1989 EDITION

Sunset

Road Atlas

UNITED STATES • CANADA • MEXICO

Copyright©1989 Creative Sales Corporation. All rights reserved. No part of this work may be copied by any process to any medium without written permission of the publisher. All data in this work are subject to change. The publisher is not responsible for loss or injury caused by erroneous data. ISBN 0-376-09018-9.

Cover photo: June Lake Road, Aspen, Colorado
Photographer: Philip M. DeRenzis

Cover design: Design Systems Group

Lane Publishing Co. • Menlo Park, California

D1738871

Proof-of-Purchase
ISBN 0-376-09018-9
Sunset

Contents

4-5	U.S. Interstate Highway Map
6-7	Mileage Chart
8-9	Time & Distance Map
10-11	National Parks & Landmarks
12-13	Canada
14-15	Mexico
16	State Map Legend
17-94	State Maps (see listing)
95	United States Demographics
96	City Demographics
97-149	Vicinity Maps (see listing)
150-176	Index to U.S. Cities (listed alphabetically by state)

State Maps

Alabama 22
Alaska 17
Arizona 20-21
Arkansas 19
California 24-27
Colorado 28-29
Connecticut 30
Delaware 31
Florida 32-33
Georgia 23
Hawaii 18
Idaho 34-35
Illinois 36-37
Indiana 38
Iowa 39
Kansas 40-41
Kentucky 43
Louisiana 44
Maine 45
Maryland 46-47
Massachusetts 30
Michigan 48-49
Minnesota 50-51
Mississippi 56
Missouri 52-53
Montana 54-55
Nebraska 58-59
Nevada 60-61
New Hampshire 57
New Jersey 31

New Mexico 62-63
New York 64-67
North Carolina 68
North Dakota 70-71
Ohio 72-73
Oklahoma 74-75
Oregon 76-77
Pennsylvania 78-79
Rhode Island 30
South Carolina 69
South Dakota 80-81
Tennessee 42
Texas 90-94
Utah 82-83
Vermont 57
Virginia 46-47
Washington 84-85
West Virginia 46-47
Wisconsin 86-87
Wyoming 88-89

Vicinity Maps

Atlanta, GA 98-99
Baltimore, MD 97
Boston, MA 100-101
Chicago, IL 102-105
Cincinnati, OH 106
Cleveland, OH 107
Dallas, TX 109
Denver, CO 112
Detroit, MI 110-111
Ft. Worth, TX 108
Houston, TX 114-115
Indianapolis, IN 113
Kansas City, MO 116-117
Los Angeles, CA 118-121
Louisville, KY 122
Memphis, TN 123
Miami, FL 124
Milwaukee, WI 125
Minneapolis, MN 126
New Orleans, LA 132
New York City, NY 128-131
Oakland, CA 146
Oklahoma City, OK 133
Orlando, FL 134
Philadelphia, PA 136-137
Phoenix, AZ 135
Pittsburgh, PA 138
Portland, OR 139
St. Louis, MO 140-141
St. Paul, MN 127

St. Petersburg, FL 142
Salt Lake City, UT 143
San Diego, CA 144
San Francisco, CA 147
San Jose, CA 145
Seattle, WA 148
Tampa, FL 142
Washington, DC 149

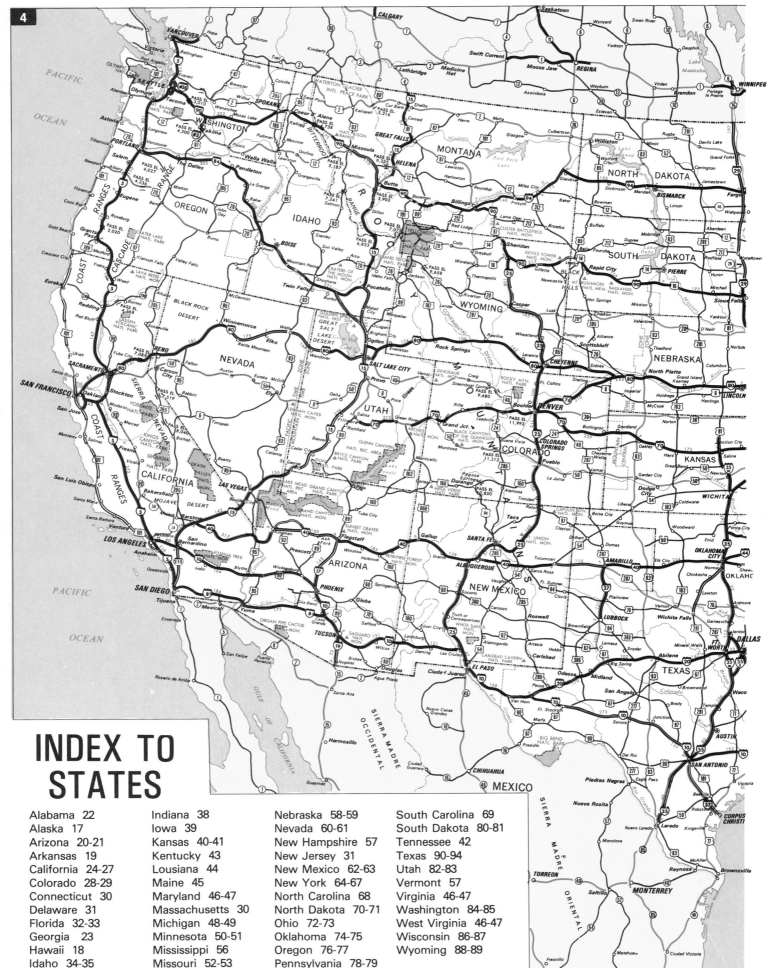

INDEX TO STATES

Alabama 22	Indiana 38	Nebraska 58-59	South Carolina 69
Alaska 17	Iowa 39	Nevada 60-61	South Dakota 80-81
Arizona 20-21	Kansas 40-41	New Hampshire 57	Tennessee 42
Arkansas 19	Kentucky 43	New Jersey 31	Texas 90-94
California 24-27	Lousiana 44	New Mexico 62-63	Utah 82-83
Colorado 28-29	Maine 45	New York 64-67	Vermont 57
Connecticut 30	Maryland 46-47	North Carolina 68	Virginia 46-47
Delaware 31	Massachusetts 30	North Dakota 70-71	Washington 84-85
Florida 32-33	Michigan 48-49	Ohio 72-73	West Virginia 46-47
Georgia 23	Minnesota 50-51	Oklahoma 74-75	Wisconsin 86-87
Hawaii 18	Mississippi 56	Oregon 76-77	Wyoming 88-89
Idaho 34-35	Missouri 52-53	Pennsylvania 78-79	
Illinois 36-37	Montana 54-55	Rhode Island 30	

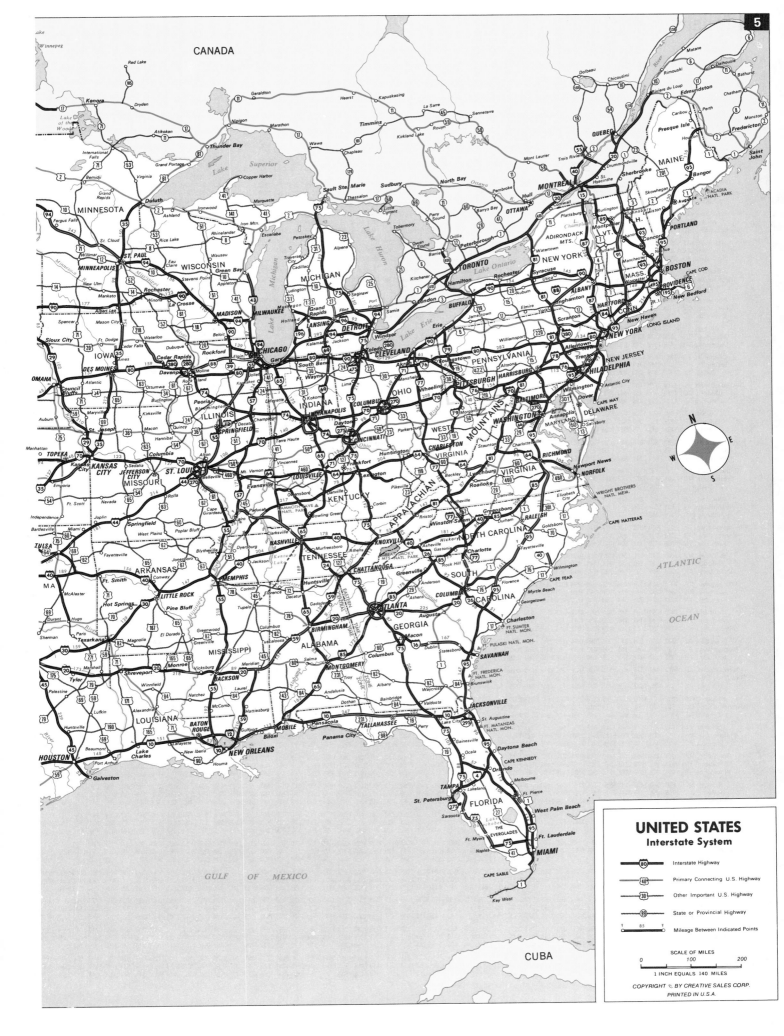

Mileage Chart

City	Albany, NY	Albuquerque, NM	Amarillo, TX	Atlanta, GA	Austin, TX	Baltimore, MD	Billings, MT	Birmingham, AL	Boise, ID	Boston, MA	Brownsville, TX	Buffalo, NY	Charleston, SC	Charleston, WV	Charlotte, NC	Chicago, IL	Cincinnati, OH	Cleveland, OH	Columbia, SC	Columbus, OH	Dallas, TX	Daytona Beach, FL	Denver, CO	Des Moines, IA	Detroit, MI	El Paso, TX	Fargo, ND	Fort Lauderdale, FL	Fort Wayne, IN	Fort Worth, TX	Grand Rapids, MI	Greensboro, NC	Hartford, CT	Houston, TX	Indianapolis, IN	Jackson, MS	Jacksonville, FL	Kansas City, MO	Knoxville, TN	Las Vegas, NV	Lincoln, NE	Little Rock, AK
Albany, NY	0	2125	1825	1007	1882	332	2073	1112	2601	170	2007	292	932	639	795	795	729	496	823	657	1679	1209	1853	1193	690	2327	1463	1403	705	1682	710	661	106	1825	836	1320	1111	1279	836	2609	1336	1370
Albuquerque, NM	2125	0	300	1387	716	1881	1022	1260	970	2214	988	1801	1695	1583	1628	1346	1394	1606	1598	1468	673	1716	446	1013	1537	267	1314	1953	1410	632	1491	1677	2084	870	1289	1087	1678	811	1407	576	837	883
Amarillo, TX	1825	300	0	1087	485	1581	1037	965	1235	1914	784	1501	1517	1304	1338	1046	1094	1306	1298	1168	363	1467	454	806	1289	508	999	1670	1109	344	1191	1377	1822	608	989	787	1378	552	1107	876	596	607
Atlanta, GA	1007	1387	1087	0	884	669	1804	160	2252	1068	1175	912	300	495	251	695	438	692	211	543	805	446	1401	924	726	1453	1364	681	612	837	749	348	969	816	543	397	329	810	204	1947	1013	540
Austin, TX	1882	716	485	884	0	1550	1449	793	1716	1930	331	1566	1247	1251	1237	1100	1127	1371	1095	1233	203	1158	1009	897	1330	583	1333	1326	1200	192	1288	1281	1867	162	1111	519	1057	680	1051	1297	851	520
Baltimore, MD	332	1881	1581	669	1550	0	1875	804	2416	409	1825	352	567	339	430	697	510	355	513	405	1347	876	1692	997	511	1997	1339	1036	550	1379	624	346	308	1409	584	1006	770	1078	503	2408	1192	1037
Billings, MT	2073	1022	1037	1804	1449	1875	0	1759	586	2232	1771	1857	2222	1762	2027	1214	1479	1662	2075	1654	1395	2173	579	959	1579	1284	625	2466	1405	1406	1396	1958	2169	1739	1400	1743	2227	1078	1723	1060	836	1439
Birmingham, AL	1112	1260	965	160	793	804	1759	0	2101	1267	1065	941	460	539	411	669	468	722	359	584	637	505	1370	838	721	1304	1311	764	610	702	739	493	1058	676	497	251	472	724	255	1822	953	394
Boise, ID	2601	970	1235	2252	1716	2416	586	2101	0	2794	1921	2271	2503	2246	2408	1777	1983	2058	2289	2069	1610	2576	870	1402	2020	1241	1245	2820	1871	1598	1917	2408	2652	1854	1890	2091	2579	1476	2022	662	1205	1781
Boston, MA	170	2214	1914	1068	1930	409	2232	1267	2794	0	2255	454	989	728	828	965	875	632	928	738	1727	1257	1953	1305	795	2376	1623	1492	847	1761	908	739	105	1878	933	1395	1167	1435	871	2765	1500	1472
Brownsville, TX	2007	988	784	1175	331	1825	1909	1065	1921	2255	0	1865	1500	1479	1426	1430	1426	1670	1360	1533	526	1353	1251	1184	1694	806	1601	1542	1455	518	1585	1480	2094	357	1427	791	1264	1008	1320	1573	1216	819
Buffalo, NY	292	1801	1501	912	1566	352	1857	941	2271	454	1865	0	947	430	666	543	438	195	822	333	1363	1069	1602	867	366	2011	1185	1400	381	1395	419	641	397	1492	512	1119	1068	1007	671	2254	1057	1046
Charleston, SC	932	1695	1517	300	1247	567	2222	460	2503	989	1500	947	0	479	203	912	628	750	113	670	1164	351	1743	1185	874	1729	1557	586	740	1116	959	271	867	1027	743	702	248	1135	368	2247	1287	814
Charleston, WV	639	1583	1304	495	1251	339	1762	539	2246	728	1479	430	479	0	276	469	178	284	376	156	1134	726	1377	761	371	1672	1126	1004	284	1056	457	227	662	1246	328	790	676	777	284	2119	960	707
Charlotte, NC	795	1628	1338	251	1237	430	2027	411	2408	828	1426	666	203	276	0	738	446	543	100	453	1054	469	1548	1029	607	1710	1414	721	602	1061	791	89	763	1053	551	640	413	940	219	2173	1151	743
Chicago, IL	795	1346	1046	695	1100	697	1214	669	1777	965	1430	543	912	469	738	0	291	348	794	340	932	1096	1037	357	284	1435	649	1348	159	945	178	729	875	1160	186	762	999	543	537	1749	527	675
Cincinnati, OH	729	1394	1094	438	1127	510	1479	468	1983	875	1426	438	628	178	446	291	0	249	502	105	924	861	1199	583	260	1472	940	1086	184	956	357	458	746	1053	105	680	786	591	246	1921	715	608
Cleveland, OH	496	1606	1306	692	1371	355	1662	722	2058	632	1670	195	750	284	543	348	249	0	627	138	1168	952	1407	672	171	1716	997	1232	211	1200	284	486	539	1297	317	924	908	803	489	2059	867	851
Columbia, SC	823	1598	1298	211	1095	513	2075	359	2289	928	1360	822	113	376	100	794	502	627	0	513	1032	381	1616	1126	745	1668	1446	622	671	1057	858	188	836	1066	625	610	290	1025	267	2162	1199	759
Columbus, OH	657	1469	1168	543	1233	405	1654	584	2069	738	1533	333	670	156	453	340	105	138	513	0	1030	901	1270	657	195	1540	989	1126	150	1062	311	381	641	1159	179	786	850	665	351	1995	776	713
Dallas, TX	1679	673	363	805	203	1347	1395	637	1610	1727	526	1363	1164	1134	1054	932	924	1168	1032	1030	0	1123	806	714	1203	648	1131	1097	1030	33	1110	1122	1664	243	908	422	1005	511	820	1249	648	317
Daytona Beach, FL	1209	1716	1467	446	1158	876	2173	505	2576	1257	1353	1069	351	726	469	1096	883	952	381	901	1123	0	1823	1329	1103	1728	1714	227	1006	1126	1143	551	1138	952	880	688	97	1209	603	2316	1401	904
Denver, CO	1853	446	454	1401	1009	1692	579	1370	870	1953	1251	1602	1743	1377	1548	1037	1199	1407	1616	1270	806	1823	0	695	1321	705	915	2067	1186	773	1201	1621	1988	1060	1091	1246	1779	608	1341	743	507	992
Des Moines, IA	1193	1013	806	924	897	997	959	838	1402	1305	1184	867	1185	761	1029	357	583	672	1126	657	714	1329	695	0	600	1114	475	1581	516	747	502	1028	1283	930	478	846	1270	203	821	1399	203	562
Detroit, MI	690	1537	1289	726	1330	511	1579	721	2020	795	1694	366	874	371	607	284	260	171	745	195	1203	1103	1321	600	0	1701	922	1346	170	1240	162	567	701	1304	293	923	1046	791	506	2011	819	850
El Paso, TX	2327	267	508	1453	583	1997	1274	1304	1241	2376	806	2011	1729	1672	1710	1435	1372	1716	1668	1540	648	1728	705	1114	1701	0	1460	1869	1573	609	1554	1783	2263	743	1460	1070	1626	915	1488	722	946	960
Fargo, ND	1463	1314	999	1364	1333	1339	625	1311	1245	1625	1601	1185	1557	1126	1414	649	940	997	1446	989	1131	1714	915	475	922	1460	0	2007	808	1072	827	1412	1531	1334	835	1335	1704	609	1195	1535	451	1091
Fort Lauderdale, FL	1435	1953	1670	681	1326	1036	2466	764	2820	1492	1542	1400	586	1004	721	1348	1086	1232	622	1126	1097	227	2067	1581	1346	1869	2007	0	1271	1129	1342	786	1403	1191	1232	883	332	1459	860	2530	1670	1184
Fort Wayne, IN	705	1410	1109	612	1200	550	1405	610	1871	847	1455	381	740	284	602	159	184	219	671	150	1030	1006	1186	516	170	1573	808	1271	0	1053	172	551	768	1176	122	784	929	648	430	1878	686	711
Fort Worth, TX	1682	632	344	837	192	1379	1406	702	1598	1761	518	1395	1116	1056	1061	945	956	1200	1057	1062	33	1126	773	747	1236	609	1072	1129	1053	0	1121	1154	1696	264	912	446	1037	513	853	1203	648	349
Grand Rapids, MI	710	1491	1191	749	1288	624	1396	739	1917	908	1585	419	959	457	791	178	357	284	858	311	1110	1143	1201	502	162	1554	827	1342	172	1121	0	707	794	1196	263	957	1071	638	573	1889	699	799
Greensboro, NC	661	1677	1377	348	1281	346	1958	493	2408	739	1480	641	271	227	89	729	458	486	188	381	1122	551	1621	1028	567	1783	1412	786	551	1154	707	0	650	1167	563	770	483	1013	283	2237	1202	778
Hartford	106	2084	1822	969	1867	308	2169	1058	2652	105	2044	397	867	662	763	875	746	339	836	641	1664	1138	1988	1283	701	2263	1531	1403	768	1696	794	650	0	1773	805	1306	1071	1297	841	2675	1378	1344
Houston, TX	1825	870	608	816	162	1409	1639	676	1854	1878	357	1492	1027	1246	1053	1160	1053	1297	1066	1159	243	952	1060	930	1304	743	1334	1191	1176	264	1196	1167	1773	0	1041	406	891	754	922	1468	892	446
Indianapolis, IN	836	1289	989	543	1111	584	1400	497	1890	933	1427	512	743	328	551	186	105	317	625	179	908	980	1091	478	293	1460	835	1232	122	912	263	563	805	1041	0	681	867	486	351	1816	673	608
Jackson, MS	1320	1087	787	397	519	1006	1743	251	2091	1395	991	1119	702	790	640	762	680	924	610	786	422	688	1246	846	923	1070	1335	883	784	446	957	770	1306	406	681	0	591	716	506	1650	874	251
Jacksonville, FL	1111	1678	1378	329	1057	770	2227	472	2579	1167	1264	1068	248	676	413	999	786	908	290	850	1005	97	1779	1270	1046	1626	1704	332	929	1037	1071	483	1071	891	867	591	0	1110	555	2238	1321	843
Kansas City, MO	1279	811	552	810	680	1078	1078	724	1476	1435	1008	1007	1135	777	940	543	591	803	1025	665	511	1209	608	203	791	916	609	1459	648	513	638	1013	1297	754	486	716	1110	0	752	1345	211	409
Knoxville, TN	836	1407	1107	204	1051	503	1723	255	2022	871	1320	671	368	284	219	537	246	489	267	351	820	603	1341	821	506	1488	1195	860	430	872	573	283	841	922	351	506	555	752	0	1983	944	523
Las Vegas, NV	2609	576	876	1947	1297	2408	1060	1822	662	2765	1573	2254	2247	2119	2173	1749	1921	2059	2162	1995	1249	2316	743	1399	2011	722	1535	2530	1878	1203	1889	2237	2675	1468	1816	1650	2238	1345	1983	0	1224	1483
Lincoln, NE	1386	837	596	1013	851	1192	836	953	1205	1500	1216	1057	1287	960	1151	527	802	867	1199	776	648	1401	507	203	819	946	451	1670	686	648	699	1202	1378	892	673	874	1321	211	944	1224	0	616
Little Rock, AK	1370	883	607	540	520	1207	1439	394	1781	1472	819	1046	814	707	743	675	608	851	759	713	317	904	992	562	850	960	1091	1184	711	349	799	778	1344	446	608	251	843	409	523	1483	616	0
Los Angeles, CA	2911	823	1095	2197	1410	2676	1254	2067	837	2993	1678	2587	2521	2394	2617	1989	2164	2392	2426	2254	1401	2407	1009	1654	2270	818	1844	2704	2137	1361	2148	2478	2829	1581	2075	1880	2402	1589	2201	275	1476	1678
Louisville, KY	868	1332	1041	421	1022	608	1550	373	1908	976	1321	543	608	258	438	300	105	349	494	211	819	801	1127	591	365	1467	949	1078	222	851	373	462	867	948	129	575	729	519	246	1861	730	502
Memphis, TN	1232	1021	721	397	658	900	1730	239	1833	1379	957	908	689	653	604	551	469	713	612	575	455	749	1151	599	712	1103	1224	989	592	487	690	640	1209	584	470	211	697	470	385	1581	647	138
Miami, FL	1439	1994	1694	665	1338	1095	2710	788	2860	1516	1580	1524	630	1046	745	1338	1086	1264	658	1210	1321	259	2131	1582	1386	1958	1986	24	1326	1353	1356	810	1427	1207	1208	907	356	1475	859	2570	1673	1208
Milwaukee, WI	933	1443	1143	784	1203	794	1143	766	1777	1078	1530	640	1032	566	835	92	388	445	891	448	1013	1180	1070	365	389	1528	576	1443	256	1059	275	826	948	1155	283	884	1067	568	643	1752	560	772
Minneapolis, MN	1215	1256	1062	1105	1120	1105	812	1088	1488	1362	1456	948	1316	874	1143	405	696	753	1276	753	1013	1458	956	251	698	1520	244	1723	564	1001	583	1135	1257	1266	591	1123	1376	459	932	1630	409	881
Mobile, AL	1322	1265	965	340	656	990	1854	269	2343	1379	851	1184	607	825	575	908	745	989	555	834	592	502	1372	954	988	1236	1413	705	839	624	1006	681	1290	478	749	178	413	819	449	1841	1039	430
Montgomery, AL	1178	1345	1042	164	804	833	1836	93	2346	1232	1041	1076	464	632	405	762	561	815	379	707	677	458	1412	1311	814	1325	1521	671	686	701	819	512	1133	709	590	255	379	867	348	2015	975	470
Nashville, TN	993	1232	932	243	869	688	1640	195	2059	1062	1168	722	576	458	399	474	283	527	458	389	666	639	1167	712	536	1314	1136	900	385	698	534	429	973	795	302	422	592	590	174	1792	770	349
New Orleans, LA	1453	1187	875	493	535	1136	1820	352	2191	1526	730	1273	727	891	721	925	820	1078	701	940	530	632	1323	978	1077	1127	1621	843	916	519	1071	810	1436	367	857	193	551	857	607	1800	1114	430
New York City, NY	146	1995	1695	855	1728	201	1742	1019	2571	203	2002	390	786	524	625	794	628	446	715	551	1525	1054	1775	1070	620	2173	1450	1289	691	1557	706	561	101	1679	730	1192	957	1192	771	2520	1274	1235
Norfolk, VA	505	1905	1632	551	1403	237	2098	711	2551	660	1735	569	454	369	341	851	681	493	412	559	1359	702	1800	1202	711	1998	1581	964	709	1382	802	227	471	1362	669	948	632	1179	412	2534	1354	1025
Oakland, CA	2982	1134	1430	2488	1786	2864	1218	2321	671	3214	2034	2745	2788	2600	2755	2098	2317	2498	2703	2391	1803	2831	1223	1742	2350	1194	1870	3041	2257	1723	2308	2809	2909	1957	2212	2270	2771	1799	2509	582	1604	1984
Oklahoma City, OK	1523	559	267	863	414	1322	1168	701	1451	1659	680	1242	1176	1031	1069	804	835	1047	1091	909	211	1257	660	576	1030	708	989	1481	852	210	932	1118	1546	454	730	575	1181	373	847	1119	433	324
Omaha, NE	1308	905	754	989	847	1143	904	904	1274	1443	1249	1005	1303	899	1135	464	721	824	1283	795	693	1402	559	146	726	1256	464	1604	634	634	640	1208	1321	949	616	914	1305	195	930	1249	57	690
Orlando, FL	1249	1751	1451	446	1142	917	2277	545	2695	1297	2034	1306	401	814	559	1127	892	1046	440	997	1078	81	1896	1363	1143	1735	1826	208	1059	1110	1188	648	1208	964	989	697	138	1266	665	2311	1452	965
Philadelphia, PA	251	1922	1622	766	1630	97	2051	880	2488	327	1954	397	688	517	522	757	559	430	627	460	1427	914	1762	662	585	2075	1370	1127	604	1459	672	438	220	1581	624	1094	859	1134	646	2449	1260	1110
Phoenix, AZ	2512	446	746	1810	1030	2311	1220	1700	1022	2644	1289	2269	2222	2045	2061	1776	1808	2045	2025	1907	1013	2101	802	1497	2019	438	1791	2244	1831	983	1906	2099	2523	1110	1719	1500	2100	1277	1845	284	1236	1379
Pittsburgh, PA	471	1654	1354	712	1412	245	1681	778	2203	584	1713	219	778	211	954	470	291	128	572	186	1209	859	1475	770	304	1833	1112	1216	336	1241	397	438	486	1345	349	965	882	851	515	2181	965	899
Portland, Or	2869	1378	1636	2763	2059	2765	867	2571	439	3149	2468	2677	2952	2615	2757	2140	2474	2416	2972	2478	2051	3018	1283	1816	2368	1661	1484	3204	2299	2003	2251	2823	2877	2369	2335	2518	3042	1809	2550	991	1641	2270
Providence, RI	178	2156	1856	1027	1898	356	2238	1189	2701	41	2222	454	956	699	785	924	790	600	888	713	1695	1201	1961	1256	746	2335	1566	1459	799	1727	859	706	73	1849	892	1362	1127	1378	935	2683	1451	1395
Raleigh, NC	656	1759	1459	397	1355	324	2273	557	2560	713	1506	721	300	297	162	802	540	559	215	453	1152	559	1694	1123	643	1800	1485	794	603	1184	754	73	624	1233	635	815	486	1086	356	2319	1263	851
Reno, NV	2763	1056	1345	2411	1775	2562	1021	2363	404	2871	2068	2433	2765	2407	2570	1897	2221	2238	2626	2295	1731	2758	1030	1606	2190	1183	1660	3008	2056	1608	2067	2591	2749	1932	2116	2143	2716	1606	2363	478	1411	1986
Richmond, VA	482	1833	1533	527	1463	155	1655	699	2594	544	1646	552	462	251	280	341	356	583	390	498	1313	713	1904	1293	609	2023	1481	946	635	1333	722	191	455	2291	907	946	693	1205	440	2406	1249	963
Rochester, NY	219	1857	1557	1015	1623	300	1922	965	2352	381	1891	81	871	495	689	608	502	268	789	397	1420	1176	1637	932	424	2036	1249	1410	464	1452	499	600	324	1555	551	1183	1102	1062	710	2371	1135	1111
Saint Louis, MO	1028	1054	754	588	806	827	1381	539	1727	1184	1216	747	884	544	689	292	340	552	737	414	641	956	859	377	585	1238	872	1208	369	698	437	762	1030	835	251	495	859	251	497	1581	462	422
Saint Paul, MN	1215	1362	1062	1105	1120	1095	812	1118	1398	892	1565	948	1316	870	1143	405	696	753	1320	745	1013	1458	956	251	698	1541	244	1753	564	987	583	1138	1257	1266	591	1160	1376	459	951	1630	408	881
Salt Lake City, UT	2290	621	1324	1900	1341	2051	579	1825	349	2417	1775	1922	2254	1896	2059	1386	1710	1727	2115	1711	1287	2283	519	1085	1679	892	1172	2578	1527	1184	1556	2059	2238	1460	1605	1742	2286	1095	1766	413	900	1054
San Antonio, TX	1986	684	530	965	81	1646	1600	895	1709	2052	300	1638	1371	1419	1272	1208	1208	1443	1180	1305	284	1175	975	1022	1500	576	1402	1378	1269	283	1353	1321	1965	203	1200	649	1086	795	1150	1273	916	592
San Diego, CA	2855	787	1078	2174	1313	2714	1309	2034	1010	2992	1574	2613	2505	2402	2423	2335	2193	2334	2389	2277	1369	2418	1054	1760	2419	721	1934	2621	2189	1332	2269	2457	2944	1484	2067	1783	2329	1627	2269	349	1573	1743
San Francisco, CA	2966	1135	1396	2511	1776	2765	1239	2371	595	3026	2044	2403	2923	2616	2756	2108	2329	2408	2738	2461	1865	2827	1233	1832	2360	1184	1886	3073	2304	1735	2318	2740	3019	1947	2224	2183	2781	1869	2549	592	1614	1994
Seattle, WA	2855	1500	1805	2656	2157	2685	815	2475	524	2961	2521	2531	2960	2748	2765	2043	2334	2336	2971	2408	2203	3070	1371	1889	2299	1775	1440	3882	2202	2071	2221	2773	2918	2498	2229	2585	3090	1884	2553	1209	1636	2368
Shreveport, LA	1599	868	568	624	340	1229	1691	474	1912	1618	644	1265	945	899	885	916	826	1070	833	592	196	903	1112	827	1069	844	1229	1144	909	228	1018	937	1519	271	827	223	814	624	729	1468	727	219
Spokane, WA	2652	1346	1563	2367	1981	2417	541	2469	369	2693	2359	2263	2700	2503	2505	1775	2066	2068	2572	2115	1978	2811	1095	1556	2020	1686	1166	3014	1921	1978	1941	2503	2650	2222	1961	2205	2822	1727	2298	1119	1460	2092
Tallahassee, FL	1249	1508	1208	268	899	932	2306	302	2512	1312	1094	1155	364	868	559	957	706	949	408	828	835	259	1727	1216	957	1492	1598	462	871	867	1022	608	1223	721	818	421	170	1045	584	2068	1227	679
Tampa, FL	1281	1759	1459	476	1150	949	2537	553	2763	1329	1345	1346	479	903	584	1143	908	1091	497	1013	1086	141	1858	1460	1200	1743	1849	268	1092	1118	1219	673	1240	972	1069	672	195	1296	730	2319	1477	1297
Toledo, OH	633	1526	1220	641	1315	454	1557	673	2020	742	1622	309	973	284	533	243	203	114	683	138	1112	1063	1272	567	65	1699	885	1289	105	1144	170	516	624	1249	243	876	989	717	449	1954	762	803
Tuscon, AZ	2442	486	656	1785	908	2246	1342	1621	1144	2571	1176	2166	2100	2039	1913	1711	1750	1971	2080	1833	964	1999	835	1484	1938	316	1897	2271	1783	932	1856	2059	2433	1079	1668	1378	2010	1281	1755	389	1246	1257
Tulsa, OK	1409	674	336	803	462	1208	1293	636	2020	1532	835	1128	1103	941	989	673	721	933	1018	795	259	1156	714	471	916	788	972	1409	738	305	818	1037	1411	504	616	510	1167	283	782	1224	416	259
Washington, DC	378	1864	1564	630	1509	41	2006	781	2441	430	1787	429	559	299	334	697	478	341	498	418	1306	802	1654	1054	506	1954	1322	1062	531	1338	608	309	341	1220	551	965	730	1046	554	2376	1184	98
West Palm Beach, FL	1396	1938	1638	632	1329	1046	2736	702	2942	1426	1524	1443	568	982	673	1289	1063	1192	586	1146	1265	195	2157	1646	1305	1922	2028	41	1157	1297	1476	737	1339	1151	1176	851	284	1052	806	2498	1598	1101
Youngstown, OH	462	1632	1346	719	1368	298	1632	738	2090	568	1695	190	709	251	502	413	275	74	561	170	1193	986	1421	737	239	1804	1038	1219	275	1225	340	482	470	1322	341	949	958	843	521	2124	923	876

Mileage Chart

Los Angeles, CA	Louisville, KY	Memphis, TN	Miami, FL	Milwaukee, WI	Minneapolis, MN	Mobile, AL	Montgomery, AL	Nashville, TN	New Orleans, LA	New York City NY	Norfolk, VA	Oakland,CA	Oklahoma City, OK	Omaha, NE	Orlando, FL	Philadelphia, PA	Phoenix, AZ	Pittsburgh, PA	Portland, OR	Providence, RI	Raleigh, NC	Reno, NV	Richmond, VA	Rochester, NY	Saint Louis, MO	Saint Paul, MN	Salt Lake City, UT	San Antonio, TX	San Diego, CA	San Francisco, CA	Seattle, WA	Shreveport, LA	Spokane, WA	Tallahassee, FL	Tampa, FL	Toledo, OH	Tucson, AZ	Tulsa, OK	Washington, DC	West Palm Beach, FL	Youngstown, OH	
2911	868	1232	1439	933	1215	1322	1178	993	1453	146	505	2982	1523	1308	1249	251	2512	471	2869	178	656	2763	482	219	1028	1215	2290	1986	2855	2966	2855	1599	2652	1249	1281	633	2442	1409	378	1396	462	Albany, NY
823	1332	1021	1994	1443	1256	1265	1345	1232	1187	1995	1905	1134	559	905	1751	1922	446	1654	1378	2156	1759	1056	1833	1857	1054	1362	621	684	787	1135	1500	868	1346	1508	1759	1526	486	674	1864	1938	1646	Albuquerque, NM
1095	1041	721	1694	1143	1062	965	1045	932	875	1695	1632	1430	267	754	1451	1622	746	1354	1636	1856	1458	1345	1533	1557	754	1062	917	530	1078	1396	1805	568	1563	1208	1459	1220	656	336	1564	1638	1346	-Amarillo, TX
2197	421	397	665	784	1105	340	164	243	493	855	551	2488	863	989	446	766	1810	712	2763	1027	397	2411	527	1015	588	1105	1900	965	2174	2511	2656	624	2367	268	476	641	1785	803	630	632	719	Atlanta, GA
1410	1022	658	1338	1203	1120	656	804	869	535	1728	1403	1786	414	847	1142	1630	1030	1412	2059	1898	1355	1775	1463	1623	806	1120	1341	81	1313	1776	2157	340	1981	899	1150	1315	908	462	1509	1329	1368	Austin, TX
2676	608	900	1095	794	1105	990	833	688	1136	201	237	2864	1322	1143	917	97	2311	245	2765	356	324	2562	155	300	827	1095	2051	1646	2714	2765	2686	1229	2417	932	949	454	2246	1208	41	1046	298	Baltimore, MD
1254	1550	1730	2710	1143	812	1854	1836	1640	1820	1742	2098	1218	1168	904	2277	2051	1220	1681	867	2238	2273	1021	1655	1922	1381	812	579	1600	1309	1239	815	1691	541	2306	2527	1557	1342	1293	2006	2736	1632	Billings, MT
2067	373	239	788	766	1088	269	93	195	352	1019	711	2321	701	904	545	880	1700	778	2571	1189	557	2363	699	965	539	1118	1825	895	2034	2371	2475	474	2469	302	553	673	1621	636	781	702	738	Birmingham, AL
837	1908	1833	2860	1777	1488	2143	2346	2059	2191	2571	2551	671	1451	1274	2695	2498	1022	2203	439	2701	2560	404	2594	2352	1727	1398	349	1709	1010	595	524	1912	369	2512	2763	2020	1144	1582	2441	2492	2090	Boise, ID
2993	976	1379	1516	1078	1362	1379	1232	1062	1525	203	560	3214	1659	1443	1297	327	2644	584	3149	41	713	2871	544	381	1184	892	2417	2052	2992	3121	2961	1618	2693	1312	1329	742	2571	1532	430	1426	568	Boston, MA
1678	1321	957	1580	1530	1456	851	1041	1168	730	2002	1735	2034	680	1249	2034	1954	1289	1713	2468	2222	1506	2068	1646	1891	1216	1565	1609	300	1574	2044	2521	644	2359	1094	1345	1622	1176	835	1787	1524	1695	Brownsville, TX
2587	543	908	1424	640	948	1184	1076	722	1273	390	569	2745	1242	1005	1306	397	2269	219	2677	454	721	2433	552	81	747	948	1922	1638	2613	2667	2531	1265	2263	1155	1346	309	2166	1128	429	1443	190	Buffalo, NY
2521	608	689	630	1032	1316	607	464	576	727	787	454	2788	1176	1303	401	688	2222	778	2952	956	300	2765	462	871	884	1316	2254	1371	2505	2923	2960	945	2700	364	479	973	2100	1103	559	568	709	Charleston, SC
2394	258	653	1046	566	874	825	632	458	891	524	369	2600	1031	899	814	517	2045	211	2615	699	297	2407	251	495	544	870	1896	1419	2402	2616	2748	899	2503	868	903	284	2039	941	299	982	251	Charleston, WV
2417	438	592	604	835	1143	575	415	399	721	625	341	2755	1069	1135	559	522	2061	487	2757	785	162	2570	280	689	689	1143	2059	1272	2423	2756	2765	885	2505	559	584	583	1913	989	334	673	502	Charlotte, NC
1989	300	551	1338	92	405	908	762	474	925	794	851	2098	804	454	1127	757	1776	470	2140	924	802	1897	341	608	292	405	1386	1208	2335	2108	2043	916	1775	957	1143	243	1711	673	697	1289	413	Chicago, IL
2164	105	469	1086	388	696	712	561	283	810	628	601	2317	835	721	892	559	1808	291	2369	790	540	2201	503	502	340	696	1671	1208	2193	2329	2334	813	2066	706	908	203	1756	721	478	1051	275	Cincinnati, OH
2392	349	713	1264	445	753	989	815	527	1078	446	493	2498	1047	824	1046	430	2045	128	2416	600	559	2238	583	268	552	753	1727	1443	2384	2408	2336	1070	2068	949	1091	114	1971	933	341	1192	74	Cleveland, OH
2426	494	612	658	891	1276	555	379	458	701	715	412	2703	1091	1283	440	627	2025	572	2972	888	215	2626	390	789	737	1320	2115	1180	2389	2738	2971	833	2572	408	497	683	2080	1018	498	586	561	Columbia, SC
2254	211	575	1210	448	753	834	707	389	940	551	559	2391	909	795	997	460	1907	186	2478	713	453	2295	498	397	414	745	1711	1305	2277	2461	2408	592	2115	828	1513	138	1833	795	418	1146	170	Columbus, OH
1401	819	455	1321	1013	1013	592	677	666	530	1525	1359	1803	211	693	1078	1427	1013	1209	2059	1695	1152	1731	1313	1420	641	1013	1287	284	1369	1865	2203	196	1978	835	1086	1112	964	259	1306	1265	1193	Dallas, TX
2407	801	749	259	1180	1458	502	458	639	632	1054	702	2831	1257	1402	81	914	2102	859	3018	1201	559	2758	713	1176	956	1458	2283	1175	2418	2827	3070	903	2811	259	141	1063	1999	1156	802	195	986	Daytona Beach, FL
1009	1127	1151	2131	1070	956	1372	1412	1167	1323	1775	1800	1223	660	559	1896	1762	802	1475	1283	1961	1694	1030	1904	1637	859	956	519	975	1054	1233	1371	1112	1095	1727	1858	1272	835	714	1654	2157	1421	Denver, CO
1654	591	599	1582	365	251	954	1131	712	978	1070	1202	1742	576	146	1363	1062	1479	770	1816	1256	1123	1606	1293	932	377	251	1085	1022	1760	1832	1889	827	1556	1216	1460	567	1484	471	1054	1646	737	Des Moines, IA
2270	365	712	1386	389	698	988	814	536	1077	620	711	2350	1030	726	1143	585	2019	304	2368	746	643	2190	609	424	535	698	1679	1500	2419	2360	2299	1069	2020	957	1200	65	1938	916	506	1305	239	Detroit, MI
818	1467	1103	1958	1528	1520	1236	1325	1314	1127	2173	1998	1194	708	1256	1735	2073	438	1833	1661	2335	1800	1185	2023	2036	1238	1541	894	576	721	1184	1775	844	1686	1492	1743	1699	316	788	1954	1922	1804	El Paso, TX
1844	949	1224	1987	576	244	1413	1525	1138	1521	1450	1581	1870	989	464	1826	1370	1791	1112	1484	1566	1485	1660	1481	1249	872	244	1172	1402	1934	1886	1440	1229	1166	1598	1849	885	1897	972	1322	2028	1038	Fargo, ND
2704	1078	989	24	1443	1723	705	671	900	843	1289	964	3041	1481	1604	208	1127	2244	1216	3204	1459	794	3008	946	1410	1208	1753	2578	1378	2621	3073	3382	1144	3014	462	268	1289	2271	1409	1062	41	1219	Fort Lauderdale
2137	222	592	1326	256	564	839	686	385	916	691	709	2257	852	634	1059	604	1831	336	2299	799	603	2056	635	464	369	564	1527	1269	2189	2304	2202	909	1921	871	1092	105	1783	738	531	1157	275	Fort Wayne, IN
1361	851	487	1353	1059	1001	624	701	698	519	1557	1382	1723	210	634	1110	1459	983	1241	2003	1727	1184	1608	1333	1452	698	987	1184	283	1332	1735	2071	228	1978	867	1118	1144	932	305	1338	1297	1225	Fort Worth, TX
2148	373	690	1356	275	583	1006	819	534	1071	706	802	2308	932	640	1188	672	1906	397	2251	859	754	2067	722	499	437	583	1556	1353	2269	2318	2221	1018	1941	1022	1219	170	1856	818	608	1476	340	Grand Rapids, MI
2478	462	640	810	826	1135	681	512	429	810	561	227	2809	1118	1208	648	438	2099	438	2823	706	73	2591	191	600	762	1138	2059	1321	2457	2740	2773	937	2503	608	673	516	2059	1037	309	737	482	Greensboro, NC
2829	867	1209	1427	948	1257	1290	1133	973	1436	101	471	2909	1546	1321	1208	220	2523	486	2877	73	624	2749	455	324	1030	1257	2238	1965	2944	3019	2918	1529	2656	1223	1240	624	2433	1411	341	1339	470	Hartford, CT
1581	948	584	1207	1155	1266	478	709	795	367	1679	1362	1957	454	949	964	1581	1110	1345	2369	1849	1233	1932	2291	1555	835	1266	1460	203	1484	1947	2498	271	2222	721	972	1249	1079	504	1420	1151	1322	Houston, TX
2075	129	470	1208	283	591	749	590	302	857	730	669	2212	730	616	989	624	1719	349	2335	892	635	2116	907	551	251	591	1605	1200	2067	2224	2229	827	1961	818	1069	243	668	616	551	1176	341	Indianapolis, IN
1880	575	211	907	884	1123	178	255	422	193	1192	948	2270	575	914	697	1094	1500	965	2518	1362	815	2143	946	1183	495	1160	1742	649	1783	2183	2585	223	2205	421	672	876	1378	510	965	851	949	Jackson, MS
2402	729	697	356	1067	1374	413	379	592	551	957	632	2771	1181	1305	138	859	2100	882	3030	1127	584	2716	693	1102	867	1374	2286	1086	2329	2781	3042	814	2822	170	195	989	2010	1167	730	284	958	Jacksonville, FL
1589	519	470	1475	568	459	819	867	590	857	1192	1179	1799	373	195	1266	1134	1277	851	1809	1378	1086	1606	1205	1062	251	459	1095	795	1627	1869	1834	624	1727	1045	1296	717	1281	283	1052	144	843	Kansas City, MO
2201	246	385	859	643	932	449	348	174	607	753	412	2509	847	930	665	646	1845	515	2550	951	356	2363	440	710	497	935	1766	1150	2269	2549	2553	729	2298	584	730	449	1755	782	554	806	521	Knoxville, TN
275	1861	1581	2570	1752	1630	1841	2015	1792	1800	2520	2534	582	1119	1249	2311	2449	284	2181	991	2683	2319	478	2406	2371	1581	1630	413	1273	349	592	1209	1468	1119	2068	2319	1954	389	1224	2376	2498	2124	Las Vegas, NV
1476	730	647	1673	560	409	1039	975	770	1014	1274	1354	1604	433	57	1452	1260	1236	965	1641	1451	1263	1411	1249	1135	462	408	900	916	1573	1614	1636	727	1460	1227	1477	762	1246	416	1184	1598	923	Lincoln, NE
1678	502	138	1208	772	881	430	470	349	430	1235	1025	1984	324	690	965	1110	1379	899	2270	1395	851	1986	963	1111	422	881	1054	592	1743	1994	2368	219	2092	679	1297	803	1257	259	98	1101	876	Little Rock, AK
0	2136	1816	2828	2238	1905	2013	2035	2027	1883	2790	2809	372	1354	1508	2585	2717	389	2449	989	2902	2554	511	2641	2400	1849	1905	672	1378	121	382	1159	1687	1406	2342	2578	2213	502	1459	2659	2772	2424	Los Angeles, CA
2136	0	364	1102	397	705	626	466	178	729	707	607	2333	738	671	907	664	1782	381	2298	421	525	2132	686	608	290	705	1638	1103	2119	2343	2343	721	2075	664	915	300	1697	633	533	1094	373	Louisville, KY
1816	364	0	1013	673	949	389	332	211	397	1095	876	2122	462	705	770	989	2948	795	2408	1257	712	2124	825	973	300	949	1613	730	1881	2132	2506	357	2230	527	778	665	1370	397	363	957	738	Memphis, TN
2828	1102	1013	0	1435	1743	729	695	908	867	1313	988	3087	1524	1670	227	1215	2448	1248	3366	1483	818	3032	988	1435	1231	1743	2602	1402	2645	3097	3406	1130	3138	486	292	1293	2326	1483	1086	65	1264	Miami, FL
2238	397	673	1435	0	332	981	859	571	1045	851	948	2171	884	503	1224	834	1873	567	2002	1021	899	1930	832	697	389	332	1419	1305	2109	2175	1970	1013	1702	1053	1240	332	1808	770	770	1386	510	Milwaukee, WI
1905	705	949	1743	332	0	1227	1281	892	1346	1160	1337	2065	803	381	1673	1126	1848	868	1670	1322	1241	1776	1237	1005	628	1	1475	1265	1975	2075	1638	985	1370	1354	1589	641	1702	705	1078	1737	794	Minneapolis, MN
2013	626	389	729	981	1227	0	176	462	146	1176	884	2351	786	1011	486	1078	1593	1050	2611	1346	738	2278	867	1273	673	1158	1903	697	1971	2361	2710	401	2342	243	494	940	1542	713	949	673	1005	Mobile, AL
2035	466	332	695	859	1281	176	0	288	322	1010	715	2295	794	1037	452	930	1655	843	2632	1191	561	2456	792	1179	632	1281	1918	904	2127	2464	2838	478	2562	209	460	766	1633	765	794	639	831	Montgomery, AL
2027	178	211	908	571	892	462	288	0	551	859	665	2374	673	735	689	761	1671	575	2376	1029	502	2189	614	770	323	892	1678	941	2092	2375	2384	669	2124	486	737	478	1581	608	632	875	543	Nashville, TN
1883	729	867	1045	1045	1346	146	322	551	0	1322	1054	2317	681	1362	624	1224	1540	1078	2505	1492	876	2278	1114	1362	681	1346	1842	551	1824	2327	2574	309	2409	381	632	1005	1419	657	1095	811	1095	New Orleans, LA
2790	707	1095	1313	851	1160	1176	1010	859	1322	0	389	2876	1436	1208	1094	101	2425	381	2330	162	510	2635	366	320	941	1160	2124	1832	2773	2886	2837	1415	2569	1109	1126	556	2360	1331	252	1223	403	New York, NY
2809	607	876	988	948	1337	884	715	665	1054	389	0	2957	1349	1362	770	268	2393	365	2968	527	197	2789	98	535	927	1337	2262	1606	2669	2967	2975	1203	2700	802	827	592	2271	1234	190	916	438	Norfolk, VA
372	2333	2122	3087	2171	2065	2351	2295	2374	2317	2876	2957	0	1660	1596	2844	2913	744	2528	614	2974	2865	201	2988	2681	2065	2065	545	1734	493	9	777	2052	979	2601	2836	2293	878	1766	2772	3055	2463	Oakland, CA
1354	738	462	1524	884	803	786	794	673	681	1436	1349	1660	0	495	1281	1363	998	1095	1946	1597	1200	1662	1287	1298	495	803	1151	478	1345	1670	2044	389	1768	1038	1289	961	908	105	1305	1468	1087	Oklahoma City, OK
1508	671	705	1670	503	381	1110	1037	735	1362	1208	1362	1596	495	0	1421	1200	1427	908	1605	1394	1281	1395	1400	1078	446	381	947	949	1641	1606	1679	729	1403	1241	1492	705	1400	401	1151	1671	875	Omaha, NE
2585	907	770	227	1224	1673	486	452	689	624	1094	770	2844	1281	1421	0	1005	2205	1025	3123	1273	608	2789	776	1200	1030	1551	2359	1159	2402	2854	3163	920	2895	243	89	1086	2083	1240	876	194	1052	Orlando, FL
2717	664	989	1215	834	1126	1078	930	761	1224	101	268	2913	1363	1200	1005	0	1025	292	2830	259	412	2627	252	324	868	1126	2209	1734	2911	2923	2780	1270	2512	1020	1037	502	2287	1249	133	1134	357	Philadelphia, PA
389	1782	1377	2448	1873	1848	1593	1655	1671	1540	2425	2393	744	998	1427	2205	1025	0	2084	1322	2586	2172	754	2285	2281	1484	1808	673	989	357	754	1492	1282	1386	1962	2213	1950	122	1103	2367	2392	2061	Phoenix, AZ
2449	381	795	1248	567	868	1050	843	575	1078	381	365	2528	1095	908	1025	292	2084	0	2538	559	519	2335	333	284	600	868	1824	1468	2441	2538	2513	1087	2190	932	932	227	2021	981	219	1111	65	Pittsburgh, PA
985	2298	2408	3366	2002	1670	2611	2632	2376	2505	2330	2968	614	1946	1605	3123	2830	1322	2538	0	3000	2903	616	3012	2676	2068	1670	764	2168	1078	624	170	2255	365	2880	3131	2311	1483	1987	2846	3310	2481	Portland, OR
2902	421	1257	1483	1021	1322	1346	1191	1029	1492	162	527	2974	1597	1394	1273	259	2586	559	3000	0	673	2814	511	381	1102	1322	2303	2002	2978	2984	2967	1585	2644	1288	1322	689	252	1483	397	1402	538	Providence, RI
2554	525	713	818	899	1241	738	561	502	876	510	197	2865	1200	1281	608	412	2172	519	2903	673	0	2716	175	600	835	1241	2205	1427	2529	2797	2911	538	2651	624	657	584	2124	1103	283	754	535	Raleigh, NC
511	2132	2124	3032	1930	1776	2278	2456	2189	2278	2635	2789	201	1662	1395	2789	2627	754	2335	616	2814	2716	0	2803	2489	1881	1776	511	1768	535	211	810	1927	770	2546	2797	2133	867	1702	2579	2976	2303	Reno, NV
2641	686	825	988	832	1237	867	792	614	1114	366	98	2988	1287	1400	776	252	2285	333	3012	511	175	2803	0	455	976	1237	2324	1555	2643	2989	3029	1173	2761	794	827	630	2425	1222	114	924	662	Richmond, VA
2400	608	973	1435	697	1005	1273	1179	770	1362	320	535	2681	1298	1078	1200	324	2281	284	2676	381	600	2489	455	0	803	1005	1978	1686	2674	2691	2660	1361	2384	1208	1257	381	2222	1398	341	1346	265	Rochester, NY
1849	290	300	1231	389	628	673	632	323	681	941	927	2065	495	446	1030	868	1484	600	2068	1102	835	1881	976	803	0	628	1370	916	1840	2075	2076	624	1881	795	1030	466	1419	381	795	1241	592	Saint Louis, MO
1905	705	949	1743	332	1	1158	1281	892	1346	1160	1337	2065	803	381	1551	1126	1808	868	1670	1322	1241	1776	1237	1005	628	0	1475	1265	1975	2075	1638	985	1370	1354	1589	641	1702	705	1078	1737	794	Saint Paul, MN
672	1638	1613	2602	1419	1475	1903	1918	1678	1842	2124	2262	545	1151	947	2359	2209	673	1824	764	2303	2205	511	2324	1978	1370	1475	0	1447	762	714	85	1563	706	2116	2318	1715	795	1223	2141	2490	1805	Salt Lake City, UT
1378	1103	730	1402	1305	1265	697	904	941	551	1832	1606	1734	478	949	1159	1734	989	1468	2168	2002	1427	1768	1555	1686	916	1265	1447	0	1274	1774	2255	474	2110	916	1151	1387	876	535	1605	1370	1477	San Antonio, TX
121	2119	1881	2645	2109	1975	1971	2127	2092	1824	2773	2669	493	1345	1641	2402	2911	357	2441	1078	2978	2529	535	2643	2674	1840	1975	762	1274	0	527	1276	1565	1403	2213	2464	2604	405	1403	2733	2072	2456	San Diego, CA
382	2343	2132	3097	2175	2075	2361	2464	2375	2327	2886	2967	9	1670	1606	2854	2923	754	2538	624	2984	2797	211	2989	2691	2075	2075	714	1774	527	0	787	2061	852	2611	2846	2415	868	1776	2949	3065	2569	San Francisco, CA
1159	2343	2506	3406	1970	1638	2710	2838	2384	2574	2837	2975	777	2044	1679	3163	2780	1492	2513	170	2967	2911	810	3029	2660	2076	1638	851	2255	1276	787	0	2335	274	2920	3155	2231	1573	2034	2684	3388	2456	Seattle, WA
1687	721	357	1130	1013	985	401	478	669	309	1415	1203	2052	389	729	920	1270	1282	1087	2255	1585	538	1927	1173	1361	624	985	1563	474	1565	2061	2335	0	2038	644	895	1099	1160	356	953	507	1081	Shreveport, LA
1406	2075	2230	3138	1702	1370	2342	2562	2124	2409	2569	2700	979	1768	1403	2895	2512	1386	2190	365	2644	2651	770	2761	2384	1816	1370	706	2110	1403	852	274	2038	0	2652	2887	2100	1496	1727	2417	3196	2125	Spokane, WA
2342	664	527	486	1053	1354	243	209	486	381	1109	802	2601	1038	1241	243	1020	1962	932	2880	1288	624	2546	794	1208	795	1354	2116	916	2213	2611	2920	644	2652	0	251	900	1840	955	891	430	1005	Tallahassee, FL
2578	915	778	292	1240	1589	494	460	737	632	1126	827	2836	1289	1492	89	1037	2213	932	3131	1322	657	2797	827	1251	1030	1589	2318	1151	2464	2846	3155	895	2887	251	0	1143	2076	1182	908	219	1088	Tampa, FL
2213	300	665	1293	332	641	940	766	478	1005	556	592	2293	961	705	1086	502	1950	227	2311	689	584	2133	630	381	466	641	1715	1387	2604	2415	2231	1099	2100	900	1143	0	2051	867	438	1248	170	Toledo, OH
502	1697	1370	2326	1808	1702	1542	1633	1581	1419	2360	2271	878	908	1400	2083	2287	122	2021	1483	252	2124	867	2425	2222	1419	1702	795	876	405	868	1573	1160	1496	1840	2076	2051	0	1038	2254	2270	2003	Tuscon, AZ
1459	633	397	1483	770	705	713	765	608	657	1331	1234	1766	105	401	1240	1249	1103	981	1987	1483	1103	1702	1222	1398	381	705	1223	535	1403	1776	2034	356	1727	955	1182	867	1038	0	1284	1167	964	Tulsa, OK
2659	533	363	1086	770	1078	949	794	632	1095	252	190	2772	1305	1151	876	133	2367	219	2846	397	283	2579	114	341	795	1078	2141	1605	2733	2949	2684	953	2417	891	908	438	2254	1284	0	1005	283	Washington, DC
2772	1094	957	65	1386	1737	673	639	875	811	1223	916	3055	1468	1671	194	1134	2392	1111	3310	1402	754	2976	924	1346	1241	1737	2490	1370	2072	3065	3388	507	3196	430	219	1248	2270	1167	1005	0	1167	West Palm Beach, FL
2424	373	738	1238	510	794	1005	831	543	1095	403	438	2463	1087	875	1052	357	2076	65	2481	527	535	2303	662	265	592	794	1877	1477	2432	2569	2456	1181	2125	1005	1088	170	2003	964	283	1167	0	Youngstown, OH

TIME & DISTANCE MAP

The Time & Distance Map has been specially prepared to help the interstate traveler estimate driving time between major cities throughout the 48 states. The mileages and times shown have been calculated giving ample consideration to normal driving conditions. Topography, congested areas, peak traffic periods, speed limits, road work and highway construction may have an unfavorable effect on these estimates. Additional time should also be allowed for rest periods, sightseeing and eating.

Limited access highways are not always the shortest distance, however, they have been used in these calculations, since they normally represent the shortest driving time and best over-all driving conditions.

Driving times have been calculated using an average speed of 50 miles per hour. Your actual driving time will depend on traffic, road conditions and your personal driving habits.

BLACK NUMERALS INDICATE MILEAGE

RED NUMERALS INDICATE DRIVING TIME ONLY

Drive safely and cautiously at all times. Obey all traffic signs.

Creative Sales Corporation
Publisher of Quality Road Maps

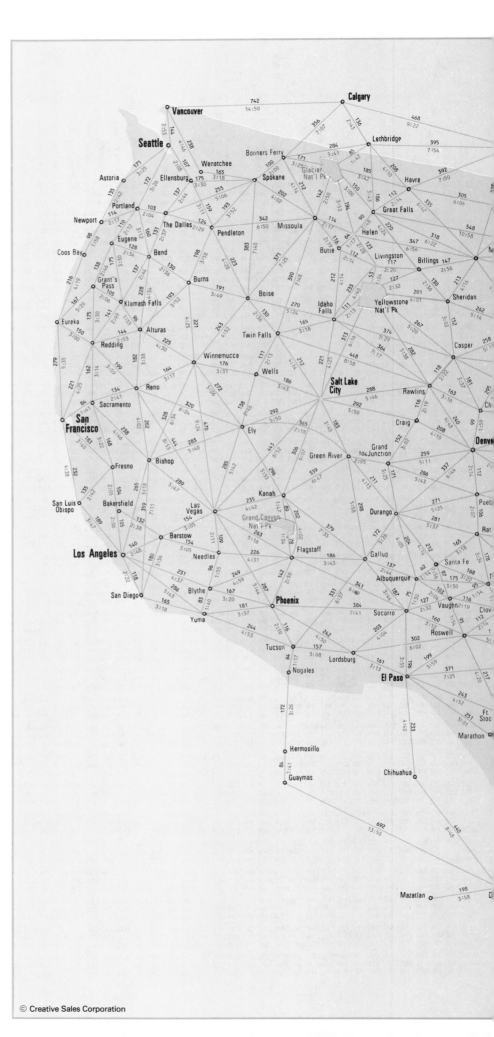

© Creative Sales Corporation

UNITED STATES
Time & Distance Map

USE ONLY FOR ORIENTATION TO NATIONAL PARKS AND LANDMARKS. FOR MORE DETAILED HIGHWAY INFORMATION SEE INTERSTATE HIGHWAY MAP, PAGES 4-5 AND STATE MAP SECTION, PAGES 17-94.

Grid columns: 1 – 7 (top), 1 – 5 (bottom)
Grid rows: A – K

Canada / Provinces & States:
BRITISH COLUMBIA, ALBERTA, SASKATCHEWAN, MANITOBA, WASHINGTON, OREGON, IDAHO, MONTANA, NORTH DAKOTA, SOUTH DAKOTA, WYOMING, NEBRASKA, NEVADA, UTAH, COLORADO, KANSAS, CALIFORNIA, ARIZONA, NEW MEXICO, OKLAHOMA, TEXAS, BAJA CALIFORNIA, SONORA, CHIHUAHUA, COAHUILA, NUEVO LEON, TAMAULIPAS, ALASKA, HAWAII, YUKON, U.S.S.R.

Cities (selected):
Vancouver, Calgary, Saskatoon, Regina, Moose Jaw, Winnipeg, Seattle, Tacoma, Olympia, Spokane, Portland, Salem, Eugene, Astoria, Pendleton, Walla Walla, Helena, Great Falls, Butte, Billings, Bismarck, Fargo, Williston, Boise, Twin Falls, Pocatello, Rapid City, Pierre, Sioux Falls, Reno, Carson City, Sacramento, San Francisco, San Jose, Oakland, Monterrey, Fresno, Bakersfield, Los Angeles, Anaheim, San Diego, Tijuana, Mexicali, Ensenada, Las Vegas, Salt Lake City, Provo, Ogden, Rock Springs, Casper, Cheyenne, Laramie, Scottsbluff, North Platte, Lincoln, Denver, Boulder, Colorado Springs, Pueblo, Durango, Dodge City, Wichita, Phoenix, Globe, Prescott, Flagstaff, Tucson, Yuma, Nogales, Douglas, Albuquerque, Santa Fe, Amarillo, Oklahoma City, Lubbock, Wichita Falls, Carlsbad, El Paso, Ciudad Juarez, Odessa, Midland, San Angelo, Abilene, Dallas, Ft. Worth, Waco, Austin, San Antonio, Corpus Christi, Brownsville, Laredo, Nuevo Casas Grandes, Piedras Negras, Nueva Rosita, Monterrey, Barrow, Wainwright, Point Hope, Prudhoe Bay, Nome, Fairbanks, College, Anchorage, Valdez, Cordova, Juneau, Ketchikan, Sitka, Kodiak, Honolulu, Pearl City, Wahiawa, Kaneohe, Lihue, Hilo, Wailuku, Laheina

National Parks, Forests & Landmarks (selected):
Pacific Rim Nat'l Park, San Juan Island Nat'l Hist. Park, Olympic Nat'l Park, Mt. Baker, N. Cascades, Okanogan Nat'l Forest, Mt. Rainier Nat'l Park, Wenatchee Nat'l Forest, Colville Nat'l Forest, Kaniksu Nat'l Forest, Waterton-Glacier Int'l Peace Park, Glacier Nat'l Park, Flathead Nat'l Forest, Lewis and Clark Nat'l Forest, Ft. Union Nat'l Hist. Site, Theodore Roosevelt Nat'l Park, Knife River Indian Village Nat'l Hist. Site, Riding Mountain Nat'l Park, Mt. Hood Nat'l Forest, Willamette Nat'l Forest, John Day Fossil Beds Nat'l Mon., Ochoco Nat'l Forest, Malheur Nat'l Forest, Wallowa Whitman Nat'l Forest, Nezperce Nat'l Forest, Clearwater Nat'l Forest, Bitterroot Nat'l Forest, Deerlodge Nat'l Forest, Gallatin Nat'l Forest, Custer Nat'l Forest, Custer Battlefield Nat'l Mon., Redwood Nat'l Seashore, Six Rivers Nat'l Forest, Klamath Nat'l Forest, Shasta Nat'l Forest, Trinity Nat'l Forest, Lassen Volcanic Nat'l Park, Modoc Nat'l Forest, Lava Beds Nat'l Mon., Crater Lake Nat'l Park, Rogue River Nat'l Forest, Siskiyou Nat'l Forest, Oregon Caves Nat'l Mon., Umpqua Nat'l Forest, Deschutes Nat'l Forest, Winema Nat'l Forest, Fremont Nat'l Forest, Sawtooth Nat'l Forest, Challis Nat'l Forest, Salmon Nat'l Forest, Payette Nat'l Forest, Boise Nat'l Forest, Targhee Nat'l Forest, Craters of the Moon Nat'l Mon., Caribou Nat'l Forest, Yellowstone Nat'l Park, Grand Teton Nat'l Park, Shoshone Nat'l Forest, Bighorn Nat'l Forest, Devil's Tower Nat'l Mon., Thunder Basin Nat'l Grassland, Black Hills Nat'l Forest, Mt. Rushmore Nat'l Mem., Badlands Nat'l Park, Wind Cave Nat'l Park, Hot Springs, Medicine Bow Nat'l Forest, Samuel R. McKelvie Nat'l Forest, Nebraska Nat'l Forest, Humboldt Nat'l Forest, Toiyabe Nat'l Forest, Tahoe Nat'l Forest, Eldorado Nat'l Forest, John Muir Nat'l Hist. Site, Pt. Reyes Nat'l Seashore, Muir Woods, Yosemite Nat'l Park, Sierra Nat'l Forest, Inyo Nat'l Forest, Kings Canyon Nat'l Park, Sequoia Nat'l Park, Devils Postpile Nat'l Mon., Pinnacles Nat'l Mon., Los Padres Nat'l Forest, Channel Islands Nat'l Park, Death Valley Nat'l Mon., Great Basin Nat'l Park, Lehman Caves Nat'l Mon., Golden Spike Nat'l Hist. Site, Wasatch Nat'l Forest, Ashley Nat'l Forest, Dinosaur Nat'l Mon., Fossil Butte Nat'l Mon., Fishlake Nat'l Forest, Manti La Sal Nat'l Forest, Dixie Nat'l Forest, Cedar Breaks Nat'l Mon., Zion Nat'l Park, Bryce Canyon Nat'l Park, Capitol Reef Nat'l Park, Canyonlands Nat'l Park, Arches Nat'l Park, Natural Bridges Nat'l Mon., Hovenweep Nat'l Mon., Mesa Verde Nat'l Park, Rainbow Bridge Nat'l Mon., Grand Canyon Nat'l Park, Lake Mead, Kaibab Nat'l Forest, Canyon De Chelly Nat'l Mon., Wupatki Nat'l Mon., Sunset Crater Nat'l Mon., Walnut Canyon Nat'l Mon., Coconino Nat'l Forest, Tonto Nat'l Forest, Prescott Nat'l Forest, Tuzigoot Nat'l Mon., Montezuma Nat'l Mon., Apache Nat'l Forest, Gila Nat'l Forest, Cibola Nat'l Forest, Petrified Forest Nat'l Park, El Morro Nat'l Mon., Chaco Canyon Nat'l Mon., Bandelier Nat'l Mon., Kiowa Nat'l Grasslands, Capulin Mtn. Nat'l Mon., Aztec Ruins Nat'l Mon., Carson Nat'l Forest, Santa Fe Nat'l Forest, Gran Quivira Nat'l Mon., Gila Cliff Dwellings Nat'l Mon., White Sands Nat'l Mon., Saguaro Nat'l Mon., Organ Pipe Cactus Nat'l Mon., Coronado Nat'l Mon., Casa Grande Ruins Nat'l Mon., Chiricahua Nat'l Mon., Tumacacori Nat'l Mon., Guadalupe Mtns. Nat'l Park, Carlsbad Caverns Nat'l Park, Lincoln Nat'l Forest, Old Ft. Davis Nat'l Hist. Site, Big Bend Nat'l Park, Platt Nat'l Park, Comanche Nat'l Grassland, Padre Island Nat'l Seashore, Pike Nat'l Forest, White River Nat'l Forest, Grand Mesa Nat'l Forest, Gunnison Nat'l Forest, San Juan Nat'l Forest, Rio Grande Nat'l Forest, Routt Nat'l Forest, Roosevelt Nat'l Forest, Arapaho Nat'l Forest, Pawnee Nat'l Grassland, Rocky Mountain Nat'l Park, Black Canyon of the Gunnison Nat'l Mon., Colorado Nat'l Mon., Pecos Nat'l Mon., Ft. Vancouver Nat'l Hist. Site and Museum, Cabrillo Nat'l Mon., Joshua Tree Nat'l Mon., Cleveland Nat'l Forest, Angeles Nat'l Forest, San Bernardino Nat'l Forest, Mojave Desert, Black Rock Desert, Great Salt Lake Desert

Alaska inset:
Noatak Nat'l Preserve, Gates of the Arctic Nat'l Park & Preserve, Bering Land Bridge Nat'l Preserve, Yukon-Charley Nat'l Preserve, Denali Nat'l Park and Preserve, Lake Clark Nat'l Park & Preserve, Katmai Nat'l Park & Preserve, Aniakchak Nat'l Park & Preserve, Kenai Fjords Nat'l Park, Chugach Nat'l Forest, Wrangell-St. Elias Nat'l Park, Glacier Bay Nat'l Park, Tongass Nat'l Forest, Kodiak Island, Aleutian Islands

Hawaii inset:
Haleakala Nat'l Park, Hawaii Volcanoes Nat'l Park, City of Refuge Nat'l Hist. Park

Water bodies / geographic features:
PACIFIC OCEAN, Arctic Ocean, Bering Sea, Bristol Bay, Gulf of Alaska, Norton Sound, Lake Sakakawea, Lake Oahe, Great Salt Lake, Lake Powell, Lake Mead, Missouri River, Yellowstone River, Snake River, Columbia River, Colorado River, Rio Grande, Pecos River, Red River, Platte River, N. Platte R., Arkansas River, SIERRA NEVADA, CASCADE RANGE, COASTAL RANGES, SIERRA MADRE OCCIDENTAL, CONTINENTAL DIVIDE

Time Zones: PACIFIC TIME ZONE, MOUNTAIN TIME ZONE, CENTRAL TIME ZONE, BERING TIME ZONE, ALASKA-HAWAII TIME ZONE, YUKON TIME ZONE, U.S.S.R. TIME ZONE, INT'L DATE LINE

CANADA / UNITED STATES / MEXICO

U.S.S.R.

B.C. — YUKON

© Creative Sales Corporation

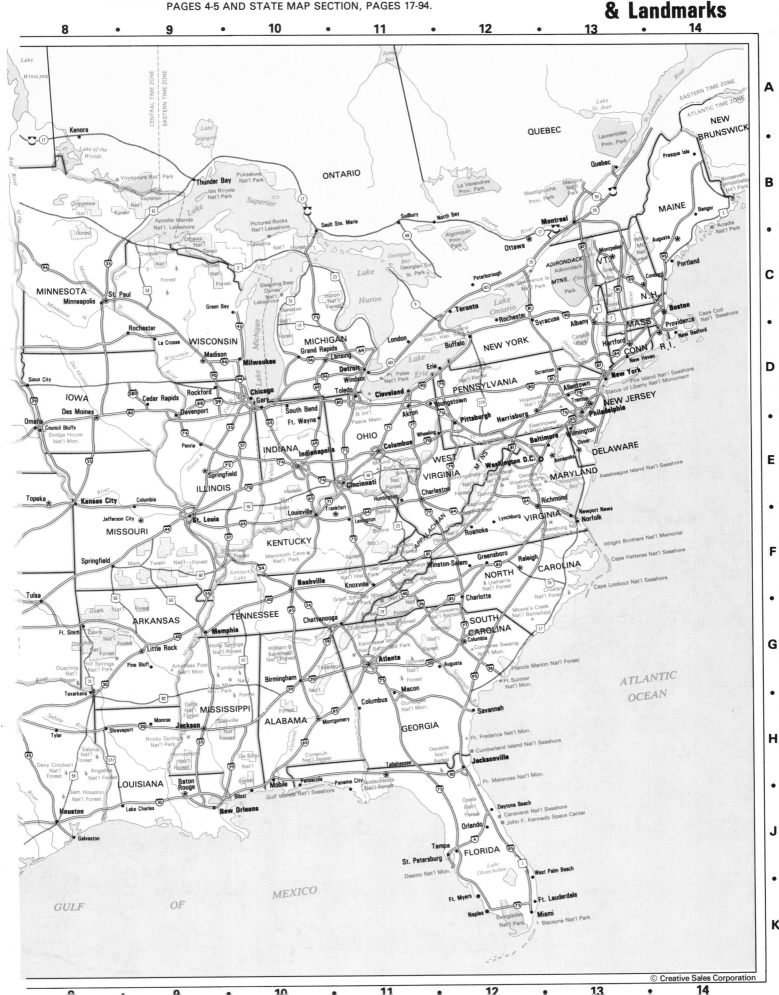

USE ONLY FOR ORIENTATION TO NATIONAL PARKS AND LANDMARKS. FOR MORE DETAILED HIGHWAY INFORMATION SEE INTERSTATE HIGHWAY MAP, PAGES 4-5 AND STATE MAP SECTION, PAGES 17-94.

National Parks & Landmarks 11

© Creative Sales Corporation

ALBERTA
Banff J-4
Calgary J-4
Camrose H-5
Cardston J-5
Drumheller H-5
Edmonton H-5
Edson H-4
Fitzgerald F-5
Fort Chipewyn . . F-4
Ft. McMurray . . . G-5
Ft. Vermillion . . . F-4

Grande Centre . . . G-5
Grande Prairie . . . G-4
Habay F-4
High Level F-4
High River J-4
Hinton H-4
Jasper H-4
Lethbridge H-5
Medicine Hat J-5
Peace Point F-4
Peace River G-4
Rainbow Lake F-4

Red Deer H-5
Rocky Mtn. House . . H-4
Waterways G-5
Wetaskiwin H-4
Whitecourt H-4

BRITISH COLUMBIA
Alert Bay J-2
Atlin F-2
Bella Coola H-2
Campbell River . . . J-2

Cassiar F-2
Courtnay J-2
Cranbrook J-4
Dawson Creek . . . G-3
Ft. Nelson F-3
Ft. St. John G-3
Hazelton G-2
Kamloops J-3
Kelowna J-3
Kimberley J-4
Kitimat H-2
Masset G-1
Nanaimo J-2

Nelson J-4
New Westminster . . . J-3
Ocean Falls H-2
Penticton J-4
Port Alberni J-2
Port Alice J-2
Powell River J-3
Prince George H-3
Prince Rupert G-2
Quesnel H-3
Revelstoke J-4
Sandpit H-1

Smithars G-2
Stewart G-2
Telegraph Creek . . . F-2
Trail J-4
Ucluelet J-2
Vancouver J-3
Vernon J-3
Victoria J-3
Ware G-3
Williams Lake H-3

MANITOBA
Berens River G-7
Brandon H-7
Brochet F-6
Churchill F-7
Dauphin H-7
Flin Flon G-6
Gillam F-7
Gods Lake G-8
Grand Rapids G-7
Island Lake G-7

Lynn Lake F-6
Norway House G-7
Portage la Prairie . . . H-7
Selkirk H-8
Snow Lake G-7
S. Indian Lake F-7
Swan River H-7
The Pas G-6
Thompson F-7
Virden J-7
Winnipeg H-8

NEW BRUNSWICK
Bathurst E-13
Campbellton E-12
Chatam E-13
Edmundston F-12
Fredricton F-13
Moncton E-13
Newcastle E-13
St. John F-13
St. Stephen F-13
Tracadie E-13

Woodstock F-13

NEWFOUNDLAND
Battle Harbour . . . C-13
Bonavista C-14
Cartwright C-12
Channel-Port aux
 Basques D-14
Churchill Falls . . . D-12
Corner Brook D-13
Gander C-14

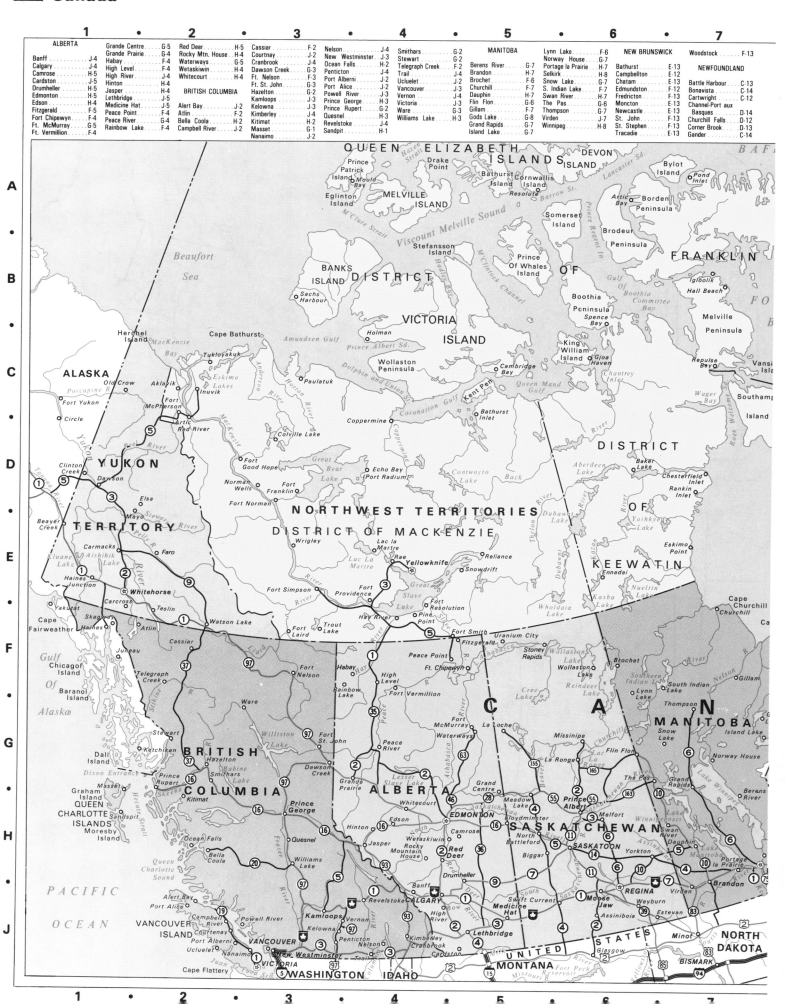

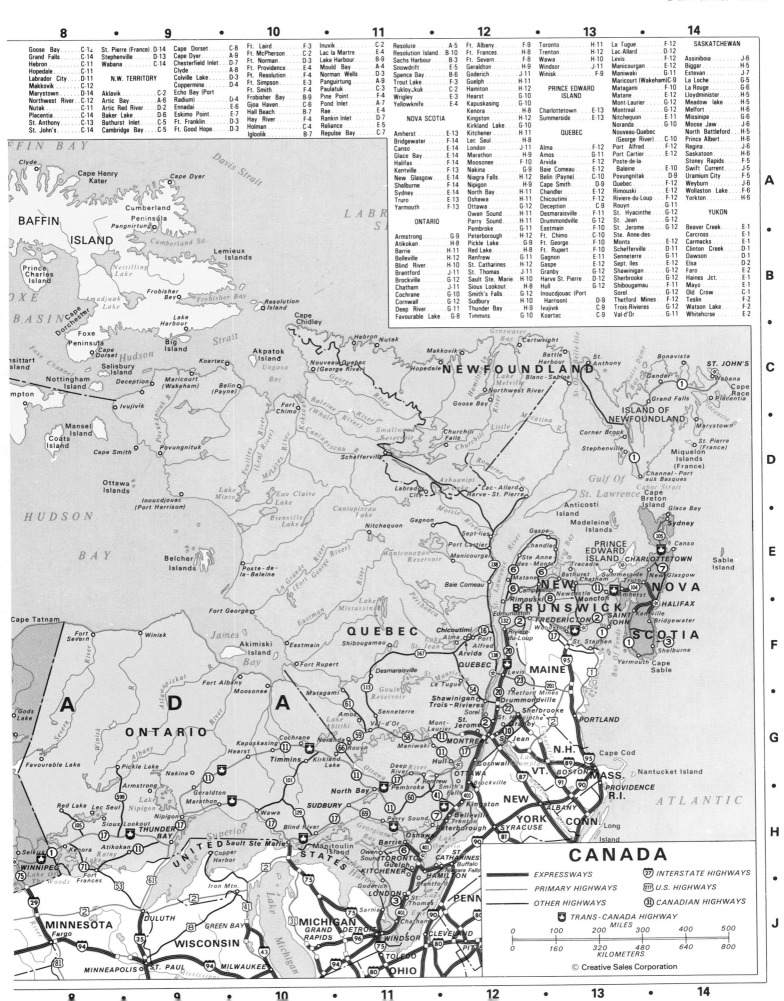

8

Goose Bay C-12
Grand Falls C-14
Hebron C-11
Hopedale C-11
Labrador City . . . D-11
Makkovik C-12
Marystown D-14
Northwest River . C-12
Nutak C-11
Placentia C-14
St. John's C-14

St. Pierre (France) . D-14
Stephenville D-13
Wabana C-14

N.W. TERRITORY

Aklavik C-2
Artic Bay A-6
Artic Red River . . D-2
Baker Lake D-6
Bathurst Inlet . . . C-5
Cambridge Bay . . C-5

9

Cape Dorset C-8
Cape Dyer A-9
Chesterfield Inlet . D-7
Clyde A-8
Colville Lake . . . D-3
Coppermine D-4
Echo Bay (Port Radium) G-4
Ennadai E-6
Eskimo Point . . . E-7
Ft. Franklin D-3
Ft. Good Hope . . D-3

10

Ft. Laird F-3
Ft. McPherson . . E-4
Ft. Norman D-3
Ft. Providence . . E-4
Ft. Resolution . . F-4
Ft. Simpson F-4
Ft. Smith F-4
Frobisher Bay . . B-9
Gjoa Haven B-7
Hall Beach B-7
Hay River E-4
Holman C-4
Igloolik B-7

11

Inuvik C-2
Lac la Martre . . . E-4
Lake Harbour . . . B-9
Mould Bay A-4
Norman Wells . . D-3
Panguirtung A-9
Paulatuk C-3
Pine Point F-4
Pond Inlet A-7
Rae E-4
Rankin Inlet . . . D-7
Reliance E-5
Repulse Bay . . . C-7

12

Resolute A-5
Resolution Island . B-10
Sachs Harbour . . B-3
Snowdrift E-5
Spence Bay B-6
Trout Lake F-3
Tukloy.kuk C-2
Wrigley E-3
Yellowknife E-4

NOVA SCOTIA

Amherst E-13
Bridgewater F-14
Canso E-14
Glace Bay E-14
Halifax F-14
Kentville F-13
New Glasgow . . . E-14
Shelburne F-14
Sydney E-14
Truro E-13
Yarmouth F-13

ONTARIO

Armstrong G-9
Atikokan H-8
Barrie H-11
Belleville H-12
Blind River H-10
Brantford J-11
Brockville G-12
Chatham J-11
Cochrane G-10
Cornwall G-12
Deep River G-11
Favourable Lake . G-8

Ft. Albany F-9
Ft. Frances H-8
Ft. Severn F-8
Geraldton H-9
Goderich J-11
Guelph H-11
Hamiton H-12
Hearst G-10
Kapuskasing . . . G-10
Kenora H-8
Kingston H-12
Kirkland Lake . . G-10
Kitchener H-11
Lec Seul H-8
London J-11
Marathon H-9
Moosonee F-10
Nakina G-9
Niagra Falls . . . H-12
Nipigon H-9
North Bay H-11
Oshawa H-11
Ottawa G-12
Owen Sound . . . H-11
Parry Sound . . . H-11
Pembroke G-11
Peterborough . . H-12
Pickle Lake . . . G-9
Red Lake G-8
Renfrew G-11
St. Catharines . . H-12
St. Thomas J-11
Sault Ste. Marie . H-10
Sioux Lookout . . H-8
Smith's Falls . . . G-12
Sudbury H-10
Thunder Bay . . . H-9
Timmins G-10

13

Toronto H-11
Trenton H-12
Wawa H-10
Windsor J-11
Winisk F-9

PRINCE EDWARD ISLAND

Charlottetown . . E-13
Summerside . . . E-13

QUEBEC

Alma F-12
Amos G-11
Arvida F-12
Baie Comeau . . E-12
Belin (Payne) . . C-10
Cape Smith . . . D-9
Chandler E-12
Chicoutimi F-12
Deception . . . .
Desmaraisville . . F-11
Drummondville . . G-12
Eastmain F-10
Ft. Chimo C-10
Ft. George F-10
Ft. Rupert F-10
Gagnon E-11
Gaspe E-12
Granby G-12
Harve St. Pierre . D-12
Hull G-12
Inoucdjouac (Port Harrison) . . . D-9
Ivujivik C-9
Koartac C-9

La Tugue F-12
Lac-Allard D-12
Levis F-12
Maniwaki G-11
Maricourt (Wakeham)C-9
Matagami F-10
Matane E-12
Mont-Laurier . . G-12
Montreal G-12
Nitchequon . . . E-11
Nouveau-Quebec (George River) . C-10
Port Alfred F-12
Port Cartier . . . E-12
Poste-de-la-Baleine . . . D-9
Povungnitak . . . D-9
Quebec F-12
Rimouski E-12
Riviere-du-Loup . F-12
Rouyn G-11
St. Hyacinthe . . G-12
St. Jean G-12
St. Jerome G-12
Ste. Anne-des-Monts E-12
Schefferville . . . D-11
Senneterre G-11
Sept. Iles E-12
Shawinigan . . . G-12
Sherbrooke . . . G-12
Shibougamau . . F-11
Sorel G-12
Thetford Mines . G-12
Trois-Rivieres . . G-12
Val-d'Or G-11

14

SASKATCHEWAN

Assiniboia J-6
Biggar H-5
Estevan J-7
La Loche G-5
La Rouge G-6
Lloydminster . . H-5
Meadow lake . . H-5
Melfort H-6
Missinipe G-6
Moose Jaw . . . J-6
North Battleford . H-5
Prince Albert . . H-6
Regina J-6
Saskatoon H-6
Stoney Rapids . . F-5
Swift Current . . J-5
Uranium City . . F-5
Weyburn J-6
Wollaston Lake . F-6
Yorkton H-6

YUKON

Beaver Creek . . E-1
Carcross E-1
Carmacks E-1
Clinton Creek . . D-1
Dawson D-1
Elsa D-2
Faro E-2
Haines Jct. . . . E-1
Mayo E-1
Old Crow C-1
Teslin F-2
Watson Lake . . F-2
Whitehorse . . . E-2

CANADA

— EXPRESSWAYS
— PRIMARY HIGHWAYS
— OTHER HIGHWAYS
TRANS-CANADA HIGHWAY

⊘27 INTERSTATE HIGHWAYS
⊙271 U.S. HIGHWAYS
⊙31 CANADIAN HIGHWAYS

MILES
0 · 100 · 200 · 300 · 400 · 500
0 · 160 · 320 · 480 · 640 · 800
KILOMETERS

© Creative Sales Corporation

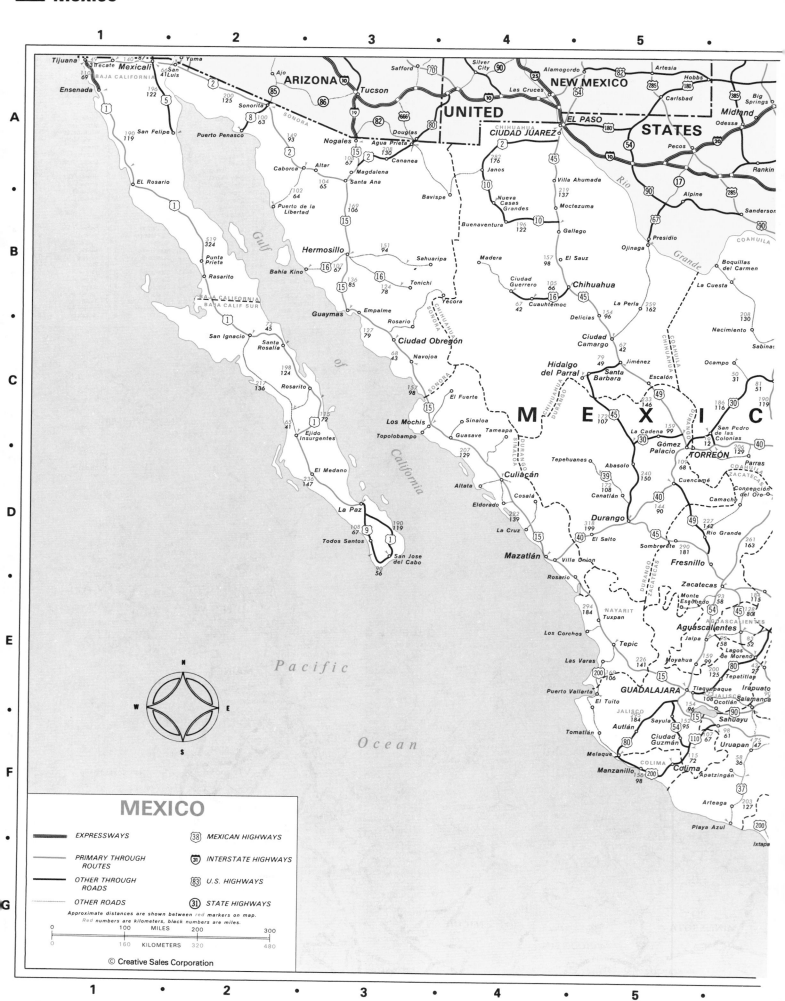

© Creative Sales Corporation

MEXICO

Cities and Towns

Abasolo	D-5	Ciudad del Maiz	E-7
Acambaro	F-6	Coatzacoalcos	F-9
Acapulco	G-6	Colima	F-5
Acatlan	F-7	Comitan	G-9
Acayucan	F-8	Conception de Oro	D-6
Agua Prieta	A-3	Cordoba	F-8
Aguascalientes	E-6	Cosala	D-4
Altar	A-2	Cuauhtemoc	B-4
Altata	D-4	Cuencame	D-5
Alvarado	F-8	Cuernavaca	F-7
Apatzingan	F-6	Culiacan	D-4
Arcelia	F-6	Delicias	B-5
Arriga	G-9	Durango	D-5
Arteaga	F-6	Dzilam de Bravo	E-10
Arlixco	F-7	Ejido Insurgentes	C-3
Autlan	F-5	Eldorado	D-4
Bahia Kino	B-2	El Fuerte	C-4
Bavispe	B-4	El Medana	D-3
Becal	E-10	El Rosario	A-1
Boquillas de		El Sauz	B-4
Carmen	B-6	El Tuito	E-5
Buenaventura	B-4	Empalme	B-3
Caborca	A-2	Ensenada	A-1
Camacho	D-6	Escalon	C-5
Campeche	E-10	Escarcega	F-10
Cananea	A-3	Fresnillo	D-5
Canatlan	D-5	Gallego	B-4
Cardenas	F-9	Gomez Palacio	C-5
Celaya	E-6	Guadalajara	E-5
Celestun	E-10	Guasave	C-4
Champoton	F-10	Guaymas	B-3
Chetumal	F-11	Hermosillo	B-3
Chihuahua	B-5	Hidalgo del Parral	C-5
Chilpancingo	F-7	Hopelchen	E-10
China	C-7	Huajuapan de Leon	F-7
Ciudad Acuna	B-6	Iguala	F-7
Ciudad Camargo	C-5	Irapuato	E-6
Ciudad Guerrero	B-4	Iturbide	F-10
Ciudad Guzman	F-5	Jalapa	F-8
Ciudad Juarez	A-4	Jalpa	E-5
Ciudad Madero	E-7	Janos	A-4
Ciudad Mante	E-7	Jimenez	C-5
Ciudad Victoria	D-7	Juchitan	G-8
Ciudad de Carmen	F-9	La Cruz	D-4
Ciudad de Valles	E-7	La Cadena	C-5
		La Cuesta	B-6
		Lagos de Morena	E-6
		La Paz	D-3
		La Perla	B-5

La Pesca	D-7	Paraiso	F-9
La Piedad	E-6	Parras	C-6
Las Varas	E-5	Peto	E-10
Leon	E-6	Piedras Negras	B-6
Linares	D-7	Pijijiapan	G-9
Los Corchos	E-5	Pinotepa Nacional	G-7
Los Mochis	C-3	Piste	E-11
Madera	B-4	Playa Azul	F-6
Magdalena	A-3	Pochutla	G-8
Malpaso	G-9	Poza Pica	E-7
Manuel	E-7	Progreso	E-10
Manzanillo	F-5	Puebla	F-7
Matamoros	C-7	Puerto de la	
Matehuala	D-6	Libertad	B-2
Matias Romero	G-8	Puerto Escondido	G-8
Mazatlan	D-4	Puerto Juarez	E-11
Melaque	F-5	Puerto Madero	G-9
Merida	E-10	Puerto Penasco	A-2
Mexicali	A-1	Punta Prieta	B-2
Mexico City	F-7	Queretaro	E-6
Miahuatlan	G-8	Rasarito	B-2
Mier	C-7	Reynosa	C-7
Minatitlan	F-8	Rio Grande	D-6
Moctezuma	B-4	Rio Lagartos	E-11
Molango	E-7	Rosario	C-3
Moncloya	C-6	Rosario	D-4
Monte Escobedo	E-5	Sabinas	C-6
Montemorelos	D-7	Sabinas Hidalgo	C-6
Monterrey	C-6	Sahuaripa	B-3
Morelia	F-6	Salamanca	E-6
Morelos	B-6	Salinas	E-6
Moyahua	E-5	Salina Cruz	G-8
Nacimiento	C-6	Saltillo	C-6
Nautla	E-8	San Andres Tuxtla	F-8
Navojoa	C-3	San Cristobal	G-9
Nogales	A-3	San Felipe	A-2
Nueva Casas		San Fernando	D-7
Grandes	B-4	San Ignacio	C-2
Nueva Rosita	C-6	San Jose del Cabo	D-3
Nuevo Laredo	C-7	San Luis	A-2
Oaxaca	G-8	San Luis Potosi	E-6
Ocampo	C-6	San Pedro de las	
Ocotlan	E-6	Colonias	C-6
Ojinaga	B-5	Santa Ana	A-3
Ometepec	G-7	Santa Barbara	C-5
Orizaba	F-8	Santa Rosalia	C-2
Pachuca	E-7	Sayula	F-5
Palenque	F-9	Sinaloa	C-4
Papantla	E-7	Sombrerete	D-5

Sonorita	A-2
Soto La Marina	D-7
Tameapa	C-4
Tampico	E-7
Tapachula	G-9
Tapanatepec	G-9
Taxco	F-7
Teapa	F-9
Tecate	A-1
Tehuacan	F-7
Tehuantepec	G-8
Temporal	E-7
Tepatitlan	E-6
Tepehuanes	D-5
Tepic	E-5
Ticul	E-10
Tijuana	A-1
Tiquicheo	F-6
Tlaciaco	G-7
Tlaxcala	F-7
Tlaxiaco	G-7
Todos Santos	D-3
Toluca	F-7
Tomatian	F-5
Tonichi	B-3
Topolobampo	C-3
Torreon	C-5
Totolapan	G-8
Tulancingo	F-7
Tulum	E-11
Tuxpan	E-5
Tuxpan	E-7
Tuxtepec	F-8
Tuxtla Gutierrez	G-9
Uruapan	F-6
Valladolid	E-11
Veracruz	F-8
Villa Ahumada	A-4
Villagran	D-7
Villahermosa	F-9
Villa Union	D-4
Xcan	E-11
Yecora	B-4
Zacatal	F-9
Zacatecas	E-6
Zamora	F-6
Zihuatanejo	F-6
Zimapan	E-7
Zitacuaro	F-6

STATE MAP LEGEND

ROAD CLASSIFICATIONS & RELATED SYMBOLS

Free Interstate Hwy.	══○90○══
Toll Interstate Hwy.	══○76○══
Divided Federal Hwy.	═○14○═
Federal Hwy.	─○20○─
Divided State Hwy.	═○31○═
State Hwy.	─○147○─
Other Connecting Road	┈[258]┈
Trans - Canada Hwy.	─⌂─
Point to Point Milage	─ 17 ─
State Boundaries	▬ ▬ ▬ ▬ ▬

LAND MARKS & POINTS OF INTEREST

Indian Reservation

National & State Forest or Wildlife Preserve

Military Installation

National & State Park or Recreation Area

Grassland

Desert

River, Lake, Ocean or other Drainage

Urban Area — **Denver**

Airport — ✈

State Capital — ✪

Park, Monument, University or other Point of Interest — ■

Roadside Table or Rest Areas — ▲

CITIES & TOWNS - Type size indicates the relative population of cities and towns

Mapleton	Kenhorst	Somerset	Butler	Auburn	Harrisburg	Madison	Chicago
under 1000	1000-5,000	5,000-10,000	10,000-25,000	25,000-50,000	50,000-100,000	100,000-500,000	500,000 and over

SCALE OF MILES
1 INCH IS APPROXIMATELY 190 MILES

0 40 80 120 160 200

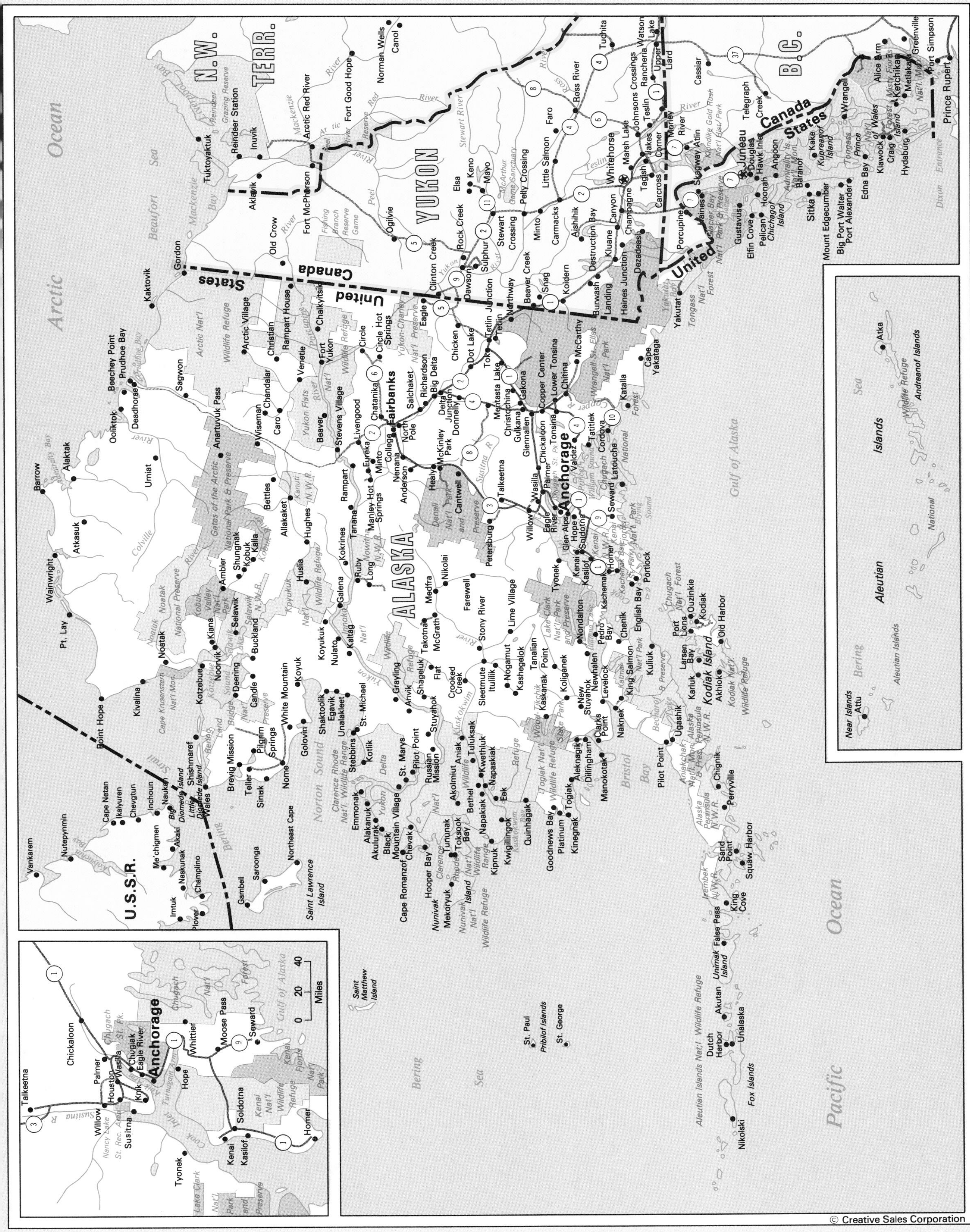

N.W. TERR.

B.C.

YUKON

Canada
United States

ALASKA

U.S.S.R.

Arctic Ocean

Beaufort Sea

Bering Sea

Gulf of Alaska

Pacific Ocean

Aleutian Islands

Near Islands

Andreanof Islands

Saint Lawrence Island

Saint Matthew Island

Pribilof Islands

Kodiak Island

Nunivak Island

Anchorage

Fairbanks

Juneau

Barrow

Nome

Kotzebue

Prudhoe Bay

Miles 0 20 40

© Creative Sales Corporation

1 INCH IS APPROXIMATELY 17.5 MILES

0 3.5 7 10.5 14 17.5

N

A B C D E F G

Maui

Kalahu Pt.
Muolea Pt.
Mokea Pt.
Kipahulu
Hana
(360)
Pukalua Pt.
Wailua
Waianapanapa St. Pk.
Kaupo
Kahakuloa Pt.
Makawao
Haleakala Crater
Haleakala Nat'l Park
Apole Pt.
(378)
Pauwela Pt.
Haiku
(37)
Pauwela
Paia
(36)
Spreckelsville
Ulupalakua
Keokea
Keoneoio
Cape Hanamanioa
Nakalele Pt.
Waihee Pt.
Kahului
Puunene
Kahului Bay
(340)
Wailuku
Iao Valley
(30)
Kihei
Nukuele Pt.
Honokahua
Mopua
Maalaea
Wailea
Kamaole Beach Park
Makena
(30)
Olowalu
Hekili Pt.
Lahaina

Pacific Ocean

4
2
0
Miles

Molokai

Lamaloa Head
Halawa
Cape Halawa
Waialua
Pauwalu
Kikipua Pt.
Pukoo
Ulapue
Kahlu Pt.
Makanalua Pen.
Kualapapa
Kalae
(450)
Kamalo
Kualapu
Kaunakakai
Kamiloloa
Lilo Pt.
(460)
Mauna Loa
Kolo
Laau Pt.

Palioolo Channel

Pacific Ocean

4
2
0
Miles

Niihau (Private)

Kaulakahi Channel
Puuwai

Kauai

Anahola
(56)
Lihue
Haena
Lawai
(50)
Mana
Waianae

Kauai Channel

Oahu

Kahuku
Kahana
(83)
Pearl City
Kaneohe
Kailua
(2)
Haleiwa
Honolulu
Makaha
Nanakuli
Waikiki

Molokai

(450)
Halawa
Kualapuu
(460)
Kamalo
Koele
Lanai City
Keomuku
Mauna Loa

Pauhi Channel
Kalohi Channel
Kealaikahiki Channel
Auau Channel

Lanai

Kahoolawe

Maui

Hana
(360)
Honokahua
Kahului
Lahaina
Ulupalakua
(36)
(37)
(31)
(30)

Kakao Pt.

Maui Co.
Hawaii Co.

Alenuihaha Channel

HAWAII

Honolulu Co.
Maui Co.

Kauai Co.
Honolulu Co.

Kauai

Hawaii

Pohoiki
Waiakea
Ophikao
Black Sands
Kaloapana
Honohina
Hakalau
Pepeekeo Pt.
Papaikou
Hilo
Pahoa
Kaimu
Kaena Pt.
(130)
Keaau
Kukuihaele
Honokaa
Ookala
(19)
Paukaa
Kurtistown
Rainbow Falls
Mountain View
Glenwood
Apua Pt.
Kunaliu Black Sand Beach
Kamuela
(200)
Mauna Kea 13,796 ft.
Hawaii Volcanoes National Park
Honuapo
(11)
Hawi
Niulii
(250)
Waiaka
Waimea
Mauna Loa 13,680 ft.
Pahala
Naalehu
Kaaluulu
Upolo Pt.
Mahukona
(270)
Kawaihae
Puako
(190)
Kalaoa
Keokea
Papa
Waiohinu
(19)
Keahole Pt.
Kailua
Keauhou
Napoopoo
Captain Cook
Kainaliu
Honokahau
Honaunau
Hookena
(11)
Miloli
Hanamalo Pt.
Kauna Pt.
Waiahukini
Ka Lae

Pacific Ocean

Hawaii

Oahu

Mokapu Pt.
Waimanalo
Kaneohe Marine Air Station
Makapuu Pt.
Sea Life Park
Kualoa Pt.
Kahana Beach
Kaneohe Bay
Kailua
Koko Head Park
Kahana
(83)
Kaaawa
(72)
Koko Head
Hawaii Kai
(3)
(61)
Pali Lookout
(1)
Kahuku
(63)
Diamond Head
Kahuku Pt.
Laie
Hauula
Kaneohe
Waikiki
Polynesian Cultural Center
Aiea
(92)
Sacred Falls
Koolau Range
Pearl City
(78)
Honolulu
Sunset Beach
Wahiawa
Wapahu
(99)
(2)
(250)
Haleiwa
Schofield Barracks
Milani Town
(750)
Ewa
(95)
Waialua
(99)
Range
Waipahu
Makakilo City
Barbers Pt.
(83)
Waianae Range
(780)
(1)
Kahuku Pt.
Mokuleia
Nanakuli
Maili
Barbers Pt. Air Sta.
Kepuhi Pt.
(930)
Dillingham Air Force Base
Makaha
Waianae
Kaena Pt.

Pacific Ocean
Kauai Channel

4
2
0
Miles

Kauai

Moloaa
Anahola
Kealia
Kapaa
Wailua
Hanamaulu
(56)
Lihue Airport
Ninini Pt.
Nawiliwili
(580)
Lihue
(583)
Puhi
(50)
Kilauea
Koloa
(56)
Hanalei
Mt. Waialeale 5243 ft.
Lawai
Makahuena Pt.
Haena Pt.
Haena
Kalaheo
Eleele
Port Allen
Koheo Pt.
Kalalau
Kaumakani
Hanapepe
(50)
Waimea
Waimea Canyon
Kokee State Park
(550)
Makaha Pt.
(550)
Mana
Kekaha

Pacific Ocean

4
2
0
Miles

© Creative Sales Corporation

SCALE OF MILES
1 INCH IS APPROXIMATELY 35 MILES

0 7 14 21 28 35

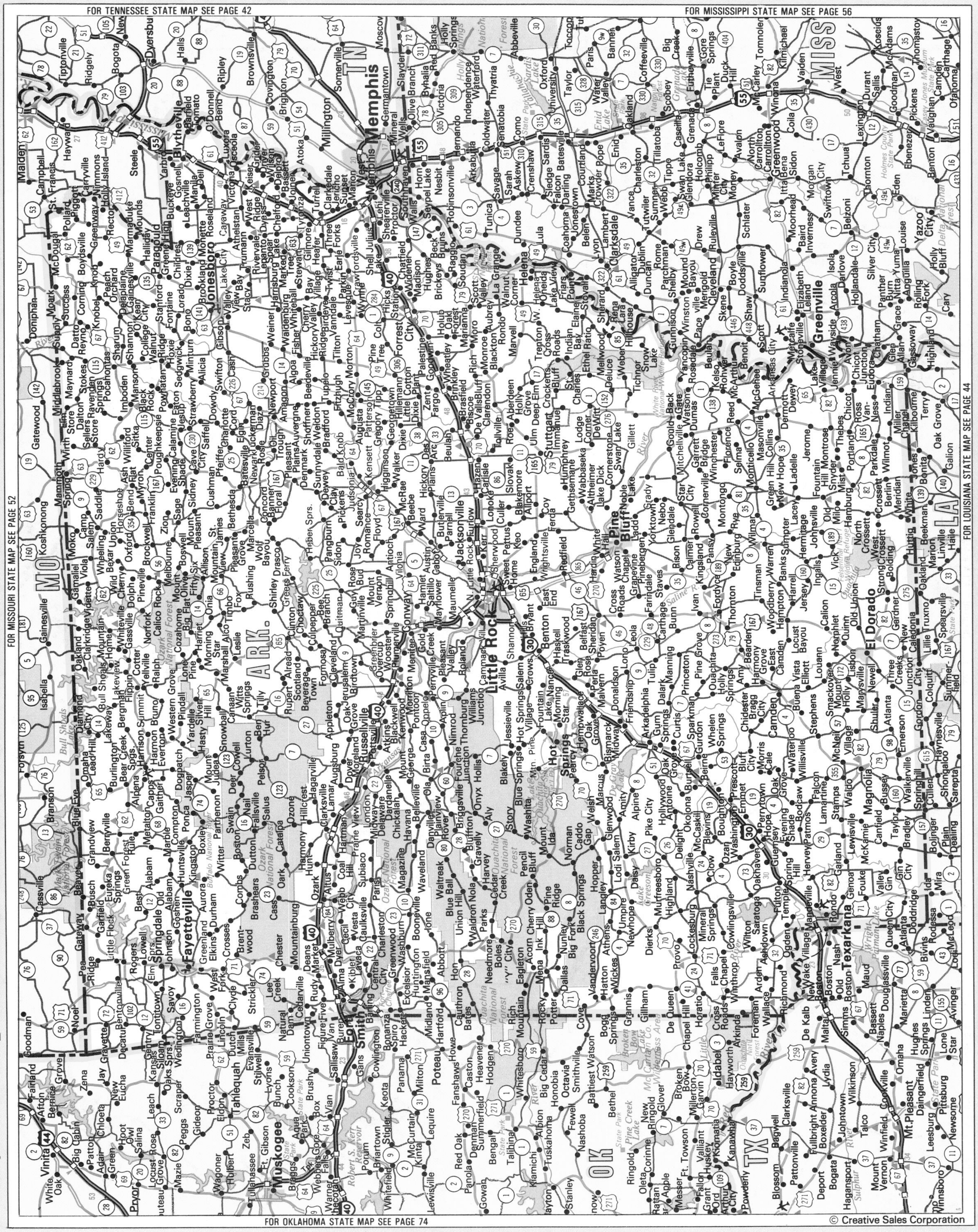

N

FOR TENNESSEE STATE MAP SEE PAGE 42

FOR MISSISSIPPI STATE MAP SEE PAGE 56

FOR MISSOURI STATE MAP SEE PAGE 52

FOR LOUISIANA STATE MAP SEE PAGE 44

FOR OKLAHOMA STATE MAP SEE PAGE 74

ARK.

MO

TN

MISS

LA

OK

TX

Memphis

Little Rock

Fayetteville

Springdale

Ft. Smith

Hot Springs

Pine Bluff

El Dorado

Texarkana

Jonesboro

Greenville

© Creative Sales Corporation

SCALE OF MILES
1 INCH IS APPROXIMATELY 35 MILES
0 7 14 21 28 35

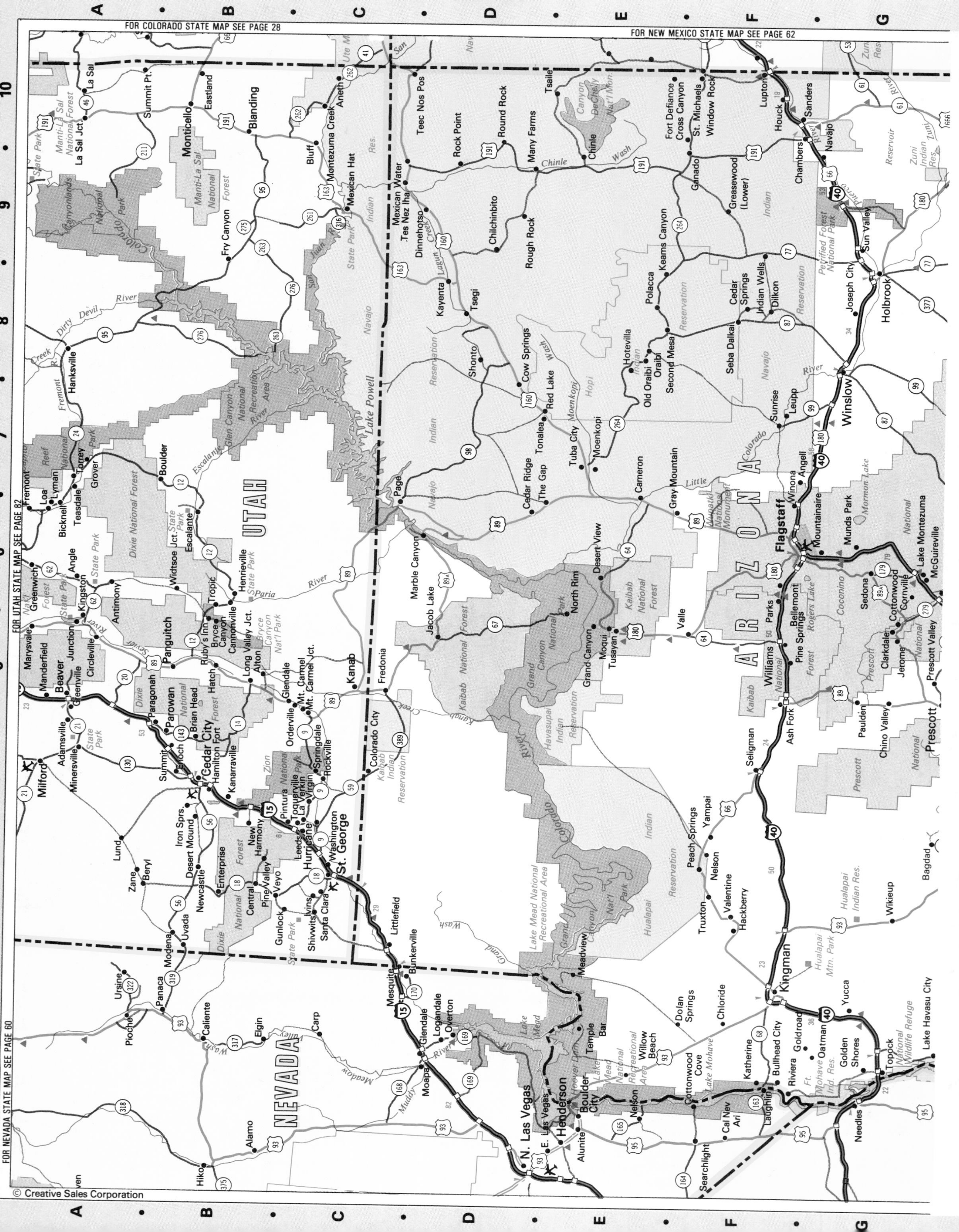

FOR COLORADO STATE MAP SEE PAGE 28

FOR NEW MEXICO STATE MAP SEE PAGE 62

FOR UTAH STATE MAP SEE PAGE 82

FOR NEVADA STATE MAP SEE PAGE 60

UTAH

NEVADA

ARIZONA

© Creative Sales Corporation

SCALE OF MILES
1 INCH IS APPROXIMATELY 35 MILES
0 7 14 21 28 35

FOR NEW MEXICO STATE MAP SEE PAGE 62

UNITED STATES
MEXICO

S O N O R A

UNITED STATES
MEXICO

FOR CALIFORNIA STATE MAP SEE PAGES 24-27

© Creative Sales Corporation

Zion Res., St. Johns, Eagar, Concho, Snowflake, Taylor, Springerville, Greer, Nutrioso, Alpine, Vernon, Clay Springs, Shumway, Show Low, Lakeside, Pinetop-Lakeside, Hawley Lake, Whiteriver, Fort Apache, Beaverhead, Hannigan Meadow, Heber, Overgaard, Christopher Creek, Cedar Creek, Carrizo, Pinedale, Star Valley, Young, Kohls Ranch, McGuireville, Camp Verde, Strawberry, Pine, Payson, Rye, Punkin Center, Tortilla Flat, Sunflower, Roosevelt, Claypool, Miami, Globe, Superior, Peridot, San Carlos, Winkelman, Dudleyville, Mammoth, San Manuel, Oracle, Oracle Jct., Catalina, Oro Valley, Tucson, South Tucson, Vail, Mt. View, Continental, Green Valley, Amado, Tubac, Carmen, Tumacacori, Sahuarita, Sasabe, Sonoita, Patagonia, Nogales

Prescott, Prescott Valley, Dewey, Humboldt, Wilhoit, Mayer, Kirkland, Kirkland Jct., Yarnell, Peeples Valley, Hillside, Congress, Aguila, Wickenburg, Morristown, Circle City, Wittmann, Surprise, Sun City, Youngtown, Peoria, Glendale, Litchfield, Goodyear, Avondale, Buckeye, Tolleson, Guadalupe, Gilbert, Chandler, Mesa, Tempe, Phoenix, Scottsdale, Paradise Valley, Fountain Hills, Carefree, Cave Creek, New River, Rock Springs, Black Canyon City, Cordes Jct.

Lake Havasu City, Parker, Parker Dam, Earp, Vidal, Vidal Jct., Rice, Midland, Blythe, Ehrenberg, Quartzsite, Bouse, Vicksburg, Brenda, Hope, Salome, Wenden, Harcuvar, Tonopah, Wintersburg, Arlington, Palo Verde, Gila Bend, Sentinel, Dateland, Tacna, Wellton, Dome, Yuma, Somerton, Gadsden, San Luis, Winterhaven

Ajo, Why, Childs, Lukeville, Covered Wells, Tracy, Santa Rosa, Anegam, Quijotoa, Sells, Artesa, Three Points (Robles Jct.), Silver Bell, Marana, Red Rock, Picacho, Eloy, Casa Grande, Coolidge, Florence, Arizona City, Stanfield, Maricopa, Francisco Grande, Chuichu, Sacaton, Olberg, Sacate, Bapchule, Ak Chin

Clifton, Morenci, Stargo, Duncan, Franklin, Solomon, Safford, Thatcher, Pima, Central, Eden, Glenbar, Ashurst, Geronimo, Bylas, Fort Thomas, Bonita, Fort Grant, Willcox, Cochise, Dragoon, Benson, Pomerene, Curtis, Fairbank, St. David, Tombstone, Bisbee, Douglas, Bowie, San Simon, Dos Cabezas, Sunizona, Sunsites, Pearce, Elfrida, McNeal, Pirtleville, Paul Spur, Huachuca City, Sierra Vista, Nicksville, Palominas, Nogales

Coronado National Forest, Saguaro National Monument, Organ Pipe Cactus National Monument, Cabeza Prieta National Wildlife Refuge, Barry M. Goldwater Air Force Range, Papago Indian Reservation, Kofa National Wildlife Refuge, Cibola Nat'l Wildlife Refuge, Imperial Nat'l Wildlife Refuge, Fort Yuma Indian Reservation, Painted Rocks State Park, Kitt Peak Nat'l Observatory, San Xavier Ind. Res., Yuma Proving Grounds, Alamo Lake State Park, Buckskin Mtn. State Pk., Lake Havasu State Park, Chemehuevi Indian Reservation

Gila River, Colorado River, San Pedro River, Santa Cruz River, Bill Williams River, Hassayampa River, Centennial Wash, Gila Bend Indian Reservation, Ak-Chin Indian Res., Gila River Indian Reservation

Aqua Prieta, Naco, Gananea, Imuris, San Ignacio, Caborca, Altar, San Luisito, El Paplo, Sonoita, Pto. Penasco, La Jovita, San Felipe

SCALE OF MILES
1 INCH IS APPROXIMATELY 35 MILES
0   7   14   21   28   35

N

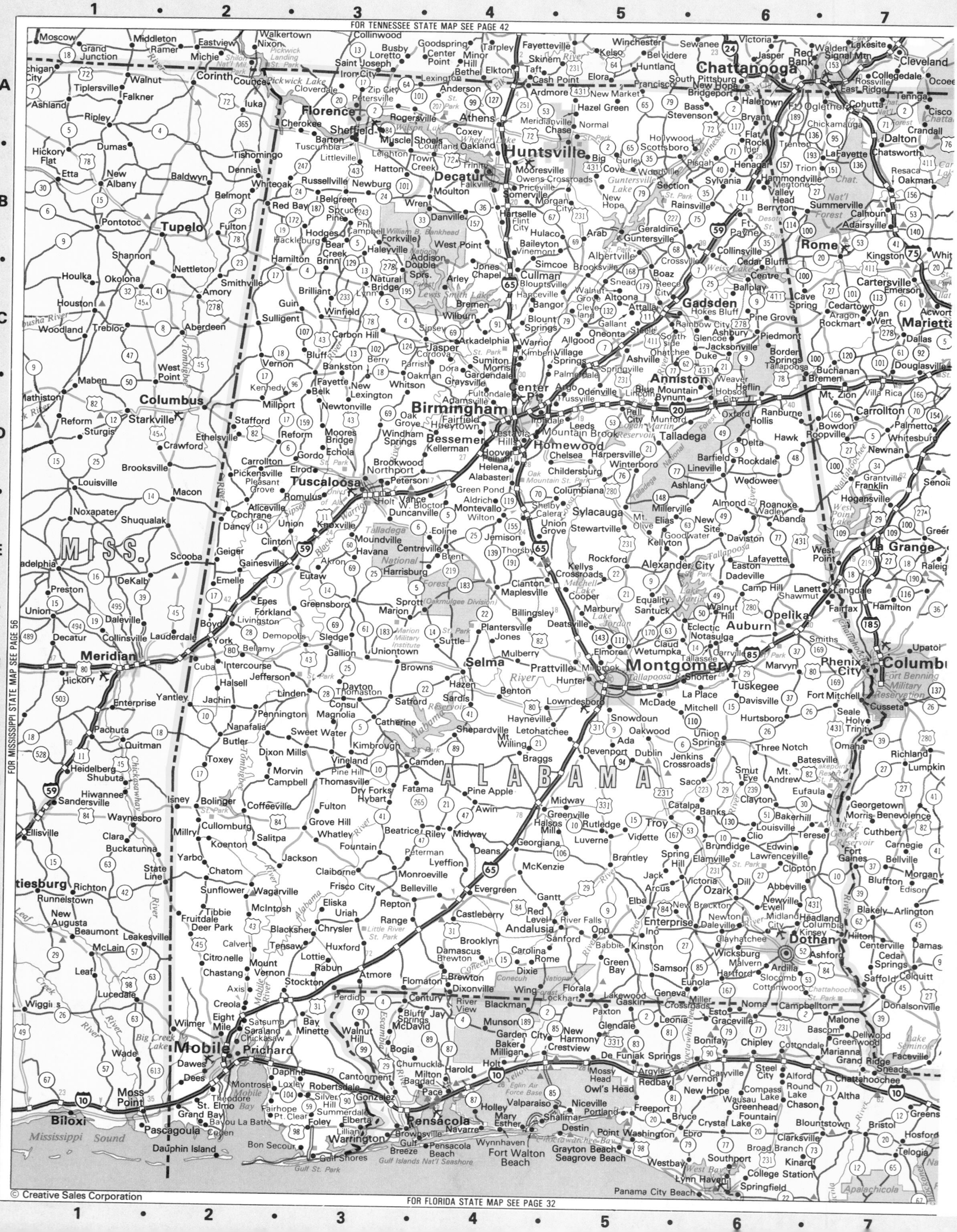

FOR TENNESSEE STATE MAP SEE PAGE 42

FOR MISSISSIPPI STATE MAP SEE PAGE 56

MISS.

ALABAMA

Major cities: Chattanooga, Dalton, Rome, Huntsville, Decatur, Florence, Sheffield, Tuscumbia, Muscle Shoals, Tupelo, Columbus, Gadsden, Anniston, Birmingham, Bessemer, Homewood, Meridian, Tuscaloosa, Selma, Montgomery, Phenix City, Columbus, Auburn, Opelika, La Grange, Dothan, Enterprise, Andalusia, Mobile, Prichard, Biloxi, Pascagoula, Pensacola, Warrington, Gulf Shores, Fort Walton Beach, Panama City Beach.

© Creative Sales Corporation

FOR FLORIDA STATE MAP SEE PAGE 32

SCALE OF MILES
1 INCH IS APPROXIMATELY 35 MILES
0 7 14 21 28 35

© Creative Sales Corporation

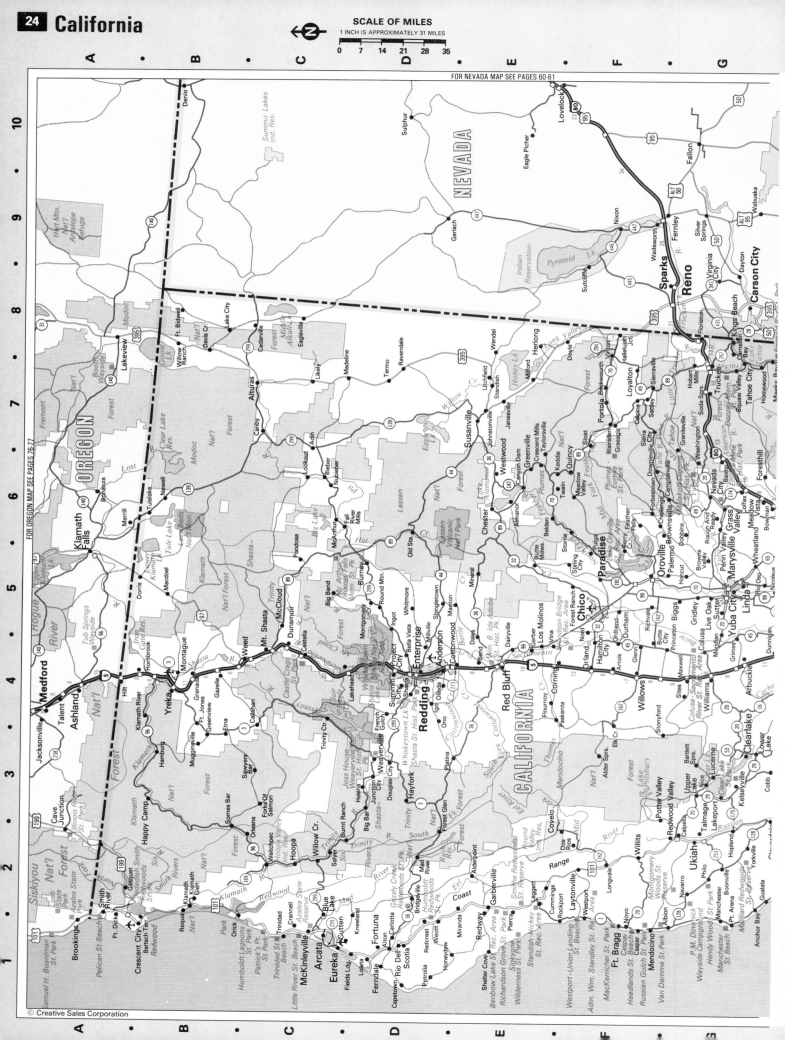

SCALE OF MILES
1 INCH IS APPROXIMATELY 31 MILES
0   7   14   21   28   35

FOR NEVADA MAP SEE PAGES 60-61

FOR OREGON MAP SEE PAGES 76-77

NEVADA

OREGON

CALIFORNIA

© Creative Sales Corporation

SCALE OF MILES
1 INCH IS APPROXIMATELY 31 MILES
0 7 14 21 28 35

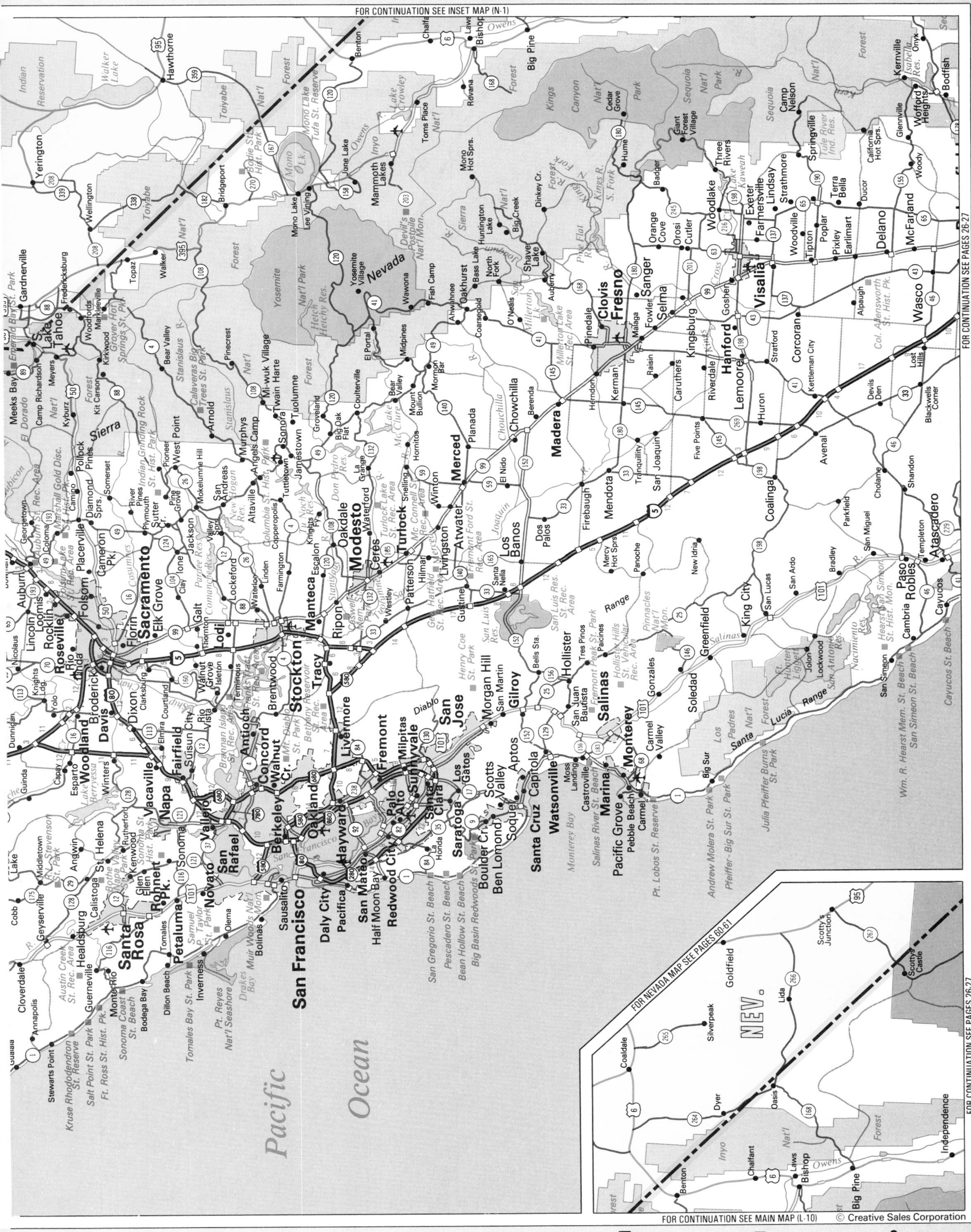

FOR CONTINUATION SEE INSET MAP (N-1)

FOR CONTINUATION SEE PAGES 26-27

FOR NEVADA MAP SEE PAGES 60-61

FOR CONTINUATION SEE PAGES 26-27

FOR CONTINUATION SEE MAIN MAP (L-10)

© Creative Sales Corporation

Pacific Ocean

Nevada

NEV.

Major cities and places: Lake Tahoe, Sacramento, Stockton, Modesto, Merced, Madera, Fresno, Clovis, Visalia, San Francisco, Oakland, Berkeley, San Jose, Santa Cruz, Monterey, Salinas, Napa, Santa Rosa, Vacaville, Fairfield, Vallejo, San Rafael, Daly City, Hayward, Fremont, Palo Alto, Sunnyvale, Gilroy, Hollister, King City, Paso Robles, Atascadero, Delano, Wasco, Hanford, Lemoore, Coalinga, Los Banos, Turlock, Manteca, Tracy, Livermore, Auburn, Placerville, Folsom, Roseville, Lodi, Galt, Oakdale, Sonora, Bishop, Big Pine, Mammoth Lakes, Bridgeport, Yosemite National Park, Sequoia National Park, Kings Canyon National Park

SCALE OF MILES
1 INCH IS APPROXIMATELY 31 MILES

0   7   14   21   28   35

N

FOR CONTINUATION SEE PAGES 24-25

3 · 4 · 5 · 6 · 7 · 8 · 9 · 10

M
N
P
Q
R
S
T
U
V

Pacific Grove
Marina   Salinas
Pebble Beach
Carmel
Monterey
Pt. Lobos St. Reserve
Carmel Valley
Gonzales
Andrew Molera St. Park
Big Sur
Soledad
Pfeiffer - Big Sur St. Park
Greenfield
Los Padres Nat'l Forest
Julia Pfeiffer Burns St. Park
King City
San Lucas
San Ardo
Ft. Hunter Liggett
Jolon
Lockwood
San Antonio Res.
Bradley
Nacimiento Res.
Parkfield
San Miguel
San Simeon
Hearst San Simeon St. Hist. Mon.
Wm. R. Hearst Mem. St. Beach
San Simeon St. Beach
Cambria
Cholame
Paso Robles
Templeton
Shandon
Cayucos St. Beach
Cayucos
Atascadero
Atascadero St. Beach
Morro Bay
Morro Bay St. Park
Santa Margarita
Montana De Oro St. Park
Baywood Pk.
San Luis Obispo
Pismo Beach
Pozo
Pismo St. Beach
Grover City
Arroyo Grande
Oceano
Nipomo
San Luis Obispo Bay
Twitchell Res.
Guadalupe
Pt. Sal St. Beach
Orcutt
Santa Maria
Casmalia
Sisquoc
Los Alamos
Surf
Los Olivos
Lompoc
La Purisima Mission St. Hist. Park
Buellton
Solvang
Lake Cachuma
Gaviota St. Park
Gaviota
El Capitan St. Beach
Montecito
Refugio St. Beach
Goleta
Summerland
Carpinteria
Ojai
Santa Barbara
Carpinteria St. Beach
Emma Wood St. Beach
Santa Paula
Fillmore
Ventura
Saticoy
Moorpark
Oxnard
Simi Valley
Port Hueneme
Agoura Hills
Pt. Mugu St. Park
Thousand Oaks
Malibu
Leo Carrillo St. Beach
Beverly Hills
Santa Monica
Los Angeles
Santa Monica Bay
Redondo Beach
Rancho Palos Verdes
Long Beach
Huntington Beach
Newport Beach
Laguna

Pacific
Ocean

San Miguel Is.
Santa Cruz Is.
Santa Rosa Is.
Santa Barbara Channel
San Miguel Passage
Santa Cruz Channel
Channel Islands Nat'l Mon.
San Nicolas Is.
Santa Barbara Is.
Channel Islands Nat'l Mon.
Santa Catalina Is.
Avalon
San Pedro Channel
Outer Santa Barbara Channel
San Clemente Is.

Mercy Hot Sprs.
Mendota
Clovis
Fresno
Kerman
Tranquillity
Panoche
San Joaquin
Raisin
Malaga
Fowler
Sanger
Selma
Caruthers
Orange Cove
Orosi
Cutler
Kingsburg
Riverdale
Woodlake
Three Rivers
Five Points
Goshen
Lemoore
Hanford
Visalia
Exeter
Farmersville
New Idria
Coalinga
Huron
Stratford
Lindsay
Strathmore
Camp Nelson
Corcoran
Woodville
Avenal
Kettleman City
Tipton
Poplar
Pixley
Earlimart
Terra Bella
Springville
Tule River Ind. Res.
Devils Den
Alpaugh
Col. Allensworth St. Hist. Pk.
Ducor
California Hot Sprs.
Lost Hills
Wasco
Delano
McFarland
Blackwells Corner
Glennville
Kernville
Wofford Heights
Onyx
Isabella Res.
Shafter
Bodfish
Buttonwillow
Green Acres
Bakersfield
McKittrick
Tule Elk St. Reserve
Edison
Fellows
Pumpkin Center
Lamont
Caliente
Ford City
Taft
Maricopa
Arvin
Keene
Tehachapi
Cuyama
New Cuyama
Frazier Pk.
Ft. Tejon St. Hist. Pk.
Willow Sprs.
Gorman
Castaic Lake St. Rec. Area
Palmdale
Castaic
Acton
San Fernando
Valencia
Glendale
Pasadena
Santa Catalina Is.

© Creative Sales Corporation

3 · 4 · 5 · 6 · 7 · 8 · 9 · 10

N

11   12   13   14   15   16   17

M
N
P
Q
R
S
T
U
V

NEVADA

Independence
Lone Pine
Keeler
Cartago
Olancha
Darwin
Little Lake
Panamint Sprs.
Stovepipe Wells
Death Valley
Salt Cr.
Furnace Cr. Ranch
Death Valley Jct.
Panamint Range
Nat'l Monument
Amargosa R.
Beatty
Armagosa Valley
Mercury
Indian Springs
Pahrump
Shoshone
Tecopa
Goodsprings
Jean
Toyabe Nat'l Forest
Las Vegas
North Las Vegas
Henderson
Boulder City
Nelson
Willow Beach
Lake Mead
Lake Mead Nat'l Rec.
Temple Bar
Mesquite
Bunkerville
Glendale
Overton
Virgin R.

Owens Lk.
Haiwee Res.
Trona
Inyokern
China Lake
Ridgecrest
Johannesburg
Randsburg
Red Mountain
Cantil
Red Rock Canyon St. Park
California City
Mojave
Boron
North Edwards
Hinkley
Barstow
Yermo
Lenwood
Helendale
Daggett
Newberry Sprs.
Oro Grande
Cady Mtns.
Ludlow
Essex
Amboy
Kelso
Goffs
Fenner
Needles
Golden Shores
Topock
Providence Mtns. St. Rec. Area
Sacramento Mtns.
Old Woman Mtns.
Sheep Hole Mtns.
Ivanpah
Cima
Nipton
Searchlight
Laughlin
Bullhead City
Black Mtns.
Mohave
Cottonwood Cove
Chloride
Kingman
McConnico
Yucca
Hualapai Mtns.
ARIZONA
Lake Havasu City
Lake Havasu State Pk.
Parker Dam
Earp
Vidal
Parker
Buckskin Mtn. State Pk.
Rice
Poston
Bouse
Quartzsite
Vicksburg
Blythe
Ehrenberg
Ripley
Palo Verde
Colorado R.
Cibola Nat'l Wildlife Refuge
Imperial Nat'l Wildlife Refuge
Stone Cabin
Martinez Lake
Picacho St. Rec. Area
Imperial Dam
Dome
Winterhaven
Yuma
Wellton
Tacna
Mohawk
Somerton
San Luis Rio Colorado
Galeana

U.S. Naval Weapons Sta.
Ft. Irwin Military Res.
Edwards Air Force Base
Rosamond
Lancaster
Pearblossom
Adelanto
Phelan
Victorville
Apple Valley
Hesperia
Lucerne Valley
Bullion Mtns.
Twentynine Palms Marine Corps Base
Joshua Tree
Twentynine Palms
Yucca Valley
Morongo Valley
Desert Hot Sprs.
Thousand Palms
Indio
Desert Center
Granite Mtns.
Chuckwalla Mtns.
Big Maria Mtns.
Saddleback Butte St. Pk.
Wrightwood
Silverwood Lake St. Rec. Area
San Bernardino Nat'l Forest
Fawnskin
Big Bear Lake
Glendora
Pomona
Ontario
San Bernardino
Riverside
Yucaipa
Beaumont
Banning
Cabazon
Fullerton
Corona
Anaheim
Santa Ana
Perris
Romoland
Sun City
Hemet
San Jacinto
Idyllwild
Palm Sprs.
Palm Desert
La Quinta
Coachella
Mecca
Salton Sea St. Rec. Area
Salton Sea
Desert Shores
Salton City
Niland
Calipatria
Glamis
Chocolate Mtns.
El Toro
Lake Elsinore
Elsinore
San Juan Capistrano
Murrieta
Temecula
Cahuilla
San Bernardino Nat'l Forest
Santa Rosa Ind. Res.
Santa Rosa Mtns.
Cahuilla Ind. Res.
Anza-Borrego Desert
Borrego Sprs.
Ocotillo Wells St. Vehicular Rec. Area
Ocotillo Wells
Westmorland
Brawley
Alamorio
Beach
San Clemente
Camp Pendleton
Fallbrook
Pala
Pauma Valley
Palomar Mtn.
Santa Ysabel
Julian
Desert
San Felipe Desert
Oceanside
Carlsbad
Vista
San Marcos
Escondido
Leucadia
Encinitas
Ramona
Del Mar
Poway
Santee
La Jolla
El Cajon
Alpine
Coronado
Chula Vista
Jamul
Campo
Boulevard
Dulzura
Tecate
Imperial Beach
Tijuana
Rosarito
Metamuco
Barrett Lk.
Pine Valley
Mt. Laguna
Cleveland Nat'l Forest
Cuyamaca Rancho St. Park
Imperial
Seeley
Ocotillo
Heber
El Centro
Holtville
Calexico
Mexicali
Algodones
Colonia Progreso
La Rumorosa
Hermosillo

U.S.
MEXICO

Cabrillo Nat'l Mon.
Silver Strand St. Beach
Border Field St. Park
San Diego
Dana Pt.
Clemente St. Beach
an Onofre St. Beach
Carlsbad St. Beach
Gulf of Catalina

Sequoia

© Creative Sales Corporation

FOR ARIZONA MAP SEE PAGES 20-21

SCALE OF MILES
1 INCH IS APPROXIMATELY 35 MILES
0 7 14 21 28 35

N

FOR WYOMING STATE MAP SEE PAGE 88

Grid columns: 1 2 3 4 5 6 7
Grid rows: A B C D E F G H J K

WYOMING

Saratoga, Medicine Bow, Centennial, Albany, Riverside, Encampment, Woods Landing, Mountain Home, Halligan, Cowdrey, Walden, Rustic, Baggs, Dixon, Savery, Medicine Bow Nat'l Forest, Hiawatha

Carter, Fort Bridger, Lyman, Urie, Mountain View, Millburne, Piedmont, Robertson, Lonetree, Burntfork, McKinnon, Green River, Quealy, Manila, Green Lake, Flaming Gorge Res., Flaming Gorge National Recreational Area, Ashley, Oak Park Res.

Routt, Lake John, State Forest, Roosevelt National Forest, Steamboat Sprs., Gould, Rand, Coalmont, Milner, Craig, Maybell, Sunbeam, Lay, Hamilton, Hayden, Routt National Forest, Oak Creek, Phippsburg, Yampa, Toponas, McCoy, Bond, State Bridge, Kremmling, Parshall, Tabernash, Fraser, Granby, Nederland, Rollinsville, Winter Park, Empire, Georgetown, Echo Lake, Silverthorne, Vail, Avon, Edwards, Minturn, Gilman, Dowd, Frisco, Dillon, Breckenridge, Blue River, Climax, Jefferson, Alma, Como, Fairplay, Leadville, Garo Park, Granite, Twin Lakes, Snowmass Village, Aspen, Marble, Redstone, Carbondale, Basalt, Snowmass, Woody Creek

Mountain Home, Whiterocks, Monarch, Neola, Altamont, Boneta, Bluebell, Cedarview, Talmage, Mt. Emmons, Upalco, Gusher, Arcadia, Ioka, Bridgeland, Myton, Duchesne, Roosevelt, Fort Duchesne, Leota, Ouray, Maeser, Vernal, Naples, Jensen, Lapoint, Leeton, Blue Mountain, Dinosaur, Elk Springs, Meeker, Buford, Rio Blanco, New Castle, Dotsero, Gypsum, Eagle, Wolcott, Rangely, Bonanza, Ashley National Forest

Wellington, Sunnyside, East Carbon City, Woodside, Green River, Thompson, Crescent Jct, Cisco, Mack, Loma, Fruita, Grand Junction, Clifton, Whitewater, Palisade, Skyway, Mesa, Molina, Collbran, Cameo, DeBeque, Parachute, Rifle, Silt, Glenwood Sprs., Redstone, Marble, Snowmass Village, Aspen, Leadville, Malta

Colorado National Monument, Orchard City, Delta, Austin, Lazear, Hotchkiss, Paonia, Bowie, Somerset, Cedaredge, Crested Butte, Buena Vista, Mount Princeton Hot Springs, Nathrop, Johnson Village, Salida, Poncha Springs, Gunnison, Parlin, Doyleville, Sargents, Garfield, Almont, Crawford, Maher, Olathe, Montrose, Cimarron, Sapinero, Powderhorn, Lake City

Moab, Gateway, Uravan, Paradox, Bedrock, La Sal Jct., La Sal, Naturita, Nucla, Vancorum, Redvale, Norwood, Ridgway, Ouray, Saguache, Mineral Hot Springs, Villa Grove, Coaldale, Howard, Arkansas River

Monticello, Eastland, Dove Creek, Cahone, Pleasant View, Yellow Jacket, Stoner, Rico, Dunton, Placerville, Saw Pit, Telluride, Ophir, Gladstone, Red Mountain, Silverton, Creede, Wagon Wheel Gap, South Fork, Del Norte, Spar City, Monte Vista, Summitville, Platoro, Center, Hooper, Mosca, Homelake, Alamosa, Great Sand Dunes National Monument

Fry Canyon, Blanding, Lewis, Arriola, Cortez, Dolores, Lebanon, Mancos, Hesperus, Rockwood, Hermosa, Durango, Bayfield, Chimney Rock, Pagosa Sprs., Capulin, La Jara, Romeo, Sanford, Manassa

Bluff, Montezuma Creek, Aneth, Towaoc, Fort Lewis, Marvel, Breen, Kline, Oxford, Ignacio, Redmesa, Allison, Arboles, Chromo, Conejos, Antonito, Mexican Hat, Navajo Indian Res.

Mexican Water, Tes Nez Iha, Dinnehotso, Teec Nos Pos, Beklabito, Flora Vista, Kirtland, Shiprock, La Plata, Cedar Hill, Aztec, Turley, Archuleta, Bloomfield, Blanco, Dulce, Lumberton, Monero, Chama, Brazos, Los Ojos, Rutheron, Ensenada, Tierra Amarilla, La Reunte, Cebolla, El Vado Lake, Canon Plaza, Vallecitos, Tres Piedra

Kayenta, Tsegi, Chilchinbito, Rock Point, Round Rock, Shiprock, Farmington, Bloomfield

UTAH — FOR UTAH STATE MAP SEE PAGE 82

FOR NEW MEXICO STATE MAP SEE PAGE 62

© Creative Sales Corporation

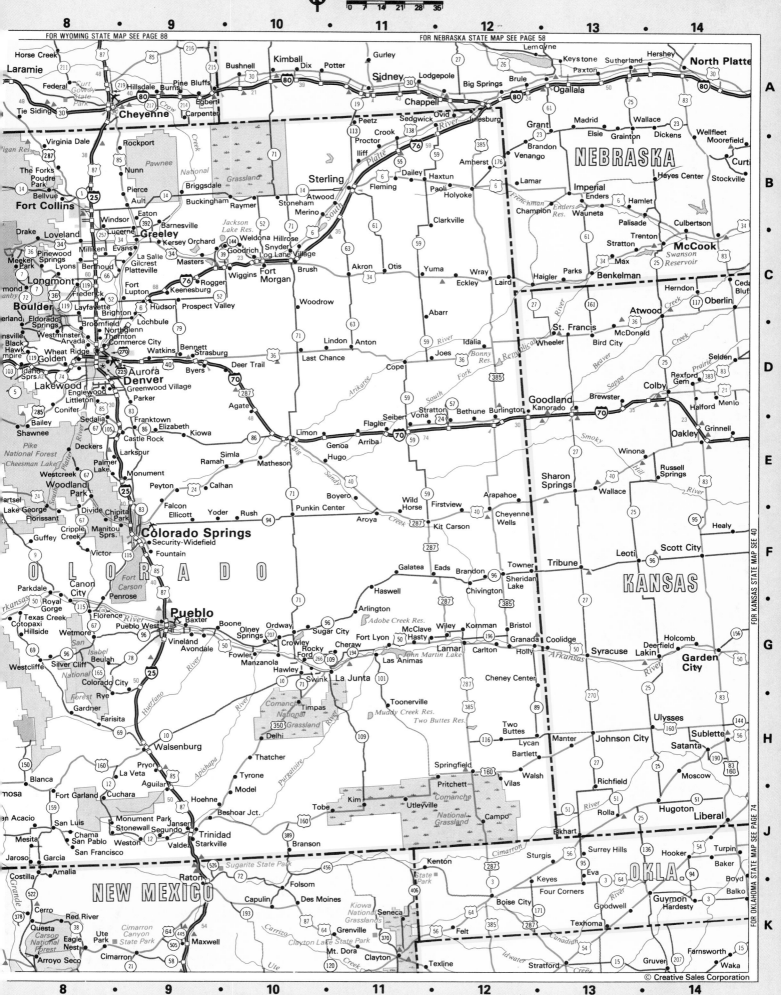

Connecticut
Massachusetts
Rhode Island

SCALE OF MILES

1 INCH IS APPROXIMATELY 20 MILES

0 4 8 12 16 20

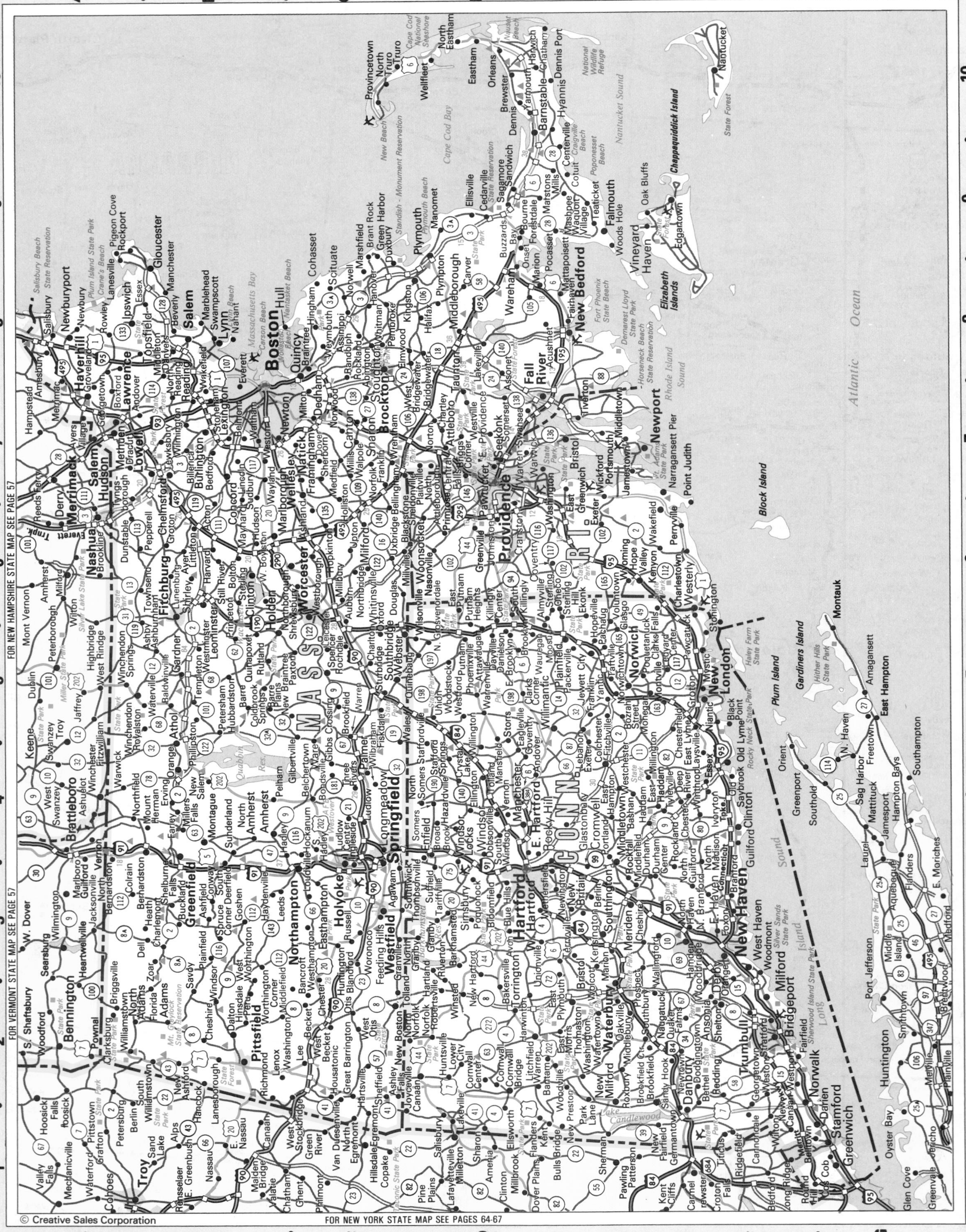

© Creative Sales Corporation

FOR NEW HAMPSHIRE STATE MAP SEE PAGE 57

FOR VERMONT STATE MAP SEE PAGE 57

FOR NEW YORK STATE MAP SEE PAGES 64-67

Atlantic Ocean

Cape Cod Bay

Nantucket Sound

Rhode Island Sound

Long Island Sound

Massachusetts Bay

Buzzards Bay

Major cities and towns shown include: Boston, Worcester, Springfield, Providence, Hartford, New Haven, Bridgeport, Fall River, New Bedford, Lowell, Lawrence, Brockton, Fitchburg, Pittsfield, Greenfield, Northampton, Holyoke, Waterbury, Danbury, Norwalk, Stamford, Norwich, New London, Newport, Nantucket, Provincetown, Gloucester, Salem, Lynn, Quincy, Plymouth, Hyannis, Falmouth, Bennington, Keene, Brattleboro.

SCALE OF MILES
1 INCH IS APPROXIMATELY 35 MILES
0   7   14   21   28   35

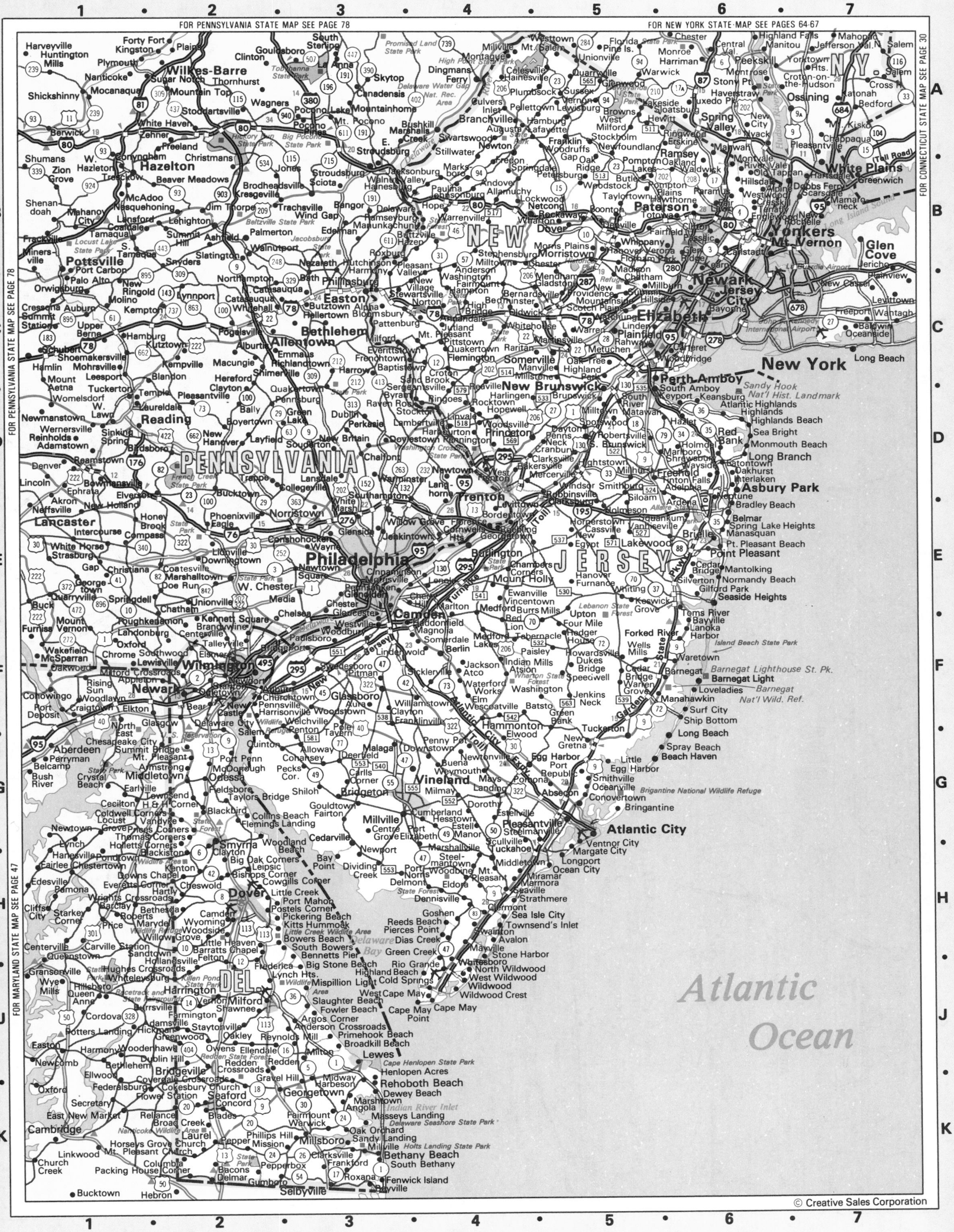

N

FOR PENNSYLVANIA STATE MAP SEE PAGE 78
FOR NEW YORK STATE MAP SEE PAGES 64-67
FOR CONNECTICUT STATE MAP SEE PAGE 30
FOR PENNSYLVANIA STATE MAP SEE PAGE 78
FOR MARYLAND STATE MAP SEE PAGE 47

NY

NEW JERSEY

PENNSYLVANIA

DEL

Atlantic Ocean

Delaware Bay

New York

Wilkes-Barre · Hazelton · Pottsville · Reading · Allentown · Bethlehem · Easton · Lancaster · Philadelphia · W. Chester · Camden · Trenton · Princeton · New Brunswick · Perth Amboy · Elizabeth · Newark · Jersey City · Paterson · Morristown · Yonkers · Mt. Vernon · White Plains · Glen Cove

Wilmington · Newark · Dover · Smyrna · Harrington · Milford · Seaford · Georgetown · Lewes · Rehoboth Beach · Bethany Beach · Fenwick Island

Atlantic City · Ventnor City · Margate City · Ocean City · Asbury Park · Long Branch · Point Pleasant · Toms River · Barnegat Light · Cape May · Wildwood

© Creative Sales Corporation

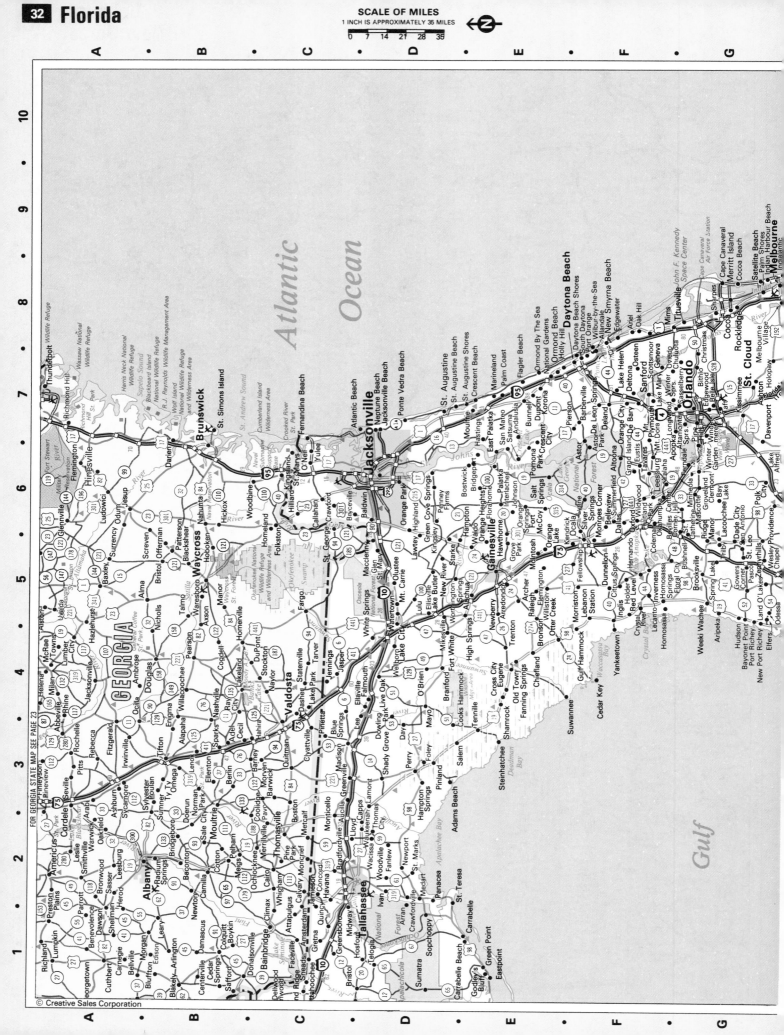

© Creative Sales Corporation

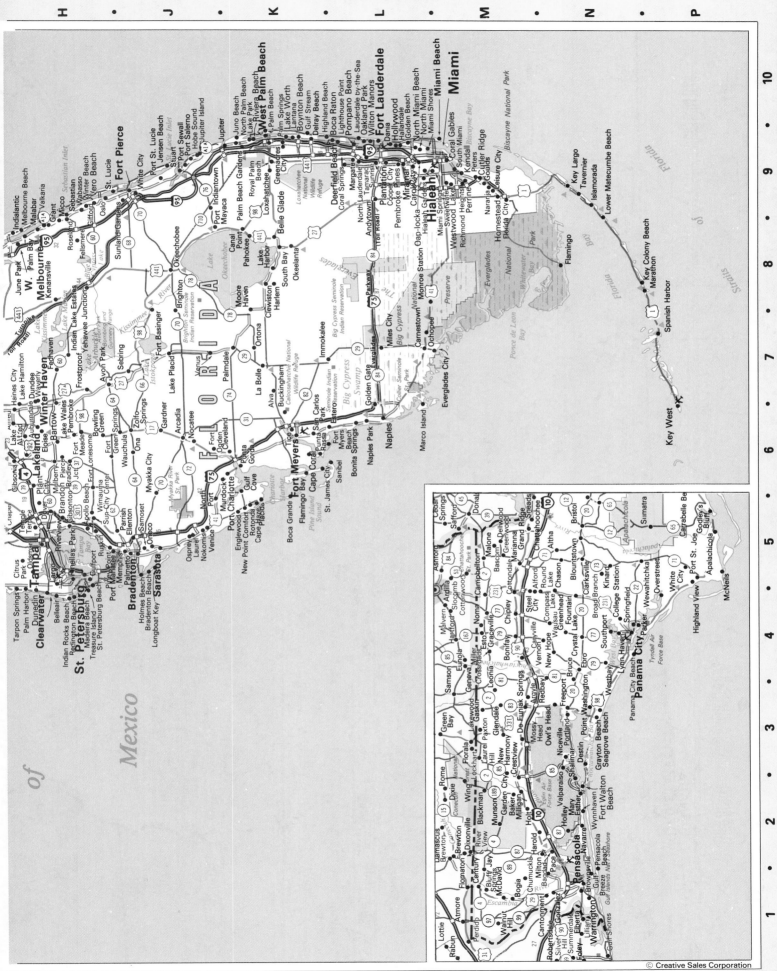

© Creative Sales Corporation

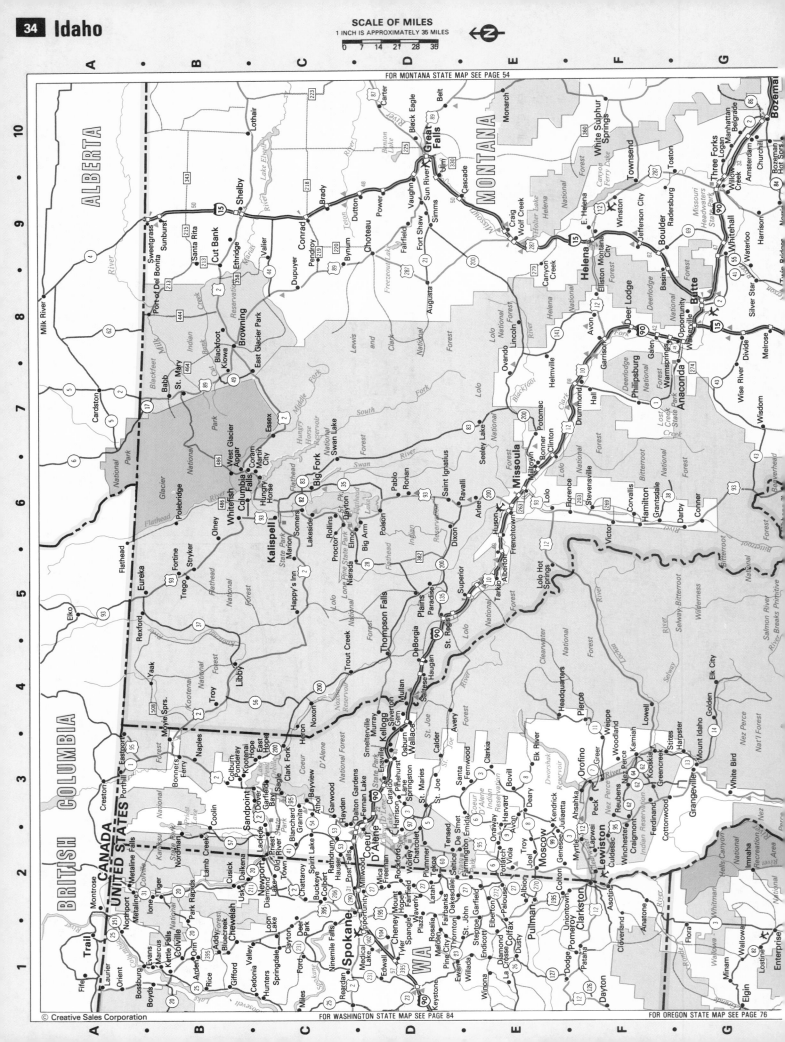

SCALE OF MILES
1 INCH IS APPROXIMATELY 35 MILES
0 7 14 21 28 35

FOR MONTANA STATE MAP SEE PAGE 54

ALBERTA

BRITISH COLUMBIA

CANADA
UNITED STATES

MONTANA

WASHINGTON

© Creative Sales Corporation

FOR WASHINGTON STATE MAP SEE PAGE 84

FOR OREGON STATE MAP SEE PAGE 76

FOR WYOMING STATE MAP SEE PAGE 88

FOR UTAH STATE MAP SEE PAGE 82

FOR NEVADA STATE MAP SEE PAGE 60

FOR OREGON STATE MAP SEE PAGE 76

© Creative Sales Corporation

SCALE OF MILES
1 INCH IS APPROXIMATELY 32 MILES
0   7   14   21   28   35

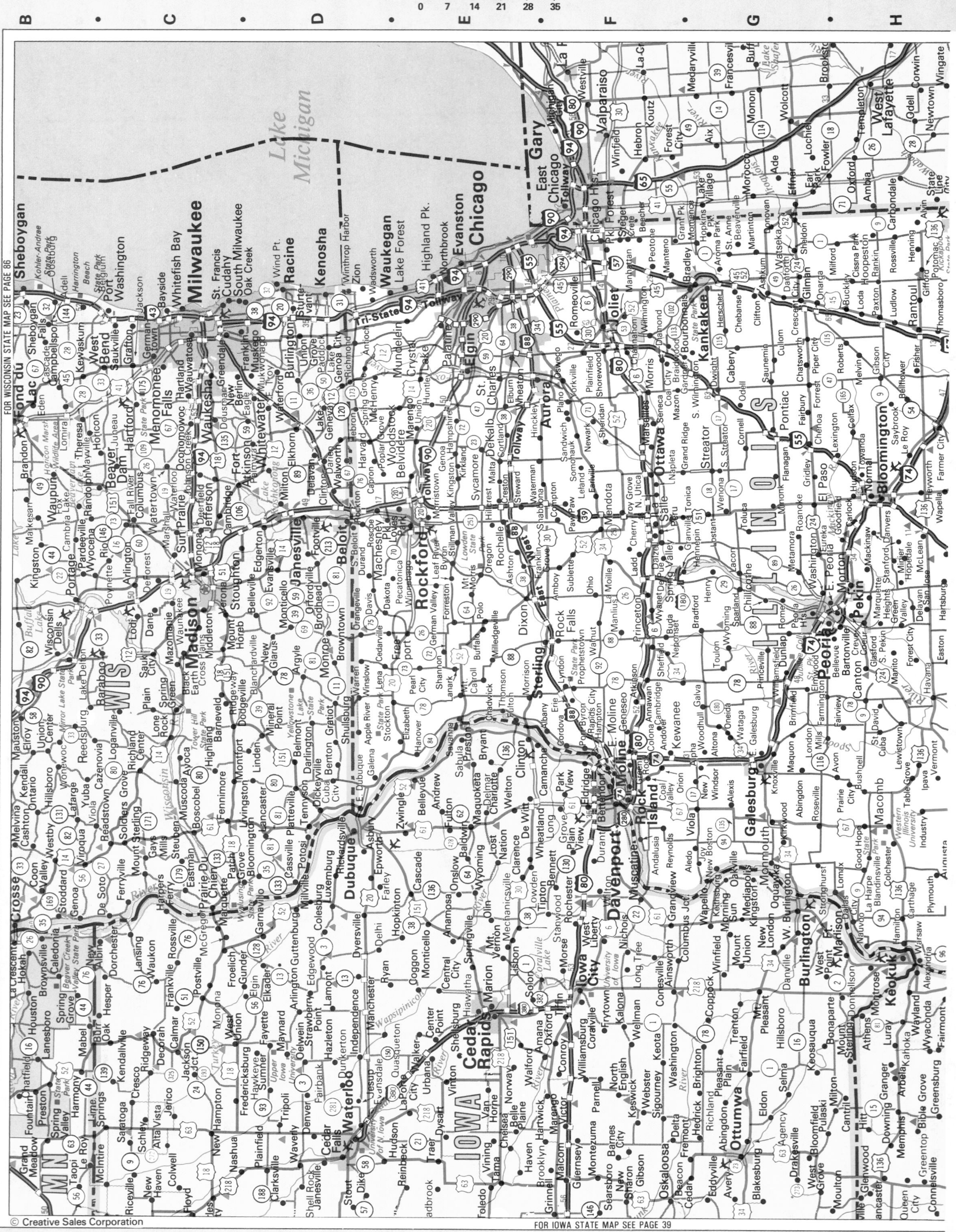

FOR WISCONSIN STATE MAP SEE PAGE 86

Lake Michigan

WIS.

MN

IOWA

ILLINOIS

Sheboygan · Fond du Lac · Milwaukee · Racine · Kenosha · Waukegan · Evanston · Chicago · East Chicago · Gary

Madison · Janesville · Beloit · Rockford · Elgin · Aurora · Joliet · Kankakee

La Crosse · Dubuque · Cedar Rapids · Waterloo · Iowa City · Davenport · Rock Island · Moline · Galesburg · Peoria · Pekin · Bloomington · Ottawa · La Salle · Streator · Pontiac

Ottumwa · Burlington · Keokuk · Macomb

Valparaiso · West Lafayette

© Creative Sales Corporation

FOR IOWA STATE MAP SEE PAGE 39

SCALE OF MILES
1 INCH IS APPROXIMATELY 32 MILES
0 7 14 21 28 35

FOR INDIANA STATE MAP SEE PAGE 38

FOR KENTUCKY STATE MAP SEE PAGE 43

FOR MISSOURI STATE MAP SEE PAGE 52

INDIANA

KENTUCKY

MISSOURI

Danville · Crawfordsville · Greencastle · Brazil · Terre Haute · Clinton · Washington · Evansville · Owensboro · Madisonville · Hopkinsville · Paducah · Mayfield

Champaign · Urbana · Decatur · Mattoon · Charleston · Effingham · Newton · Mount Vernon · Centralia · West Frankfort · Carbondale · Cape Girardeau · Sikeston · Poplar Bluff

Springfield · Jacksonville · Quincy · Hannibal · Bowling Green · Mexico · Columbia · Jefferson City · Rolla · West Plains

St. Louis · E. St. Louis · Belleville · Alton · St. Charles

© Creative Sales Corporation

SCALE OF MILES
1 INCH IS APPROXIMATELY 35 MILES
0 7 14 21 28 35

N

FOR MICHIGAN STATE MAP SEE PAGE 48

The Interstate Highway System in and around the Chicago area is confusing to many people. It is helpful to remember that, in most cases, Interstate Highways running north to south have odd numbers, and Interstate Highways running east and west have even numbers

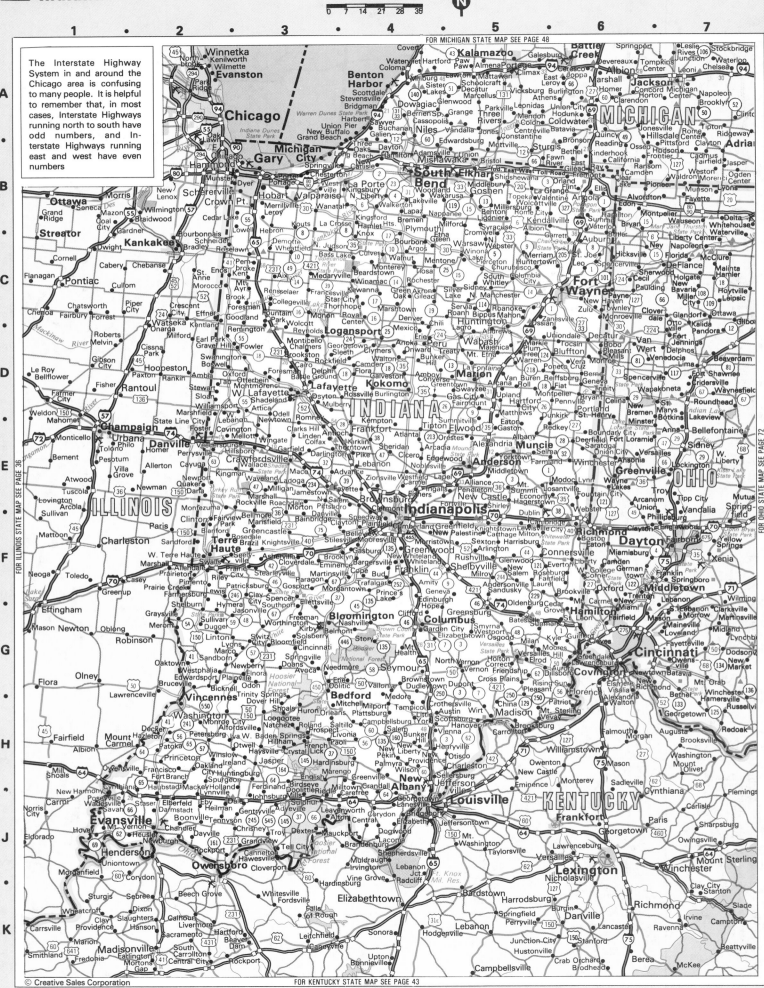

© Creative Sales Corporation

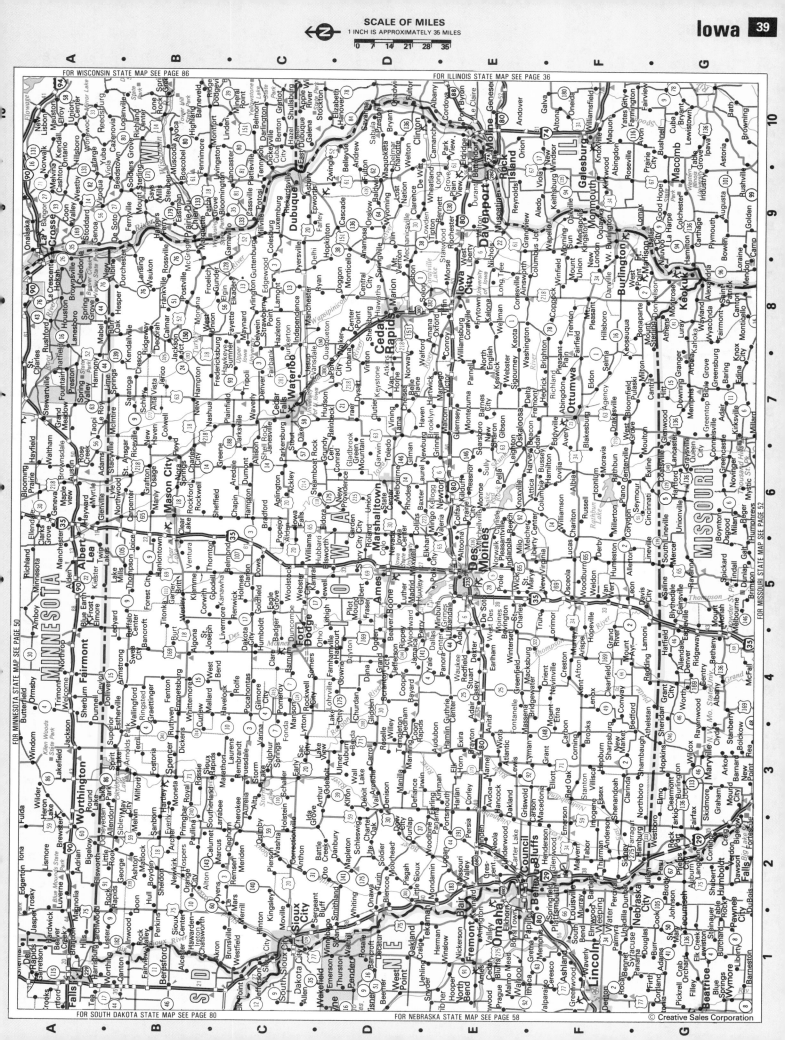

SCALE OF MILES
1 INCH IS APPROXIMATELY 35 MILES
0 7 14 21 28 35

FOR WISCONSIN STATE MAP SEE PAGE 86

FOR ILLINOIS STATE MAP SEE PAGE 36

FOR MINNESOTA STATE MAP SEE PAGE 50

FOR MISSOURI STATE MAP SEE PAGE 52

FOR SOUTH DAKOTA STATE MAP SEE PAGE 80

FOR NEBRASKA STATE MAP SEE PAGE 58

© Creative Sales Corporation

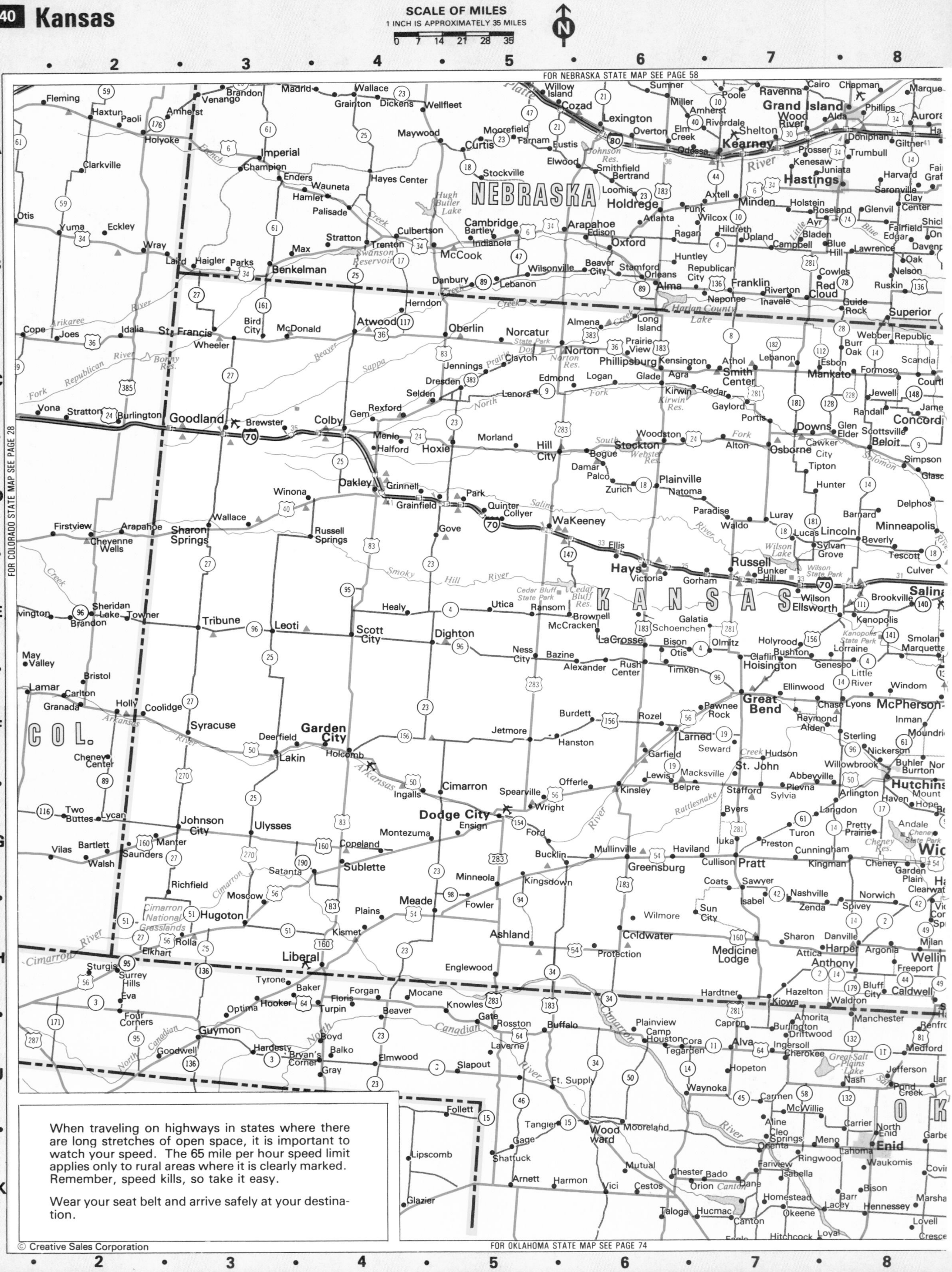

SCALE OF MILES
1 INCH IS APPROXIMATELY 35 MILES
0 7 14 21 28 35

N

FOR NEBRASKA STATE MAP SEE PAGE 58

FOR COLORADO STATE MAP SEE PAGE 28

NEBRASKA

KANSAS

COL.

When traveling on highways in states where there are long stretches of open space, it is important to watch your speed. The 65 mile per hour speed limit applies only to rural areas where it is clearly marked. Remember, speed kills, so take it easy.

Wear your seat belt and arrive safely at your destination.

© Creative Sales Corporation

FOR OKLAHOMA STATE MAP SEE PAGE 74

SCALE OF MILES
1 INCH IS APPROXIMATELY 35 MILES

FOR NEBRASKA STATE MAP SEE PAGE 58
FOR IOWA STATE MAP SEE PAGE 39
FOR MISSOURI STATE MAP SEE PAGE 52
FOR OKLAHOMA STATE MAP SEE PAGE 74

© Creative Sales Corporation

SCALE OF MILES
1 INCH IS APPROXIMATELY 35 MILES
0 7 14 21 28 35

N

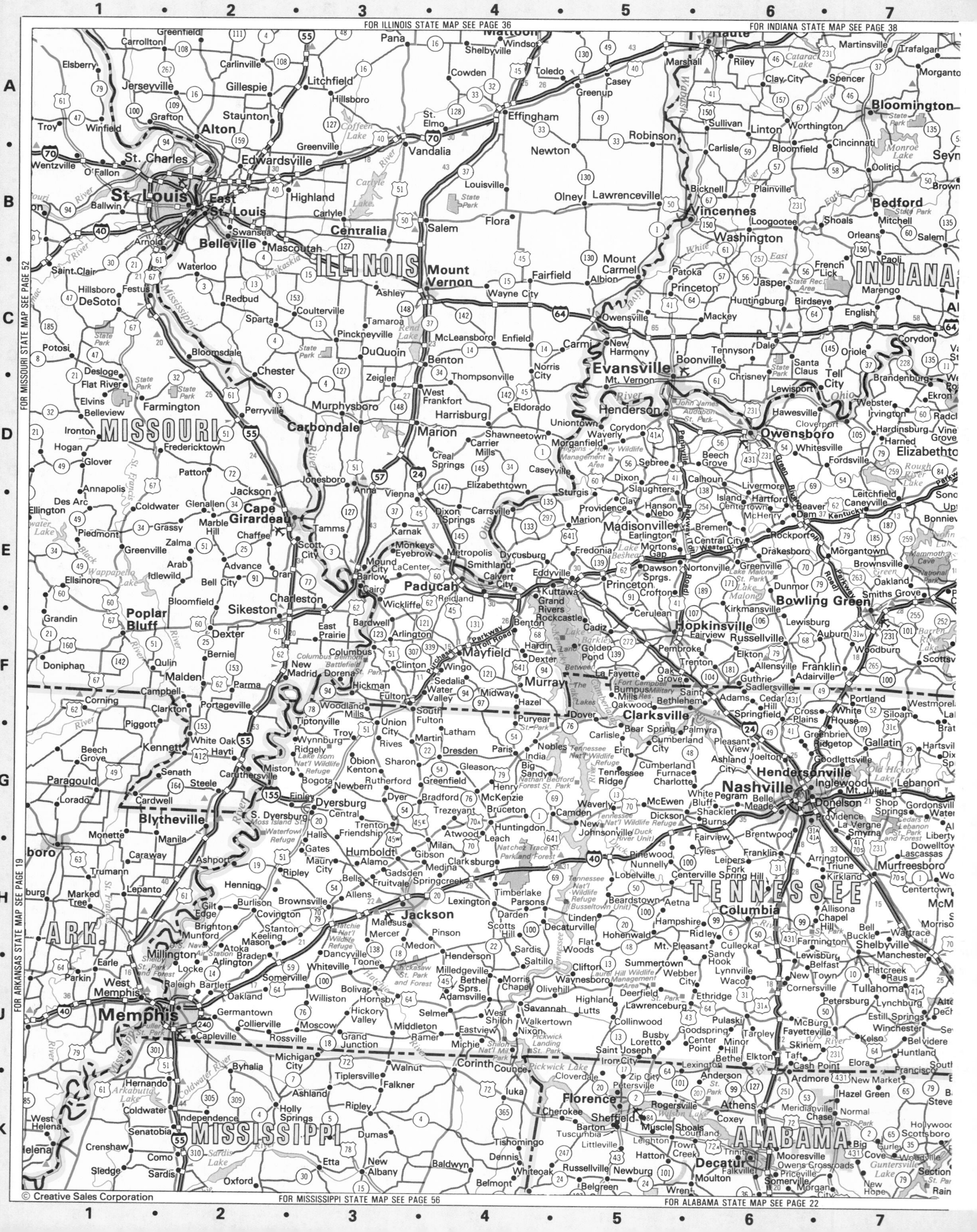

FOR ILLINOIS STATE MAP SEE PAGE 36
FOR INDIANA STATE MAP SEE PAGE 38
FOR MISSOURI STATE MAP SEE PAGE 52
FOR ARKANSAS STATE MAP SEE PAGE 19
FOR MISSISSIPPI STATE MAP SEE PAGE 56
FOR ALABAMA STATE MAP SEE PAGE 22
ILLINOIS
INDIANA
MISSOURI
TENNESSEE
ARK.
MISSISSIPPI
ALABAMA
St. Louis
East St. Louis
Alton
St. Charles
Edwardsville
Belleville
Centralia
Mount Vernon
Carbondale
Marion
Cape Girardeau
Paducah
Evansville
Henderson
Owensboro
Vincennes
Washington
Bloomington
Bedford
Bowling Green
Hopkinsville
Clarksville
Nashville
Hendersonville
Murfreesboro
Columbia
Jackson
Dyersburg
Blytheville
Memphis
West Memphis
Florence
Sheffield
Decatur
Athens
Poplar Bluff
Sikeston
Madisonville
Mayfield
Murray
Union City
Tullahoma
Shelbyville
Mississippi River
Ohio River
Wabash River
Kentucky Lake
Lake Barkley
© Creative Sales Corporation

FOR OHIO STATE MAP SEE PAGE 72

FOR WEST VIRGINIA STATE MAP SEE PAGE 46
FOR VIRGINIA STATE MAP SEE PAGE 47
FOR NORTH CAROLINA STATE MAP SEE PAGE 68
FOR SOUTH CAROLINA STATE MAP SEE PAGE 69
FOR ALABAMA STATE MAP SEE PAGE 22
FOR GEORGIA STATE MAP SEE PAGE 23

OHIO

WEST VIRGINIA

VIRGINIA

KENTUCKY

NC

GA

© Creative Sales Corporation

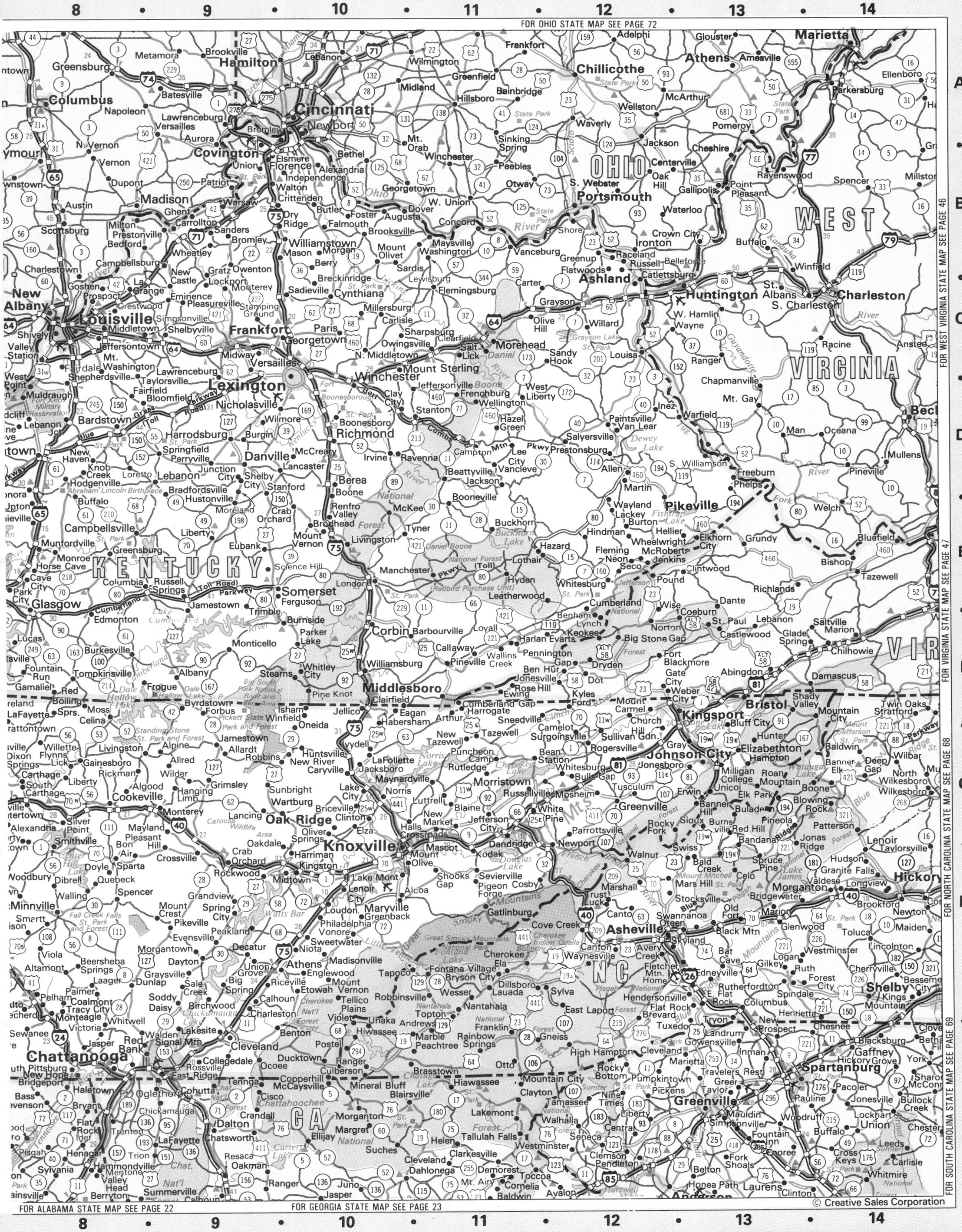

SCALE OF MILES
1 INCH IS APPROXIMATELY 35 MILES
0 7 14 21 28 35

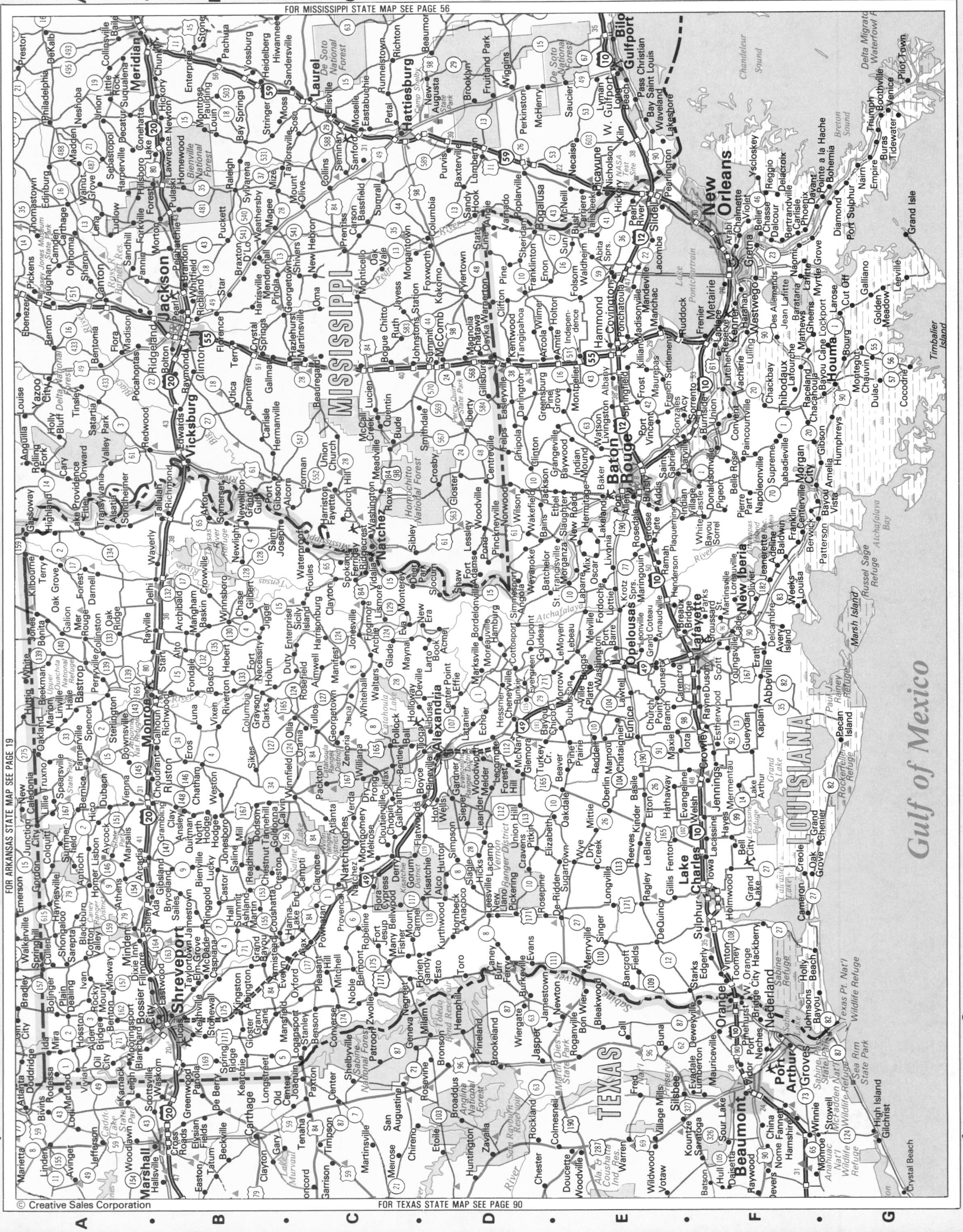
FOR MISSISSIPPI STATE MAP SEE PAGE 56
FOR ARKANSAS STATE MAP SEE PAGE 19
FOR TEXAS STATE MAP SEE PAGE 90
MISSISSIPPI
LOUISIANA
TEXAS
Gulf of Mexico
Shreveport
Monroe
Alexandria
Lafayette
Lake Charles
Baton Rouge
New Orleans
Houma
New Iberia
Opelousas
Jackson
Vicksburg
Natchez
Hattiesburg
Laurel
Meridian
Gulfport
Biloxi
Beaumont
Port Arthur
Orange
Marshall
Lake Pontchartrain
Chandeleur Sound
Breton Sound
Marsh Island
Timbalier Island
Grand Isle

© Creative Sales Corporation

SCALE OF MILES
1 INCH IS APPROXIMATELY 35 MILES
0 7 14 21 28 35
N

1 2 3 4 5 6 7

Passports or visas are not required of Canadian citizens or British subjects residing in Canada entering the United States for a period of six months or less; however, evidence of citizenship is rigidly controlled. Check with customs officials for complete regulations and requirements.

United States Citizens Visiting Canada

All persons entering Canada must report to the Canadian Immigration and Customs Office at the Port of Entry and secure the necessary permits for admission of their person and possessions. The transportation of plants and produce is rigidly controlled. Check with customs officials for complete regulations and requirements.

QUEBEC

CANADA
UNITED STATES

MAINE

NEW BRUNSWICK

NH

Atlantic

Ocean

FOR NEW HAMPSHIRE STATE MAP SEE PAGE 57

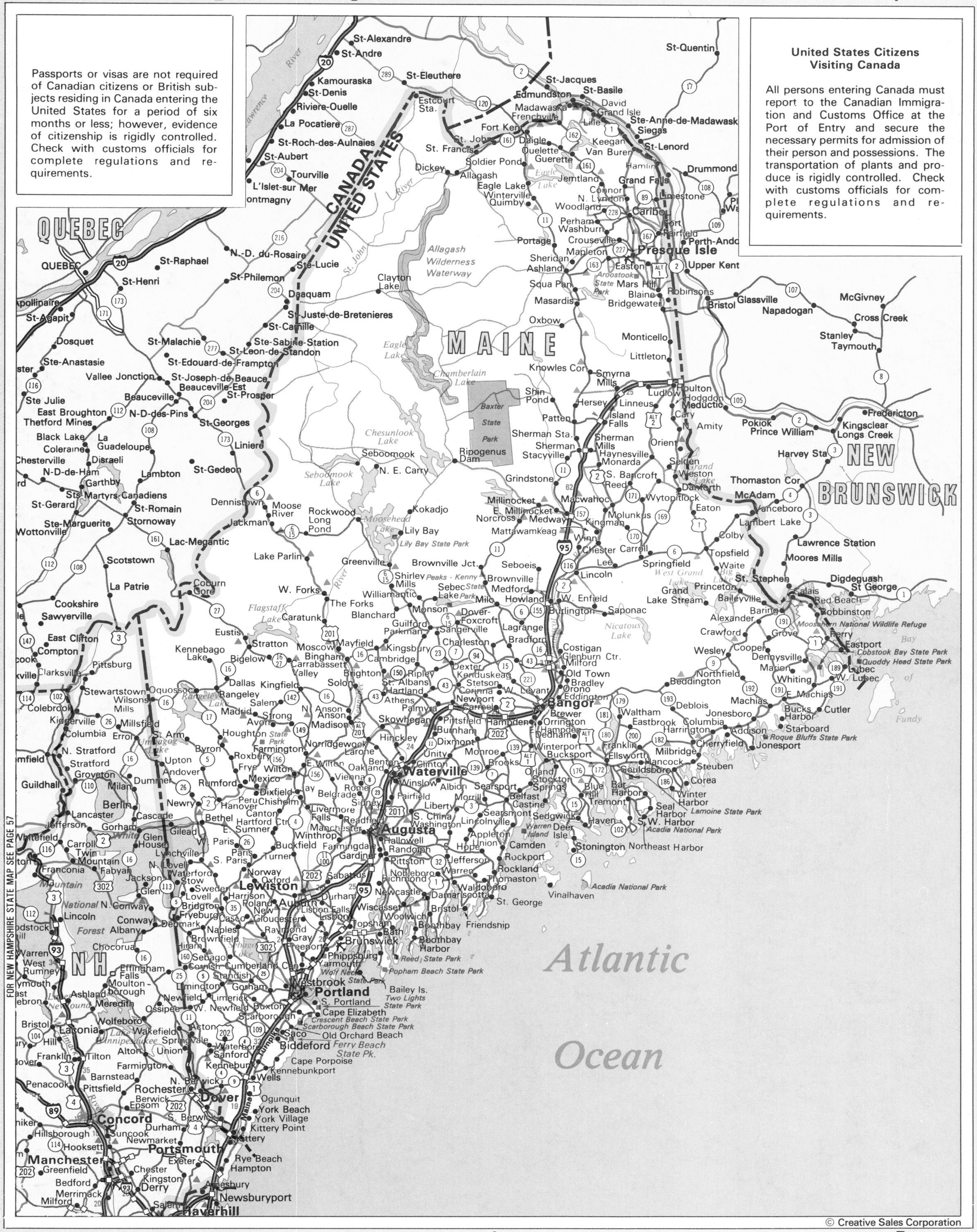

© Creative Sales Corporation

SCALE OF MILES
1 INCH IS APPROXIMATELY 35 MILES
0 7 14 21 28 35

N

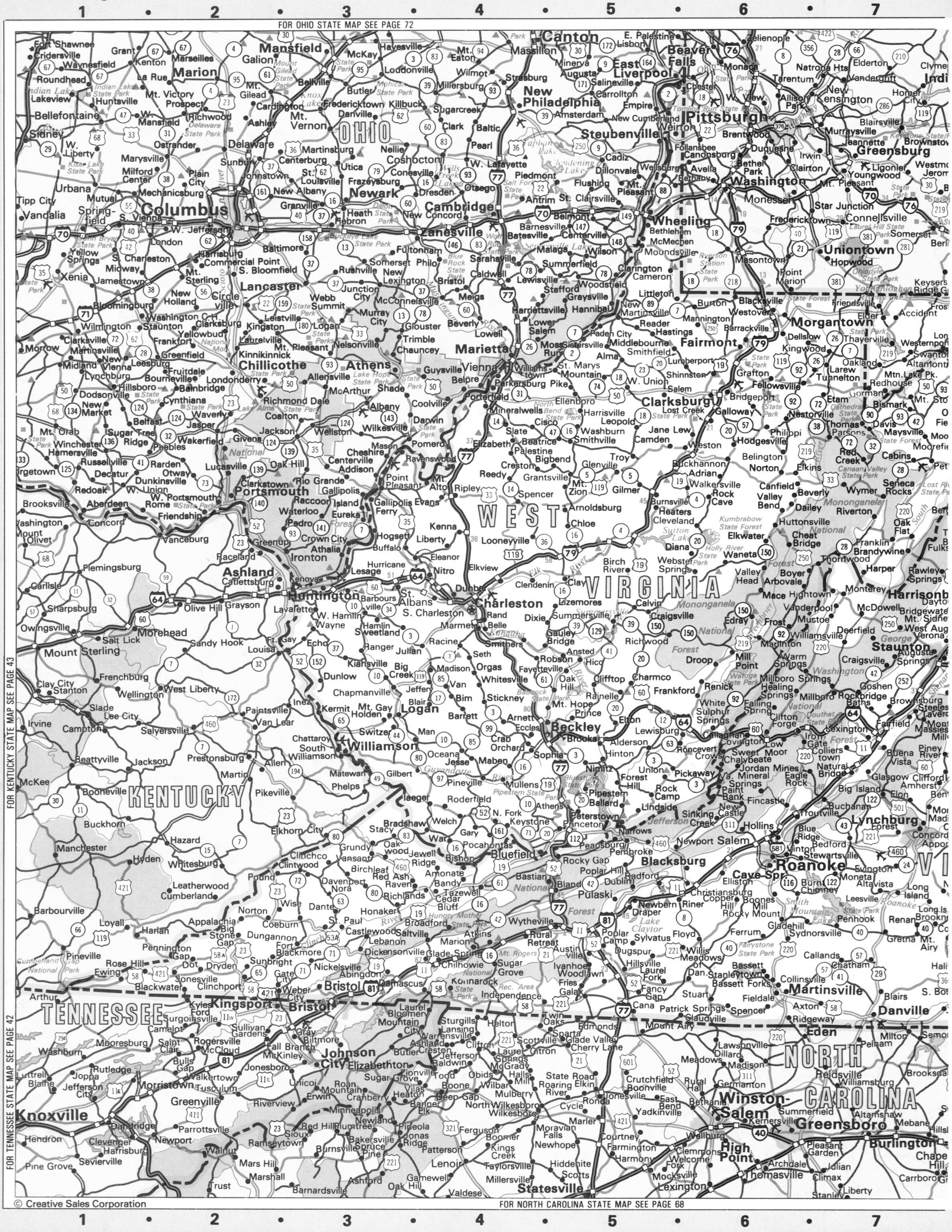

FOR OHIO STATE MAP SEE PAGE 72

(Grid coordinates: columns 1–7 across top and bottom; rows A–K down the sides.)

States / regions labeled: OHIO, WEST VIRGINIA, KENTUCKY, TENNESSEE, NORTH CAROLINA

Major cities: Columbus, Mansfield, Marion, Newark, Cambridge, Zanesville, Lancaster, Chillicothe, Athens, Marietta, Pittsburgh, Steubenville, Wheeling, Washington, Uniontown, Morgantown, Fairmont, Clarksburg, Parkersburg, Charleston, Huntington, Ashland, Portsmouth, Beckley, Bluefield, Logan, Williamson, Staunton, Lynchburg, Roanoke, Pulaski, Blacksburg, Martinsville, Danville, Bristol, Johnson City, Kingsport, Knoxville, Winston-Salem, Greensboro, High Point, Burlington

FOR KENTUCKY STATE MAP SEE PAGE 43

FOR TENNESSEE STATE MAP SEE PAGE 42

FOR NORTH CAROLINA STATE MAP SEE PAGE 68

© Creative Sales Corporation

SCALE OF MILES
1 INCH IS APPROXIMATELY 35 MILES
0 7 14 21 28 35

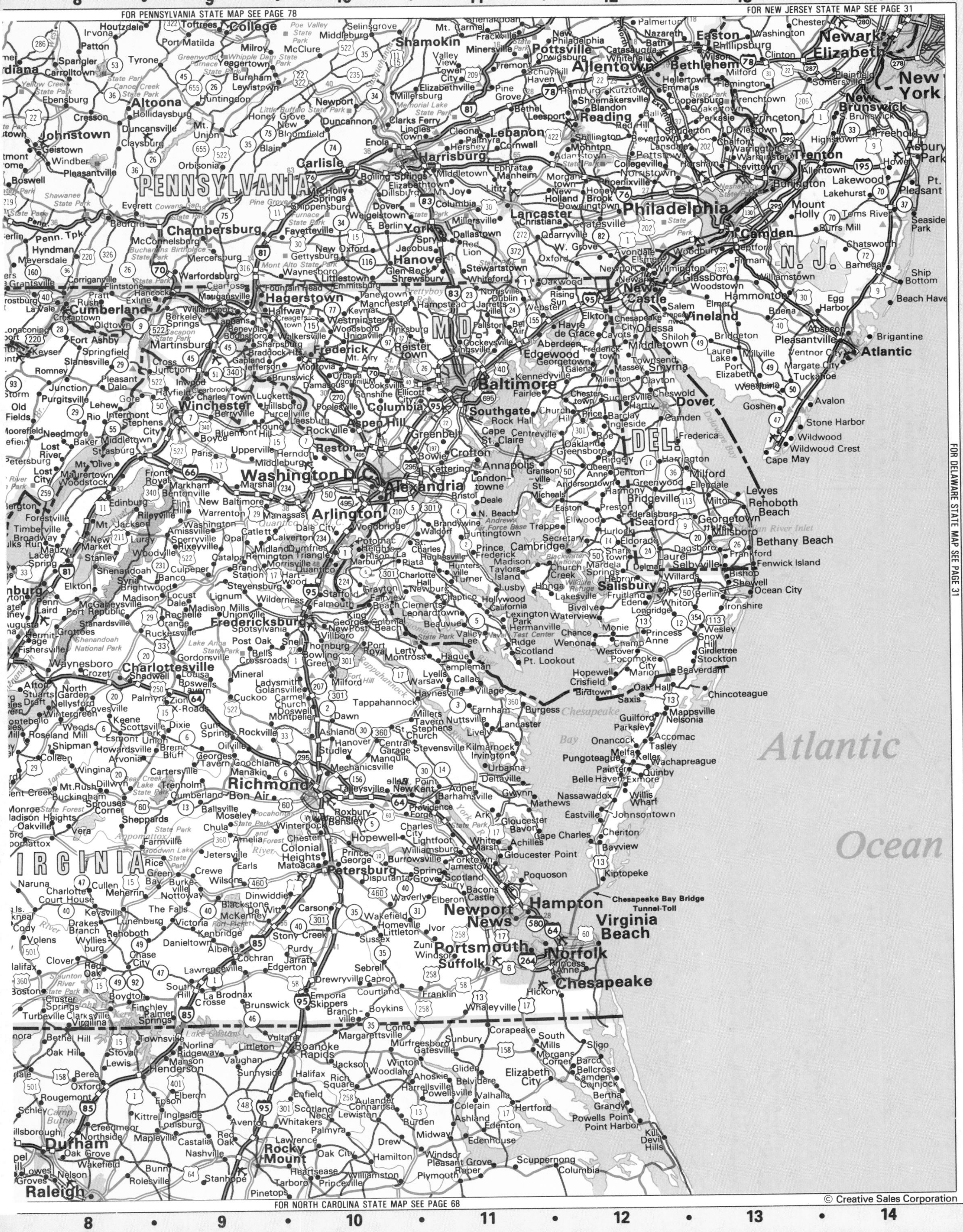

**Maryland
Virginia
West Virginia**

47

FOR PENNSYLVANIA STATE MAP SEE PAGE 78

FOR NEW JERSEY STATE MAP SEE PAGE 31

FOR DELAWARE STATE MAP SEE PAGE 31

FOR NORTH CAROLINA STATE MAP SEE PAGE 68

Atlantic

Ocean

Chesapeake
Bay

PENNSYLVANIA

N. J.

MD.

DEL.

VIRGINIA

Washington D.C.

Baltimore

Philadelphia

Richmond

Norfolk

Virginia Beach

Chesapeake Bay Bridge Tunnel-Toll

© Creative Sales Corporation

SCALE OF MILES
1 INCH IS APPROXIMATELY 35 MILES
0   7   14   21   28   35

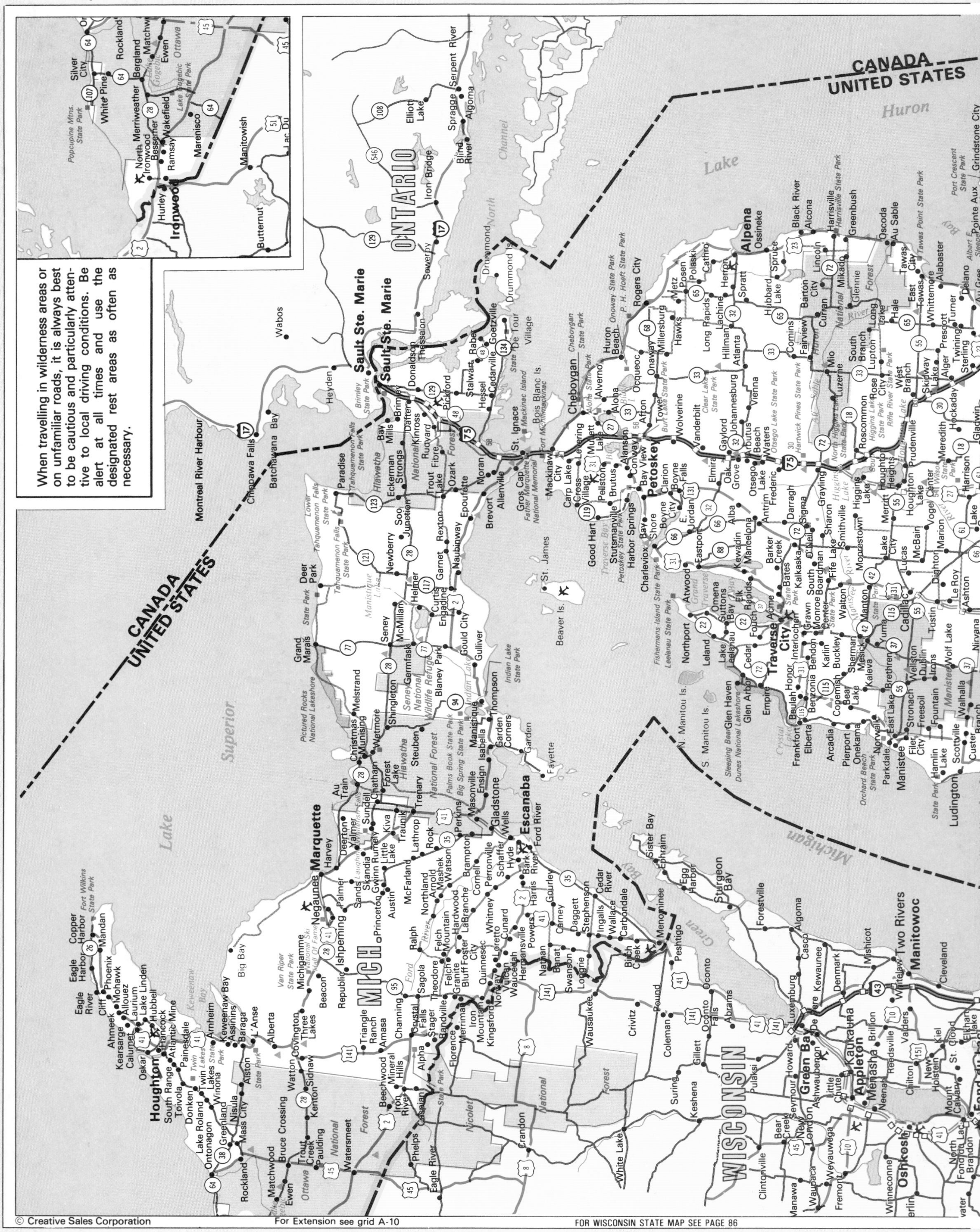

When travelling in wilderness areas or on unfamiliar roads, it is always best to be cautious and particularly attentive to local driving conditions. Be alert at all times and use the designated rest areas as often as necessary.

CANADA
UNITED STATES

ONTARIO

Sault Ste. Marie
Sault Ste. Marie

Lake Superior

Lake Huron

Lake Michigan

MICH

WISCONSIN

Marquette
Negaunee
Escanaba
Houghton
Traverse City
Cheboygan
Alpena
Petoskey
Cadillac
Manistee
Ludington
Manitowoc
Two Rivers
Green Bay
Appleton
Menasha
Neenah
Oshkosh

© Creative Sales Corporation

For Extension see grid A-10

FOR WISCONSIN STATE MAP SEE PAGE 86

SCALE OF MILES
1 INCH IS APPROXIMATELY 35 MILES

0 7 14 21 28 35

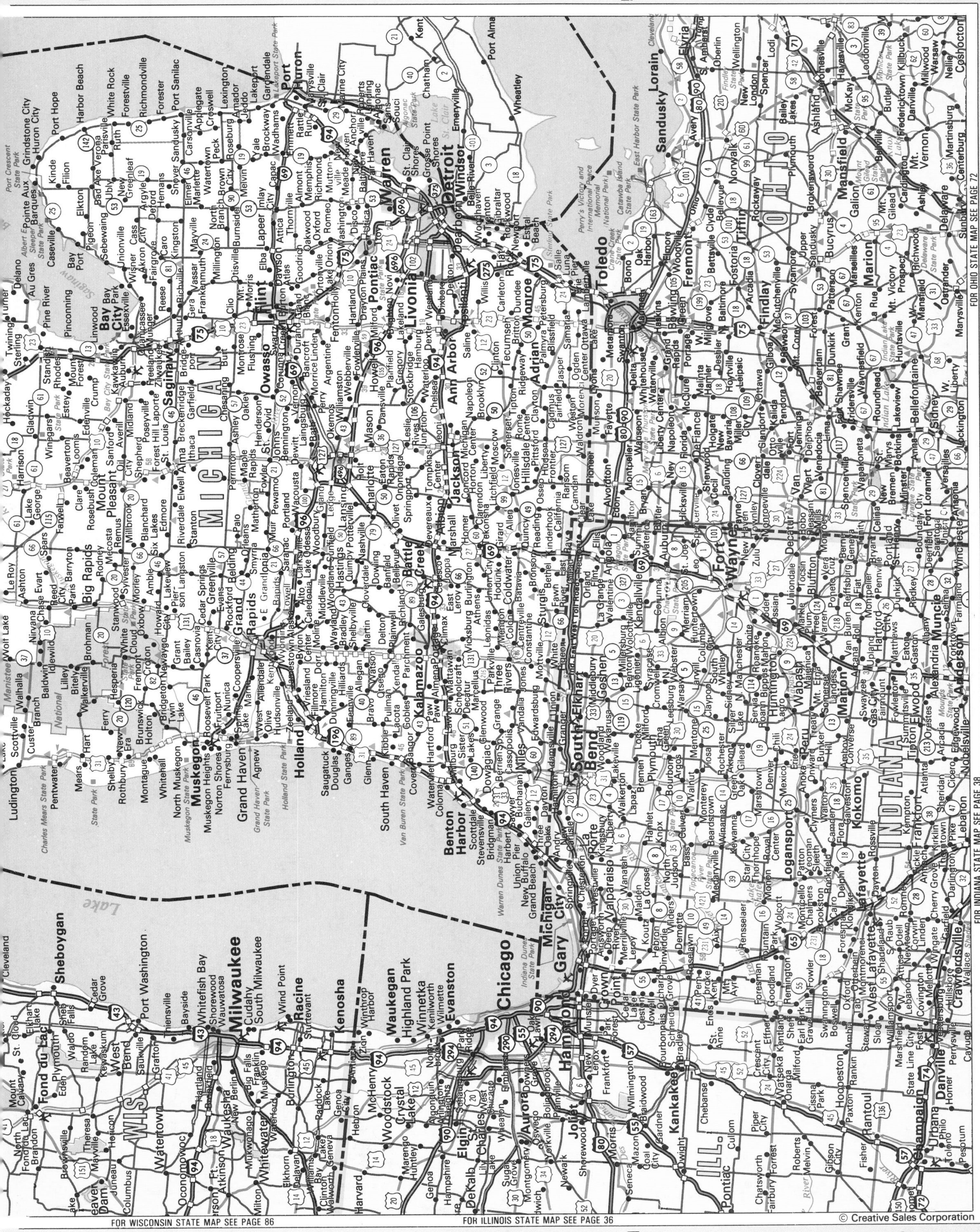

N

MICHIGAN

Detroit
Windsor
Flint
Saginaw
Bay City
Pontiac
Livonia
Warren
Ann Arbor
Monroe
Adrian
Jackson
Battle Creek
Kalamazoo
Grand Rapids
Holland
Benton Harbor
Muskegon
Grand Haven
Big Rapids
Mount Pleasant
Ludington
Owosso
Port Huron

OHIO
Toledo
Sandusky
Lorain
Elyria
Norwalk
Fremont
Findlay
Tiffin
Mansfield
Marion
Bucyrus
Galion
Bellefontaine

INDIANA
South Bend
Elkhart
Goshen
Fort Wayne
Kokomo
Logansport
Lafayette
West Lafayette
Muncie
Anderson
Marion
Michigan City

ILLINOIS
Chicago
Gary
Hammond
Joliet
Aurora
Elgin
Evanston
Waukegan
Highland Park

WISCONSIN
Milwaukee
Racine
Kenosha
Sheboygan
Fond du Lac
West Bend

Lake Michigan
Lake Huron
Lake St. Clair
Lake Erie

FOR INDIANA STATE MAP SEE PAGE 38
FOR OHIO STATE MAP SEE PAGE 72

© Creative Sales Corporation

SCALE OF MILES
1 INCH IS APPROXIMATELY 35 MILES
0 7 14 21 28 35

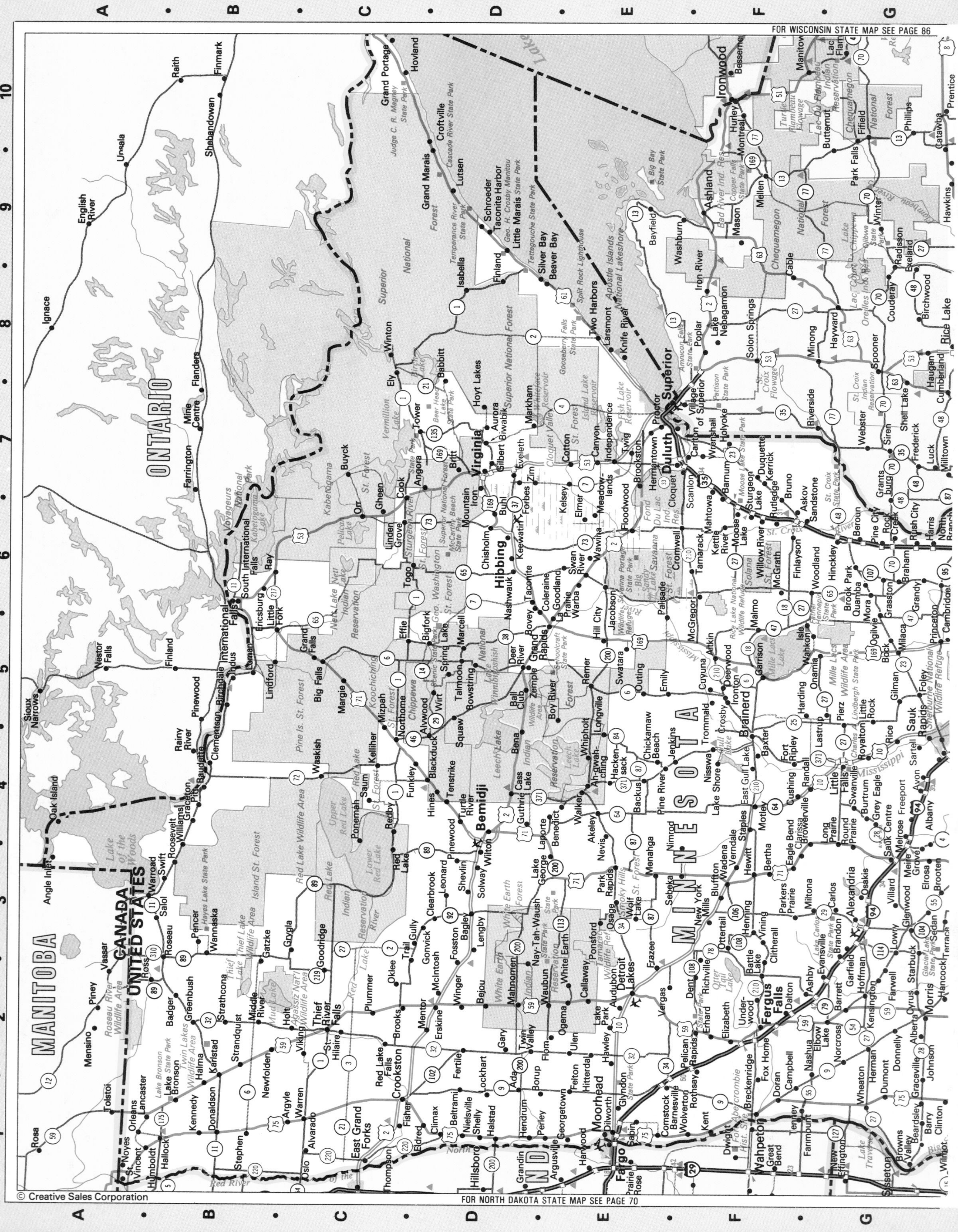

N

FOR WISCONSIN STATE MAP SEE PAGE 86

FOR NORTH DAKOTA STATE MAP SEE PAGE 70

ONTARIO

MANITOBA

CANADA
UNITED STATES

MINNESOTA

Lake Superior

Lake of the Woods

Voyageurs National Park

Superior National Forest

Chippewa National Forest

Red Lake

Upper Red Lake

Lower Red Lake

Leech Lake

Mille Lacs Lake

Duluth · Superior · Proctor

Bemidji · Grand Rapids · Hibbing · Virginia · Ely

Brainerd · Fergus Falls · Alexandria · Detroit Lakes

Moorhead · Fargo · Wahpeton · Breckenridge

International Falls · Baudette · Roseau · Warroad

Grand Marais · Two Harbors · Silver Bay

Crookston · Thief River Falls · East Grand Forks

© Creative Sales Corporation

SCALE OF MILES
1 INCH IS APPROXIMATELY 35 MILES
0 7 14 21 28 35

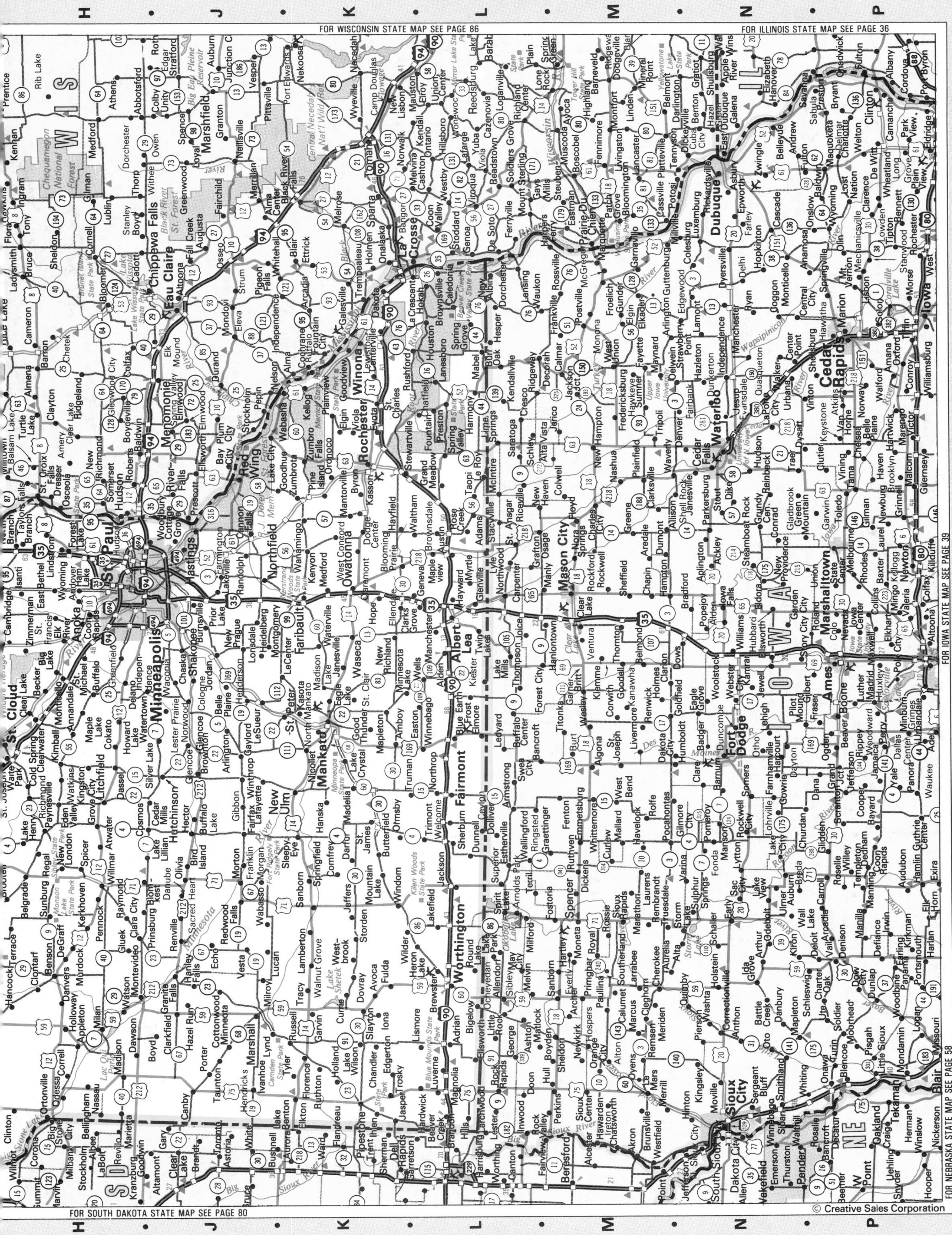
FOR WISCONSIN STATE MAP SEE PAGE 86
FOR ILLINOIS STATE MAP SEE PAGE 36
FOR IOWA STATE MAP SEE PAGE 39
FOR NEBRASKA STATE MAP SEE PAGE 58
FOR SOUTH DAKOTA STATE MAP SEE PAGE 80
H J K L M N P
1 2 3 4 5 6 7 8 9 10
Minneapolis
St. Paul
Rochester
Mankato
Worthington
Fairmont
Albert Lea
Winona
Eau Claire
La Crosse
Dubuque
Waterloo
Cedar Rapids
Ames
Fort Dodge
Mason City
Sioux City
Marshalltown
© Creative Sales Corporation

SCALE OF MILES
1 INCH IS APPROXIMATELY 35 MILES

0   7   14   21   28   35

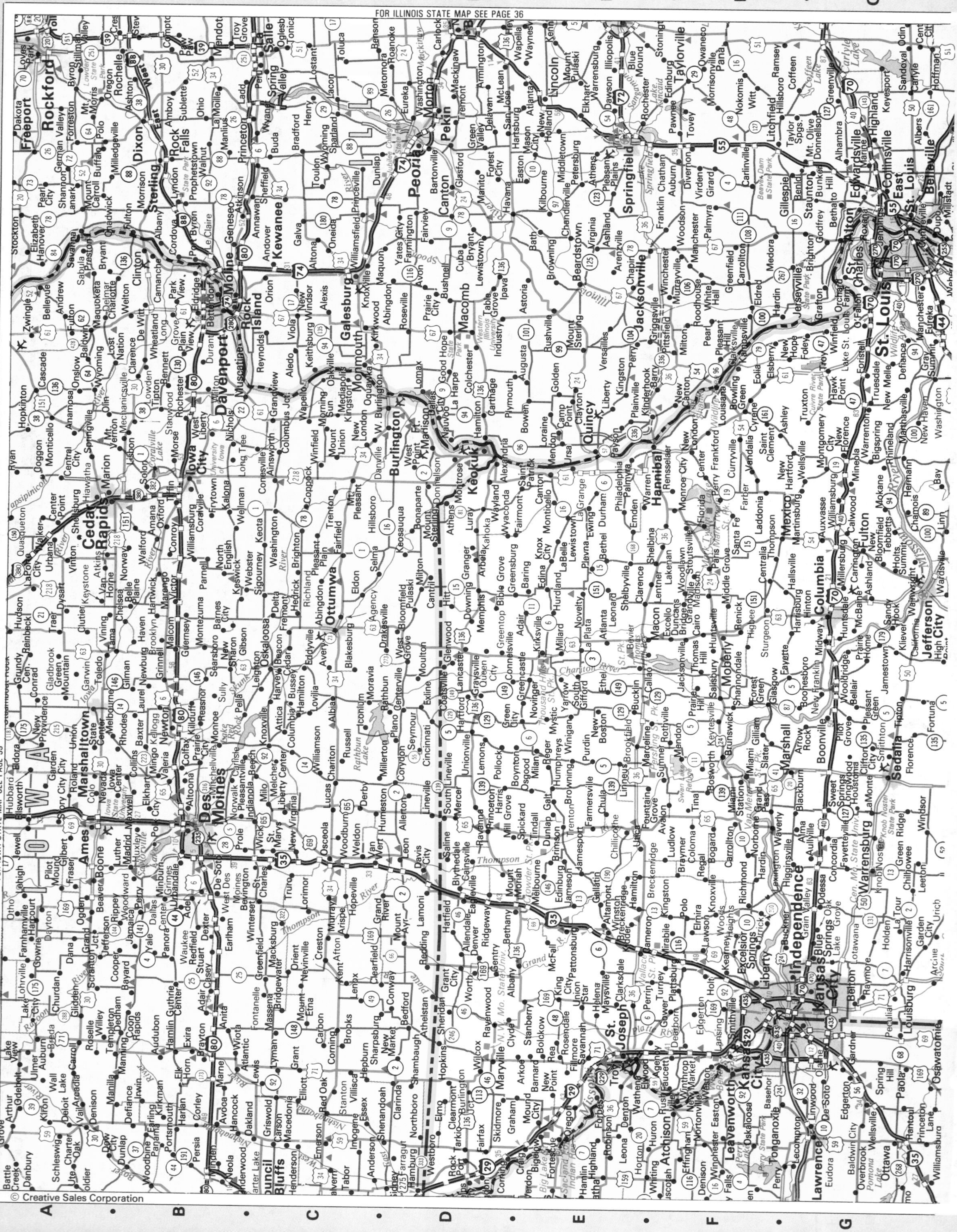

N

FOR ILLINOIS STATE MAP SEE PAGE 36

FOR IOWA STATE MAP SEE PAGE 39

© Creative Sales Corporation

SCALE OF MILES
1 INCH IS APPROXIMATELY 35 MILES
0 7 14 21 28 35

FOR ILLINOIS STATE MAP SEE PAGE 36
FOR TENNESSEE STATE MAP SEE PAGE 42
FOR MISSISSIPPI STATE MAP SEE PAGE 56
FOR ARKANSAS STATE MAP SEE PAGE 19
FOR OKLAHOMA STATE MAP SEE PAGE 74
FOR KANSAS STATE MAP SEE 40
ILLINOIS
MISSOURI
KANSAS
OKLAHOMA
ARKANSAS
TENN.
MISS.
Memphis
Springfield
Joplin
Little Rock
Jonesboro
Fort Smith
Fayetteville
Rolla
Cape Girardeau
Poplar Bluff
West Plains
© Creative Sales Corporation

SCALE OF MILES
1 INCH IS APPROXIMATELY 35 MILES
0 7 14 21 28 35
N

1 2 3 4 5 6 7

A

BRITISH COLUMBIA   ALBERTA

Milk River

Elko

Cardston
River

CANAD
UNITED STA

Flathead

Eastport

Rexford Eureka

Yaak

Sweetgrass
Port of Del Bonita Sunburst

Moyie Sprs.

Fortine

National
Park

Blackfeet
Milk

Babb
St. Mary

Santa Rita

Naples

Troy

Trego
Stryker

Olney

Polebridge

Glacier

National
Park

Indian

Blackfoot
Kiowa

Cut Bank

Inverness Rudyard Gildford
Joplin Hingham Kremlin

Libby

Kootenai
National
Forest

Whitefish
Columbia
Falls

West Glacier
Apgar

Browning

Ethridge

Shelby

Lothair Chester

Box Elder

Hope
East Hope

Happy's Inn
Marion

Coram
Martin City

Essex

East Glacier Park

Valier

Rocky Boys
Indian Reservati

Fork

Kalispell
State Park

Somers

Hungry
Horse

Reservation

Dupuyer

Conrad

Big Sandy

Heron
Noxon

Lakeside

Big Fork

Hungry
Horse
Reservoir

Pendroy

Brady

Loma

MO

Trout Creek

Proctor
Rollins

Swan Lake

National

Bynum

Dutton

Fort Benton

Thompson Falls
Hot Springs

Niarada
Elmo
Big Arm

Dayton

Lewis

Choteau

Power

Carter
Montague

Geraldine

Murray
Kellogg
Silverton
Gem Mullan

Polson

Pablo
Ronan

and

Freezeout Lake

Fairfield

Vaughn
Fort Shaw

Black Eagle

Belt

Osburn
Wallace

DeBorgia

Plains
Paradise

Clark

Augusta

Simms

Sun River
Ulm

Great
Falls

Raynesford

Calder
Avery

Saltese
Haugan

St. Regis

Dixon

Saint Ignatius

National

Fort Shaw

Cascade

Geyser

Denton

Superior

Arlee

Ravalli

Craig
Wolf Creek

Monarch

Stanford

Moccasin
Hobson

Tarkio
Alberton

Huson
Frenchtown

Seeley Lake

Lolo

Ovando
Lincoln

Lewis

Neihart

Coffee

Headquarters

Missoula

Milltown
Bonner Clinton

National

Helmville

Helena

Canyon
Creek

Holter Lake

and

National

Pierce

Lolo Hot
Springs

Lolo

Florence

Potomac

Clark

Avon

Helena

E. Helena

White Sulphur
Springs

Judith Ga

Checkerboard

Weippe
Woodland

Stevensville

Drummond

Hall

Garrison

Elliston Montana
City

Winston

Canyon
Ferry Lake

Townsend

Martinsdale

Twodot

Kamiah
Lowell

Victor

Corvallis

Deerlodge

Philipsburg
Galen

Deer Lodge

Jefferson City

Ringling

Lewis and Clark
National Forest

Stites
Harpster

Hamilton
Grantsdale

Warmsprings

Deerlodge

Basin

Boulder

Radersburg

Toston

Melville

Mount Idaho

Darby

Conner

Anaconda
Opportunity
Walkerville

Butte

Whitehall

Three Forks
Logan

Manhattan
Belgrade

Wilsall

Golden Elk City

Wise River
Divide

Silver Star
Waterloo

Willow
Creek

Amsterdam
Churchill

Clyde Park

Big Timber

Nez Perce
Nat'l Forest

Wisdom

Melrose

Harrison

Bozeman
Hot Sprs.

Livingston

Springdale

Gibbonsville

Jackson

Glen

Twin Bridges
Sheridan

Norris

Gallatin Gateway

Bozeman

McLeod

North Fork

Laurin

Ennis

Big Sky

Emigrant

Pray

Ny

Salmon
Baker

Alder
Virginia City

Cameron

Chico
Hot Springs

Custer

Tendoy
Lemhi

Bannack
State Park

Grant

Dillon

Gardiner

Warm Lake

Challis

Leadore

Dell

Mammoth
Springs Jct.

Tower Jct.

IDAHO

May

Patterson

Lima

West Yellowstone

Norris Jct.

Canyon Jct.

Cape Horn

Blue Dome

Spencer

Dubois

Madison Jct.

Lake Jct.

Sunbeam
Stanley

Clayton

MacKay

Island
Park

Old
Faithful

W. Thumb Jct.

Pioneerville
Centerville
Idaho City

Sawtooth
National
Recreational
Area

Mud Lake Hamer

St. Anthony
Parker

Ashton
Chester

Warm
River

Heart
Lake

© Creative Sales Corporation

SCALE OF MILES
1 INCH IS APPROXIMATELY 35 MILES
0 7 14 21 28 35

N

SASKATCHEWAN

Claydon Frontier Val Marie Rock Glen Minton Lake Alma Oungre
Climax Bracken Orkney Coronach Big Beaver Regway Port of Oungre
DA STATES Port of Climax Monchy W. Poplar Port of Coronoch Port of Raymond Westby Fortuna
Port of Opheim Port of Whitetail Whitetail Raymond Alkabo
Ophiem Four Buttes Flaxville Redstone Westby
Turner Port of Morgan Richland 248 Peerless Scobey Plentywood Antelope
Loring Four Buttes Reserve Grenora
232 233 241 242 Saco Hinsdale Medicine Lake Zahl
Lohman Chinook Zurich Harlem Medicine Lakes Medicine Nat'l Wildlife Refuge
Havre Bearpaw State Recreational Area Fort Belknap Agency Dodson Homestead Froid Williston
Chief Joseph Battleground State Mon. Wagner Malta Nashua Wolf Point Poplar Brockton Culbertson Bainville Fort Union Nat'l Hist. Site
Lloyd Cleveland Glasgow Oswego Fort Buford State Hist. Site Cartwright
NTANA Fort Belknap Indian Reservation Frazer Fairview Charbonneau
Fort Peck Bear Creek Recreational Area Vida Sidney
Missouri U.S. Bend Nat'l Wildlife Refuge Fort Peck Lake Rock Creek Recreational Area Richey Lambert Crane
Charles M. Russel Nat'l Wildlife Refuge James Kipp Recreational Area Devils Creek Recreational Area Nelson Creek Recreational Area Savage
Winifred Charles M. Russel Nat'l Wildlife Refuge Bloomfield
Suffolk Crooked Creek Recreational Area Circle Intake
Christina Roy Jordan Brockway Lindsay Glendive ND
Hilger Lewistown Teigen Cohagen Makoshika State Park Wibaux
Grassrange Winnett Rock Springs Terry Fallon Beach
Moore Forestgrove Flatwillow Angela Golva
Lewis and Clark National Forest Garneill Melstone Sumatra Ingomar Plevna Baker
Gap Harlowtown Roundup Musselshell Cartersville Miles City Marma
Shawmut Klein Hysham Hathaway Rosebud Medicine Rocks State Park
Ryegate Lavina Forsyth
Broadview Bighorn Custer Colstrip Volborg Ekalaka Ladner
Rapelje Worden Powderville Custer National Forest
Acton Huntley Pompeys Pillar Lame Deer Ashland Broadus Camp C
Billings Hardin Busby Capitol Custer National Forest
Columbus Laurel Crow Agency Northern Cheyenne Indian Reservation Birney Hammond Alzada SD
Reedpoint Park City Silesia Garryowen Crow Indian Reservation Custer National Forest
Absarokee Joliet Edgar Lodge Grass Ford Biddle Lightning Flat
Nye Boyd Fromberg Wyola Decker Colony
Dean Fishtail Roberts Bighorn Canyon Nat'l Recreation Area Parkman Acme Rockypoint New Haven Hulett Alva Belle Fourche
Red Lodge Bearcreek Bridger Ranchester Recluse Devils Tower Nat. Monument Aladdin Beulah
Cooke City Belfry Clark Elk Basin Deaver Dayton Sheridan Leiter Spotted Horse Weston Devil's Tower Jct. Spearfish
Powell Frannie Cowley Burgess Jct. Beckton Big Horn Arvada Oshoto Sundance Lea
Lovell Garland Byron Banner Clearmont Thunder Basin Nat'l Grassland Black Hills
Cody Ralston Emblem Story Ucross Buffalo Bill Shoshone State Park Keyhole Res.
Buffalo Bill Reservoir Otto Shell Lake DeSmet Rozet Central
Greybull Basin Big Horn National Forest WYOMING Gillette Moorcroft
Meeteetse Butlington Hyattville Buffalo Upton Four Corners Deerfie
Pitchfork Manderson Ten Sleep Osage
Valley Worland Savageton Newcastle
Grass Creek

FOR WYOMING STATE MAP SEE PAGE 88

FOR NORTH DAKOTA STATE MAP SEE PAGE 70

FOR SOUTH DAKOTA STATE MAP SEE PAGE 80

© Creative Sales Corporation

SCALE OF MILES
1 INCH IS APPROXIMATELY 35 MILES
0 7 14 21 28 35

N

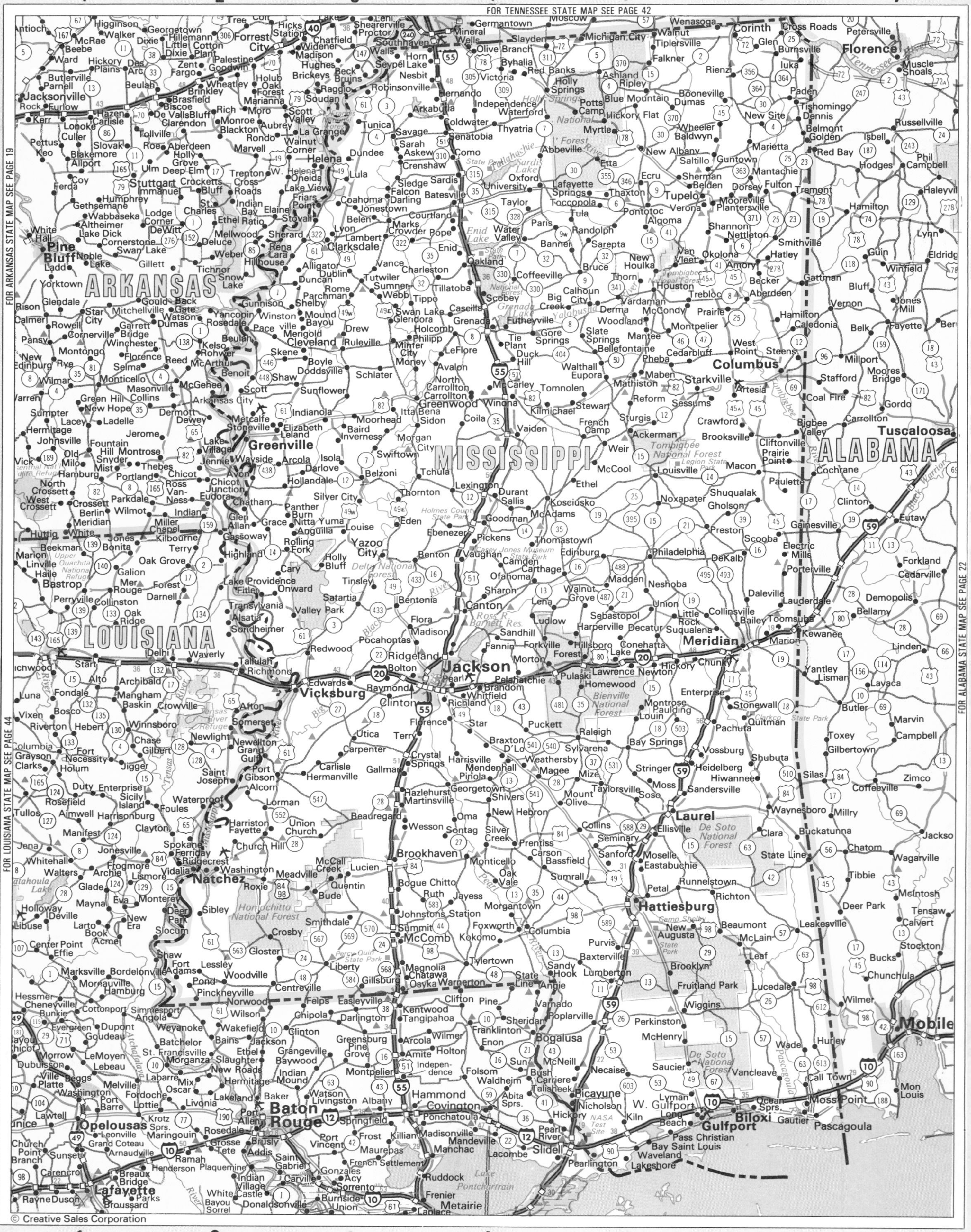

FOR TENNESSEE STATE MAP SEE PAGE 42
FOR ARKANSAS STATE MAP SEE PAGE 19
FOR LOUISIANA STATE MAP SEE PAGE 44
FOR ALABAMA STATE MAP SEE PAGE 22
ARKANSAS
LOUISIANA
MISSISSIPPI
ALABAMA
1 2 3 4 5 6 7
A B C D E F G H J K
Jackson
Vicksburg
Greenville
Clarksdale
Cleveland
Greenwood
Yazoo City
Natchez
McComb
Brookhaven
Hattiesburg
Laurel
Meridian
Columbus
Starkville
Tupelo
Corinth
Oxford
Biloxi
Gulfport
Pascagoula
Baton Rouge
Mobile
Florence
Tuscaloosa
Pine Bluff
Lafayette
Opelousas
Helena
© Creative Sales Corporation

SCALE OF MILES
1 INCH IS APPROXIMATELY 20 MILES
0 4 8 12 16 20

N

QUEBEC

CANADA
UNITED STATES

MAINE

VERMONT

NEW HAMPSHIRE

Lake Memphremagog

Connecticut Lakes State Forest

Coleman State Park

Lake Champlain

Plattsburgh
Burlington
South Burlington
Montpelier
Barre
Rutland
Springfield
Brattleboro
Bennington

Berlin
Gorham
White Mountain National Forest
Mount Washington
Concord
Manchester
Nashua
Laconia
Lake Winnipesaukee
Portsmouth
Dover
Rochester
Keene
Claremont
Lebanon
Hanover

Lake Francis State Park

Mooselookmeguntic Lake
Rangeley Lake
Kennebago Lake

Lawrence
Lowell
Fitchburg
Haverhill
Newburyport
Gloucester
Ipswich

© Creative Sales Corporation

FOR NEW YORK STATE MAP SEE PAGES 64-67
FOR MAINE STATE MAP SEE PAGE 45
FOR MASSACHUSETTS STATE MAP SEE PAGE 30

SCALE OF MILES
1 INCH IS APPROXIMATELY 35 MILES
0 7 14 21 28 35

N

FOR SOUTH DAKOTA STATE MAP SEE PAGE 80

W Y

Mule Cr. Jct.
Redbird
Lance Creek
Lost Springs
Manville Lusk
Keeline
Node
Jay Em
Veteran
Yoder
Huntley
...ugwater
Hawk Sprs.
La Grange
Albin

Provo
Edgemont
Buffalo Gap National Grassland
Angostura Res.
Oglala National Grassland
Harrison
Crawford
Fort Robinson State Park
Chadron State Park
Whitney Chadron
Hemingford
Nebraska National Forest
Alliance
Antioch Lakeside
Ellsworth Bingham
Angora

Hot Springs
Gap
Cheyenne
Oelrichs
Pine Ridge Indian Reservation
Oglala
Wounded Knee
Pine Ridge
Hay Springs Rushville
Clinton
Gordon
Niobrara River
Snake River

Kyle
Allen
Wounded Knee Battle Site
Batesland
Martin
Merriman
Eli
Samuel R. McKelvie Nat'l Forest
Merritt Res.

Long Valley
White River
Wood
Rosebud
Parmelee
Okreek Mission
Saint Francis
Rosebud Indian Reservation
Cody Nenzel Kilgore Crookston
Valentine
Norden
Wood Lake
Johnstown
Ainsworth Long Pine
Brownlee
Elsmere
Purdum

Winner
Sparks
Springv...
Wewala

Fort Laramie
Lingle
Torrington
Morrill
Mitchell
Scottsbluff
Gering Minatare
Terrytown
Melbeta Bayard
McGrew
Harrisburg
Bridgeport
Broadwater
Lisco
Dalton
Gurley
Kimball Dix Potter
Bushnell
Sidney
Lodgepole
Chappell
Big Springs
Sedgwick Ovid
Julesburg
Proctor Crook
Briggsdale
Buckingham Raymer
Stoneham
Willard
Sterling
Fleming
Haxtun Paoli
Atwood
Merino
Snyder
Fort Morgan Brush
Akron Otis
Yuma Eckley
Woodrow
Wray

Hyannis
Whitman
Ashby
Mullen
Seneca
Thedford
Halsey
Dunning
Arthur
Tryon
Oshkosh
Lewellen
Lemoyne
Keystone
Lake C.W. McConaughy
Brule Ogallala
Paxton Sutherland
Hershey
North Platte
Maxwell Brady
Sutherland Res.
Grant
Brandon
Venango
Madrid
Elsie Wallace
Grainton Dickens
Wellfleet
Maywood
Curtis
Moorefield
Farnam
Eustis
Smithfield Bertrand
Loomis
Holdrege

NEBRASKA

Brewster
Dunning
Halsey
Nebraska National Forest
Middle Loup River
Stapleton
Gandy
Anselmo
Merna
Arnold
Broken Bow
Callaway
Oconto
Eddyville
Gothenburg
Willow Island Cozad
Lexington Overton
Johnson Res.
Elwood

Imperial
Champion
Enders
Wauneta
Hamlet
Palisade
Hayes Center
Hugh Butler Lake
Culbertson
Stratton
Max
Trenton
Swanson Reservoir
Indianola
McCook
Bartley
Cambridge
Arapahoe
Edison
Oxford
Wilsonville
Danbury Lebanon
Beaver City
Stamford
Orleans
Aln...

COLORADO

Strasburg
Byers
Deer Trail
Last Chance
Lindon Anton
Cope
Joes
Idalia
Wheeler
St. Francis
Bird City
McDonald
Atwood
Oberlin
Norcatur
Norton
Jennings
Clayton
Dresden
Selden
Rexford
Gem
Menlo
Hoxie
Halford
Morland
Hill City
Bogue
Damar Palco
Zurich
Stockton
Woodsto...
Almena
Long Island
Phillipsburg
Prairie View
Logan Glade
Edmond
Lenora

Agate
River Bend Limon
Hugo
Punkin Center
Aroya
Wild Horse
Firstview
Cheyenne Wells
Kit Carson
Eads
Genoa
Arriba
Seibert Vona
Stratton
Burlington
Goodland
Brewster
Colby
Oakley
Grinnell
Grainfield
Park
Gove
Winona
Wallace
Sharon Springs
Russell Springs
Quinter
Collyer
WaKeeney
Hays
Victoria
Ellis

I-70
I-80

FOR WYOMING STATE MAP SEE PAGE 88
FOR COLORADO STATE MAP SEE PAGE 28
FOR KANSAS STATE MAP SEE 40

© Creative Sales Corporation

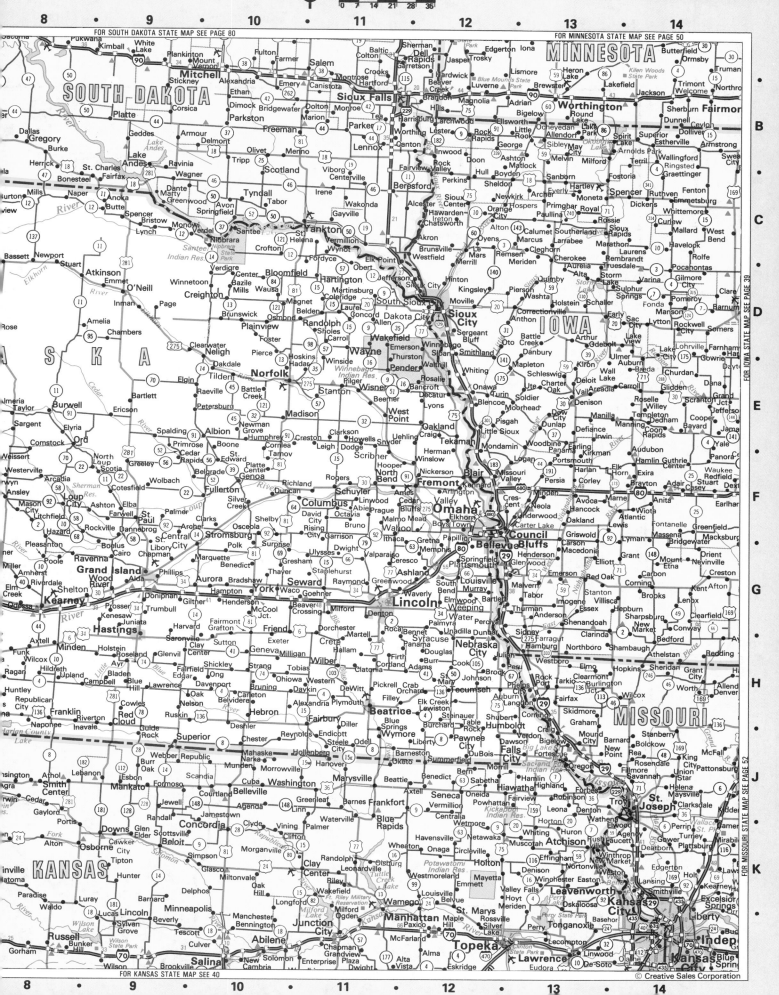

SCALE OF MILES
1 INCH IS APPROXIMATELY 35 MILES
0 7 14 21 28 35

FOR UTAH STATE MAP SEE PAGE 82

FOR IDAHO STATE MAP SEE PAGE 34

FOR OREGON STATE MAP SEE PAGE 76

UTAH

IDAHO

OREGON

NEVADA

CAL.

Twin Falls, Burley, Murtaugh, Oakley, Rock Creek, Hansen, Marion, Hollister, Rogerson, Three Creek, Riddle, Grasmere, Owyhee, Mountain City, Jarbidge, Jack Creek, Tuscarora, Midas, Willow Rock, Tuscarora Mts.

Grouse Creek Mts., Lucin, Montello, Oasis, Thousand Springs, Contact, Wells, Deeth, Halleck, Lamoille, Elko, Carlin, Beowawe, Beowawe Geysers, Battle Mountain, Golconda, Valmy, Winnemucca, Paradise Valley, McDermitt, Orovada, Sulphur, Gerlach, Empire

Wendover, Bonneville Speedway, Toana Range, White Horse Pass, Currie, Lage's, Cherry Creek, McGill, Ruth, East Ely, Ely, Preston, Lund, Kimberly, Duckwater, Currant, Eureka, Austin, Ione, Gabbs, Luning

East Humboldt Range, Ruby Mountain Scenic Area, Ruby Valley, Harrison Pass, Jiggs, Shantytown, Railroad Pass, Crescent Valley, Garden Pass, Simpson Park Mts., Toquima Range, Round Mt., Manhattan, Toiyabe Range, Toiyabe National Forest, Berlin-Ichthyosaur State Park

Humboldt National Forest, Lehman Caves Nat'l Monument, Sacramento, Wheeler Peak Scenic Area, Baker, Major's Place, Connors Pass, Garrison, Snake Range

Lovelock, Rye Patch Reservoir, Rye Patch Dam, Imlay, Mill City, Unionville, Oreana, Eagle Picher Mine, Humboldt Range, Sonoma Range, Santa Rosa Range, Little Humboldt River

Carson Sink, Stillwater, U.S. Naval Air Station, Fallon, Silver Springs, Hazen, Fernley, Wadsworth, Nixon, Pyramid Lake, Sutcliffe, Carson Lake, Lahontan Reservoir, Frenchman, Cold Spring, Middle Gate, Dixie Valley, Alpine, Clan Alpine Mts., Stillwater Range

Reno, Sparks, Verdi, Virginia City, Gold Hill, Silver City, Dayton, Carson City, Stewart, Genoa, Minden, Gardnerville, Wellington, Yerington, Mason, Smith, Schurz, Walker Lake, Wassuk Range, Babbitt, Walker River Indian Reservation

Lake Tahoe, Incline Village, Crystal Bay, Zephyr Cove, Stateline, Glenbrook, Meyers, Markleeville, Woodfords, Coleville, Topaz

Lakeview, Adel, Cedarville, Eagleville, Ravendale, Litchfield, Wendel, Doyle, Beckwourth, Chilcoot Pass, Loyalton, Truckee

Honey Lake, Pyramid Lake Indian Reservation, Smoke Creek Desert, Granite Range, Black Rock Desert, High Rock Canyon, Massacre Lake, Charles Sheldon Ind. Res., Antelope Range, Pine Forest Range, Jackson Mts., Quinn River, Fort McDermitt Indian Reservation, Summit Lake Indian Reservation

Blue Mtn. Pass, Trout Creek Mts., Crooked River, Owyhee River, Crump Lake, Abert Lake, Alkali Lake, Pueblo Mts., Fields, Denio, Denio Junction

U.S. Route 80, 93, 95, 50, 6, 40

© Creative Sales Corporation

FOR CALIFORNIA STATE MAP SEE PAGES 24-27

SCALE OF MILES
1 INCH IS APPROXIMATELY 35 MILES
0 7 14 21 28 35

FOR UTAH STATE MAP SEE PAGE 82

FOR ARIZONA STATE MAP SEE PAGE 20

FOR CALIFORNIA STATE MAP SEE PAGES 24-27

FOR CALIFORNIA STATE MAP SEE PAGES 24-27

© Creative Sales Corporation

SCALE OF MILES
1 INCH IS APPROXIMATELY 35 MILES
0 7 14 21 28 35

N

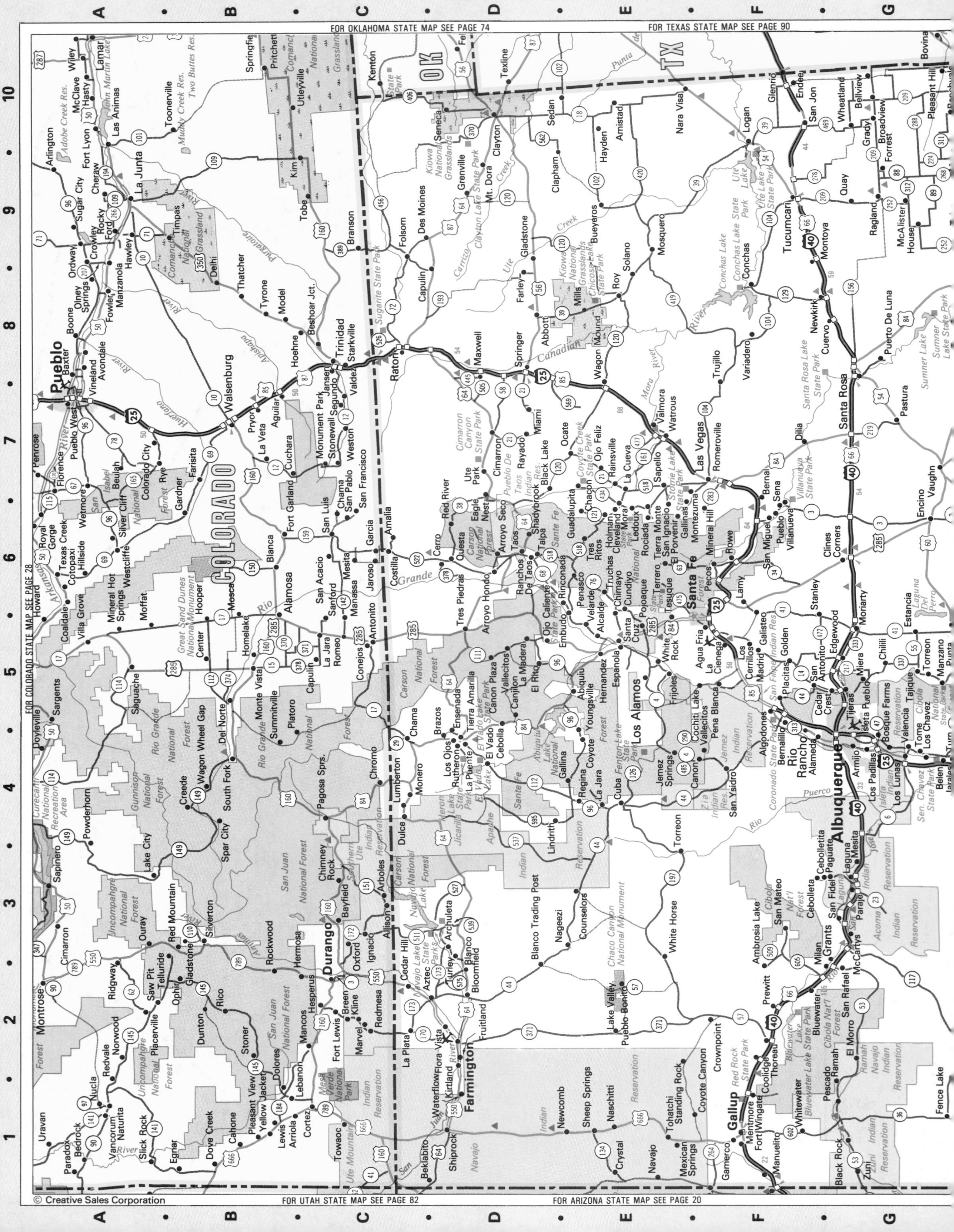
FOR OKLAHOMA STATE MAP SEE PAGE 74
FOR TEXAS STATE MAP SEE PAGE 90
FOR COLORADO STATE MAP SEE PAGE 28
FOR UTAH STATE MAP SEE PAGE 82
FOR ARIZONA STATE MAP SEE PAGE 20
COLORADO
OK
TX
Pueblo
Trinidad
Raton
Santa Fe
Los Alamos
Las Vegas
Albuquerque
Rio Rancho
Santa Rosa
Tucumcari
Clayton
Durango
Farmington
Gallup
Grants
Alamosa
Walsenburg
La Junta
Taos
Espanola
© Creative Sales Corporation

SCALE OF MILES
1 INCH IS APPROXIMATELY 35 MILES
0 7 14 21 28 35

FOR TEXAS STATE MAP SEE PAGE 90

NEW MEXICO

TEXAS

CHIHUAHUA

Clovis
Portales
Ranchvale
St. Vrain
Melrose
Farwell
Goo
Arch
Rogers
Causey
Lingo
Pleasant Hill
Floyd
Dora
Pep
Milnesand
Tolar
Taiban
Elida
Kenna
Crossroads
Caprock
Tatum
McDonald
Hilburn City
Knowles
Hobbs
Lovington
Humble City
Monument
Eunice
Oil Center
Nadine
Jal
Kermit
Wink
Pyote
Barstow
Saragosa
Balmorhea
Toyahvale
Pecos
Mentone
Kent
Van Horn
Allamoore
Sierra Blanca
Lobo
Fort Sumner
Yeso
Mesa
Ramon
Elkins
Roswell
Dexter
Greenfield
Hagerman
Lake Arthur
Riverside
Artesia
Atoka
Hope
Seven Rivers
Loco Hills
Malaga
Loving
Carlsbad
Whites City
Black River Village
Carlsbad Caverns
El Paso Gap
Pine Springs
Salt Flat
Cornudas
Duran
Cedarvale
Corona
Ancho
Jicarilla
White Oaks
Carrizozo
Capitan
Lincoln
Fort Stanton
Hondo
Picacho
Tinnie
Arabela
Dunken
Mayhill
Weed
Pinon
Ruidoso Downs
Ruidoso
Alto
Angus
Mescalero
Bent
Elk
High Rolls
Cloudcroft
Sunspot
Sacramento
Orogrande
Tularosa
La Luz
Alamogordo
Three Rivers
White Sands National Monument
White Sands Missile Range
Fort Bliss
Organ
Las Cruces
University Pk.
Mesquite
Vado
Berino
Chaparral
Newman
Anthony
El Paso
Juarez
Horizon City
Fabens
Clint
Socorro
Acala
McNary
El Porvenir
Ahumada
Samalayuca
Ascension
Janos
Magdalena
Datil
Pie Town
Quemado
Omega
Salt Lake
Red Hill
Luna
Reserve
Apache Creek
Cruzville
Aragon
Alma
Glenwood
Mogollon
Pleasanton
Buckhorn
Cliff
Gila
Red Rock
Mule Creek
Virden
Silver City
Pinos Altos
Central
Bayard
Hurley
Santa Rita
Hanover
Mimbres
San Lorenzo
Kingston
Hillsboro
Winston
Monticello
Placita
Cuchillo
Williamsburg
Truth or Consequences
Las Palomas
Caballo
Arrey
Derry
Garfield
Salem
Hatch
Rincon
Radium Springs
Leasburg
Hill
Dona Ana
Fairacres
Mesilla
San Miguel
La Mesa
Chamberino
La Union
Anapra
Deming
Sunshine
Columbus
Hachita
Antelope Wells
Rodeo
Animas
Cotton City
Separ
Lordsburg
Tyrone
Gage
Datil
Socorro
Magdalena
Bernardo
La Joya
San Acacia
Abeytas
Escondida
San Antonio
Bingham
Oscuro
Bernardo
Jarales
Sabinal
Veguita
Las Nutrias
Luis Lopez
Laborcita
Abo
Mountainair
Willard
Claunch
Gran Quivira

Lincoln National Forest
Cibola National Forest
Gila National Forest
Apache National Forest

Trinity Site
World's First Atomic Explosion (July 16, 1945 - Closed to Public)

Rio Grande
Elephant Butte Lake
Caballo Res.

Pecos River
Red Bluff Lake
Salt Lake
Bottomless Lakes State Park
Living Desert State Park
Guadalupe Mtns. National Park
Carlsbad Caverns National Park

FOR ARIZONA STATE MAP SEE PAGE 20

© Creative Sales Corporation

SCALE OF MILES
1 INCH IS APPROXIMATELY 18 MILES
0 3 6 9 12 15 18
N

SCALE OF MILES
1 inch equals 10.25 miles
0 2 4 6 8 10

Atlantic Ocean

Long Island Sound

Block Island Sound

Lake Ontario

CONN.

NEW YORK

NEW JERSEY

CANADA
UNITED STATES

FOR CONNECTICUT STATE MAP SEE PAGE 30

FOR NEW JERSEY STATE MAP SEE PAGE 31

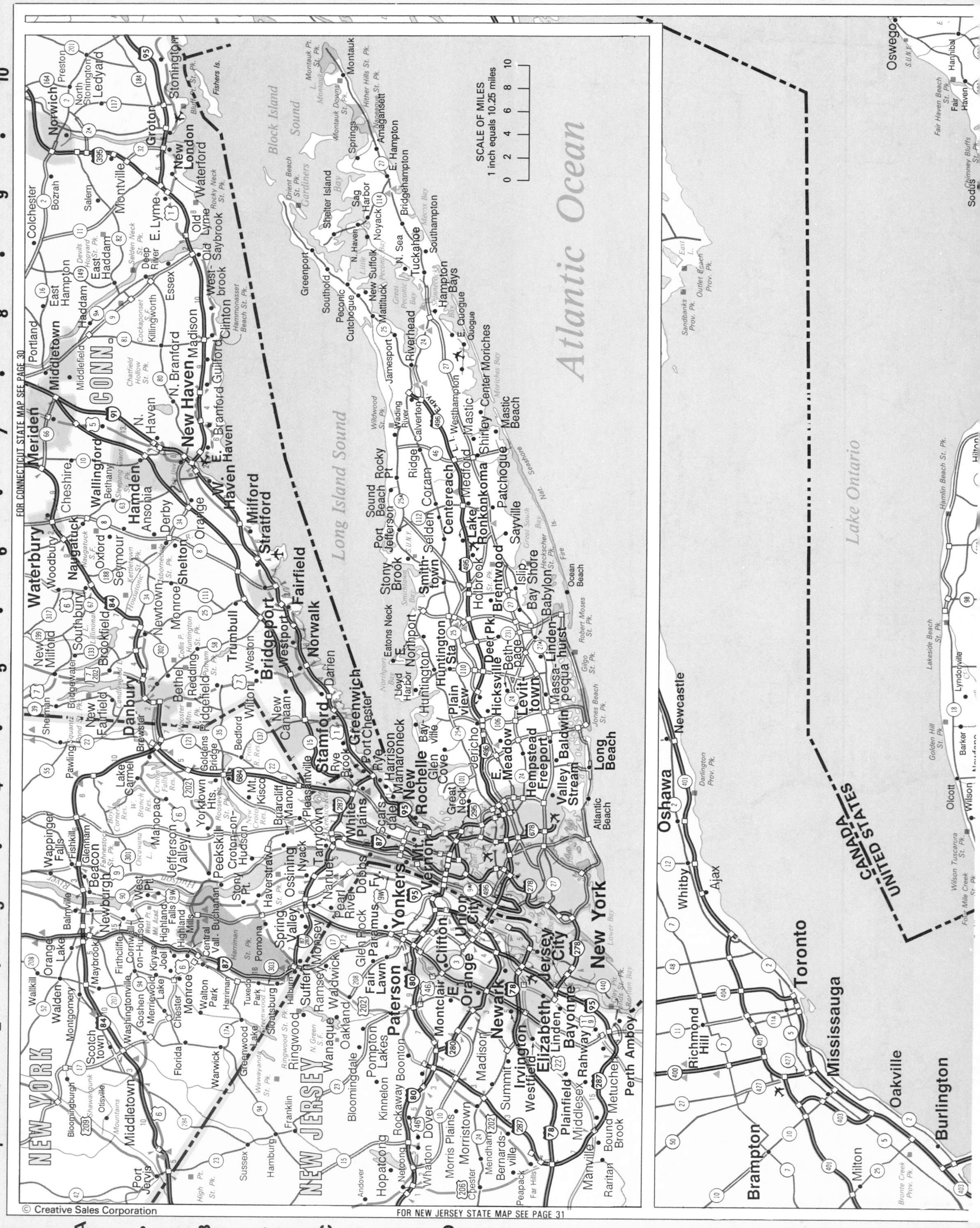

© Creative Sales Corporation

New York, Yonkers, Jersey City, Newark, Paterson, Elizabeth, Bayonne, Clifton, Union City, Perth Amboy, Mt. Vernon, New Rochelle, White Plains, Stamford, Greenwich, Norwalk, Westport, Fairfield, Bridgeport, Stratford, Milford, New Haven, Hamden, Waterbury, Meriden, Wallingford, Naugatuck, Danbury, Bethel, Ridgefield, Brookfield, Peekskill, Ossining, Tarrytown, Nyack, Spring Valley, Suffern, Monroe, Middletown, Newburgh, Beacon, Wappinger Falls, Port Jervis

Hempstead, Freeport, Long Beach, Levittown, Hicksville, Massapequa, Bay Shore, Babylon, Islip, Brentwood, Huntington, Northport, Smithtown, Patchogue, Sayville, Center Moriches, Mastic, Shirley, Medford, Coram, Selden, Port Jefferson, Stony Brook, Riverhead, Mattituck, Southold, Greenport, Southampton, Bridgehampton, E. Hampton, Montauk, Sag Harbor, Shelter Island, Gardiners

Toronto, Mississauga, Oakville, Burlington, Brampton, Milton, Oshawa, Whitby, Ajax, Newcastle, Richmond Hill, Oswego, Sodus

SCALE OF MILES
1 INCH IS APPROXIMATELY 18 MILES
0 3 6 9 12 15 18

H J K L M N O P

NEW YORK

PENNSYLVANIA

ONTARIO

CANADA
UNITED STATES

Lake Erie

Lake Ontario

Burlington
Hamilton
Dundas
Stoney Creek
Grimsby
St. Catherines
Niagara-on-the-Lake
Niagara Falls
Welland
Thorold
Port Colborne
Dunnville

Youngstown
Lewiston
Ransomville
Niagara Falls
Fort Erie
Tonawanda
N. Tonawanda
Kenmore
Buffalo
Lackawanna
Blasdell
Hamburg
Orchard Pk.
Eden
N. Collins
Angola
Angola-on-the-Lake
Lake Erie Beach
Farnham
Silver Creek
Dunkirk
Fredonia
Brocton
Westfield
Mayville
Sherman
Panama
Findley L.
Sugar Grove
Lakewood
Jamestown
Celoron
Falconer
Frewsburg
Bemus Pt.
Long Pt. on Lake Chautauqua
Cassadaga
Sinclairville
Cherry Creek
Forestville
Perrysburg
Gowanda
Cattaraugus
Little Valley
Ellicottville
Salamanca
Allegany
Olean
Westons Mills
Portville
Bolivar
Richburg
Cuba
Friendship
Belmont
Wellsville
Stannards
Andover
Alfred
Angelica
Belfast
Caneadea
Houghton
Fillmore
Pike
Castile
Silver Springs
Perry
Warsaw
Gainesville
Wyoming
Attica
Alexander
Darien Lakes St. Pk.
Alden
Lancaster
Depew
Williamsville
Clarence Center
Akron
Cowlesville
Corfu
Batavia
LeRoy
Bergen
Churchville
Spencerport
Brockport
Holley
Hilton
Greece
Rochester
E. Rochester
Pittsford
Fairport
Victor
Honeoye Falls
Lima
Avon
Caledonia
Scottsville
Geneseo
Leicester
Mt. Morris
Nunda
Dansville
Wayland
Cohocton
Naples
Avoca
Bath
Savona
Painted Post
Corning
Riverside
Gang Mills
S. Corning
Flats
Southport
Elmira
Elmira Hts.
Horseheads
Montour Falls
Watkins Glen
Odessa
Millport
Burdett
Penn Yan
Dundee
Hammondsport
Kanona
Canisteo
Hornell
N. Hornell
Almond
Arkport
Canaseraga
Dansville
Springwater
Livonia
Lakeville
Conesus
Honeoye
Bristol
Canandaigua
Clifton Spgs.
Manchester
Shortsville
Palmyra
Macedon
Newark
Lyons
Phelps
Waterloo
Seneca Falls
Geneva
Ovid
Interlaken
Trumansburg
Dresden
Willard
Lodi
Rushville
Holcomb
Bloomfield
Webster
Fairport

Auburn
Cayuga
Union Springs
Aurora
Weedsport
Port Byron
Cato
Seneca Falls
Clyde
Red Creek
Wolcott
Sodus
Sodus Pt.
Fair Haven
Fair Haven Beach St. Pk.
Chimney Bluffs St. Pk.

Olcott
Wilson
Barker
Newfane
Middleport
Medina
Albion
Oakfield
Oakfield
Lyndonville
Lockport
Gasport
Middleport

Bradford
Warren
Youngsville
Corry
Union City
N. East
Clarendon
Kane
Mt. Jewett
Smethport
Port Allegany
Coudersport
Galeton
Westfield
Shinglehouse
Eldred
Bradford
Emporium
Johnsonburg
Wellsboro
Mansfield
Blossburg
Canton
Troy
Liberty
Tioga
Elkland
Addison
Woodhull
Jasper
Canton
Allegheny National Forest
Tionesta
Titusville
Pleasantville
Hydetown

FOR PENNSYLVANIA STATE MAP SEE PAGE 78

© Creative Sales Corporation

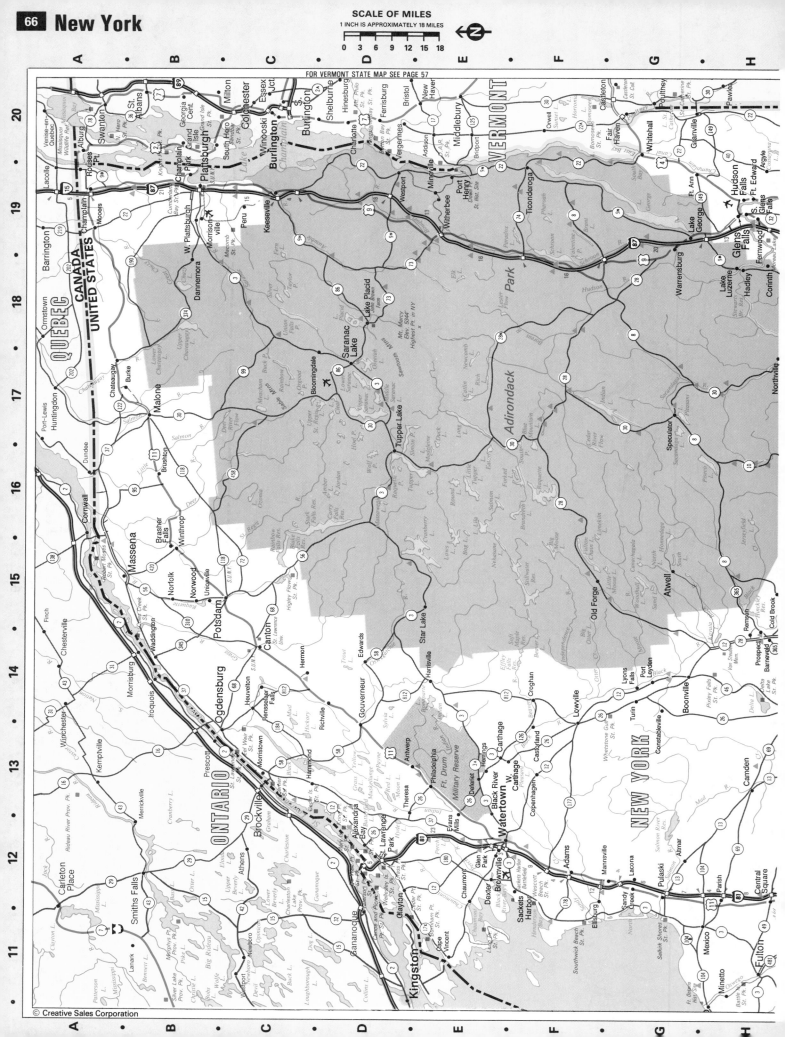

SCALE OF MILES
1 INCH IS APPROXIMATELY 18 MILES
0 3 6 9 12 15 18

FOR VERMONT STATE MAP SEE PAGE 57

VERMONT

QUEBEC

CANADA
UNITED STATES

ONTARIO

NEW YORK

Adirondack Park

© Creative Sales Corporation

SCALE OF MILES
1 INCH IS APPROXIMATELY 18 MILES
0 3 6 9 12 15 18

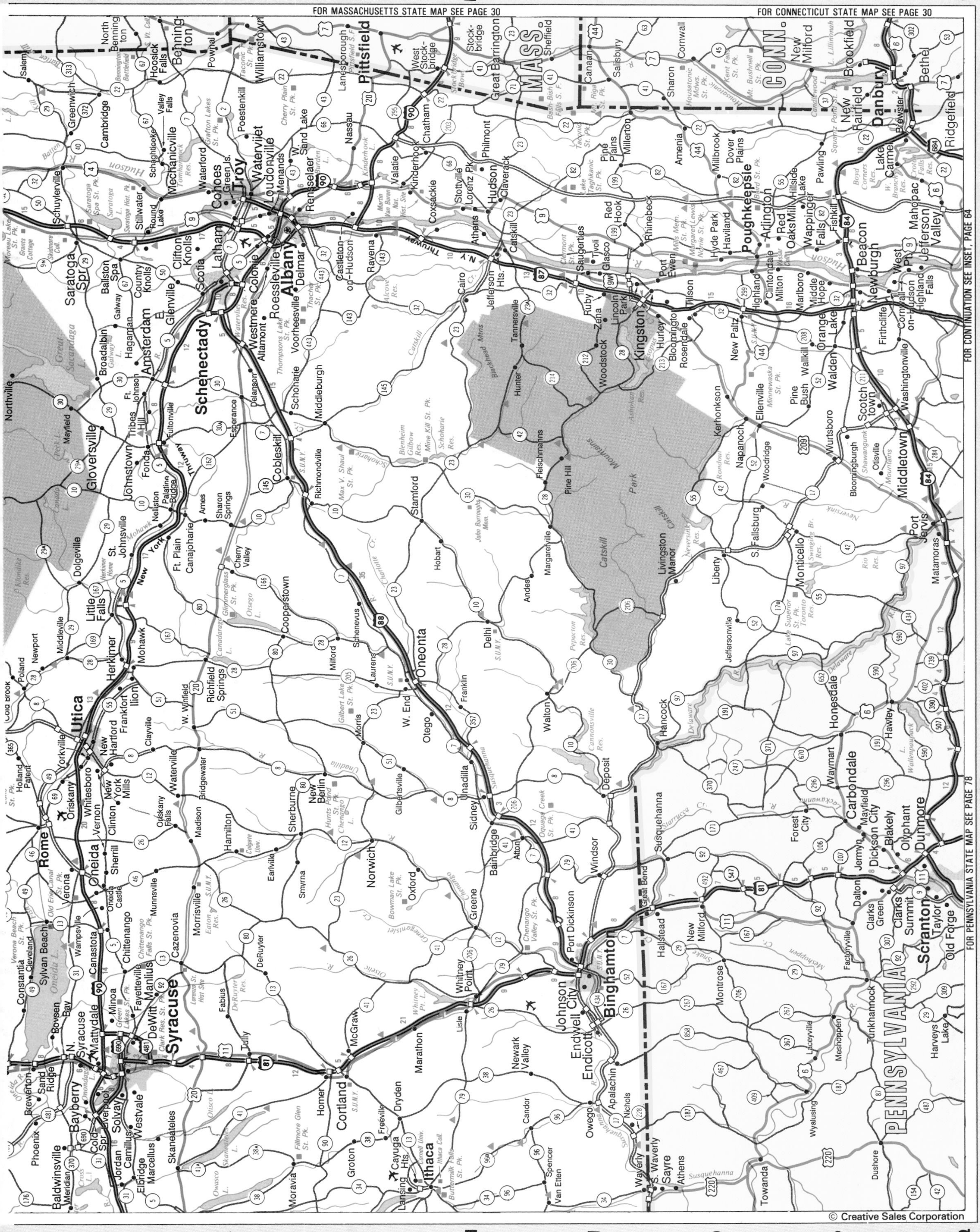

FOR MASSACHUSETTS STATE MAP SEE PAGE 30

FOR CONNECTICUT STATE MAP SEE PAGE 30

FOR CONTINUATION SEE INSET, PAGE 64

FOR PENNSYLVANIA STATE MAP SEE PAGE 78

MASS.

CONN.

PENNSYLVANIA

Albany, Schenectady, Troy, Watervliet, Cohoes, Saratoga Spa, Amsterdam, Gloversville, Johnstown, Utica, Rome, Oneida, Syracuse, Cortland, Ithaca, Binghamton, Endicott, Endwell, Johnson City, Oneonta, Delhi, Walton, Hancock, Monticello, Liberty, Middletown, Port Jervis, Matamoras, Poughkeepsie, Newburgh, Beacon, Kingston, Catskill, Hudson, Scranton, Carbondale, Danbury, Pittsfield

N.Y. Thruway 90 87 88 81 84

Catskill Mountains, Catskill Park, Ashokan Res., Pepacton Res., Cannonsville Res., Neversink Res., Schoharie Res.

© Creative Sales Corporation

SCALE OF MILES
1 INCH IS APPROXIMATELY 35 MILES
0 7 14 21 28 35

N

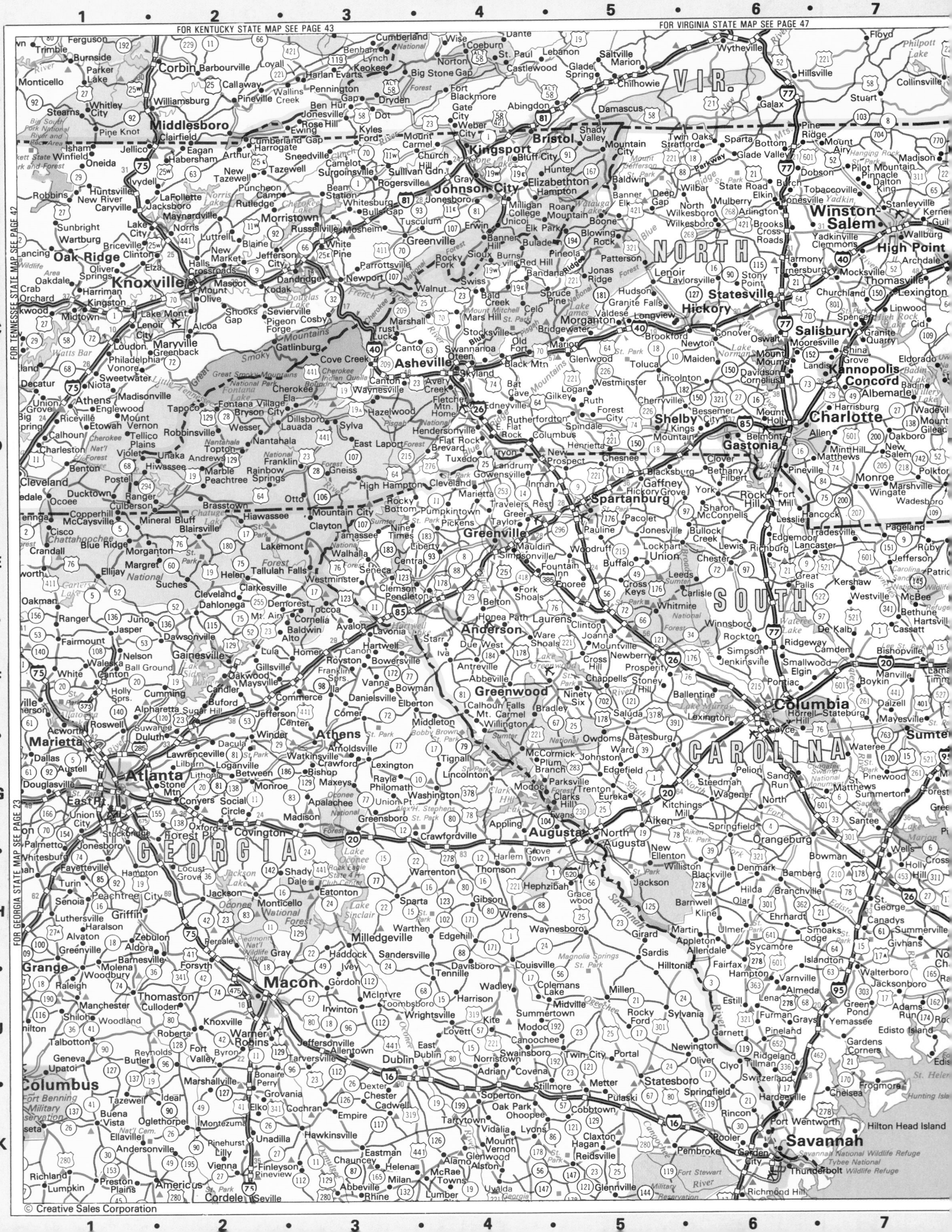

FOR KENTUCKY STATE MAP SEE PAGE 43

FOR VIRGINIA STATE MAP SEE PAGE 47

FOR TENNESSEE STATE MAP SEE PAGE 42

FOR GEORGIA STATE MAP SEE PAGE 23

VIR.

NORTH CAROLINA

SOUTH CAROLINA

GEORGIA

© Creative Sales Corporation

SCALE OF MILES
1 INCH IS APPROXIMATELY 35 MILES
0 7 14 21 28 35

N

FOR VIRGINIA STATE MAP SEE PAGE 47

Atlantic Ocean

Callands, Chatham, Clover, Chase City, South Hill, Lawrenceville, Courtland, Suffolk, Chesapeake

Martinsville, Danville, Eden, Halifax, S. Boston, Clarksville, Wise, Roanoke Rapids, Emporia, Franklin, Roduco, Gates, Corapeake, Morgans Corner, Barco, Coinjock, Bertha, Grandy, Jarvisburg

Mayodan, Reidsville, Bethel Hill, Roxboro, Lewis, Oxford, Stovall, Middleburg, Liberia, Rheasville, Garysburg, Weldon, Jackson, Halifax, Conway, Winton, Ahoskie, Colerain, Hertford, Elizabeth City, Camden, Powells Point, Harbinger

Yanceyville, Leasburg, Brooksdale, Picks, Surl, Henderson, Arcola, Essex, Centerville, Enfield, Rich Square, Spring Hill, Edenton, Point Harbor, Mamie, Nags Head, Whalebone

Greensboro, Durham, Chapel Hill, Raleigh, Cary, Wake Forest, Zebulon, Wilson, Rocky Mount, Tarboro, Greenville, Williamston, Plymouth, Columbia, Manteo, Wanchese

CAROLINA

Asheboro, Sanford, Goldsboro, Benson, Walnut Creek, Snow Hill, Ayden, Winterville, Chocowinity, Wilmar, New Holland, Avon

Fayetteville, Dunn, Newton Grove, Clinton, Warsaw, Kenansville, Kinston, New Bern, Trenton, Maysville, Newport, Morehead City, Beaufort, Ocracoke, Portsmouth

Aberdeen, Southern Pines, Spring Lake, Fort Bragg, Hope Mills, Lumber Bridge, Roseboro, Magnolia, Jacksonville, Swansboro, Fort Macon

Rockingham, Hamlet, Laurinburg, Maxton, Pembroke, Lumberton, Elizabethtown, Burgaw, Holly Ridge, Surf City, Topsail Beach

Cheraw, Bennettsville, McColl, Clio, Tatum, Rowland, Fairmont, Council, Bolton, Maco, Freeman, Scotts Hill, Wrightsville Beach

Society Hill, Darlington, Dillon, Latta, Mullins, Nichols, Whiteville, Chadbourn, Brunswick, Wilmington, Seabreeze, Carolina Beach, Kure Beach

Florence, Marion, Green Sea, Tabor City, Loris, Supply, Shallotte, Southport, Long Beach

Conway, Aynor, Bayboro, Centenary, Nixonville, Little River, Cresent Beach

Manning, Kingstree, Rhems, Myrtle Beach, Surfside Beach, Garden City, Litchfield Beach, Pawleys Island

Andrews, Georgetown, Debidue Beach

Charleston, North Charleston, Mt. Pleasant, Isle of Palms, Folly Beach

Atlantic Ocean

Albemarle Sound, Pamlico Sound, Raleigh Bay, Onslow Bay, Long Bay, Bulls Bay

Cape Hatteras National Seashore, Cape Lookout National Seashore, Cape Romain National Wildlife Refuge

© Creative Sales Corporation

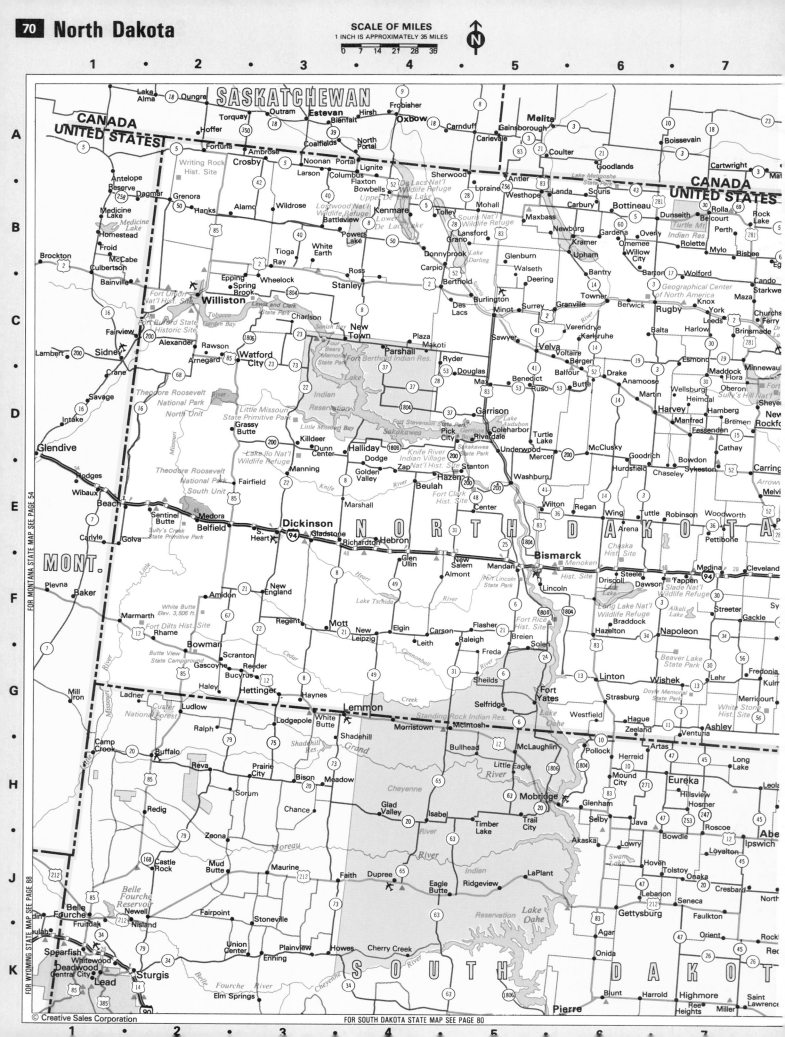

SCALE OF MILES
1 INCH IS APPROXIMATELY 35 MILES
0 7 14 21 28 35

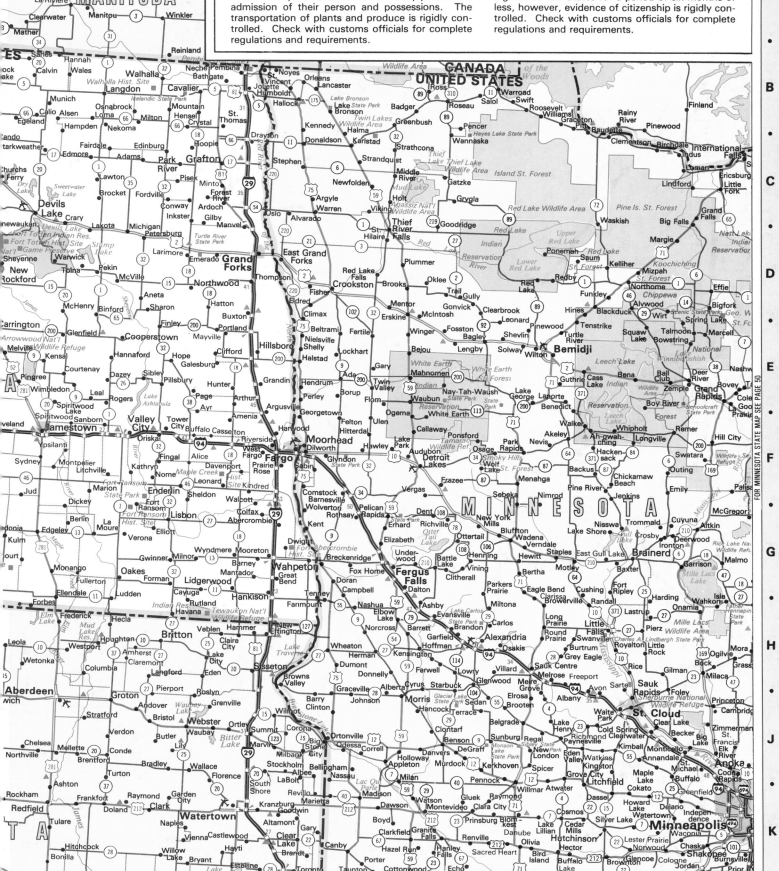

United States Citizens Visiting Canada

All persons entering Canada must report to the Canadian Immigration and Customs Office at the Port of Entry and secure the necessary permits for admission of their person and possessions. The transportation of plants and produce is rigidly controlled. Check with customs officials for complete regulations and requirements.

Canadian Citizens Visiting the United States

Passports or visas are not required of Canadian citizens or British subjects residing in Canada entering the United States for a period of six months or less, however, evidence of citizenship is rigidly controlled. Check with customs officials for complete regulations and requirements.

FOR MINNESOTA STATE MAP SEE PAGE 50

FOR SOUTH DAKOTA STATE MAP SEE PAGE 80

© Creative Sales Corporation

SCALE OF MILES
1 INCH IS APPROXIMATELY 23.5 MILES
0 5 10 15 20 25
N

FOR PENNSYLVANIA STATE MAP SEE PAGE 78

FOR MICHIGAN STATE MAP SEE PAGE 48

CANADA
UNITED STATES

ONTARIO

Lake Huron

Lake Erie

MICHIGAN

OHIO

London
St. Thomas
Sarnia
Port Huron
Chatham
Windsor
Detroit
Warren
Sterling Hts.
Troy
Royal Oak
Southfield
Livonia
Dearborn
Westland
Taylor
Pontiac
Farmington Hills
Flint
Lansing
E. Lansing
Ann Arbor
Jackson

Toledo
Bowling Green
Sylvania
Oregon
Maumee
Perrysburg
Findlay
Lima
Sandusky
Lorain
Elyria
Vermilion
Avon Lake
Westlake
Lakewood
Cleveland
Parma
Euclid
Cleveland Hts.
Shaker Hts.
Garfield Hts.
Strongsville
Brunswick
Medina
Wooster
Ashland
Mansfield
Bucyrus
Marion
Tiffin
Fremont
Fostoria
Norwalk
Conneaut
Ashtabula
Geneva
Painesville
Mentor
Willoughby
Eastlake
Willowick
Cuyahoga Falls
Stow
Kent
Ravenna
Akron
Barberton
Wadsworth
Massillon
Canton
Alliance
Salem
Youngstown
Niles
Warren
Struthers
Campbell
Hubbard
Sharon
Dover
New Philadelphia

FOR INDIANA STATE MAP SEE PAGE 38

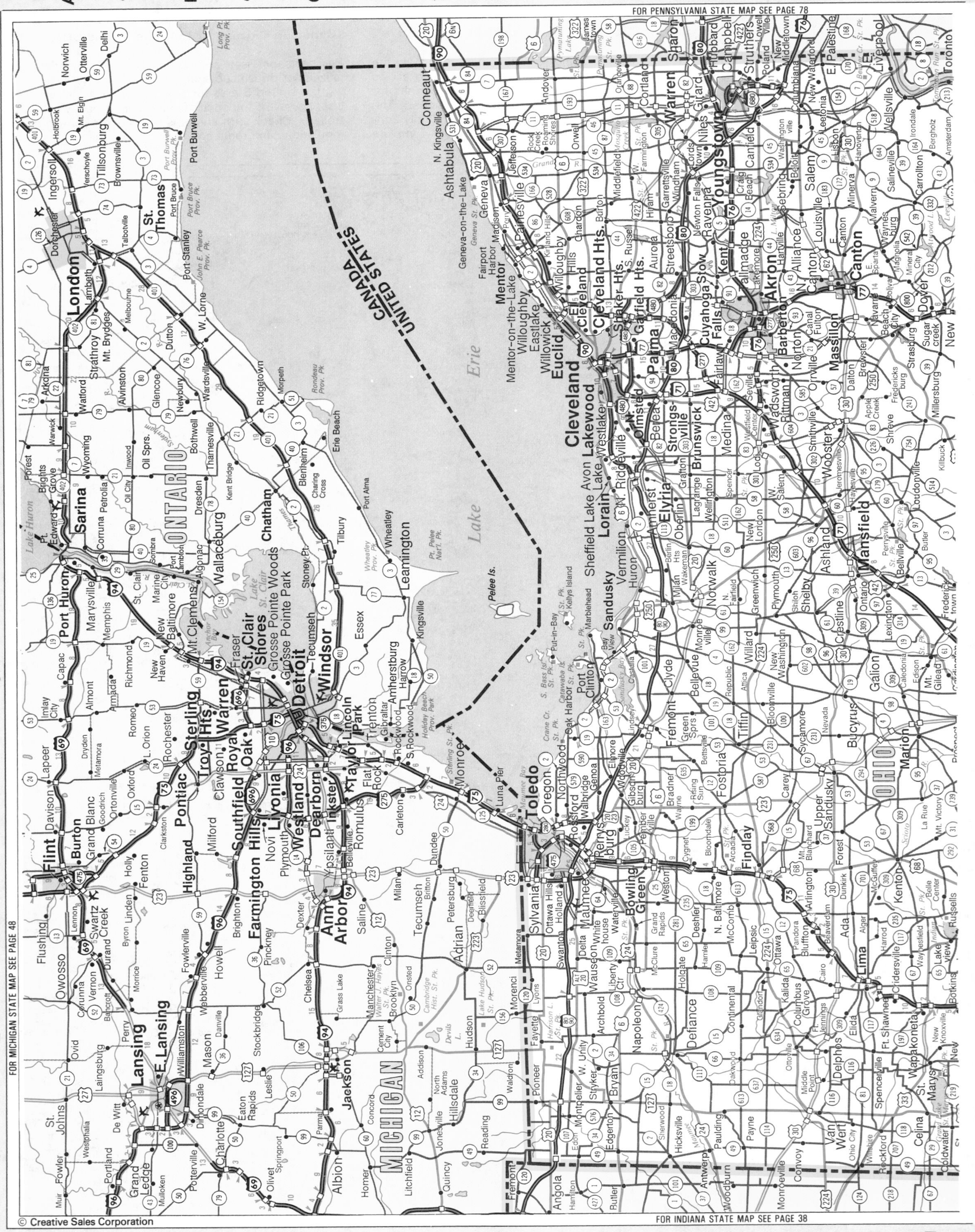

© Creative Sales Corporation

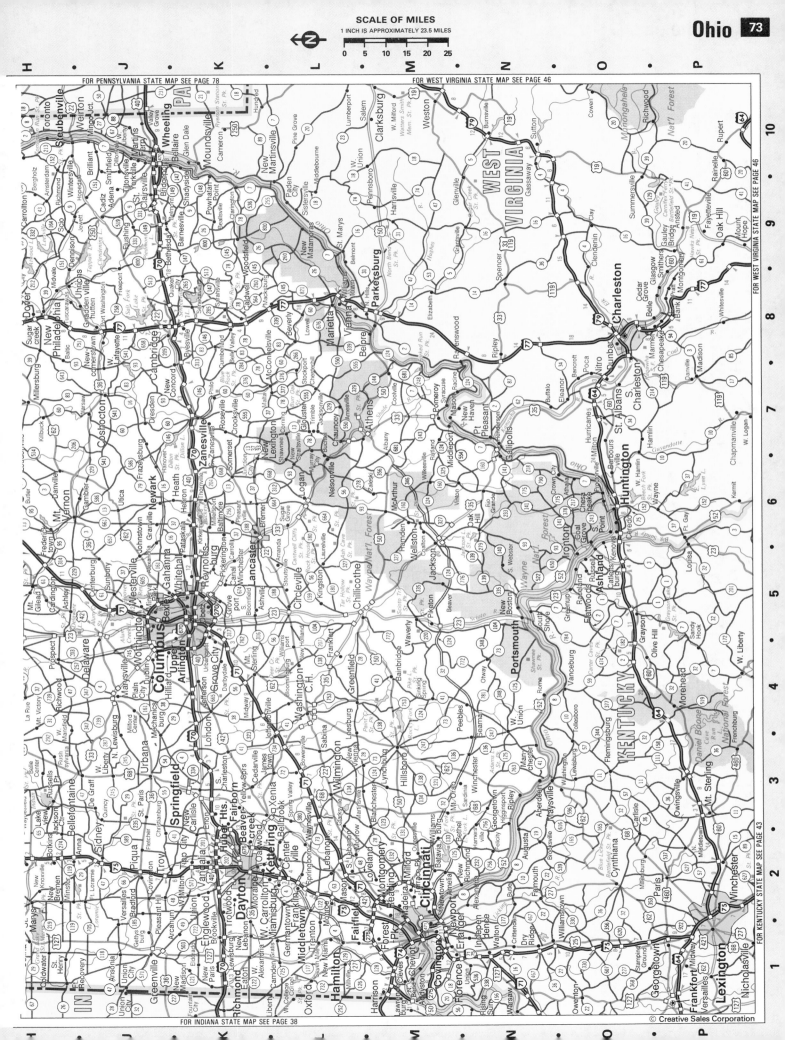

SCALE OF MILES
1 INCH IS APPROXIMATELY 35 MILES
0 7 14 21 28 35

N

FOR COLORADO STATE MAP SEE PAGE 28   FOR KANSAS STATE MAP SEE PAGE 40

COLORADO

KANSAS

NM

TEXAS

Dodge City

Cimarron • Belpre
Greensburg • Cullison
Johnson • Montezuma • Bucklin
Sublette • Minneola • Coldwater
Meade • Ashland
Liberal • Englewood • Lookout

Pritchett • Bartlett
Springfield
Vilas
Utleyville
Campo
Comanche National Grassland
Elkhart • Richfield • Hugoton
Tyrone
Forgan • Mocane • Knowles • Buffalo • Plainview
Floris • Turpin • Rosston • Gate • Edith • Tegarden • Cora
Kenton • Sturgis • Surrey Hills • Hooker • Guymon • Optima
Keyes • Eva • Four Corners • Goodwell • Hardesty
Boise City • Griggs
Wheeless
Kiowa National Grasslands
Grenville • Mt. Dora • Clayton • Felt • Texline
Texhoma
Boyd • Balko • Slapout • May • Ft. Supply
Bryan's Corner • Elmwood • Catesby
Perryton • Booker • Darrouzett
Farnsworth • Follett
Woodward
Sedan • Amistad • Dalhart • Stratford • Gruver • Waka • Lipscomb • Higgins
Shattuck • Goodwin
Gage • Sharon
Hayden • Cactus • Etter • Spearman • Glazier
Logan • Hartley • Sunray • Morse
Dumas • Stinnett • Canadian
Durham • Crawford • Angora
Channing • Masterson • Sanford • Borger • Phillips • Miami
Reydon • Strong City • Cheyenne
Boys Ranch • Tascosa • Fritch • Bunavista • Skellytown • Allison
Dempsey • Rankin • Hammon • Butler • Custer City • Clinton • Weatherford
New Mobeetie • Briscoe • Elk City • Arapaho
White Deer • Pampa • Mobeetie • Wheeler • Foss
Kings Mill • Lefors • Kellerville • Sweetwater • Mayfield • Sayre • Doxey • Dill City • Cordell
Panhandle • McLean • Lela • Twitty • Erick • Texola • Carter • Retrop • Rocky • Sentinel
Amarillo • Bushland • Lark • Conway • Groom • Alanreed • Shamrock • Willow • Cloud Chief
San Jon • Endee • Adrian • Vega • Wildorado • Claude • Ashtola • Dozier • Samnorwood • Lone Wolf • Hobart • Lake Valley
Glenrio • Canyon • Goodnight • Clarendon • Lutie • Vinson • Reed • Brinkman • Granite • Babbs
Umbarger • Dawn • Lelia Lake • Hedley • Madge • Wellington • Mangum • Lugert • Cooperton
Hereford • Summerfield • Wayside • Quail • Blair • Martha • Warren • Roosevelt
Black • Friona • Happy • Brice • Memphis • Hollis • Gould • Duke • Altus • Snyder
Dimmitt • Nazareth • Tulia • Silverton • Lakeview • Newlin • McQueen • Olustee • Headrick • Indiahoma
Bovina • Hart • Kress • Parnell • Estelline • Lincoln • Creta • Elmer • Tipton • Chattanooga
Clovis • Farwell • Texico • Lariat • Edmonson • Quitaque • Turkey • Childress • Goodlett • Quanah • Frederick
Portales • Arch • Muleshoe • Springlake • Olton • Plainview • Aiken • South Plains • Flomot • Northfield • Tell • Kirkland • Acme • Chillicothe • Vernon
Rogers • Needmore • Earth • Fieldton • Hale Center • Lockney • Whiteflat • Medicine Mound • Lockett • Oklaunion • Burkburnett
Dora • Pep • Goodland • Enochs • Amherst • Cotton Center • Floydada • Matador • Paducah • Rayland • Thalia • Electra • Grayback • Iowa Park
Lingo • Maple • Morton • Bula • Littlefield • Spade • Anton • Petersburg • Dougherty • Crowell • Kamay • Mankins
Milnesand • Bledsoe • Lehman • Whitharral • Shallowater • Idalou • Abernathy • Glenn • Dumont • Finney • Gilliland • Vera • Seymour • Westover
McDonald • Whiteface • Reese Vill. • New Deal • Ralls • McAdoo • Guthrie • Benjamin • Red Springs • Dundee
Hilburn City • Levelland • Hurlwood • Lorenzo • Crosbyton • Dickens • Knox City • Munday • Goree • Bomarton • Archer City
Allred • Smyer • Wolfforth • Lubbock • Posey • Spur • O'Brien • Rochester • Weinert • Seymour
Tatum • Bronco • Sundown • New Home • Slaton • Kalgary • Girard • Jayton • Swenson • Old Glory • Haskell • Throckmorton • Olney • Elbert • Newcastle
Plains • Ropesville • Meadow • Wilson • Post • Clairemont • Aspermont • Sagerton
Denver City • Brownfield • Tahoka • Grassland • Double • Salt • Woodson • South Bend
Knowles • Wellman • Loop • Welch • Draw • O'Donnell • Justiceburg • Fluvanna • Rotan • Hamlin • Stamford • Avoca • Fort Griffin
Hobbs • Seminole

Palo Duro Canyon State Park
Caprock Canyons State Park
Copper Breaks State Park
Mackenzie State Park
North Canadian River
Canadian River
Arkansas River
Red River
Wichita River
Brazos River
Lake Kemp
Lake Stamford

© Creative Sales Corporation

FOR NEW MEXICO STATE MAP SEE PAGE 62
FOR TEXAS STATE MAP SEE PAGES 90-94

SCALE OF MILES
1 INCH IS APPROXIMATELY 35 MILES
0 7 14 21 28 35

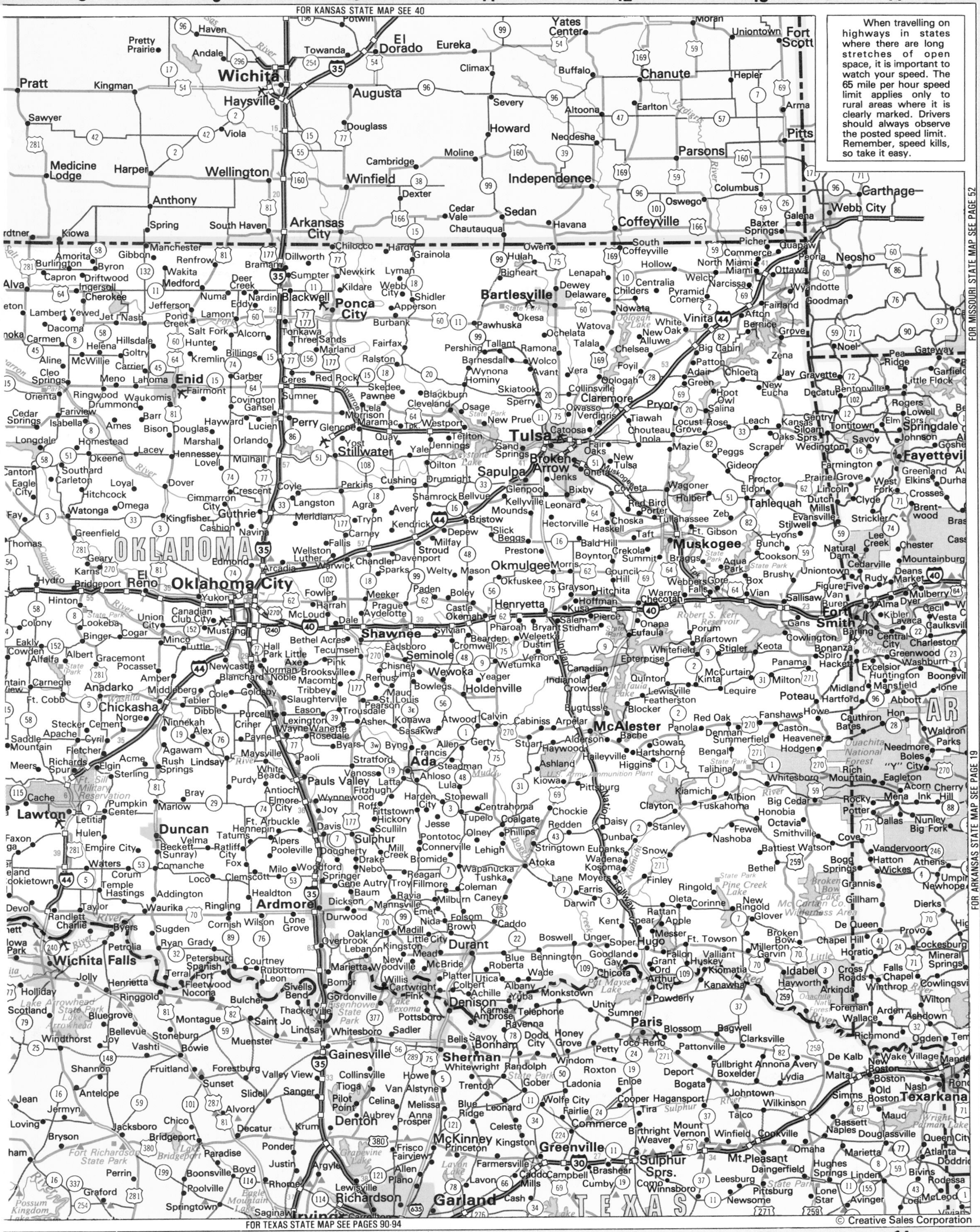

N

FOR KANSAS STATE MAP SEE PAGE 40

When travelling on highways in states where there are long stretches of open space, it is important to watch your speed. The 65 mile per hour speed limit applies only to rural areas where it is clearly marked. Drivers should always observe the posted speed limit. Remember, speed kills, so take it easy.

FOR MISSOURI STATE MAP SEE PAGE 52

FOR ARKANSAS STATE MAP SEE PAGE 19

FOR TEXAS STATE MAP SEE PAGES 90-94

OKLAHOMA

TEXAS

AR

© Creative Sales Corporation

SCALE OF MILES
1 INCH IS APPROXIMATELY 35 MILES
0 7 14 21 28 35

FOR WASHINGTON STATE MAP SEE PAGE 84

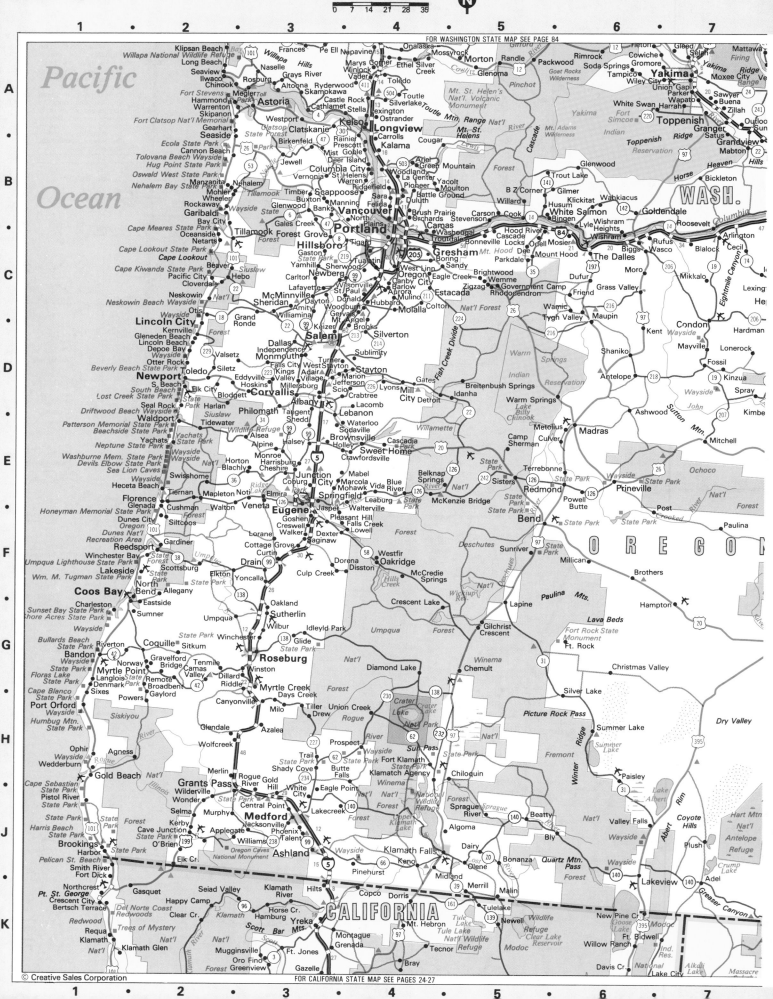

© Creative Sales Corporation

FOR CALIFORNIA STATE MAP SEE PAGES 24-27

SCALE OF MILES
1 INCH IS APPROXIMATELY 35 MILES
0 7 14 21 28 35

N

FOR WASHINGTON STATE MAP SEE PAGE 84

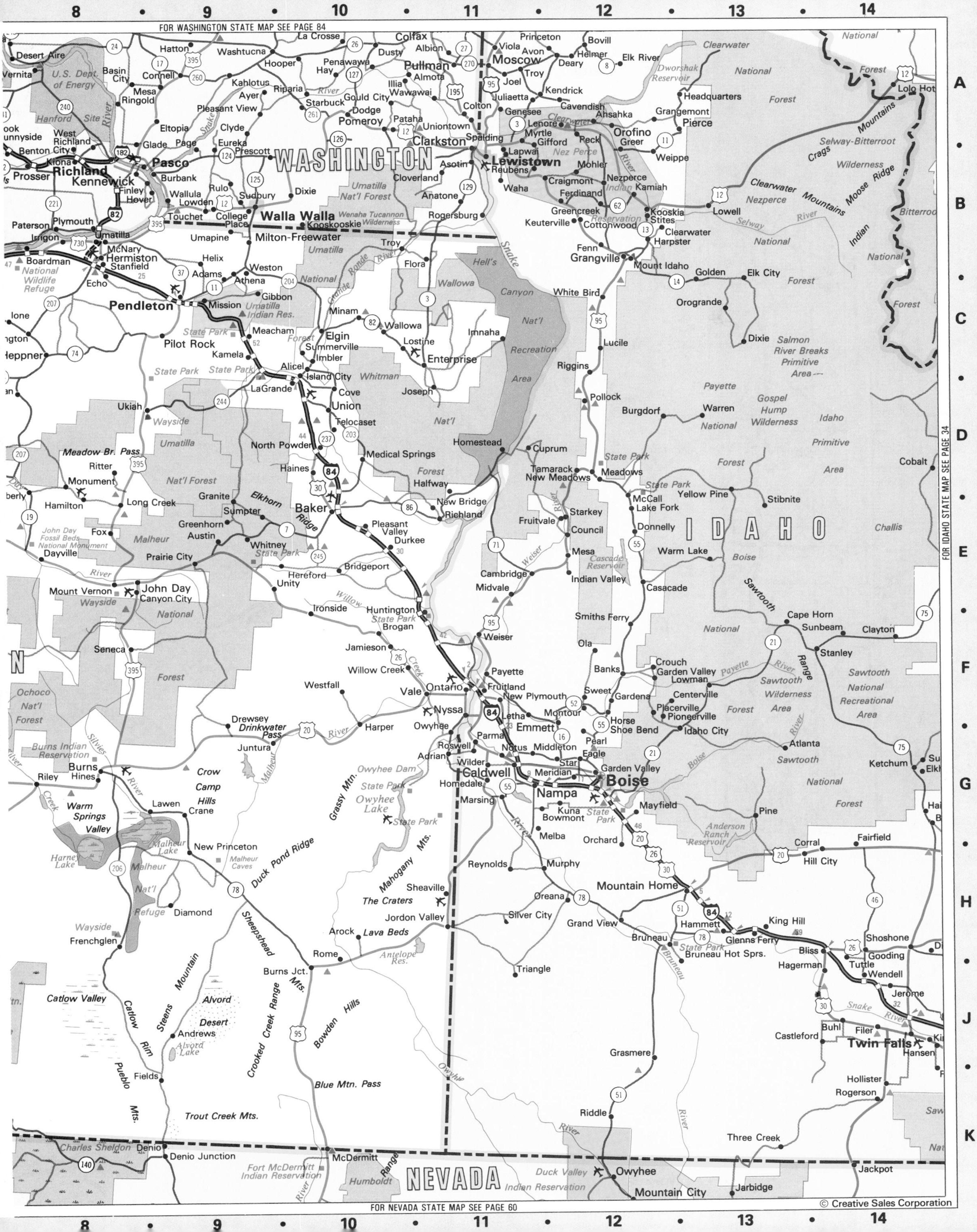

WASHINGTON

OREGON

IDAHO

NEVADA

FOR IDAHO STATE MAP SEE PAGE 34

FOR NEVADA STATE MAP SEE PAGE 60

© Creative Sales Corporation

SCALE OF MILES
1 INCH IS APPROXIMATELY 21.5 MILES
0 4 8 12 16 20

N

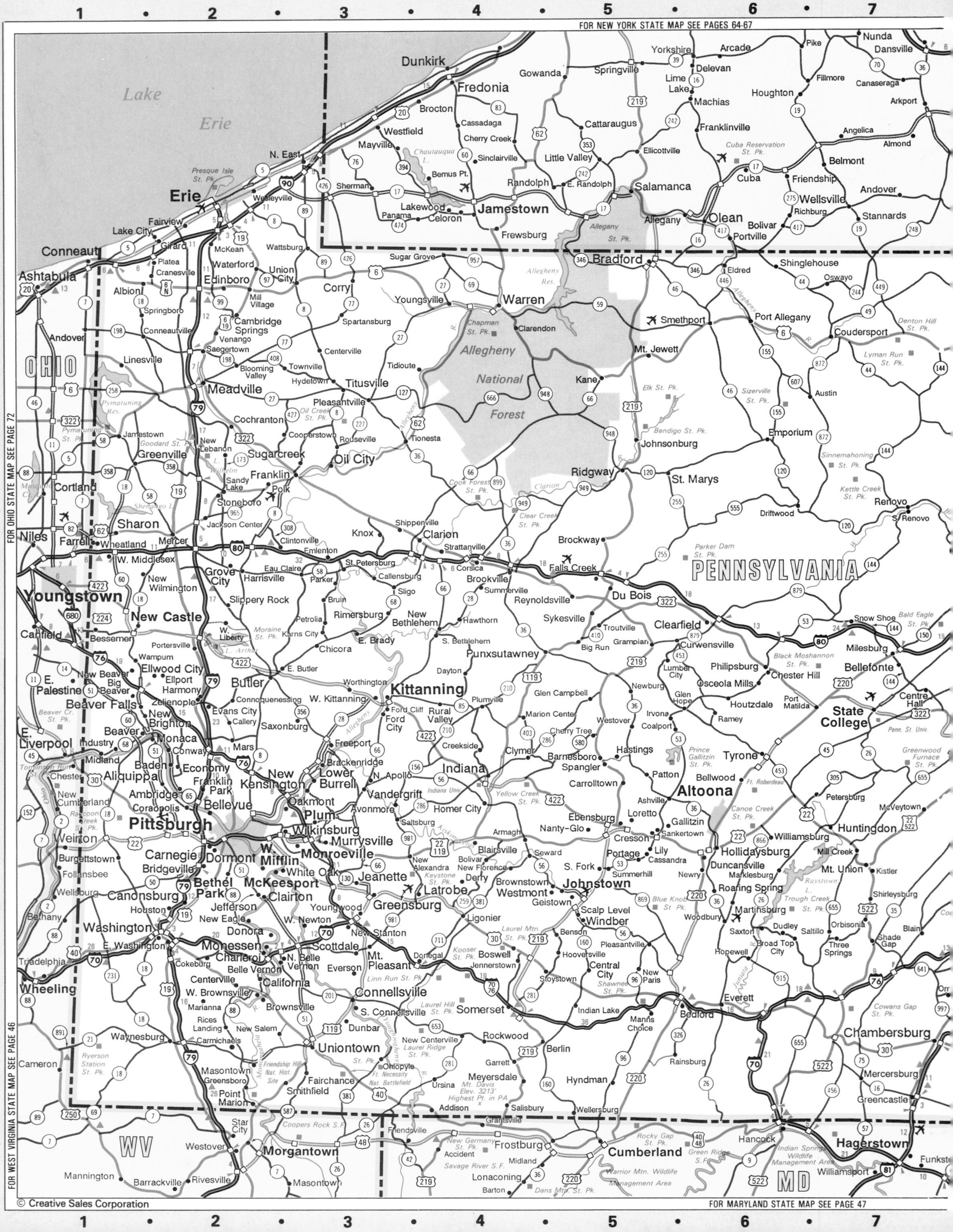

FOR NEW YORK STATE MAP SEE PAGES 64-67

Lake Erie

PENNSYLVANIA

OHIO

WV

MD

Allegheny National Forest

Erie **Pittsburgh** **Jamestown** **Youngstown** **New Castle** **Altoona** **Johnstown** **Bradford** **Warren** **State College** **Chambersburg** **Hagerstown** **Cumberland** **Morgantown** **Wheeling**

Conneaut Ashtabula Andover Albion Edinboro Cambridge Springs Meadville Greenville Sharon Farrell Niles Canfield Grove City Slippery Rock Butler Kittanning Indiana Clearfield Du Bois Clarion Brookville Ridgway St. Marys Emporium Coudersport Port Allegany Kane Mt. Jewett Smethport Punxsutawney Tyrone Hollidaysburg Huntingdon Bedford Somerset Uniontown Connellsville Greensburg Latrobe Washington Monessen Charleroi California Brownsville Waynesburg

Dunkirk Fredonia Westfield N. East Corry Titusville Oil City Franklin Youngsville Clarendon

© Creative Sales Corporation

FOR OHIO STATE MAP SEE PAGE 72
FOR WEST VIRGINIA STATE MAP SEE PAGE 46
FOR MARYLAND STATE MAP SEE PAGE 47

SCALE OF MILES
1 INCH IS APPROXIMATELY 21.5 MILES
0 4 8 12 16 20

N

FOR NEW YORK STATE MAP SEE PAGES 64-67

NEW YORK

Wayland · Cohocton · Avoca · Bath · Hammondsport · Watkins Glen · Montour Falls · Trumansburg · Lansing · Dryden · Marathon · Norwich · Oxford · Oneonta · W. End · Otego · Stamford · Hobart

Painted Post · Corning · Horseheads · Spencer · Candor · Newark Valley · Owego · Endwell · Endicott · Binghamton · Greene · Sidney · Unadilla · Delhi · Andes · Margaretville

Addison · Woodhull · S. Corning · Elmira Hts. · Elmira · Southport · Waverly · S. Waverly · Athens · Sayre · Apalachin · Nichols · Johnson City · Windsor · Deposit · Livingston Manor

Catskill Park · Pepacton Res.

Elkland · Knoxville · Westfield · Tioga · Roseville · Rome · Le Raysville · Montrose · New Milford · Susquehanna · Hancock · Liberty · Jeffersonville · Monticello

Galeton · Wellsboro · Mansfield · Sylvania · Troy · Burlington · Towanda · Wyalusing · Hop Bottom · Union Dale · Forest City · Honesdale

Canton · Alba · Monroe · New Albany · Dushore · Meshoppen · Nicholson · Lackawanna · Factoryville · Waymart · Carbondale · Archbald

Liberty · Forksville · Laporte · Eagles Mere · Tunkhannock · Dalton · Clarks Summit · Dickson City · Blakely · Olyphant · Hawley · Matamoras · Milford

Williamsport · Montoursville · S. Williamsport · Hughesville · Picture Rocks · Harveys Lake · Scranton · Taylor · Old Forge · Moosic · Moscow · Promised Land St. Pk.

Avis · Jersey Shore · Duboistown · Watsontown · Benton · Stillwater · W. Pittston · Kingston · Pittston · Edwardsville · Wilkes-Barre · Mt. Pocono · E. Stroudsburg · Newton

Lock Haven · Flemington · Mill Hall · Beech Creek · Logantown · Milton · Orangeville · Berwick · Briar Creek · Nanticoke · Shickshinny · Nuangola · Ashley · White Haven · Stroudsburg · Delaware Water Gap · Portland · E. Stroudsburg

Lewisburg · Danville · Bloomsburg · Catawissa · W. Hazleton · Hazleton · Freeland · Jeddo · Weatherly · Jim Thorpe · Roseto · Bangor · Hackettstown

Mifflinburg · Riverside · Northumberland · Ringtown · Shenandoah · Lansford · Lehighton · Palmerton · Bowmanstown · Wind Gap · Pen Argyl · Tatamy · Belvidere · Washington

Millheim · New Berlin · Shamokin Dam · Mt. Carmel · Mahanoy City · Tamaqua · Summit Hill · Slatington · Northampton · Nazareth · Easton · Phillipsburg · Hampton · High Bridge

Sunbury · Selinsgrove · Shamokin · Kulpmont · Frackville · New Philadelphia · Port Carbon · Walnutport · Bath · Catasauqua · Wilson · W. Easton · Clinton · Lebanon

Beavertown · Middleburg · Freeburg · Herndon · Gordon · St. Clair · Pottsville · Orwigsburg · Whitehall · Fountain Hill · Freemansburg · Milford · Flemington

Burnham · Millerstown · Pillow · Tremont · Minersville · Cressona · Hamburg · Emmaus · Allentown · Bethlehem · Coopersburg · Frenchtown

Lewistown · Mifflintown · Port Royal · Newport · Liverpool · Millersburg · Berrysburg · Gratz · Tower City · Williamstown · Pine Grove · Schuylkill Haven · Shoemakersville · Strausstown · Leesport · Kutztown · Topton · Alburtis · Macungie · Richlandtown · Quakertown · Stockton

Duncannon · Dauphin · Elizabethville · Halifax · New Buffalo · Jonestown · Myerstown · Womelsdorf · Robesonia · W. Reading · Reading · Fleetwood · E. Greenville · Pennsburg · Bally · Red Hill · Telford · Sellersville · Perkasie · Doylestown · Lambertville · Pennington

Landisburg · Marysville · Harrisburg · Lemoyne · Hershey · Hummelstown · Palmyra · Lebanon · Richland · Cleona · Cornwall · Wyomissing · Shillington · Laureldale · Temple · Boyertown · Pottstown · Souderton · N. Wales · Lansdale · Hatboro · Ambler · Chalfont · Yardley · Trenton

Carlisle · Mechanicsburg · New Cumberland · Middletown · Elizabethtown · Mt. Joy · Manheim · Denver · Lincoln · Adamstown · Birdsboro · Spring City · Royersford · Trappe · Collegeville · Phoenixville · Norristown · Conshohocken · Langhorne · Morrisville · Levittown · Bristol

Newville · Newburg · Mt. Holly Springs · Dillsburg · Manchester · Marietta · Mountville · Lititz · Terre Hill · New Holland · Honey Brook · Downingtown · Philadelphia · Narberth · Lansdowne · Camden · Palmyra

Shippensburg · Orrstown · Wellsville · York Springs · Dover · Wrightsville · Columbia · Lancaster · Millersville · Strasburg · Parkesburg · Coatesville · W. Chester · Media · Yeadon · Darby · Collingswood · Medford

Bendersville · Arendtsville · E. Berlin · York · Dallastown · Spring Grove · Jacobus · Red Lion · Quarryville · Kennett Square · S. Coatesville · Chester Hts. · Brookhaven · Chester · Gloucester · Bellmawr · Stratford

Gettysburg · Fairfield · New Oxford · Hanover · Glen Rock · Winterstown · Oxford · W. Grove · Wilmington · Elsmere · Marcus Hook · Trainer · Runnemede · Pitman · Berlin

Waynesboro · Mont Alto · Littlestown · McSherrystown · Shrewsbury · New Freedom · Fawn Grove · Delta · Newport · New Castle · Penns Grove · Linwood · Lindenwold · Chesilhurst · Hammonton

Emmitsburg · Taneytown · Manchester · Hampstead · Rising Sun · Port Deposit · Perryville · Elkton · North East · Salem · Woodstown · Clayton · Newfield · Buena

Thurmont · Westminster · Bel Air · Charlestown · New Castle

DE · NJ

FOR MARYLAND STATE MAP SEE PAGE 47 FOR DELAWARE STATE MAP SEE PAGE 31 FOR NEW JERSEY STATE MAP SEE PAGE 31

© Creative Sales Corporation

SCALE OF MILES
1 INCH IS APPROXIMATELY 35 MILES
0   7   14   21   28   35

N

FOR NORTH DAKOTA STATE MAP SEE PAGE 70

FOR MONTANA STATE MAP SEE PAGE 54

FOR WYOMING STATE MAP SEE PAGE 88

MONT.

WY

NORTH DAKOTA

SOUTH DAKOTA

NEBRASKA

Beach, Sentinel Butte, Medora, Belfield, S. Heart, Dickinson, Gladstone, Richardton, Hebron, Marshall, Center, Wilton, Regan, Wing, Tuttle, Robinson, Woodworth, Carlyle, Golva, Sully's Creek State Primitive Park, Glen Ullin, New Salem, Mandan, Bismarck, Chaska Hist Site, Steele, Arena, Pettibone

Plevna, Baker, Amidon, New England, Mott, New Leipzig, Elgin, Carson, Flasher, Breien, Solen, Hazelton, Napoleon, Almont, Lincoln, Fort Lincoln State Park, Long Lake Nat'l Wildlife Refuge, Braddock, Driscoll, Dawson, Tappen, Medina, Streeter

Marmarth, Rhame, Bowman, Scranton, Gascoyne, Reeder, Bucyrus, Haley, Hettinger, Haynes, Regent, Raleigh, Freda, Sheilds, Fort Yates, Linton, Strasburg, Wishek, Lehr, Braddock, Beaver Lake State Park

Ekalaka, Mill Iron, Ladner, Ludlow, Lemmon, Camp Crook, Buffalo, Ralph, Lodgepole, White Butte, Shadehill, Morristown, McIntosh, Selfridge, Westfield, Hague, Zeeland, Venturia, Ashley, White Butte Elev. 3,506 ft., Fort Dilts Hist. Site, Butte View State Campground, Custer National Forest

Reva, Prairie City, Bison, Meadow, Bullhead, McLaughlin, Pollock, Herreid, Mound City, Eureka, Hillsview, Hosmer, Long Lake, Redig, Sorum, Chance, Glad Valley, Isabel, Timber Lake, Trail City, Mobridge, Glenham, Selby, Java, Roscoe, Bowdle, Loyalton

Alzada, Colony, Zeona, Maurine, Faith, Dupree, Eagle Butte, Ridgeview, LaPlant, Akaska, Lowry, Hoven, Tolstoy, Onaka, Cresbard, Castle Rock, Mud Butte

Belle Fourche Reservoir, Newell, Nisland, Fairpoint, Stoneville, Howes, Cherry Creek, Reservation, Lake Oahe, Gettysburg, Agar, Seneca, Faulkton, Orient, Beulah, Spearfish, Whitewood, Deadwood, Central City, Lead, Sturgis, Union Center, Plainview, Enning, Onida, Newcastle, Horton, Black Hills National Forest, Hill City, Rapid City, Black Hawk, Box Elder, Ellsworth A.F. Base, Creighton, Wasta, Blunt, Harrold, Highmore, Ree Heights, Miller, Pierre, Fort Pierre, Stephan, Fort Thompson

Mt. Rushmore, Crazy Horse Mon., Keystone, Rockerville, Farmingdale, Wall, Quinn, Philip, Midland, Cottonwood, Draper, Vivian, Presho, Kennebec, Reliance, Murdo, Chamberlain, Oacoma, Pukwana, Kimball, Custer, Hermosa, Scenic, Interior, Kadoka, Belvidere

Pringle, Fairburn, Buffalo Gap, Hot Springs, Badlands, Kyle, Long Valley, White River, Wood, Mule Cr. Jct., Edgemont, Provo, Oelrichs, Oglala, Allen, Parmelee, Okreek, Mission, Winner, Wounded Knee, Wounded Knee Battle Site, Batesland, Martin, Rosebud, Saint Francis, Dallas, Gregory, Burke, Herrick, St. Charles, Bonesteel, Fairf

Pine Ridge, Chadron, Whitney, Crawford, Fort Robinson State Park, Hay Springs, Clinton, Rushville, Gordon, Merriman, Cody, Nenzel, Kilgore, Crookston, Wewala, Harrison, Nebraska National Forest, Valentine, Sparks, Norden, Burton, Mills, Naper, Springview, Ft. Niobrara Nat'l Wildlife Refuge

Hemingford, Samuel R. McKelvie Nat'l Forest, Valentine Migratory Waterfowl Refuge, Merritt Res., Wood Lake, Johnstown, Bassett, Newport, Stuart, Atkinson

Morrill, Mitchell, Scottsbluff, Alliance, Antioch, Lakeside, Ellsworth, Bingham, Whitman, Long Pine, Ainsworth, Brownlee, Elsmere, Rose, Amelia

© Creative Sales Corporation

FOR NEBRASKA STATE MAP SEE PAGE 58

SCALE OF MILES
1 INCH IS APPROXIMATELY 35 MILES
0 7 14 21 28 35

N

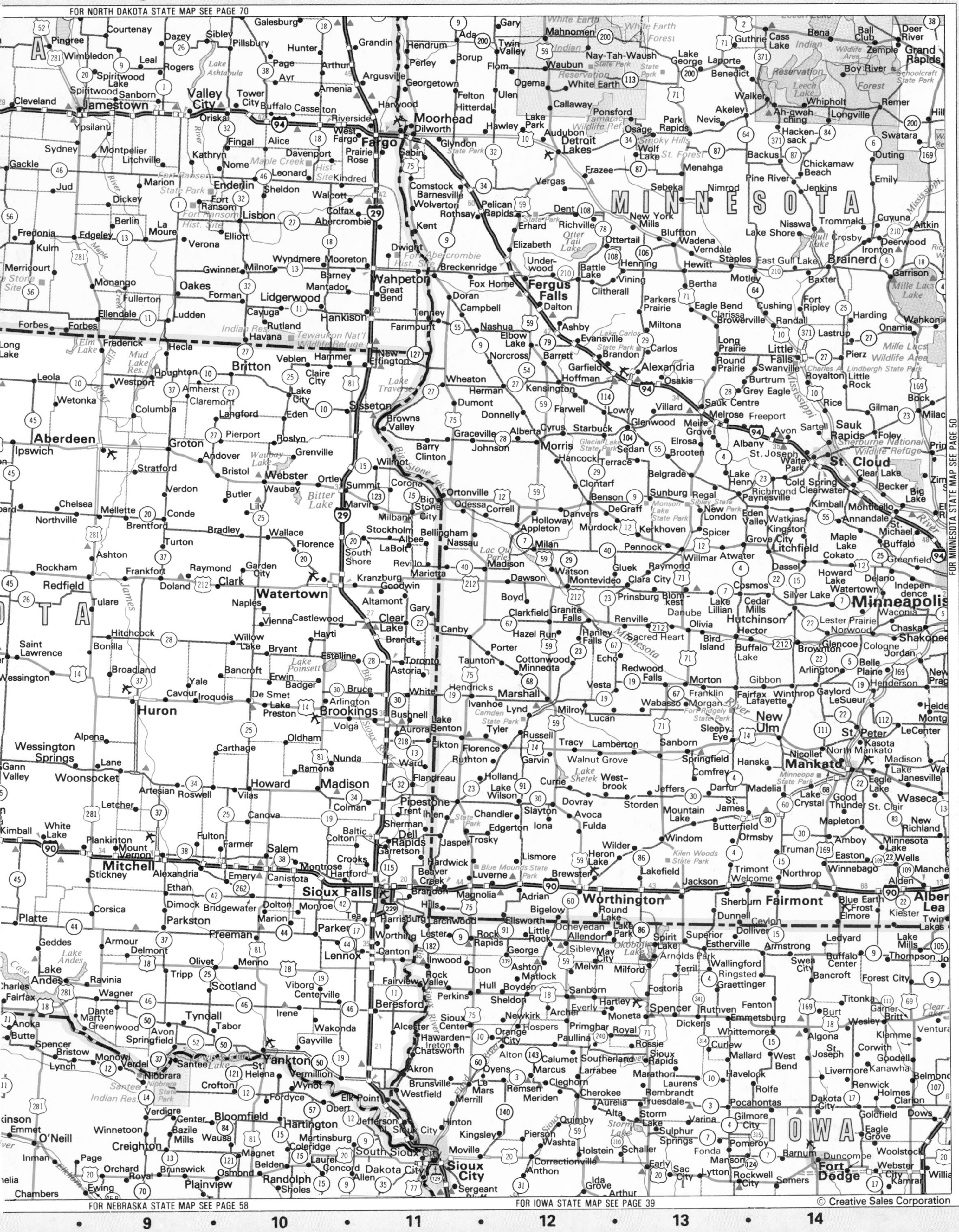

FOR NORTH DAKOTA STATE MAP SEE PAGE 70

MINNESOTA

IOWA

FOR NEBRASKA STATE MAP SEE PAGE 58

FOR IOWA STATE MAP SEE PAGE 39

FOR MINNESOTA STATE MAP SEE PAGE 50

© Creative Sales Corporation

SCALE OF MILES
1 INCH IS APPROXIMATELY 35 MILES
0 7 14 21 28 35
N

FOR COLORADO STATE MAP SEE PAGE 28
FOR WYOMING STATE MAP SEE PAGE 88
FOR IDAHO STATE MAP SEE PAGE 34
FOR NEVADA STATE MAP SEE PAGE 60

WYOMING

IDAHO

NV

COL

Salt Lake City
Ogden
Provo
Orem
Logan
Brigham City
Bountiful
Clearfield
Layton
Nephi
Tooele
Pocatello
Blackfoot
Twin Falls
Soda Springs
Wendover
Evanston
Rock Springs
Green River
Lander
Duchesne
Roosevelt
Vernal
Dinosaur
Rangely

Great Salt Lake
Utah Lake
Bear Lake
Flaming Gorge Reservoir
Wasatch National Forest
Ashley National Forest
Uinta National Forest
Caribou National Forest
Sawtooth National Forest
Uintah and Ouray Indian Reservation
Great Salt Lake Desert
Dugway Proving Grounds
Hill Air Force Range
Bonneville Speedway

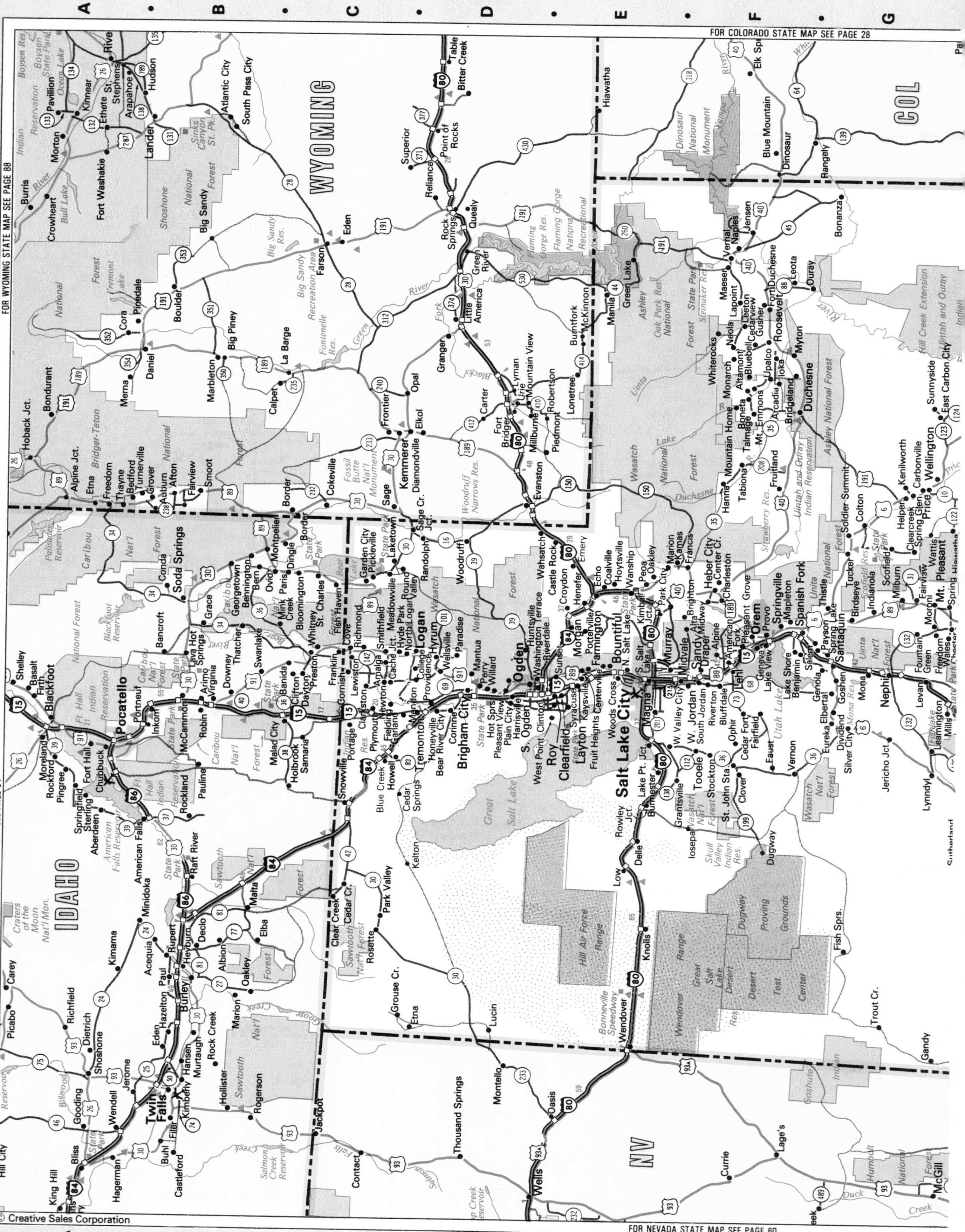

© Creative Sales Corporation

SCALE OF MILES
1 INCH IS APPROXIMATELY 35 MILES

0   7   14   21   28   35

H   J   K   L   M   N   P

U T A H

ARIZONA

Grand Junction
Moab
Green River
Crescent Jct.
Thompson
Cisco
Mack
Loma
Fruita
Clifton
White
Cameo
DeBeq
Palisade

Price
East Carbon City
Woodside
Cleveland
Huntington
Castle Dale
Elmo
Clawson
Orangeville
Ferron
Molen
Emery
Moore

Mt. Pleasant
Spring City
Ephraim
Manti
Sterling
Mayfield
Gunnison
Centerfield
Axtell
Salina
Aurora
Sigurd
Glenwood
Redmond
Fayette

Delta
Oasis
Deseret
Hinckley
Abraham
Sutherland
Oak City
Holden
Scipio
Fillmore
Flowell
Meadow
Kanosh
Hatton

Milford
Minersville
Beaver
Adamsville
Greenville
Manderfield
Marysvale
Junction
Circleville
Kingston
Antimony
Angle
Koosharem
Burrville
Greenwich

Richfield
Elsinore
Monroe
Joseph
Sevier
Annabella
Venice
Central
Austin

Cedar City
Enoch
Summit
Parowan
Paragonah
Brian Head
Hamilton Fort
Kanarraville
New Harmony
Pintura
Leeds
Toquerville
La Verkin
Hurricane
Virgin
Springdale
Rockville
Veyo
Gunlock
Central
Pine Valley
Santa Clara
Ivins
Shivwits
St. George
Washington
Bloomington

Cannonville
Tropic
Henrieville
Escalante
Boulder
Widtsoe Jct.
Panguitch
Hatch
Long Valley Jct.
Alton
Glendale
Orderville
Mt. Carmel
Mt. Carmel Jct.
Kanab
Fredonia

Torrey
Grover
Teasdale
Bicknell
Lyman
Loa
Fremont
Fruita

Hanksville
Fry Canyon

Monticello
Blanding
Eastland
Summit Pt.
La Sal
La Sal Jct.
Bluff
Mexican Hat
Montezuma Creek
Aneth
Bland

Nucla
Naturita
Naturita
Slick Rock
Egnar
Dove Creek
Paradox
Bedrock
Vancorum
Uravan
Gateway
Redvale

Pleasant View
Yellow Jacket
Lewis
Arriola
Cortez
Towaoc
Dolores
Lebanon

Farmington
Kirtland
Flora Vista
Newcomb
Sheep Springs
Naschitti
Shiprock
Beklabito
Crystal
Navajo
Mexican Springs
Coyote Canyon
Tohatchi
Standing Rock

Teec Nos Pos
Tes Nez Iha
Mexican Water
Dinnehotso
Kayenta
Tsegi
Rock Point
Round Rock
Many Farms
Chinle
Chilchinbito
Rough Rock
Cross Canyon
St. Michaels
Window Rock
Ganado
Keams Canyon
Polacca
Old Oraibi
Oraibi
Second Mesa
Cow Springs
Red Lake
The Gap
Cedar Ridge
Tonalea
Tuba City
Page
Marble Canyon
Jacob Lake
North Rim
Grand Canyon
Desert View
Tusayan
Moqui
Cameron
Gray Mountain
Valle

Colorado City
Littlefield
Beaver Dam
Mesquite
Bunkerville
Glendale
Overton
Logandale

Lake Powell
Glen Canyon National Recreation Area
Canyonlands National Park
Arches National Park
Capitol Reef National Park
Bryce Canyon National Park
Zion National Park
Grand Canyon National Park
Lake Mead National Recreational Area

Manti-La Sal National Forest
Dixie National Forest
Fishlake National Forest
Kaibab National Forest
Navajo Indian Reservation
Hopi Indian Reservation
Kaibab Indian Reservation
Havasupai Indian Reservation
Hualapai Indian Reservation

© Creative Sales Corporation

SCALE OF MILES
1 INCH IS APPROXIMATELY 35 MILES
0 7 14 21 28 35

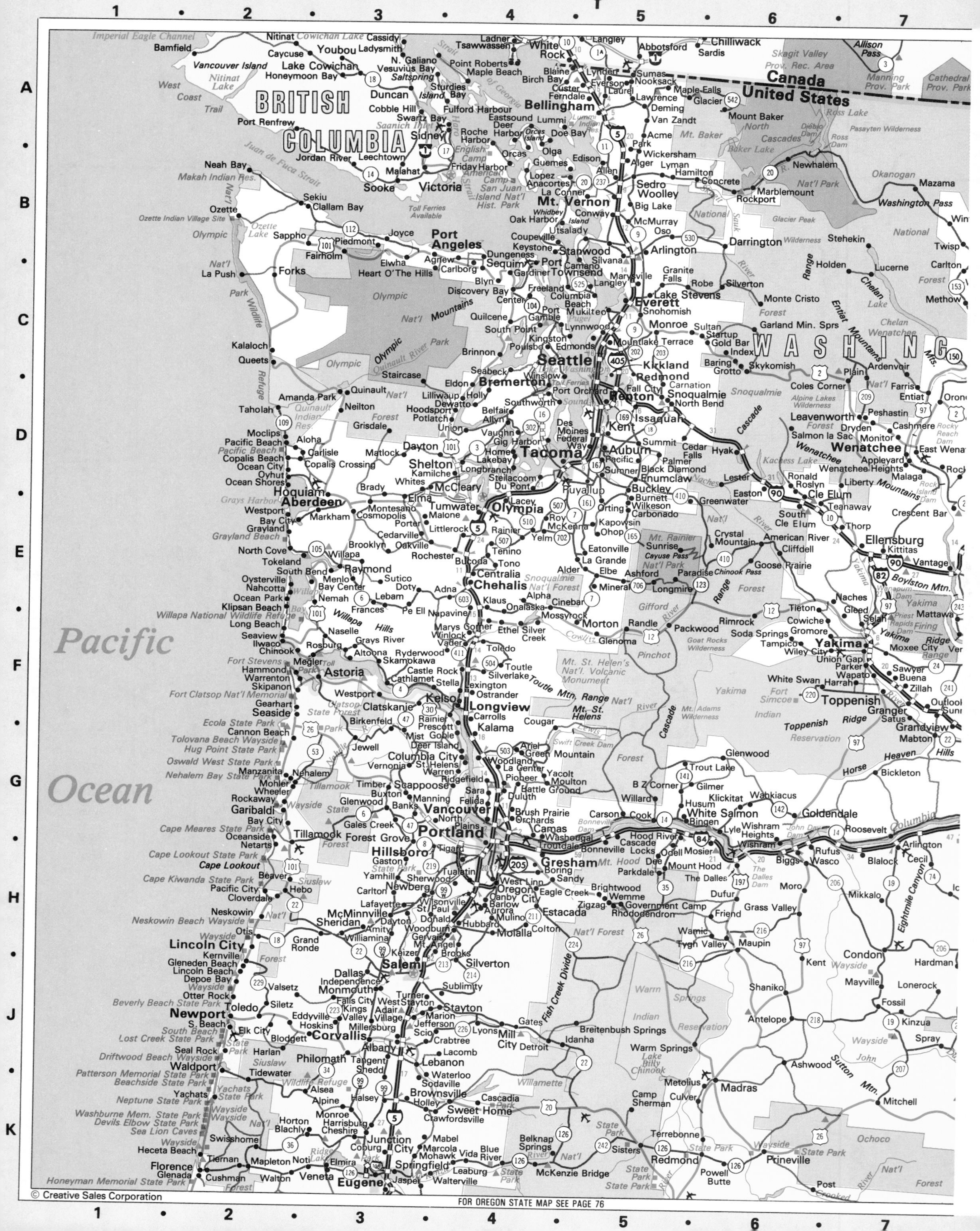

N

Pacific

Ocean

BRITISH COLUMBIA

Canada
United States

WASHINGTON

Vancouver Island

Juan de Fuca Strait

Strait of Georgia

Olympic Nat'l Park

Olympic Mountains

Seattle
Tacoma
Olympia
Bellingham
Mt. Vernon
Everett
Bremerton
Kirkland
Redmond
Renton
Kent
Auburn
Puyallup
Aberdeen
Hoquiam
Centralia
Chehalis
Longview
Kelso
Vancouver
Portland
Salem
Corvallis
Albany
Eugene
Newport
Lincoln City
Astoria
Yakima
Ellensburg
Wenatchee
Cle Elum
Leavenworth
Toppenish
Goldendale
The Dalles

Mt. Rainier Nat'l Park
Mt. St. Helen's Nat'l Volcanic Monument
Gifford Pinchot Nat'l Forest
Snoqualmie Nat'l Forest
Wenatchee Nat'l Forest

Columbia River

Willapa National Wildlife Refuge

© Creative Sales Corporation

FOR OREGON STATE MAP SEE PAGE 76

SCALE OF MILES
1 INCH IS APPROXIMATELY 35 MILES
0 7 14 21 28 35

N

8 9 10 11 12 13 14

A

B

C

D

E

F

G

H

J

K

CANADA
UNITED STATES

MONTANA

OREGON

IDAHO

Keremeos, Cawston, Olalla, Oliver, Westbridge, Bonanza Pass, Robson, Thrums, Castlegar, Ymir, Salmo, Sanca, Moyie, Jaffray, Wardner, Elko
Richter Pass, Rock Creek, Greenwood, Christina Lake, Rossland, Erie, Wynndel, Sirdar, Kitchener, Kingsgate
Nighthawk, Osoyoos, Oroville, Molson, Chesaw, Ferry, Danville, Laurier, Orient, Northport, Boundary, Porthill, Eastport, Newgate, Grasmere, Roosville
Loomis, Ellisford, Wauconda, Old Toroda, Curlew, Malo, Metaline, Spirit, Metaline Falls, Copeland, Meadow Creek, Rexford, Eureka, Flathead
Conconully, Tonasket, Havillah, Toroda, Republic, Boyds, Bossburg, Evans, Marcus, Echo, Aladdin, Tiger, Ione, Bonners Ferry, Nordman, Yaak, Kootenai, Trego, Stryker
Omak, Riverside, Aeneas, Keller, Kettle Falls, Colville, Rice, Arizina, Park Rapids, Lamb Creek, Coolin, Elmira, Naples, Troy, Libby, Whitefish, Olney
Malott, Okanogan, Disautel, Inchelium, Gifford, Orin, Arden, Addy, Cusick, Usk, Colburne, Kootenai, Hope, East Hope, Kalispell
Brewster, Nespelem, Covada, Bluecreek, Chewelah, Valley, Dalkena, Sandpoint, Dover, Sagle, Heron, Noxon, Marion, Kila, Happy's Inn
Bridgeport, Del Rio, Elmer City, Monahans, Ninemile, Cedonia, Hunters, Springdale, Diamond Lake, Newport, Oldtown, Laclede, Priest River, Westmond, Clark Fork, Trout Creek
Withrow, St. Andrews, Hartline, Almira, Wilbur, Creston, Govan, Davenport, Reardan, Loon Lake, Clayton, Deer Park, Ford, Camden, Blanchard, Careywood, Granite, Bayview, Niarada, Elmo, Camas
Waterville, Coulee City, Wilson Creek, Stratford, Krupp, Harrington, Spokane, Coeur D'Alene, Athol, Rathdrum, Hauser, Post Falls, Hayden, Dalton Gardens, Fernan Lake, Prichard, Murray, Belknap, Thompson Falls, Hot Springs
Quincy, Soap Lake, Ephrata, Odessa, Lamona, Keystone, Airway Heights, Opportunity, Mica, Freeman, Rose Lake, Cataldo, Enaville, Kellogg, Silverton, Gem, Mullan, Plains, Perma
Moses Lake, George, Wheeler, Marcellus, Sprague, Cheney, Four Lakes, Mt. Hope, Rockford, Springston, Medimont, Lane, Pinehurst, Wardner, Wallace, DeBorgia, Paradise
Low Gap, Warden, Schrag, Tokio, Lamont, Malden, Pine City, Fairfield, Waverly, Latah, Harrison, St. Maries, Calder, Avery, Haugan, Saltese, St. Regis, Superior
Royal City, Roxboro, Ritzville, Paha, Ralston, Rosalia, Thornton, Tekoa, Chatcolet, St. Joe, Tarkio, Albert
Beverley, Othello, Lind, Benge, Winona, Oakesdale, Farmington, Worley, Fernwood, Emida, Clarkia, Dixon
Smyrna, Corfu, Hatton, Cunningham, Endicott, Diamond, St. John, Garfield, Plummer, Tensed, Santa, Harvard, Bovill
Desert Aire, Vernita, Basin City, Washtucna, La Crosse, Colfax, Palouse, De Smet, Sanders, Potlatch, Princeton, Viola, Helmer, Elk River, Headquarters
Mesa, Ringold, Connell, Hooper, Dusty, Albion, Pullman, Moscow, Avon, Deary, Troy, Pierce, Grangemont
Mattawa, Eltopia, Kahlotus, Penawawa, Hay, Illia, Wawawai, Almota, Joel, Juliaetta, Kendrick, Cavendish, Ahsahka, Weippe
West Richland, Benton City, Glade, Page, Ayer, Riparia, Starbuck, Gould City, Dodge, Colton, Genesee, Lenore, Myrtle, Reck, Gifford, Orofino, Greer
Richland, Kennewick, Pasco, Burbank, Clyde, Eureka, Prescott, Pomeroy, Pataha, Uniontown, Spalding, Lapwai, Mohler, Weippe
Prosser, Finley, Hover, Wallula, Lowden, Sudbury, Dixie, Dayton, Clarkston, Lewiston, Reubens, Nezperce, Kamiah
Kiona, Plymouth, Paterson, Touchet, College Place, Waitsburg, Asotin, Cloverland, Waha, Craigmont, Ferdinand, Kooskia, Stites, Clearwater, Harpster
Umatilla, McNary, Walla Walla, Kooskooskie, Milton-Freewater, Anatone, Rogersburg, Greencreek, Cottonwood, Keuterville, Fenn, Lowell
Irrigon, Boardman, Hermiston, Stanfield, Umapine, Milton-Freewater, Weston, Athena, Troy, Flora, White Bird, Grangeville, Mount Idaho, Golden, Elk City
Echo, Adams, Helix, Gibbon, Wallowa, Minam, Lucile, Orogrande, Dixie
Pendleton, Mission, Meacham, Elgin, Summerville, Imbler, Lostine, Enterprise, Imnaha, Riggins, Pollock, Burgdorf, Warren
Pilot Rock, Kamela, Alicel, Island City, Cove, Union, Joseph, Homestead, Tamarack, New Meadows, Meadows, Yellow Pine, Stinbite
La Grande, Telocaset, North Powder, Medical Springs, Cuprum, McCall, Lake Fork, Cobalt
Ukiah, Haines, Halfway, New Bridge, Richland, Fruitvale, Starkey, Council, Donnelly, Challis
Ritter, Monument, Long Creek, Granite, Sumpter, Baker, Pleasant Valley, Durkee, Cambridge, Mesa, Warm Lake, Boise
Fox, Greenhorn, Austin, Whitney, Indian Valley, Midvale, Cascade
Hamilton, Dayville, Mount Vernon, John Day, Canyon City, Prairie City, Hereford, Unity, Ironside, Huntington, Smiths Ferry, Cape Horn

Columbia River, Christina Lake, Staglepp Prov. Park, Kaniksu National Forest, Colville Nat'l Forest, Okanogan Nat'l Forest, Omak Lake, Banks Lake, Franklin D. Roosevelt Lake, Grand Coulee Dam, Coeur D'Alene, Priest Lake, Lake Pend Oreille, Pend Oreille, Kootenai Lake, Koocanusa, Purcell Mts., Cabinet Mts., Selkirk Mts., Bitterroot Range, Rocky Mts.
Moses Lake, Potholes Res., Snake River, Lower Monumental Dam, Ice Harbor Dam, U.S. Dept. of Energy Hanford Site, National Wildlife Refuge, Hell's Canyon Nat'l Recreation Area, Dworshak Reservoir, Clearwater National Forest, Nezperce Indian Reservation, Selway-Bitterroot Wilderness, Clearwater Mountains, Moose Ridge, Gospel Hump Wilderness, Salmon River Breaks Primitive Area, Payette National Forest, Lolo Hot Springs, Whitefish Lake
Umatilla Nat'l Forest, Wallowa, Whitman Nat'l Forest, Eagle Cap Wilderness, Elkhorn Ridge, Malheur National Forest, John Day Fossil Beds National Monument, Cascade Reservoir, Weiser River, Willow, Sawtooth

FOR MONTANA STATE MAP SEE PAGE 54
FOR IDAHO STATE MAP SEE PAGE 34

© Creative Sales Corporation

SCALE OF MILES
1 INCH IS APPROXIMATELY 35 MILES
0 7 14 21 28 35

FOR MICHIGAN STATE MAP SEE PAGE 48

FOR MINNESOTA STATE MAP SEE PAGE 50

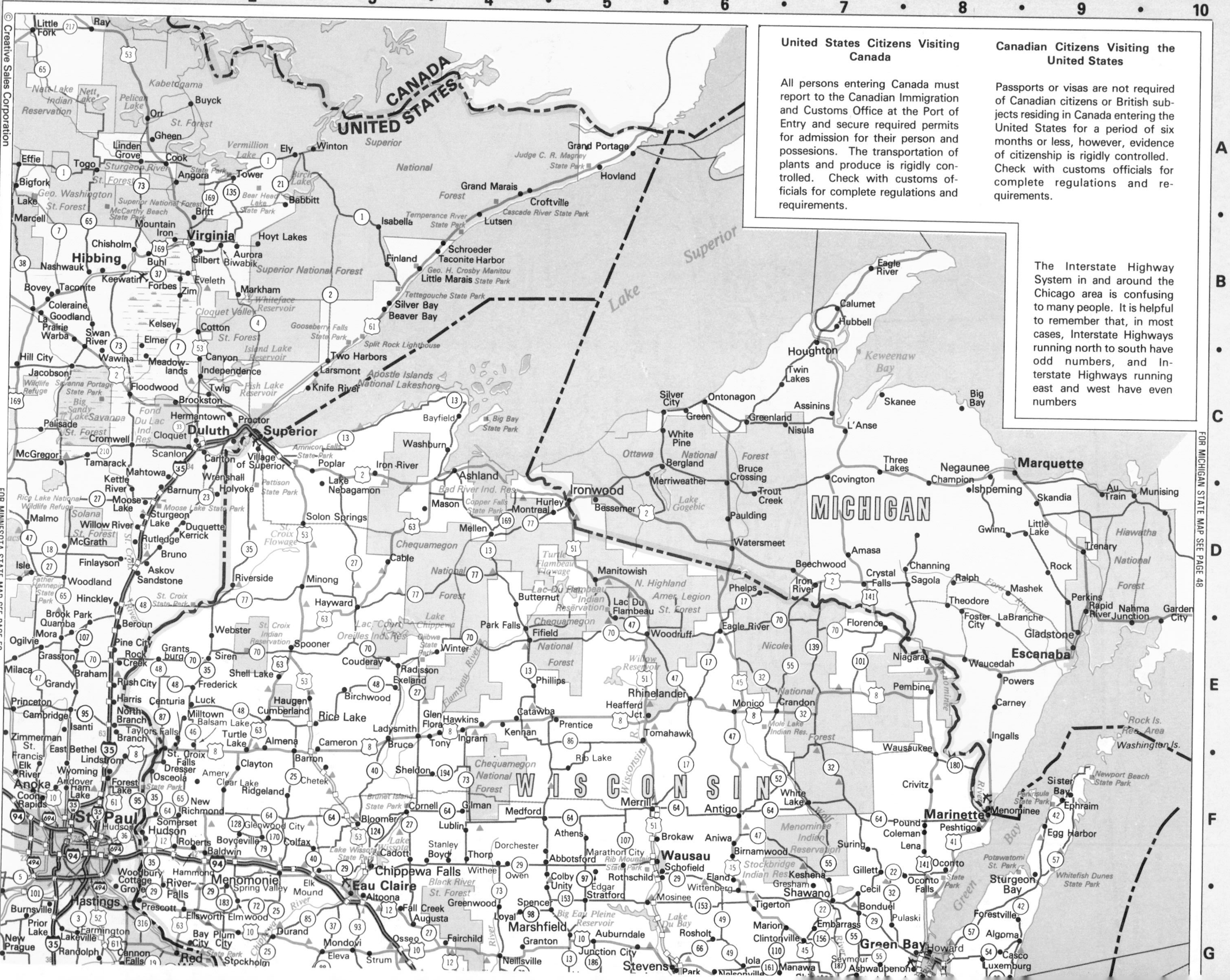

© Creative Sales Corporation

United States Citizens Visiting Canada

All persons entering Canada must report to the Canadian Immigration and Customs Office at the Port of Entry and secure required permits for admission for their person and possesions. The transportation of plants and produce is rigidly controlled. Check with customs officials for complete regulations and requirements.

Canadian Citizens Visiting the United States

Passports or visas are not required of Canadian citizens or British subjects residing in Canada entering the United States for a period of six months or less, however, evidence of citizenship is rigidly controlled. Check with customs officials for complete regulations and requirements.

The Interstate Highway System in and around the Chicago area is confusing to many people. It is helpful to remember that, in most cases, Interstate Highways running north to south have odd numbers, and Interstate Highways running east and west have even numbers

SCALE OF MILES
1 INCH IS APPROXIMATELY 35 MILES
0 7 14 21 28 35

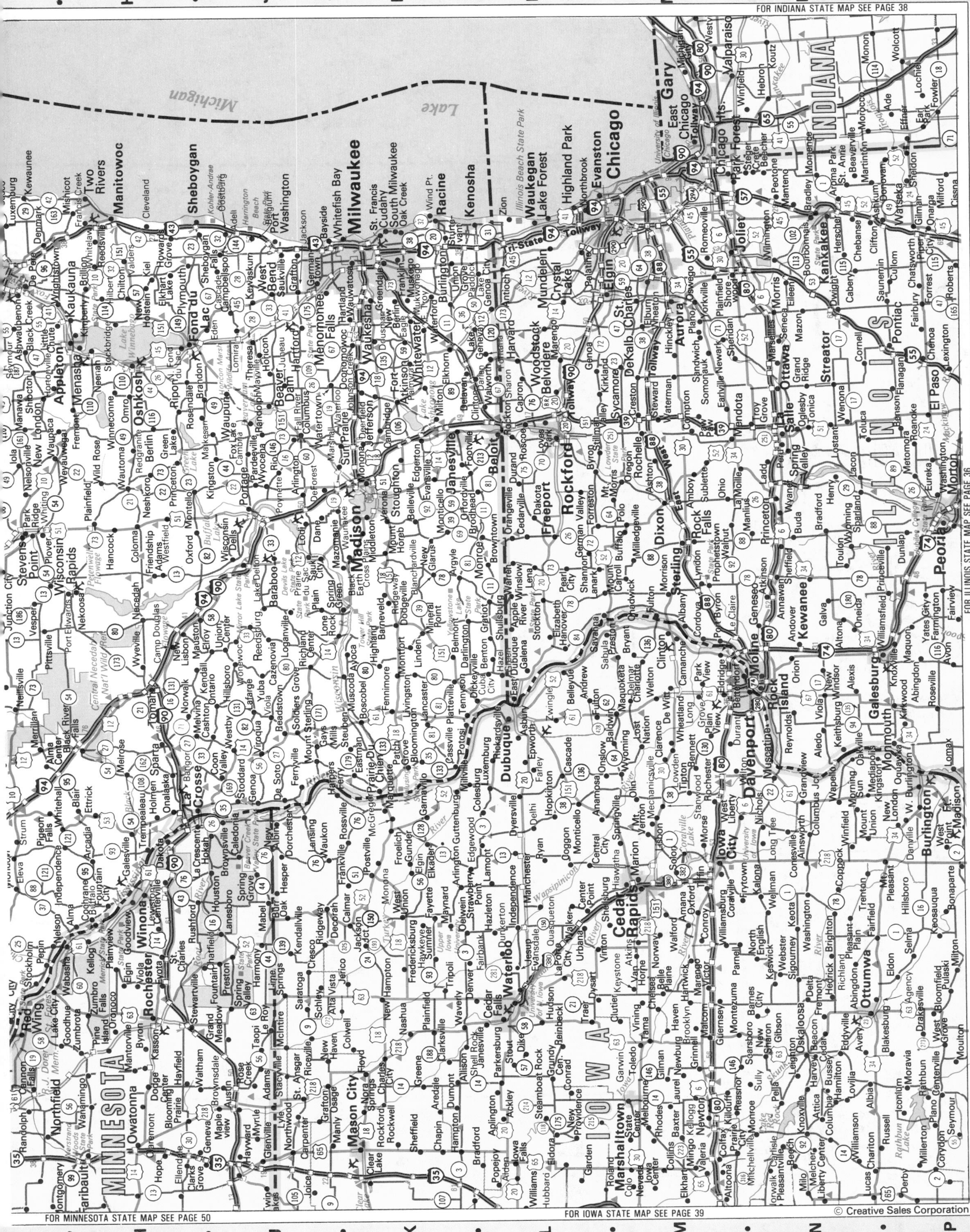

FOR INDIANA STATE MAP SEE PAGE 38
FOR ILLINOIS STATE MAP SEE PAGE 36
FOR MINNESOTA STATE MAP SEE PAGE 50
FOR IOWA STATE MAP SEE PAGE 39

Lake Michigan

INDIANA
ILLINOIS
IOWA
MINNESOTA

Major cities: Manitowoc, Sheboygan, Milwaukee, Racine, Kenosha, Waukegan, Evanston, Chicago, Gary, East Chicago, Highland Park, Appleton, Menasha, Oshkosh, Fond du Lac, Stevens Point, Wisconsin Rapids, Madison, Janesville, Beloit, Rockford, Freeport, La Crosse, Winona, Rochester, Waterloo, Cedar Rapids, Iowa City, Davenport, Rock Island, Moline, Dubuque, Marshalltown, Mason City, Ottumwa, Galesburg, Monmouth, Burlington, Peoria, Kankakee, Joliet, Aurora, Elgin, DeKalb, Dixon, Sterling, Kewanee, Streator

© Creative Sales Corporation

SCALE OF MILES
1 INCH IS APPROXIMATELY 35 MILES
0 7 14 21 28 35

N

FOR MONTANA STATE MAP SEE PAGE 54

FOR IDAHO STATE MAP SEE PAGE 34

IDAHO

WYOMING

UTAH

COLORADO

Glen, Twin Bridges, Norris, Hot Sprs., Gallatin Gateway, Absarokee, Park City, Silesia, Crow Agency, Garryowen
Sheridan, Laurin, Emigrant, Joliet, Boyd, Edgar, Nye, Fishtail, Dean, Roberts, Fromberg
Ennis, Alder, Virginia City, Pray, Chico Hot Springs, Bridger, Bearcreek
Dillon, Big Sky, Cameron, Red Lodge, Belfry, 310, Custer National Forest
Grant, Gardiner, Cooke City, Clark, Elk Basin, Frannie, Deaver, Cowley, Burgess Jct.
Dell, Mammoth Springs Jct., Tower Jct., Powell, Garland, Byron, Lovell
Lima, Norris Jct., Canyon Jct., Ralston, 310
Blue Dome, Madison Jct., Lake Jct., Cody, Emblem, Otto, Greybull, Shell
Spencer, Old Faithful, W. Thumb Jct., Yellowstone Lake, Burlington, Basin
Dubois, Macks Inn, Island Park, Shoshone Lake, Lewis Lake, Buffalo Bill Reservoir, Manderson
Ashton, Warm River, Heart Lake, Meeteetse, Worland
St. Anthony, Parker, Chester, Grand Teton National Park, Pitchfork, Grass Creek
Hamer, Sugar City, Newdale, Jenny Lake, Moran Jct., Hamilton Dome, Kirby, Lucerne
Mud Lake, Rexburg, Teton, Tetonia, Moose, Thermopolis, E. Thermopolis
Thornton, Menan, Lorenzo, Driggs, Teton Village, Kelly, Dubois, Burris, Crowheart, Morton, Pavillion, Kinnear
Roberts, Lewisville, Rigby, Ririe, Victor, Wilson, Jackson, Hoback Jct., Bondurant, Fort Washakie, Ethete, St. Stephens, Riverton
Idaho Falls, Iona, Ammon, Swan Valley, Irwin, Alpine Jct., Arapahoe, Hudson, Lander
Shelley, Basalt, Firth, Blackfoot, Palisades Reservoir, Etna, Freedom, Thayne, Merna, Cora, Pinedale, Daniel
Moreland, Rockford, Pingree, Bancroft, Conda, Bedford, Turnerville, Grover, Auburn, Afton, Marbleton, Big Piney, Boulder
Pocatello, Inkom, Soda Springs, Fairview, Smoot, Calpet, La Barge, Big Sandy
Portneuf, Lava Hot Springs, Grace, Marbleton, Big Piney, Atlantic City, South Pass City, Jeffrey City
Rockland, McCammon, Virginia, Downey, Thatcher, Georgetown, Bennington, Bern, Border, Cokeville, Eden, Bairoil
Pauline, Robin, Arimo, Swanlake, Ovid, Paris, Dingle, Farson
Malad City, Banida, Clifton, Dayton, Mink Creek, Bloomington, Border, Sage, Frontier, Kemmerer, Diamondville, Elkol, Opal, Superior, Creston, Wamsutter
Holbrook, Samaria, Preston, Whitney, St. Charles, Fish Haven, Cokeville, Sage Jct., Granger, Little America, Green River, Rock Springs, Reliance, Point of Rocks, Table Rock, Bitter Creek, Red Desert
Snowville, Franklin, Gornish, Cove, Garden City, Pickleville, Laketown, Randolph, Carter, Fort Bridger, Lyman, Mountain View, Qualey
Portage, Clarkston, Lewiston, Richmond, Amalga, Smithfield, Meadowville, Round Valley, Sage Cr., Millburne, Urie, Robertson
Blue Ctr, Howell, Fielding, Beaver Dam, Riverside, Newton, Hyde Park, Benson, Collinston, Cache Jct., Woodruff, Evanston, Piedmont, Lonetree, Burntfork, McKinnon
Cedar Springs, Tremonton, Penrose, Elwood, Deweyville, College Ward, Logan, Hyrum, Wellsville, Paradise, Woodruff Narrows Res.
Brigham City, Corinne, Mantua, Perry, Willard, Liberty, Manila, Green Lake, Hiawatha, Dixon, Baggs
Ogden, Huntsville, N. Ogden, Plain City, Harrisville, Riverdale, Kanesville, Hooper, Washington Terrace, Wahsatch, Echo, Coalville, Hoytsville
Roy, Clinton, Clearfield, Layton, Kaysville, Syracuse, Centerville, Devils Slide, Castle Rock, Croydon, Henefer, Emory
Salt Lake City, Bountiful, Woods Cross, Morgan, Milton, Potterville, Farmington, Wanship, Peoa, Kamas, Francis, Marion
Magna, West Valley City, Holladay, Murray, Midvale, Crescent, Snyderville, Kimball, Brighton, Oakley
Tooele, Grantsville, Mills Jct., Burmester, Lake Pt. Jct.

Great Salt Lake

Wind River, Indian Reservation, Boysen Res., Boysen State Park, Ocean Lake, Bull Lake, Shoshoni, Lysite, Moneta

Shoshone National Forest, Fremont Lake, Sinks Canyon St. Pk.

Bridger-Teton National Forest

Caribou National Forest

Fontenelle Res., Fossil Butte Nat'l Monument, Big Sandy Recreation Area, Big Sandy Res.

Flaming Gorge Res., Flaming Gorge National Recreational Area, Ashley National Forest, Uinta National Forest, Wasatch National Forest

Dinosaur National Monument, Sunbeam, Maybell, Craig

Bighorn Canyon Nat'l Recreational Area, Bighorn Lake, Custer National Forest, Crow Indian Reservation, Little Bighorn River

Buffalo Bill State Park

Hot Springs State Park

© Creative Sales Corporation

FOR UTAH STATE MAP SEE PAGE 82 FOR COLORADO STATE MAP SEE PAGE 28

SCALE OF MILES
1 INCH IS APPROXIMATELY 35 MILES
0 7 14 21 28 35

N

FOR MONTANA STATE MAP SEE PAGE 54

MONTANA

SOUTH DAKOTA

NEBRASKA

Busby
Birney
Lodge Grass
Wyola
Decker
Ford
Biddle
Broadus
Hammond
Alzada
Capitol
Sorum
Zeona
Hoover
Usta
Maurine
Mud Butte
Castle Rock
Faith
kman Parkman
Ranchester
Dayton
Beckton
Big Horn
Story
Banner
Sheridan
Acme
Leiter
Clearmont
Arvada
Spotted Horse
Recluse
New Haven
Weston
Oshoto
Hulett
Alva
Aladdin
Beulah
Belle Fourche
Fruitdale
Newell
Nisland
Fairpoint
Stoneville
Marcus
Plainvie
White Owl
Union Center
Creigh
Lightning Flat
Rockypoint
Devils Tower Nat'l Monument
Devil's Tower Jct.
Spearfish
Central City
Whitewood
Sturgis
Deadwood
Lead
Pluma
Elm Springs
Ucross
Buffalo
Gillette
Rozet
Moorcroft
Sundance
Rochford
Deerfield
Silver City
Hisega
Caputa
Farmingdale
Box Elder
Wasta
Wall
Ten Sleep
Savageton
Upton
Four Corners
Osage
Newcastle
Hill City
Keystone
Hermosa
Mt. Rushmore Nat'l Mon.
Crazy Horse Mon.
Custer
Game Lodge
Scenic
Badlands National Park
Mayoworth
Big Trails
Kaycee
Sussex
Linch
Wright
Reno Jct.
Rochelle
Hampshire
Clareton
Morrisey
Pringle
Blue Bell
Fairburn
Wind Cave National Park
Hot Springs
Oral
Smithwick
Manderson
Porcup
Wounded Knee Battle Site
Denby
Barnum
Edgerton
Midwest
Bill
Mule Cr. Jct.
Edgemont
Redbird
Igloo
Provo
Ardmore
Oelrichs
Oglala
Wounded Knee
Whiteclay
Lost Cabin
Arminto
Waltman
Natrona
Hiland
Powder River
Bar Nunn
Mills
Casper
Evansville
Glenrock
Orpha
Lost Springs
Shawnee
Keeline
Lusk
Node
Van Tassell
Harrison
Crawford
Chadron
Whitney
Clinton
Rushville
Hay Springs
Alliance
Ellsworth
Antioch
Lakesid
Douglas
Orin
Glendo Res.
Glendo
Hartville
Sunrise
Jay Em
Marsland
Hemingford
Boxelder
Edness Kimball Wilkins State Park
Medicine Bow National Forest
Lamont
Alcova
Shirley Basin
Medicine Bow
Fort Laramie
Guernsey
Lingle
Torrington
Morrill
Mitchell
Scottsbluff
Minatare
Bayard
Bridgeport
Broadwater
Lisco
Rawlins
Sinclair
Walcott
Hanna
Elmo
Rock River
Wheatland
Veteran
Yoder
Huntley
Lyman
Terrytown
Gering
Melbeta
McGrew
Dalton
Gurley
Elk Mountain
McFadden
Bosler
Iron Mountain
Chugwater
Slater
Hawk Sprs.
La Grange
Harrisburg
Bushnell
Kimball
Dix
Potter
Sidney
Lodgepole
Arlington
Saratoga
Laramie
Horse Creek
Albin
Chappell
Peetz
Crook
Proctor
Sedgwick
Ovid
Riverside
Centennial
Albany
Encampment
Woods Landing
Federal
Tie Siding
Hillsdale
Burns
Pine Bluffs
Egbert
Carpenter
Cheyenne
Iliff
Sterling
Dixon
Savery
Mountain Home
Virginia Dale
Rockport
Nunn
Briggsdale
Sidney
Atwood
Fleming
Dailey
Haxtun
Paoli
Cowdrey
Walden
Gould
Rustic
The Forks
Poudre Park
Bellvue
Fort Collins
Pierce
Ault
Eaton
Buckingham
Raymer
Stoneham
Merino
Windsor
Barnesville
Clarkvil
Milner
Steamboat Sprs.

Ellsworth A.F. Base

Belle Fourche Reservoir

Thunder Basin National Grassland

Black Hills National Forest

Buffalo Gap National Grassland

Wounded Knee Indian Pine Ridge Reservation

Oglala National Grasslands

Nebraska National Forest

Pathfinder Reservoir

Seminoe Reservoir

Wheatland Res. No. 2

Curt Gowdy State Park

Roosevelt National Forest

Routt National Forest

Lake John

Medicine Bow Nat'l Forest

Northern Cheyenne Indian Reservation

Custer National Forest

Big Horn National Forest

Tongue River Res.

Lake DeSmet

Keyhole Res.

Angostura Reservoir

Black Hills

Missouri River
Powder River
Little
Belle Fourche
White River
Niobrara River
North Platte River
Laramie River
South Platte River
Cheyenne River

FOR SOUTH DAKOTA STATE MAP SEE PAGE 80
FOR NEBRASKA STATE MAP SEE PAGE 58
FOR COLORADO STATE MAP SEE PAGE 28

© Creative Sales Corporation

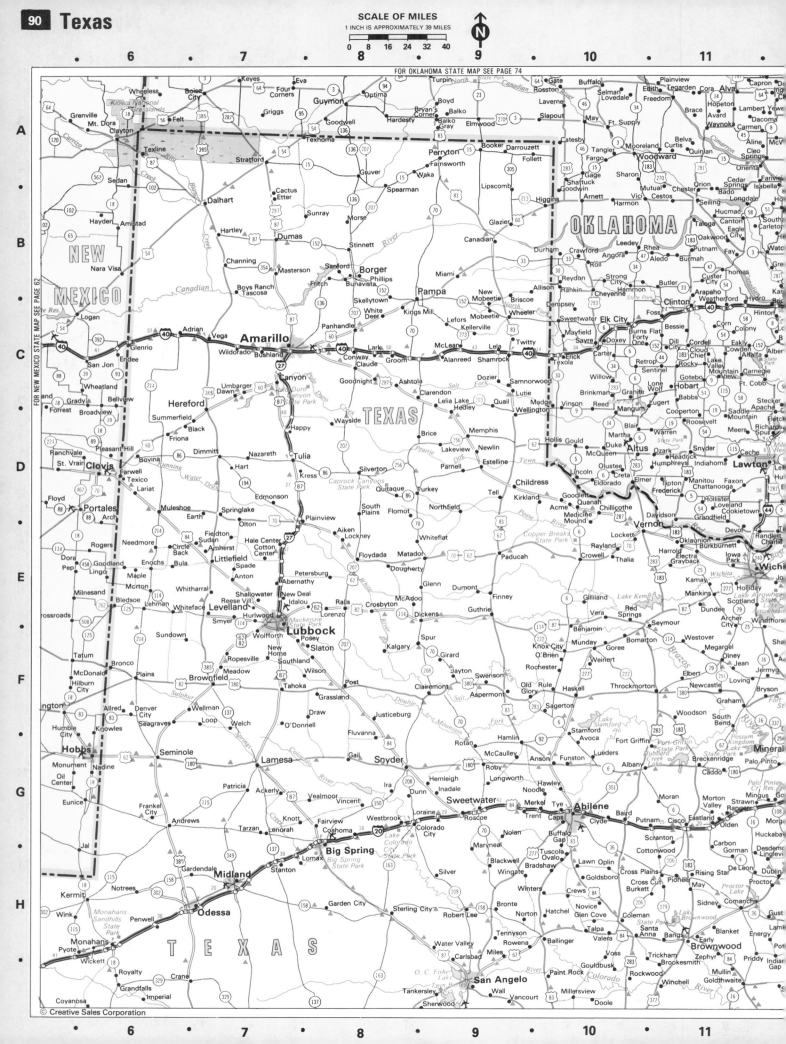

SCALE OF MILES
1 INCH IS APPROXIMATELY 39 MILES
0 8 16 24 32 40

N

FOR OKLAHOMA STATE MAP SEE PAGE 74

OKLAHOMA

NEW MEXICO

TEXAS

TEXAS

FOR NEW MEXICO STATE MAP SEE PAGE 62

© Creative Sales Corporation

N

12 • 13 • 14 • 15 • 16 • 17 • 18

FOR OKLAHOMA STATE MAP SEE PAGE 74   FOR MISSOURI STATE MAP SEE PAGE 52

A

B

C

D

E

F

G

H

OKLAHOMA

ARKANSAS

TEXAS

LOU

FOR ARKANSAS STATE MAP SEE PAGE 19

FOR LOUISIANA STATE MAP SEE PAGE 44

Enid
Tulsa
Broken Arrow
Sapulpa
Muskogee
Oklahoma City
Shawnee
Fort Smith
Fayetteville
Springdale
McAlester
Ada
Ardmore
Sulphur
Duncan
Chickasha
Denton
Fort Worth
Arlington
Grand Prairie
Dallas
Mesquite
Garland
Irving
Richardson
Weatherford
Cleburne
Waco
Corsicana
Waxahachie
Ennis
McKinney
Greenville
Sherman
Denison
Paris
Gainesville
Sulphur Sprs.
Mt. Pleasant
Texarkana
Marshall
Longview
Tyler
Kilgore
Henderson
Shreveport
Nacogdoches
Lufkin
Palestine
Jacksonville

© Creative Sales Corporation

12 • 13 • 14 • 15 • 16 • 17 • 18

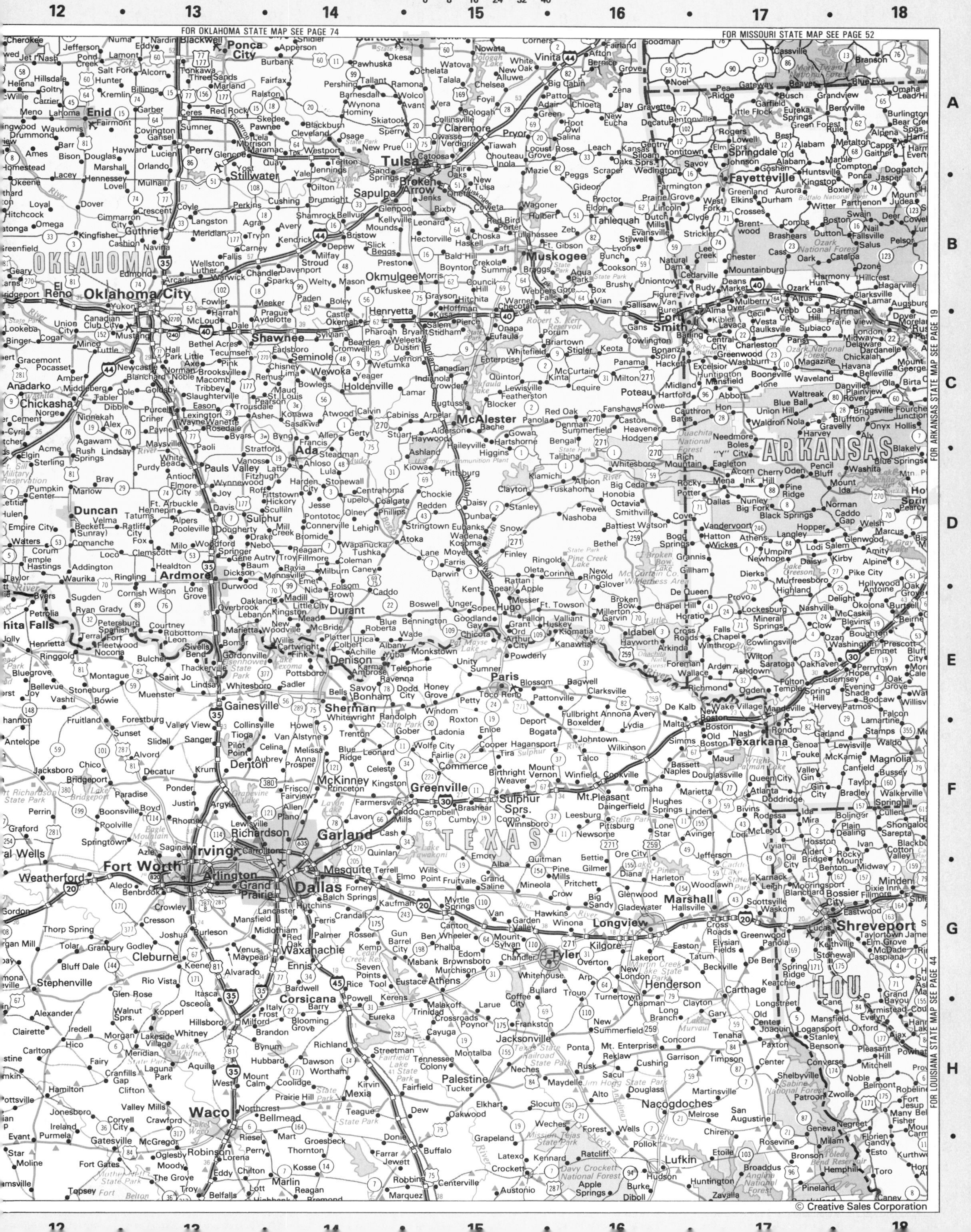

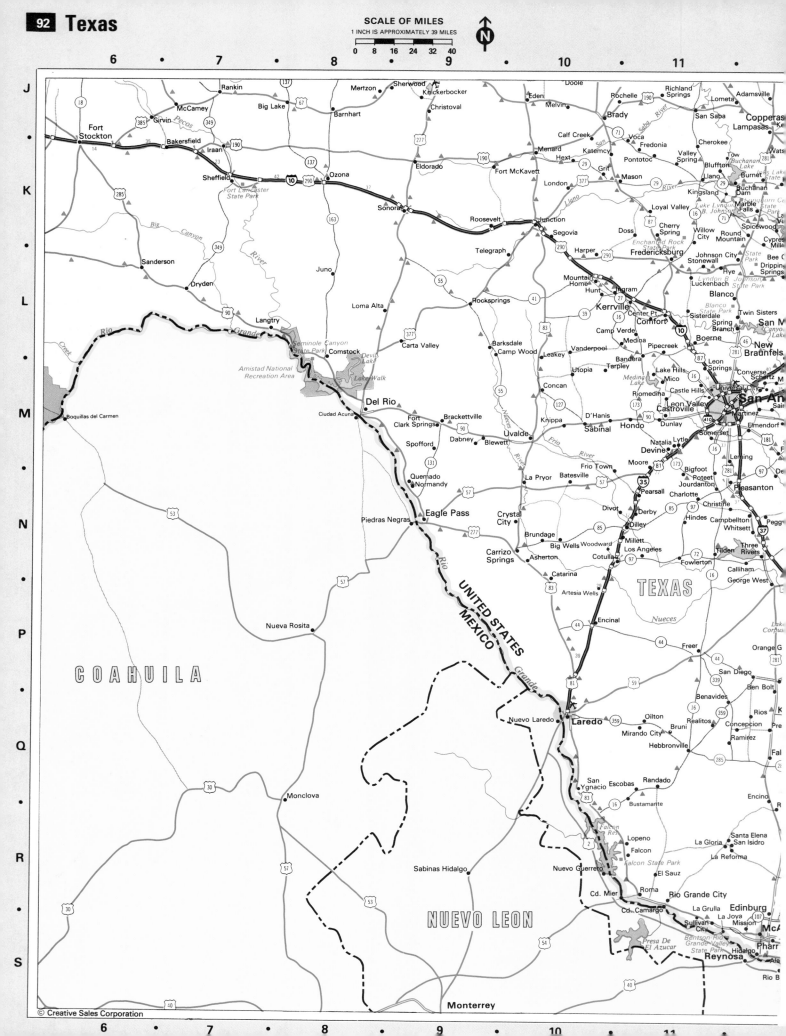

SCALE OF MILES
1 INCH IS APPROXIMATELY 39 MILES

0   8   16   24   32   40

N

6   7   8   9   10   11

J
K
L
M
N
P
Q
R
S

Rankin
Mertzon
Sherwood
Knickerbocker
Doole
Eden
Rochelle
Richland Springs
Lometa
Adamsville

McCamey
Big Lake
Christoval
Melvin
Brady
San Saba
Copperas
Lampasas

Girvin
Barnhart
Calf Creek
Voca
Cherokee
Tow
Buchanan Dam
Watts

Fort Stockton
Bakersfield
Iraan
Eldorado
Menard
Hext
Katemcy
Fredonia
Valley Spring
Bluffton
Llano
Burnet

Sheffield
Ozona
Fort McKavett
London
Grit
Mason
Kingsland
Buchanan Dam
Marble Falls

Fort Lancaster State Park
Sonora
Roosevelt
Junction
Doss
Loyal Valley
Cherry Spring
Willow City
Round Mountain
Spicewood

Sanderson
Segovia
Telegraph
Harper
Fredericksburg
Johnson City
Stonewall
Dripping Springs

Dryden
Juno
Rocksprings
Mountain Home
Hunt
Ingram
Luckenbach
Blanco

Langtry
Loma Alta
Carta Valley
Kerrville
Center Pt.
Comfort
Sisterdale
Spring Branch
San M

Comstock
Seminole Canyon State Park
Devils Lake
Barksdale
Camp Wood
Camp Verde
Medina
Pipecreek
Boerne
New Braunfels

Amistad National Recreation Area
Lake Walk
Leakey
Vanderpool
Bandera
Tarpley
Lake Hills
Mico
Leon Springs
Converse
Schertz

Boquillas del Carmen
Del Rio
Ciudad Acuña
Fort Clark Springs
Brackettville
Utopia
Concan
Riomedina
Castle Hills
Universal City
San An

Spofford
Dabney
Blewett
Uvalde
Knippa
D'Hanis
Sabinal
Hondo
Dunlay
Leon Valley
Castroville
Martinez
Elmendorf

Quemado
Normandy
Frio Town
Batesville
La Pryor
Moore
Natalia
Lytle
Devine
Somerset
Leming

Piedras Negras
Eagle Pass
Crystal City
Divot
Derby
Pearsall
Charlotte
Christine
Bigfoot
Poteet
Jourdanton
Pleasanton

Carrizo Springs
Brundage
Big Wells
Woodward
Dilley
Millett
Los Angeles
Hindes
Campbellton
Whitsett
Pegg

Nueva Rosita
Asherton
Catarina
Cotulla
Fowlerton
Tilden
Three Rivers

Artesia Wells
TEXAS
Calliham
George West

Encinal
Nueces
Lake Corpus

Freer
Orange G

San Diego
Ben Bolt

Nuevo Laredo
Laredo
Oilton
Benavides
Rios

Mirando City
Bruni
Realitos
Concepcion
Ramirez

Hebbronville
San Ygnacio
Escobas
Randado
Encino

COAHUILA
Monclova
Bustamante
Fal

Sabinas Hidalgo
Falcon Res.
Lopeno
La Gloria
Santa Elena
San Isidro

Falcon
La Reforma

Nuevo Guerrero
Falcon State Park
El Sauz

Cd. Mier
Roma
Rio Grande City

NUEVO LEON
Cd. Camargo
La Grulla
La Joya
Edinburg
Mission
McA

Presa De El Azucar
Sullivan City
Bentsen-Rio Grande Valley State Park
Pharr
Hidalgo

Monterrey
Reynosa

UNITED STATES
MEXICO
Rio Grande

© Creative Sales Corporation

SCALE OF MILES
1 INCH IS APPROXIMATELY 39 MILES

0 8 16 24 32 40

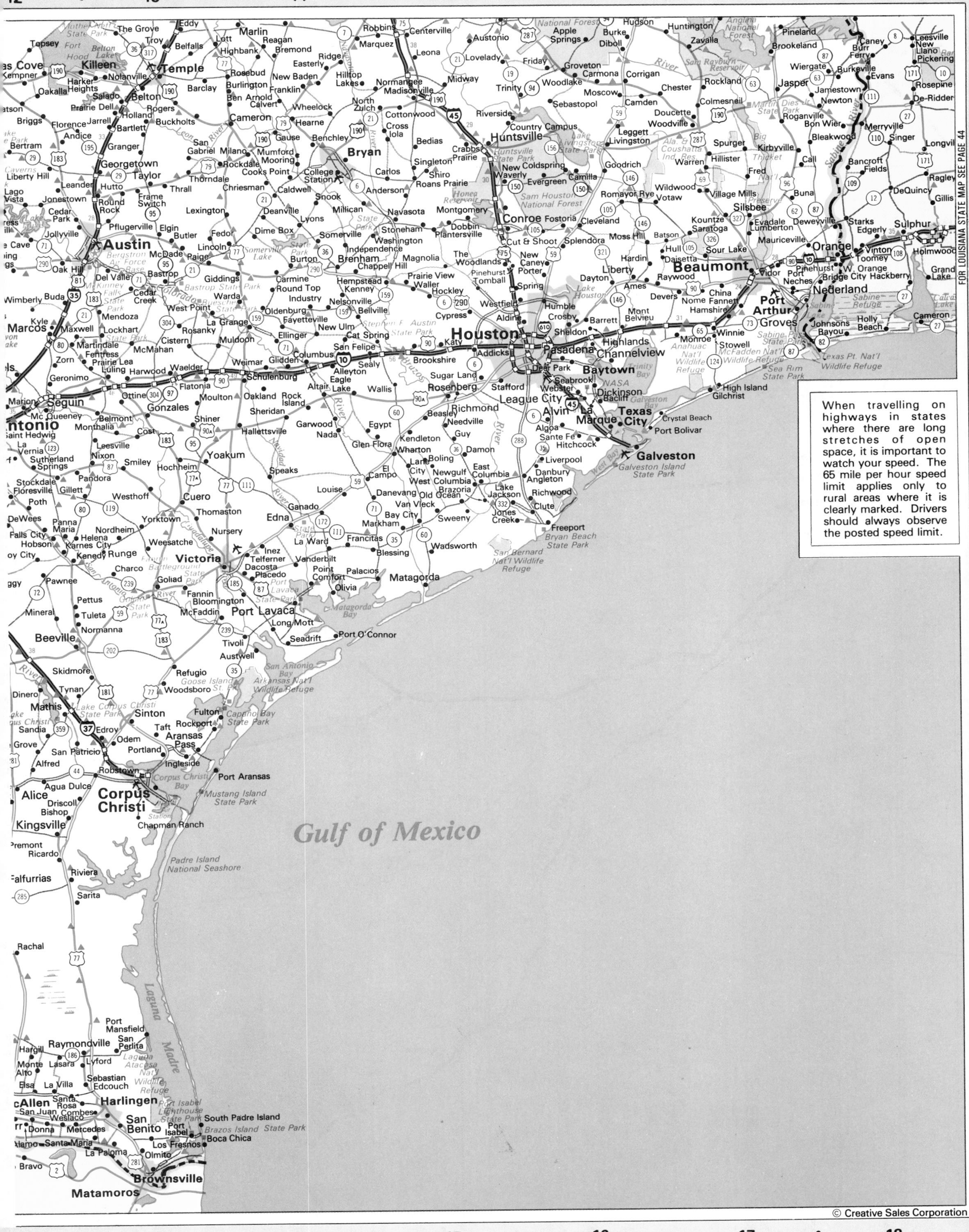

N

12 · 13 · 14 · 15 · 16 · 17 · 18

J K L M N P Q R S

When travelling on highways in states where there are long stretches of open space, it is important to watch your speed. The 65 mile per hour speed limit applies only to rural areas where it is clearly marked. Drivers should always observe the posted speed limit.

FOR LOUISIANA STATE MAP SEE PAGE 44

Gulf of Mexico

Killeen, Temple, Belton, Austin, Marlin, Bryan, College Station, Huntsville, Conroe, Houston, Pasadena, Baytown, Channelview, Beaumont, Port Arthur, Nederland, Groves, Orange, Vidor, Galveston, Texas City, La Marque, Alvin, League City, Richmond, Rosenberg, Sugar Land, Victoria, Port Lavaca, Seguin, Gonzales, Cuero, Yoakum, Beeville, Corpus Christi, Alice, Kingsville, Robstown, Sinton, Aransas Pass, Rockport, Fulton, Portland, Ingleside, Mathis, Harlingen, Brownsville, Matamoros, San Benito, Raymondville, McAllen, Weslaco, Falfurrias, Kenedy, Jasper, Silsbee, Livingston, Woodville, Newton, De Ridder, Merryville, DeQuincy, Sulphur, Padre Island National Seashore, Laguna Madre, Mustang Island State Park, Galveston Island State Park, South Padre Island, Boca Chica, Freeport, Angleton, Bay City, Matagorda, Port O'Connor, Port Aransas

© Creative Sales Corporation

SCALE OF MILES
1 INCH IS APPROXIMATELY 39 MILES
0 8 16 24 32 40

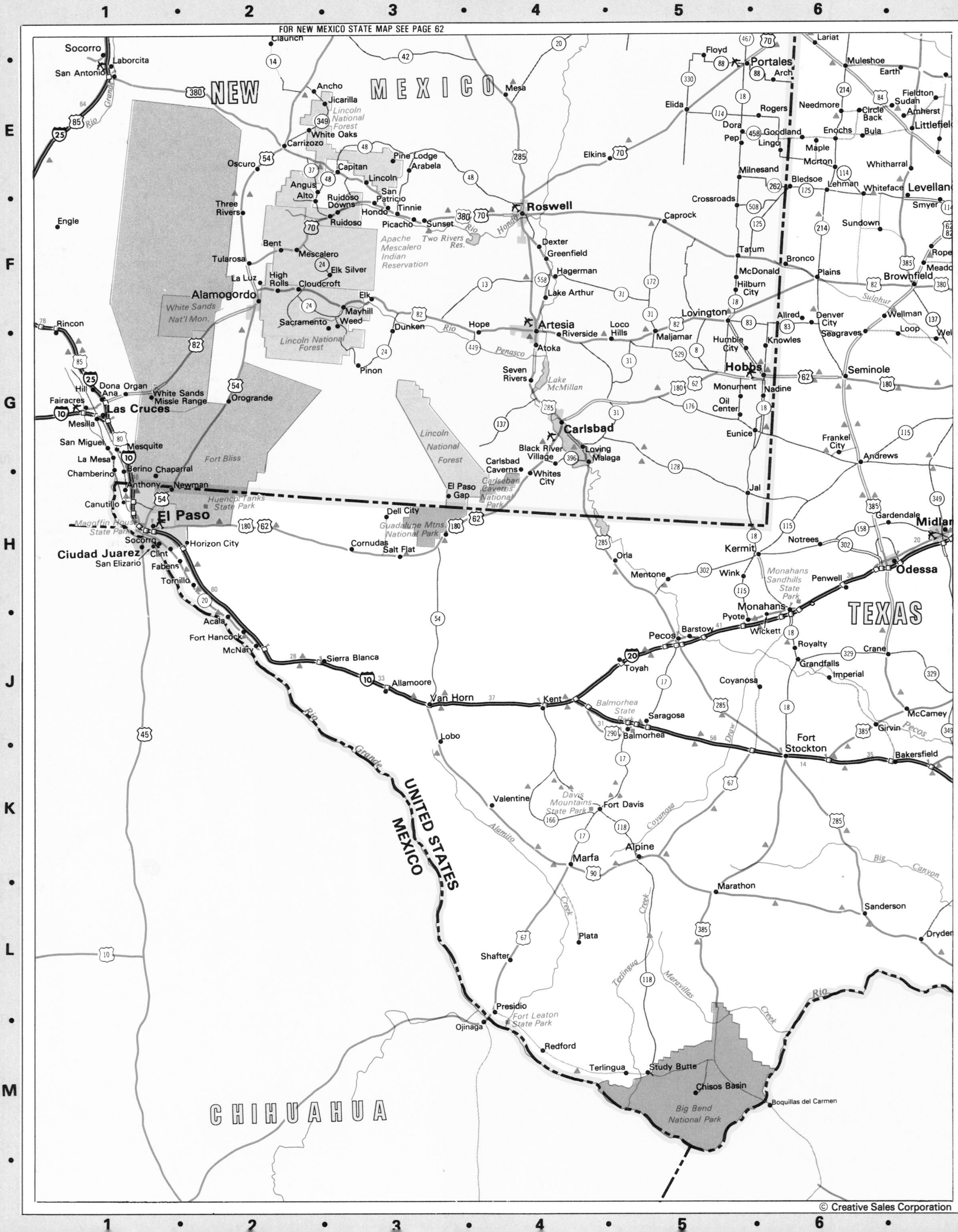

N

FOR NEW MEXICO STATE MAP SEE PAGE 62

NEW MEXICO

Socorro
Laborcita
San Antonio
Claunch
Ancho
Jicarilla
Mesa
White Oaks
Carrizozo
Oscuro
Three Rivers
Capitan
Pine Lodge
Arabela
Lincoln
Angus
Alto
San Patricio
Ruidoso Downs
Tinnie
Hondo
Picacho
Sunset
Ruidoso
Engle
Bent
Mescalero
Elk Silver
Tularosa
La Luz
High Rolls
Cloudcroft
Elk
Alamogordo
White Sands Nat'l Mon.
Sacramento
Mayhill
Weed
Dunken
Hope
Pinon
Rincon
Apache Mescalero Indian Reservation
Lincoln National Forest
Two Rivers Res.
Roswell
Dexter
Greenfield
Hagerman
Lake Arthur
Elkins
Caprock
Riverside
Artesia
Atoka
Loco Hills
Seven Rivers
Lake McMillan
Lincoln National Forest
Carlsbad
Black River Village
Carlsbad Caverns
Whites City
Loving
Malaga
Hill
Dona Ana
Organ
White Sands Missle Range
Orogrande
Las Cruces
Fairacres
Mesilla
San Miguel
La Mesa
Mesquite
Chamberino
Berino Chaparral
Anthony Newman
Canutillo
El Paso
Socorro
Ciudad Juarez
Clint
San Elizario
Fabens
Tornillo
Horizon City
Magoffin House State Park
Huenco Tanks State Park
El Paso Gap
Dell City
Cornudas
Salt Flat
Carlsbad Caverns National Park
Guadalupe Mtns National Park
Acala
Fort Hancock
McNary
Sierra Blanca
Allamoore
Van Horn
Kent
Lobo
Balmorhea State Park
Saragosa
Balmorhea
Toyah
Coyanosa
UNITED STATES
MEXICO
Valentine
Davis Mountains State Park
Fort Davis
Alpine
Marfa
Marathon
Plata
Shafter
Presidio
Fort Leaton State Park
Ojinaga
Redford
Terlingua
Study Butte
Chisos Basin
Boquillas del Carmen
Big Bend National Park
CHIHUAHUA

Lariat
Floyd
Portales
Arch
Muleshoe
Earth
Elida
Rogers
Needmore
Fieldton
Sudah
Amherst
Dora
Pep
Goodland
Enochs
Bula
Littlefield
Milnesand
Morton
Whitharral
Levelland
Crossroads
Lehman
Whiteface
Smyer
Bledsoe
Sundown
Tatum
Bronco
Plains
Brownfield
McDonald
Hilburn City
Lovington
Maljamar
Humble City
Allred
Denver City
Wellman
Seagraves
Loop
Hobbs
Monument
Oil Center
Knowles
Nadine
Seminole
Eunice
Frankel City
Andrews
Jal
Gardendale
Midland
Kermit
Notrees
Odessa
Monahans Sandhills State Park
Wink
Penwell
Monahans
Pyote
Pecos
Barstow
Wickett
Royalty
Crane
Grandfalls
Imperial
Fort Stockton
Girvin
McCamey
Bakersfield
Sanderson
Dryden

TEXAS

© Creative Sales Corporation

Alabama
Population: 3,893,888
Capital: Montgomery (F-5)
Largest City: Birmingham 284,413 (D-4)
Highest Elevation: Cheaha Mtn. 2,407 ft.
Land Area (sq. miles): 50,767

Alaska
Population: 401,851
Capital: Juneau (G-9)
Largest City: Anchorage 174,431 (E-6)
Highest Elevation: Mt. McKinley 20,320 ft.
Land Area (sq. miles): 570,833

Arizona
Population: 2,718,215
Capital: Phoenix (J-6)
Largest City: Phoenix 789,704 (J-6)
Highest Elevation: Humphreys Peak 12,633 ft.
Land Area (sq. miles): 113,508

Arkansas
Population: 2,286,435
Capital: Little Rock 158,461 (D-5)
Largest City: Little Rock 158,461 (D-5)
Highest Elevation: Magazine Mtn. 2,753 ft.
Land Area (sq. miles): 52,078

California
Population: 23,667,902
Capital: Sacramento (J-5)
Largest City: Los Angeles 2,966,850 (T-8)
Highest Elevation: Mount Whitney 14,494 ft.
Land Area (sq. miles): 156,299

Colorado
Population: 2,889,964
Capital: Denver (D-9)
Largest City: Denver 492,365 (D-9)
Highest Elevation: Mount Elbert 14,433 ft.
Land Area (sq. miles): 103,595

Connecticut
Population: 3,107,576
Capital: Hartford (D-3)
Largest City: Bridgeport 142,546 (F-2)
Highest Elevation: Mount Frissell 2,380 ft.
Land Area (sq. miles): 4,872

Delaware
Population: 594,338
Capital: Dover (H-2)
Largest City: Wilmington 70,195 (F-2)
Highest Elevation: Near Brandywine 442 ft.
Land Area (sq. miles): 1,932

District of Columbia
Population: 638,333
Capital: Washington (F-5)
Largest City: Washington 638,333 (F-5)
Highest Elevation: Tenleytown 410 ft.
Land Area (sq. miles): 63

Florida
Population: 9,746,324
Capital: Tallahassee (D-2)
Largest City: Jacksonville 540,920 (D-6)
Highest Elevation: Near Gaskin 345 ft.
Land Area (sq. miles): 58,560

Georgia
Population: 5,463,105
Capital: Atlanta (C-8)
Largest City: Atlanta 425,022 (C-8)
Highest Elevation: Brasstown Bald 4,784 ft.
Land Area (sq. miles): 54,153

Hawaii
Population: 964,691
Capital: Honolulu (C-5)
Largest City: Honolulu 365,048 (C-5)
Highest Elevation: Mauna Kea 13,796 ft.
Land Area (sq. miles): 6,425

Idaho
Population: 943,935
Capital: Boise (M-3)
Largest City: Boise 102,451 (M-3)
Highest Elevation: Borah Peak 12,662 ft.
Land Area (sq. miles): 82,413

Illinois
Population: 11,426,518
Capital: Springfield (J-5)
Largest City: Chicago 3,005,072 (E-8)
Highest Elevation: Charles Mound 1,235 ft.
Land Area (sq. miles): 55,645

Indiana
Population: 5,490,224
Capital: Indianapolis (F-4)
Largest City: Indianapolis 700,807 (F-4)
Highest Elevation: Near Fountain City 1,257 ft.
Land Area (sq. miles): 35,932

Iowa
Population: 2,913,808
Capital: Des Moines (E-5)
Largest City: Des Moines 191,003 (E-5)
Highest Elevation: Near Allendorf 1,670 ft.
Land Area (sq. miles): 55,965

Kansas
Population: 2,363,679
Capital: Topeka (D-11)
Largest City: Wichita 279,272 (G-9)
Highest Elevation: Mount Sunflower 4,039 ft.
Land Area (sq. miles): 81,781

Kentucky
Population: 3,660,777
Capital: Frankfort (C-9)
Largest City: Louisville 298,451 (C-8)
Highest Elevation: Black Mtn. 4,145 ft.
Land Area (sq. miles): 39,669

Louisiana
Population: 4,205,900
Capital: Baton Rouge (E-7)
Largest City: New Orleans 557,515 (F-9)
Highest Elevation: Driskill Mtn. 535 ft.
Land Area (sq. miles): 44,521

Maine
Population: 1,124,660
Capital: Augusta (G-3)
Largest City: Portland 61,572 (J-3)
Highest Elevation: Mt. Katahdin 5,268 ft.
Land Area (sq. miles): 30,995

Maryland
Population: 4,216,975
Capital: Annapolis (D-11)
Largest City: Baltimore 786,775 (D-11)
Highest Elevation: Backbone Mtn. 3,360 ft.
Land Area (sq. miles): 9,837

Massachusetts
Population: 5,737,037
Capital: Boston (B-8)
Largest City: Boston 562,994 (B-8)
Highest Elevation: Mt. Greylock 3,491 ft.
Land Area (sq. miles): 7,824

Michigan
Population: 9,262,078
Capital: Lansing (K-6)
Largest City: Detroit 1,203,339 (L-9)
Highest Elevation: Mt. Curwood 1,980 ft.
Land Area (sq. miles): 56,954

Minnesota
Population: 4,075,970
Capital: St. Paul (H-6)
Largest City: Minneapolis 370,951 (J-5)
Highest Elevation: Eagle Mtn. 2,301 ft.
Land Area (sq. miles): 79,548

Mississippi
Population: 2,520,638
Capital: Jackson (F-4)
Largest City: Jackson 202,895 (F-4)
Highest Elevation: Woodall Mtn. 806 ft.
Land Area (sq. miles): 47,233

Missouri
Population: 4,916,686
Capital: Jefferson City (G-5)
Largest City: St. Louis 453,085 (G-9)
Highest Elevation: Taum Sauk Mtn. 1,772 ft.
Land Area (sq. miles): 68,945

Montana
Population: 786,690
Capital: Helena (F-5)
Largest City: Billings 66,798 (G-9)
Highest Elevation: Granite Peak 12,799 ft.
Land Area (sq. miles): 145,398

Nebraska
Population: 1,569,825
Capital: Lincoln (G-12)
Largest City: Omaha 314,255 (F-12)
Highest Elevation: near Bushnell 5,426 ft.
Land Area (sq. miles): 76,644

Nevada
Population: 800,493
Capital: Carson City (F-2)
Largest City: Las Vegas 164,674 (M-8)
Highest Elevation: Boundary Peak 13,143 ft.
Land Area (sq. miles): 109,893

New Hampshire
Population: 920,610
Capital: Concord (H-5)
Largest City: Manchester 90,936 (J-5)
Highest Elevation: Mt. Washington 6,288 ft.
Land Area (sq. miles): 8,993

New Jersey
Population: 7,364,823
Capital: Trenton (D-4)
Largest City: Newark 329,248 (C-6)
Highest Elevation: High Point 1,803 ft.
Land Area (sq. miles): 7,468

New Mexico
Population: 1,302,894
Capital: Santa Fe (F-6)
Largest City: Albuquerque 331,767 (G-4)
Highest Elevation: Wheeler Peak 13,161 ft.
Land Area (sq. miles): 121,335

New York
Population: 17,558,072
Capital: Albany (L-19)
Largest City: New York 7,071,639 (E-3)
Highest Elevation: Mount Marcy 5,344 ft.
Land Area (sq. miles): 47,377

North Carolina
Population: 5,881,766
Capital: Raleigh (B-10)
Largest City: Charlotte 314,447 (D-7)
Highest Elevation: Mt. Mitchell 6,684 ft.
Land Area (sq. miles): 48,843

North Dakota
Population: 652,717
Capital: Bismark (F-6)
Largest City: Fargo 61,383 (F-10)
Highest Elevation: White Butte 3,506 ft.
Land Area (sq. miles): 69,300

Ohio
Population: 10,797,630
Capital: Columbus (J-4)
Largest City: Cleveland 573,822 (E-7)
Highest Elevation: Campbell Hill 1,550 ft.
Land Area (sq. miles): 41,004

Oklahoma
Population: 3,025,487
Capital: Oklahoma City (E-9)
Largest City: Oklahoma City 403,213 (E-9)
Highest Elevation: Black Mesa 4,973 ft.
Land Area (sq. miles): 68,655

Oregon
Population: 2,633,105
Capital: Salem (D-3)
Largest City: Portland 366,383 (C-4)
Highest Elevation: Mt. Hood 11,239 ft.
Land Area (sq. miles): 96,184

Pennsylvania
Population: 11,863,895
Capital: Harrisburg (H-9)
Largest City: Philadelphia 1,688,210 (J-13)
Highest Elevation: Mt. Davis 3,213 ft.
Land Area (sq. miles): 44,888

Rhode Island
Population: 947,154
Capital: Providence (D-6)
Largest City: Providence 156,804 (D-6)
Highest Elevation: Jerimoth Hill 812 ft.
Land Area (sq. miles): 1,055

South Carolina
Population: 3,121,820
Capital: Columbia (F-6)
Largest City: Columbia 101,208 (F-6)
Highest Elevation: Sassafrass Mtn. 3,560 ft.
Land Area (sq. miles): 30,203

South Dakota
Population: 690,768
Capital: Pierre (F-6)
Largest City: Sioux Falls 81,343 (H-10)
Highest Elevation: Harney Peak 7,242 ft.
Land Area (sq. miles): 75,952

Tennessee
Population: 4,591,120
Capital: Nashville (G-6)
Largest City: Memphis 646,356 (J-1)
Highest Elevation: Clingmans Dome 6,643 ft.
Land Area (sq. miles): 41,155

Texas
Population: 14,229,191
Capital: Austin (L-13)
Largest City: Houston 1,595,138 (L-15)
Highest Elevation: Guapalupe Peak 8,749
Land Area (sq. miles): 262,017

Utah
Population: 1,461,037
Capital: Salt Lake City (E-5)
Largest City: Salt Lake City 163,033 (E-5)
Highest Elevation: Kings Peak 13,528 ft.
Land Area (sq. miles): 82,073

Vermont
Population: 511,456
Capital: Montpelier (D-3)
Largest City: Burlington 37,712 (D-1)
Highest Elevation: Mt. Mansfield 4,393 ft.
Land Area (sq. miles): 9,273

Virginia
Population: 5,346,818
Capital: Richmond (G-10)
Largest City: Norfolk 266,979 (H-12)
Highest Elevation: Mt. Rogers 5,729 ft.
Land Area (sq. miles): 39,703

Washington
Population: 4,132,156
Capital: Olympia (E-4)
Largest City: Seattle 493,846 (C-5)
Highest Elevation: Mt. Rainier 14,410 ft.
Land Area (sq. miles): 66,511

West Virginia
Population: 1,949,644
Capital: Charleston (E-4)
Largest City: Charleston 63,968 (E-4)
Highest Elevation: Spruce Knob 4,863 ft.
Land Area (sq. miles): 24,119

Wisconsin
Population: 4,705,767
Capital: Madison (K-6)
Largest City: Milwaukee 636,212 (K-8)
Highest Elevation: Timms Hill 1,951 ft.
Land Area (sq. miles): 54,426

Wyoming
Population: 469,559
Capital: Cheyenne (J-11)
Largest City: Casper 51,016 (F-9)
Highest Elevation: Gannet Peak 13,804 ft.
Land Area (sq. miles): 96,989

Atlanta, GA
Land Area (sq. miles): 131.0
City Population: 425,022
Metropolitan Population: 1,613,357

Baltimore, MD
Land Area (sq. miles): 80.3
City Population: 786,775
Metropolitan Population: 1,755,477

Boston, MA
Land Area (sq. miles): 47.2
City Population: 562,994
Metropolitan Population: 2,678,762

Chicago, IL
Land Area (sq. miles): 228.1
City Population: 3,005,072
Metropolitan Population: 6,779,799

Cincinnati, OH
Land Area (sq. miles): 78.1
City Population: 385,457
Metropolitan Population: 1,123,412

Cleveland, OH
Land Area (sq. miles): 79.0
City Population: 573,822
Metropolitan Population: 1,752,424

Dallas, TX
Land Area (sq. miles): 333.0
City Population: 904,078
Metropolitan Population: including Ft. Worth 2,451,390

Denver, CO
Land Area (sq. miles): 110.6
City Population: 492,365
Metropolitan Population: 1,352,070

Detroit, MI
Land Area (sq. miles): 135.6
City Population: 1,203,339
Metropolitan Population: 3,809,327

Fort Worth, TX
Land Area (sq. miles): 240.2
City Population: 385,164
Metropolitan Population: including Dallas 2,451,390

Houston, TX
Land Area (sq. miles): 556.4
City Population: 1,595,138
Metropolitan Population: 2,412,644

Indianapolis, IN
Land Area (sq. miles): 352.0
City Population: 700,807
Metropolitan Population: 836,472

Kansas City, MO
Land Area (sq. miles): 316.3
City Population: 448,159
Metropolitan Population: 1,097,793

Los Angeles, CA
Land Area (sq. miles): 464.7
City Population: 2,966,850
Metropolitan Population: 9,479,436

Louisville, KY
Land Area (sq. miles): 60.0
City Population: 298,451
Metropolitan Population: 761,002

Memphis, TN
Land Area (sq. miles): 264.1
City Population: 646,305
Metropolitan Population: 774,551

Miami, FL
Land Area (sq. miles): 34.3
City Population: 346,865
Metropolitan Population: 1,608,159

Milwaukee, WI
Land Area (sq. miles): 95.8
City Population: 636,212
Metropolitan Population: 1,207,008

Minneapolis, MN
Land Area (sq. miles): 55.1
City Population: 370,951
Metropolitan Population: including St. Paul 1,787,564

New Orleans, LA
Land Area (sq. miles): 199.4
City Population: 557,515
Metropolitan Population: 1,078,299

New York City, NY
Land Area (sq. miles): 301.5
City Population: 7,071,639
Metropolitan Population: 15,590,274

Oakland, CA
Land Area (sq. miles): 59.3
City Population: 339,337
Metropolitan Population: including San Francisco 3,190,698

Oklahoma City, OK
Land Area (sq. miles): 603.6
City Population: 388,599
Metropolitan Population: 674,322

Omaha, NE
Land Area (sq. miles): 90.9
City Population: 314,255
Metropolitan Population: 512,438

Philadelphia, PA
Land Area (sq. miles): 136.0
City Population: 1,688,210
Metropolitan Population: 4,112,933

Phoenix, AZ
Land Area (sq. miles): 324.0
City Population: 789,704
Metropolitan Population: 1,437,392

Pittsburgh, PA
Land Area (sq. miles): 55.4
City Population: 423,938
Metropolitan Population: 1,810,038

Portland, OR
Land Area (sq. miles): 103.3
City Population: 366,383
Metropolitan Population: 1,026,144

Saint Louis, MO
Land Area (sq. miles): 61.4
City Population: 453,085
Metropolitan Population: 1,848,590

Saint Paul, MN
Land Area (sq. miles): 52.4
City Population: 270,230
Metropolitan Population: including Minneapolis 1,787,564

Saint Petersburg, FL
Land Area (sq. miles): 55.5
City Population: 238,647
Metropolitan Population: 833,337

Salt Lake City, UT
Land Area (sq. miles): 75.2
City Population: 163,033
Metropolitan Population: 674,201

San Diego, CA
Land Area (sq. miles): 320.0
City Population: 875,538
Metropolitan Population: 1,704,352

San Francisco, CA
Land Area (sq. miles): 46.4
City Population: 678,974
Metropolitan Population: including Oakland 3,190,698

Seattle, WA
Land Area (sq. miles): 144.6
City Population: 493,846
Metropolitan Population: 1,391,535

Tampa, FL
Land Area (sq. miles): 84.4
City Population: 271,523
Metropolitan Population: 520,912

Washington, DC
Land Area (sq. miles): 62.7
City Population: 638,333
Metropolitan Population: 2,763,105

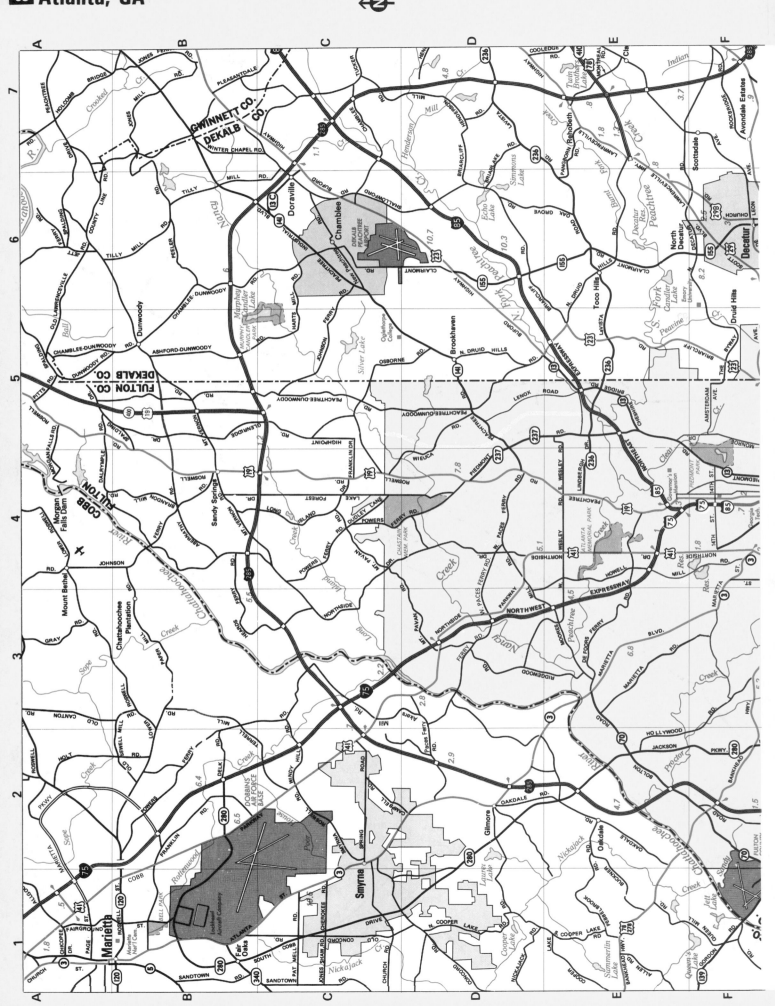

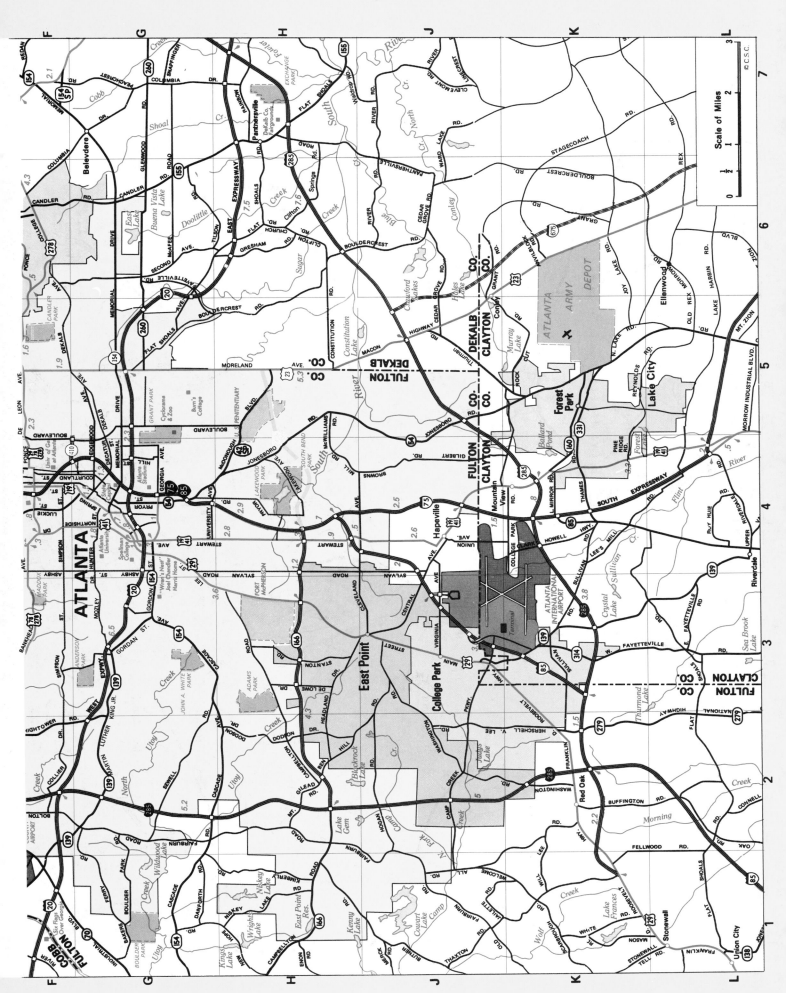

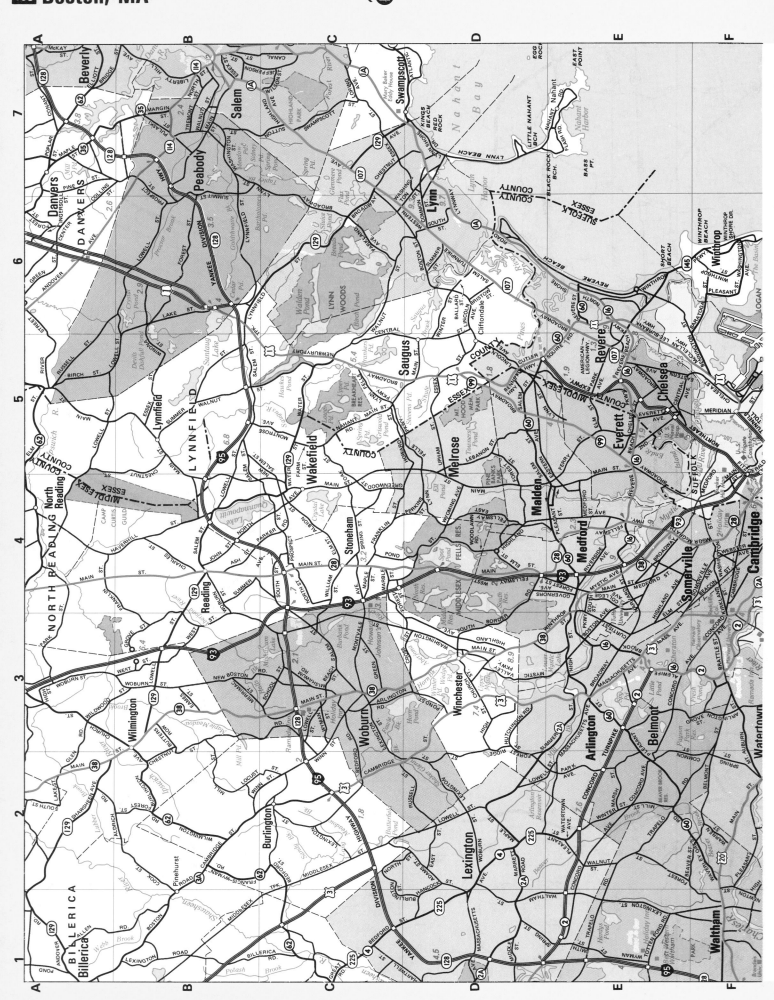

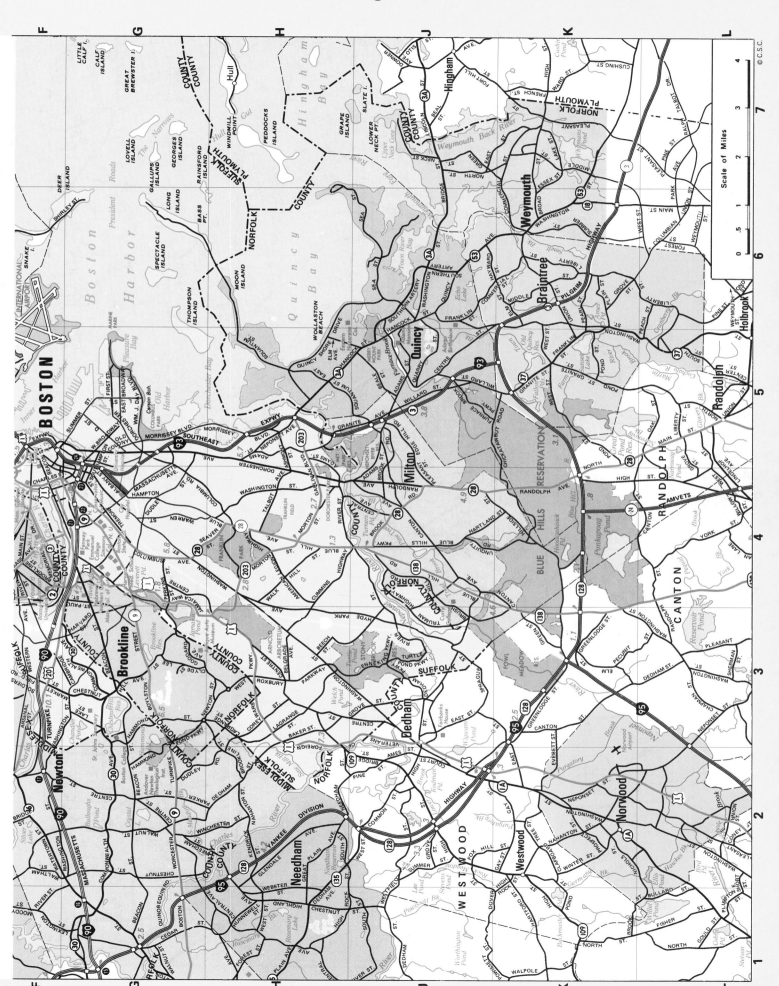

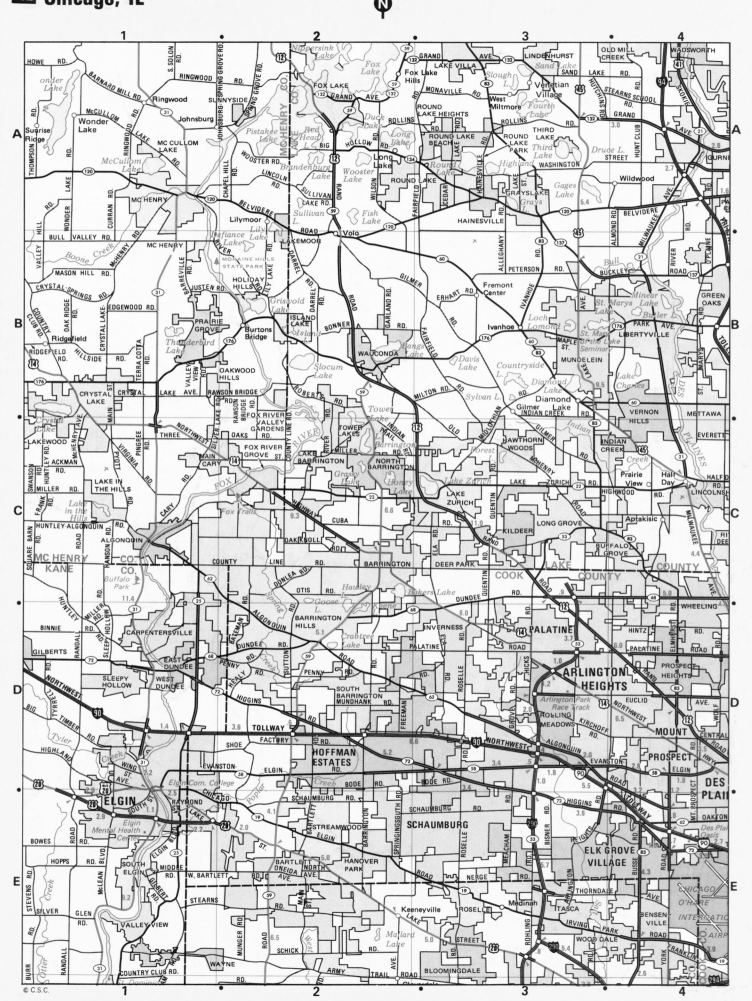

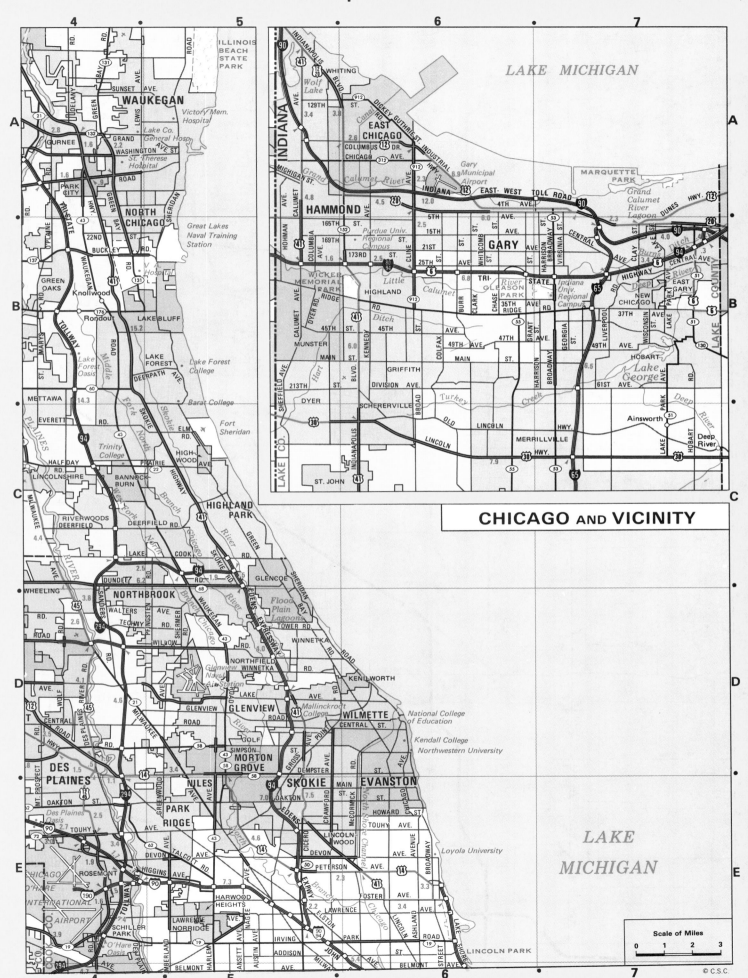

N

CHICAGO AND VICINITY

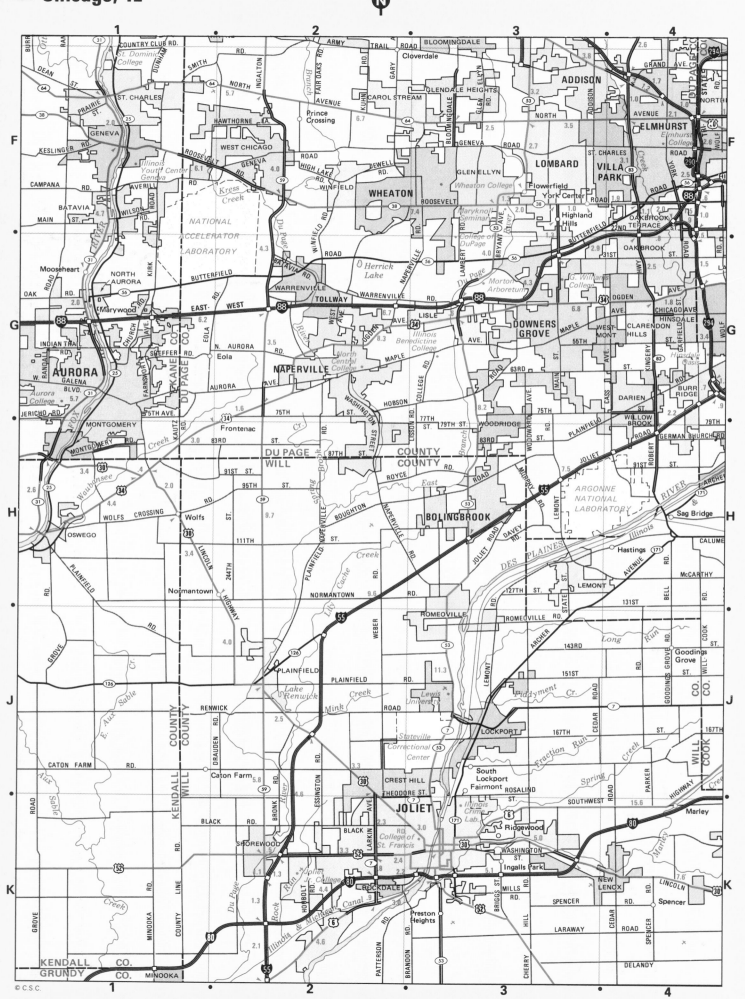

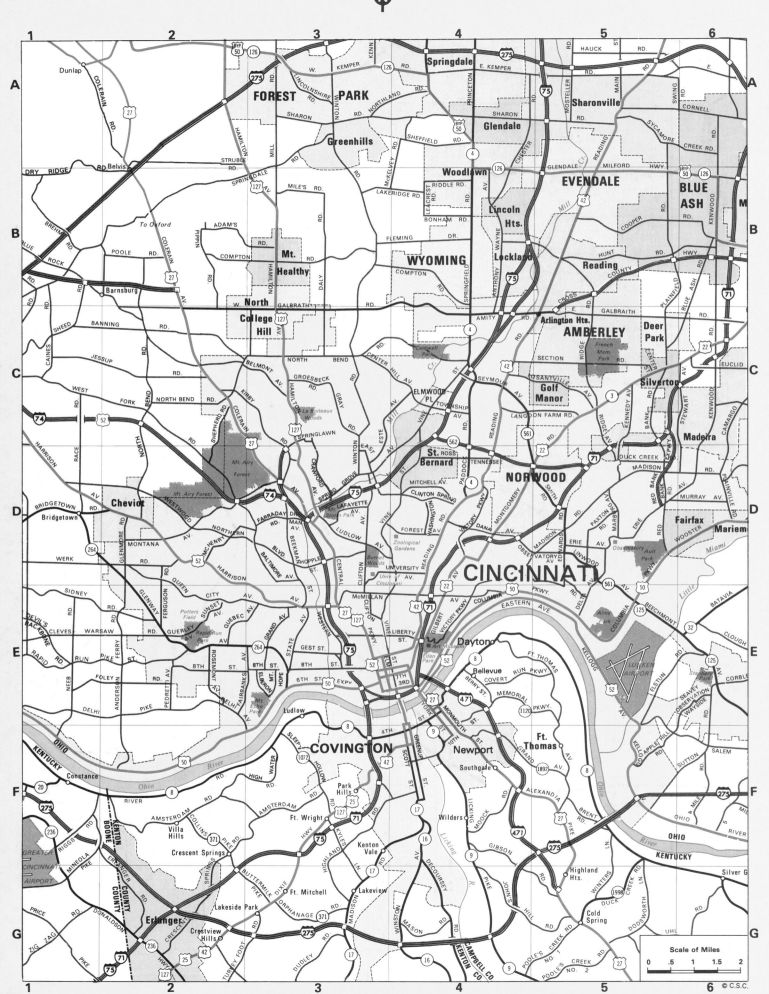

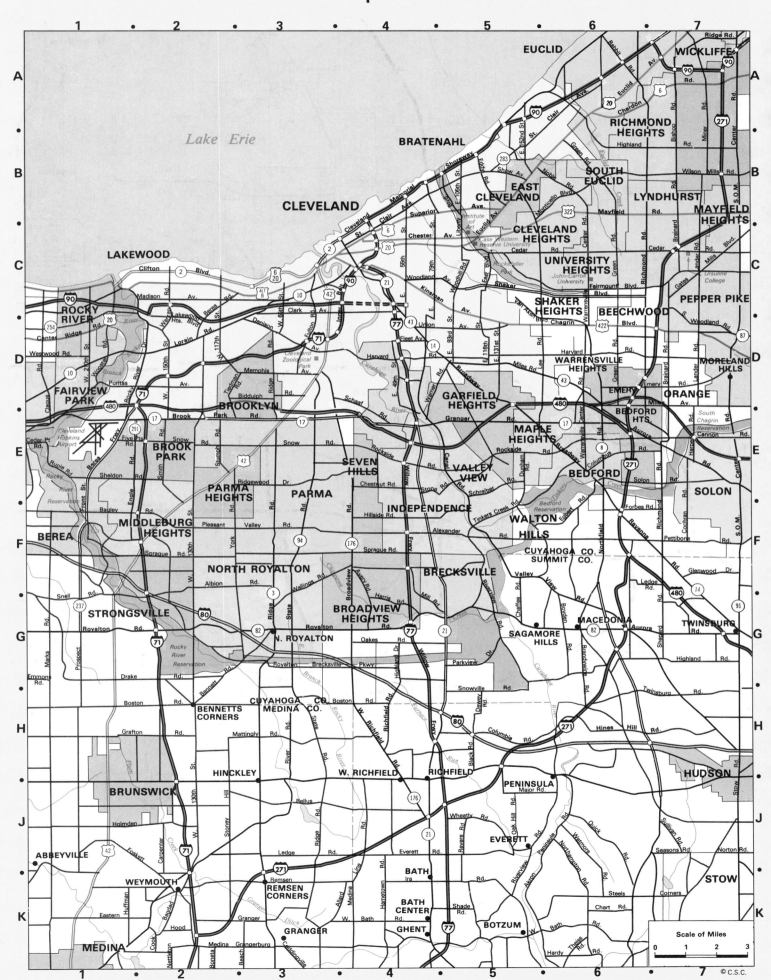

Cleveland, OH 107

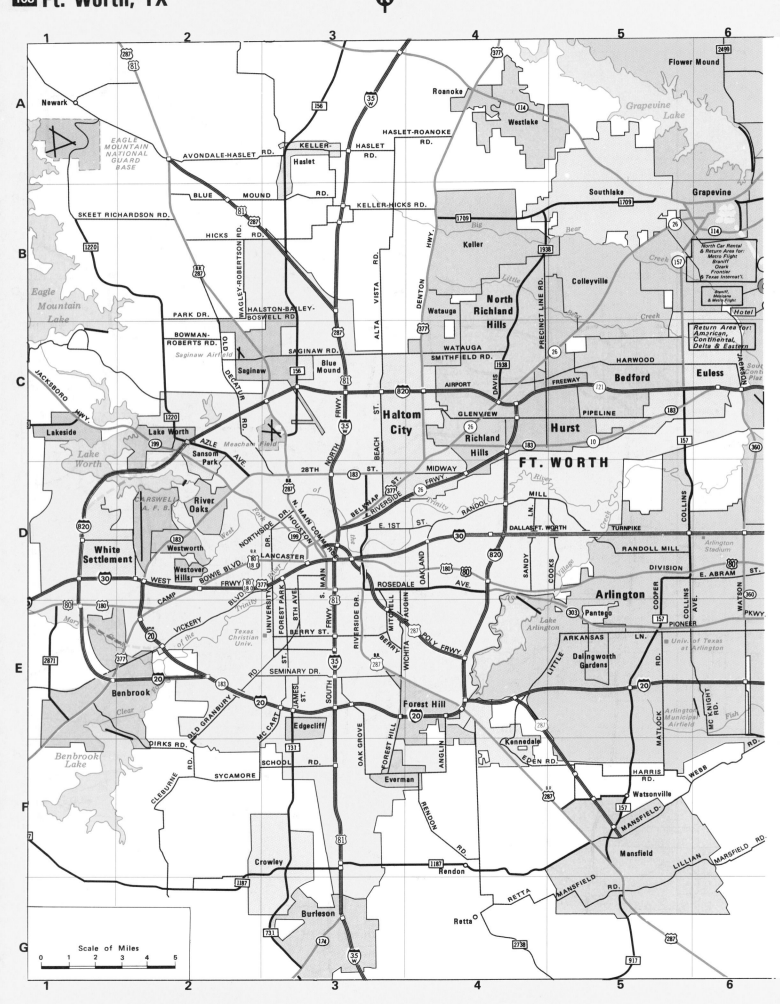

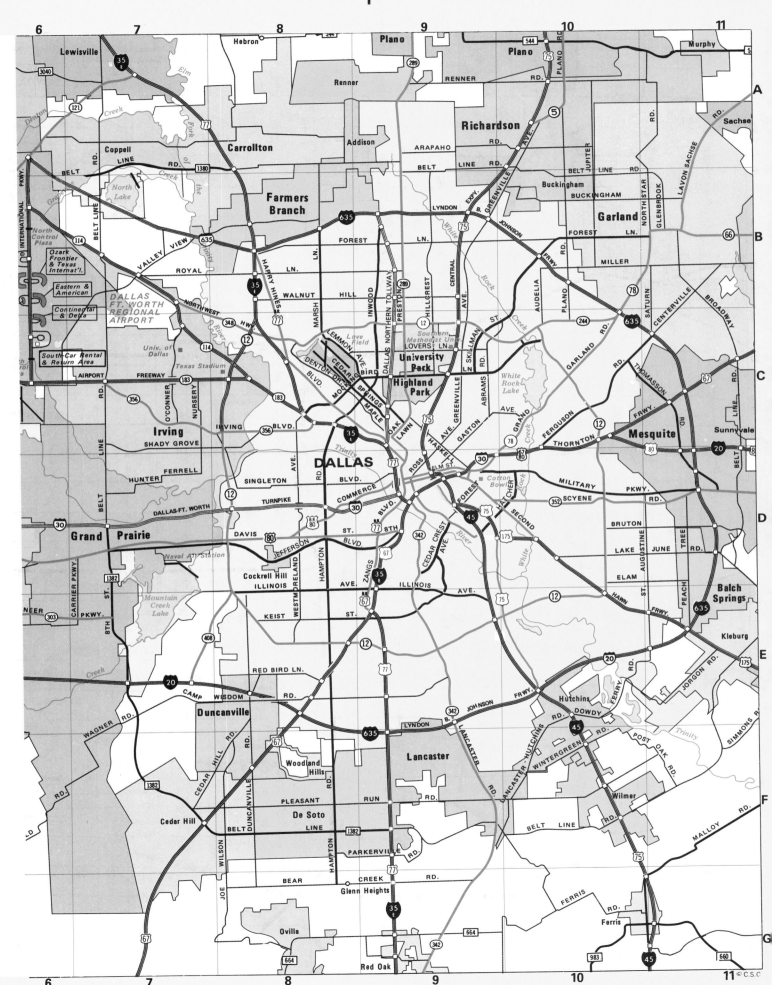

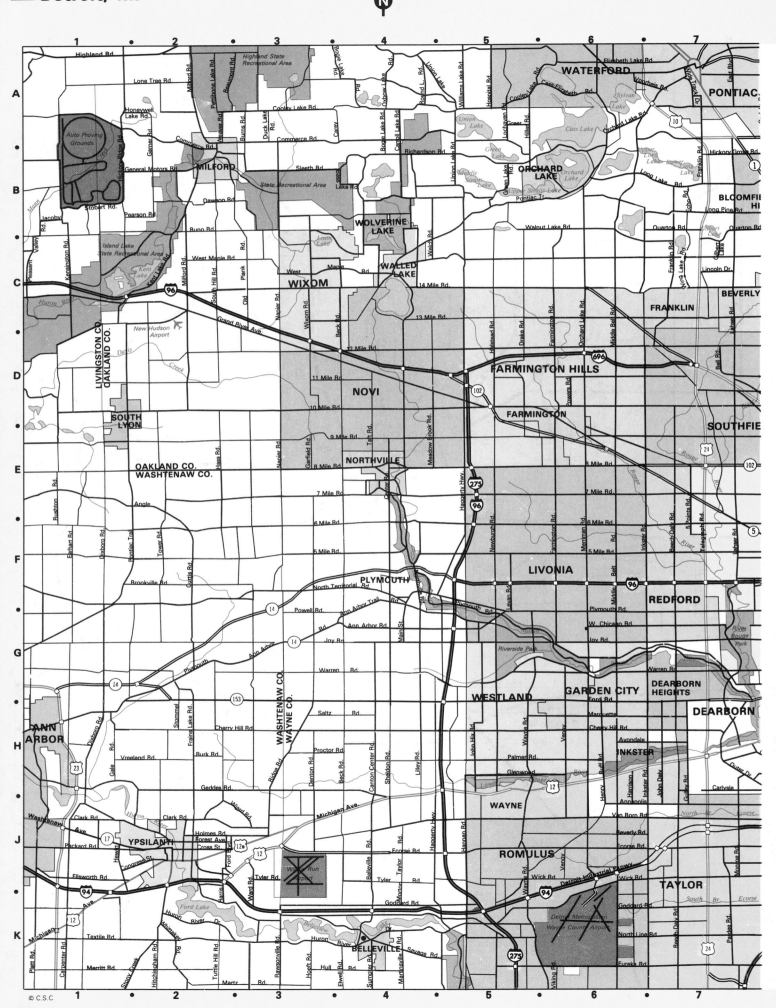

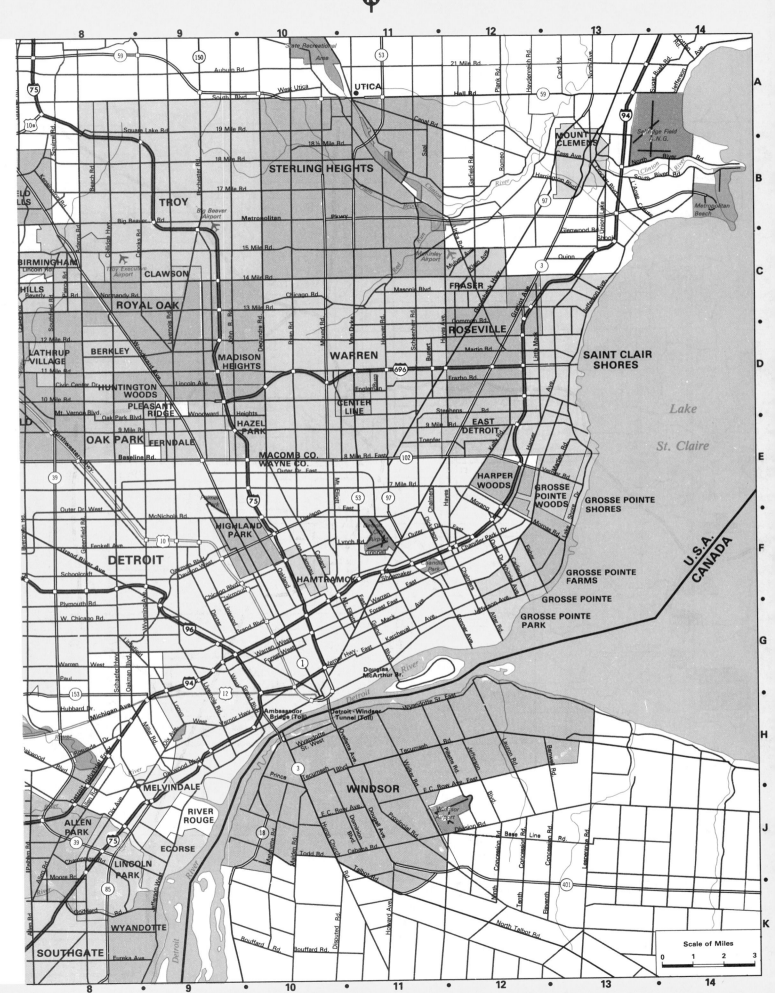

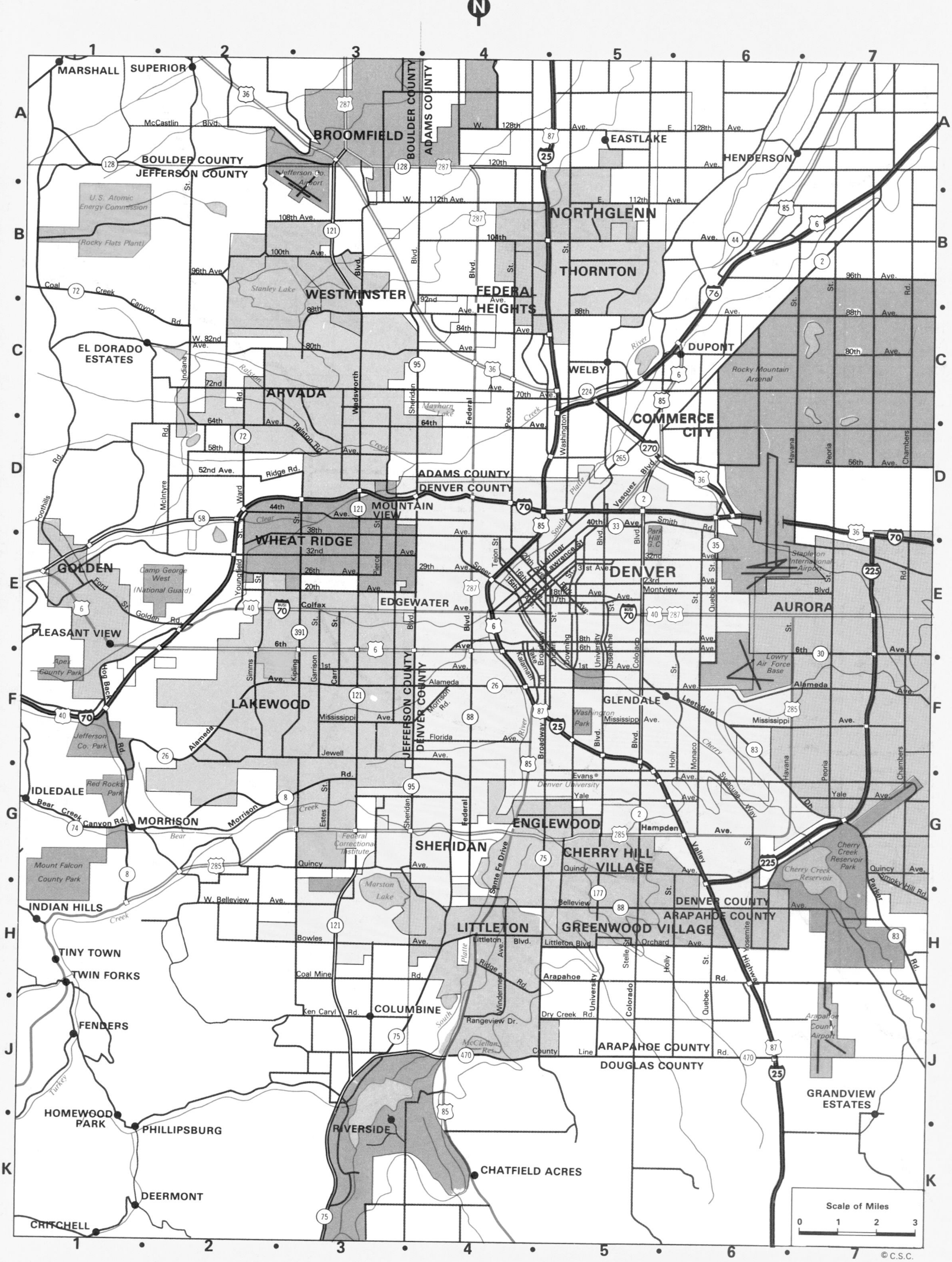

N

MARSHALL SUPERIOR

BROOMFIELD

BOULDER COUNTY
JEFFERSON COUNTY

McCaslin Blvd.

128

U.S. Atomic Energy Commission

Rocky Flats Plant

BOULDER COUNTY
ADAMS COUNTY

EASTLAKE

HENDERSON

W. 128th Ave.   E. 128th Ave.

120th

W. 112th Ave.   E. 112th Ave.

NORTHGLENN

108th Ave.

Jefferson Co. Airport

100th Ave.

104th

96th Ave.

88th

WESTMINSTER

Coal Creek Canyon Rd.

72

Stanley Lake

W. 82nd Ave.

88th

THORNTON

FEDERAL HEIGHTS

92nd

84th

EL DORADO ESTATES

80th

ARVADA

72nd

64th Ave.

58th Ave.

Mayhorn Lake

Ralston Rd.

64th

WELBY

DUPONT

80th Ave.

56th Ave.

Rocky Mountain Arsenal

COMMERCE CITY

52nd Ave.

Ridge Rd.

ADAMS COUNTY
DENVER COUNTY

44th

MOUNTAIN VIEW

GOLDEN

Camp George West (National Guard)

WHEAT RIDGE

38th

32nd

40th Ave.

32nd

DENVER

Park Hill G.C.

Stapleton International Airport

PLEASANT VIEW

26th

20th

Colfax

EDGEWATER

29th

Montview

AURORA

6th

LAKEWOOD

Mississippi

Florida

Jewell

Alameda

GLENDALE

Lowry Air Force Base

6th Ave.

Alameda

IDLEDALE

Red Rocks Park

MORRISON

JEFFERSON COUNTY
DENVER COUNTY

Morrison Rd.

Washington Park

Denver University

Mississippi Ave.

Mississippi

Bear Creek

Mount Falcon County Park

Federal Correctional Institute

SHERIDAN

ENGLEWOOD

Hampden

Evans

Yale

CHERRY HILL VILLAGE

Cherry Creek Reservoir Park

INDIAN HILLS

W. Belleview Ave.

Marston Lake

Quincy

Belleview

Quincy

DENVER COUNTY
ARAPAHOE COUNTY

Cherry Creek Reservoir

TINY TOWN

TWIN FORKS

Bowles

LITTLETON

GREENWOOD VILLAGE

Littleton Blvd.

Orchard

Coal Mine Rd.

Arapahoe

FENDERS

Ken Caryl Rd.

COLUMBINE

Rangeview Dr.

Dry Creek Rd.

ARAPAHOE COUNTY
DOUGLAS COUNTY

Arapahoe County Airport

GRANDVIEW ESTATES

HOMEWOOD PARK

PHILLIPSBURG

RIVERSIDE

McClellan Res.

CHATFIELD ACRES

DEERMONT

CRITCHELL

Scale of Miles
0 1 2 3

© C.S.C.

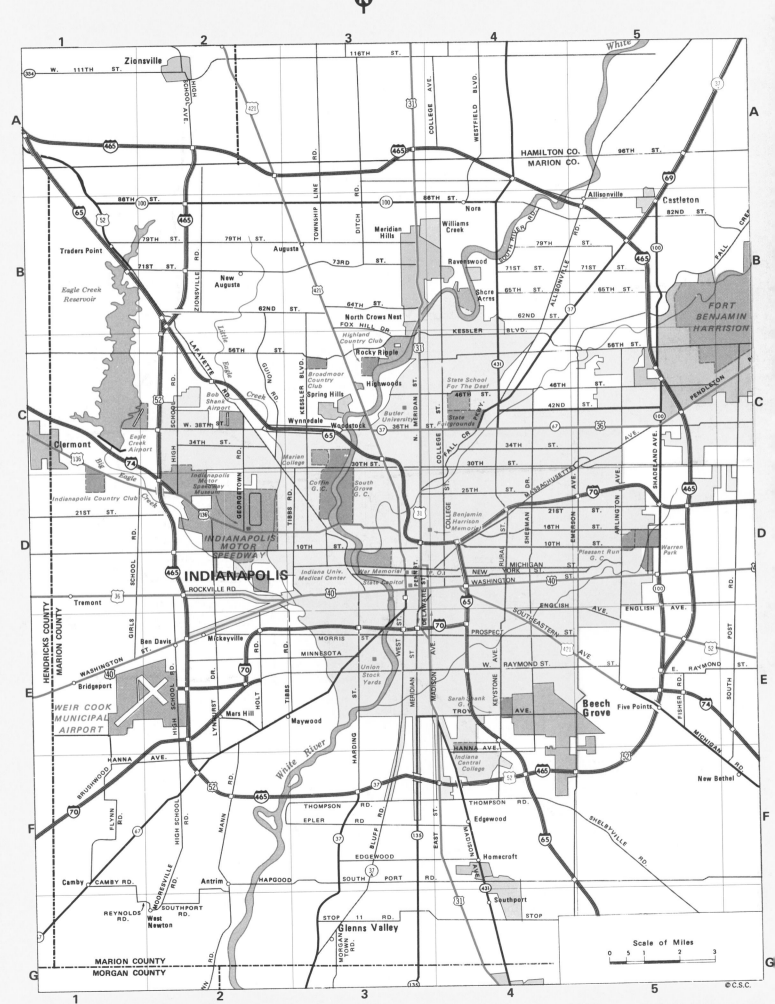

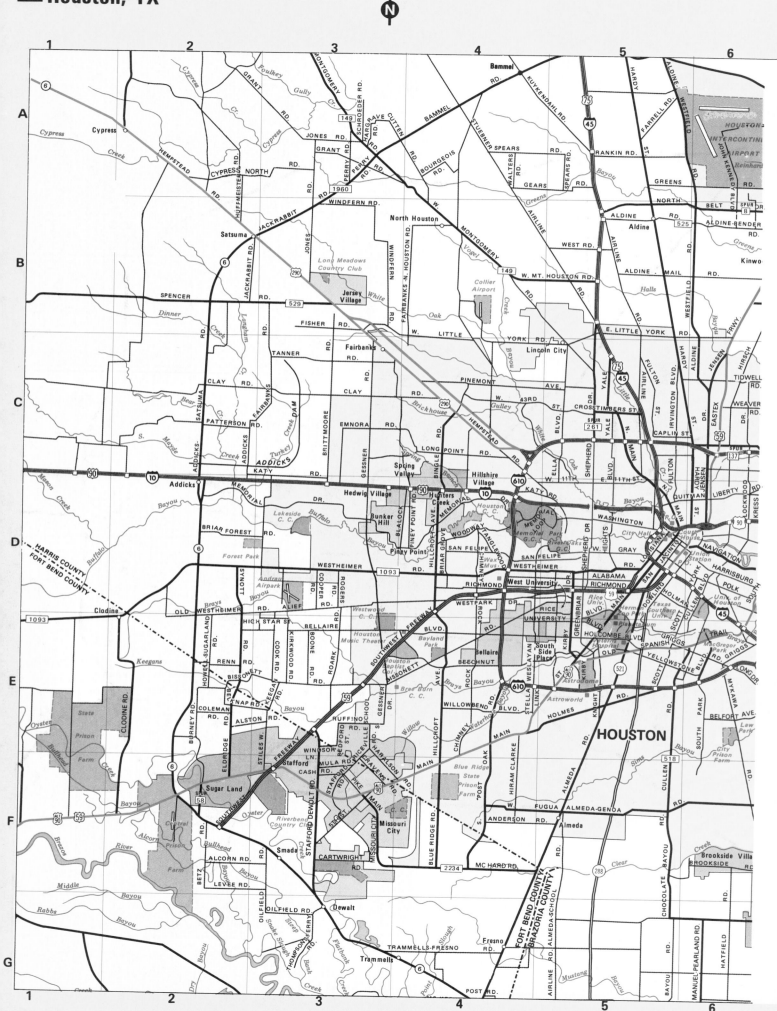

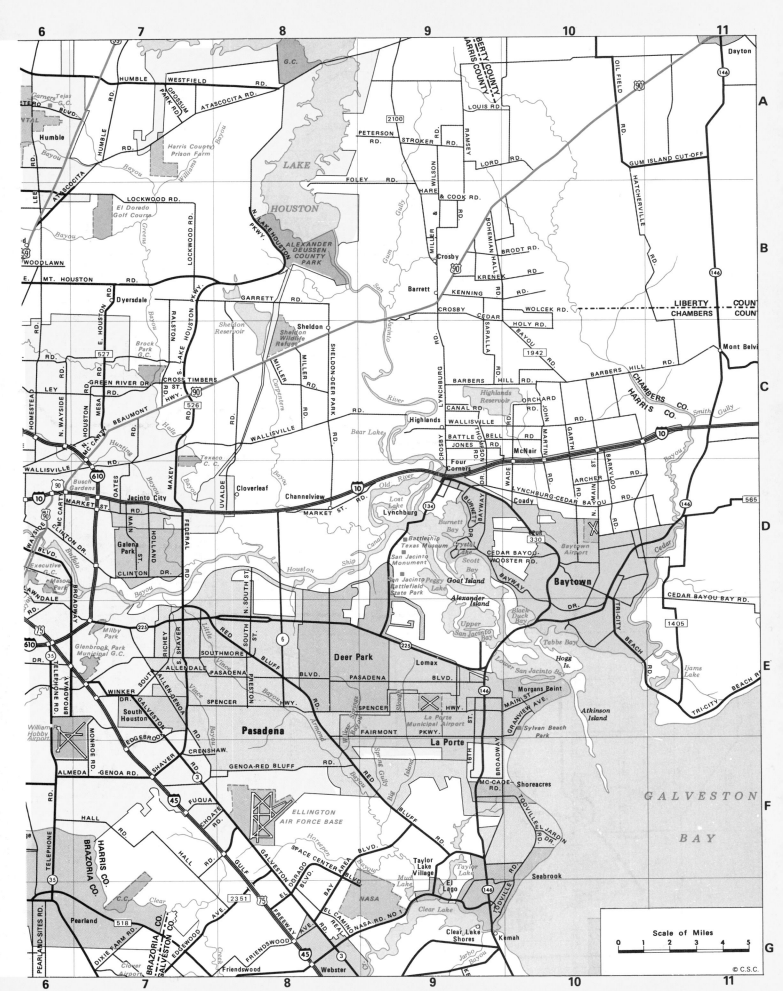

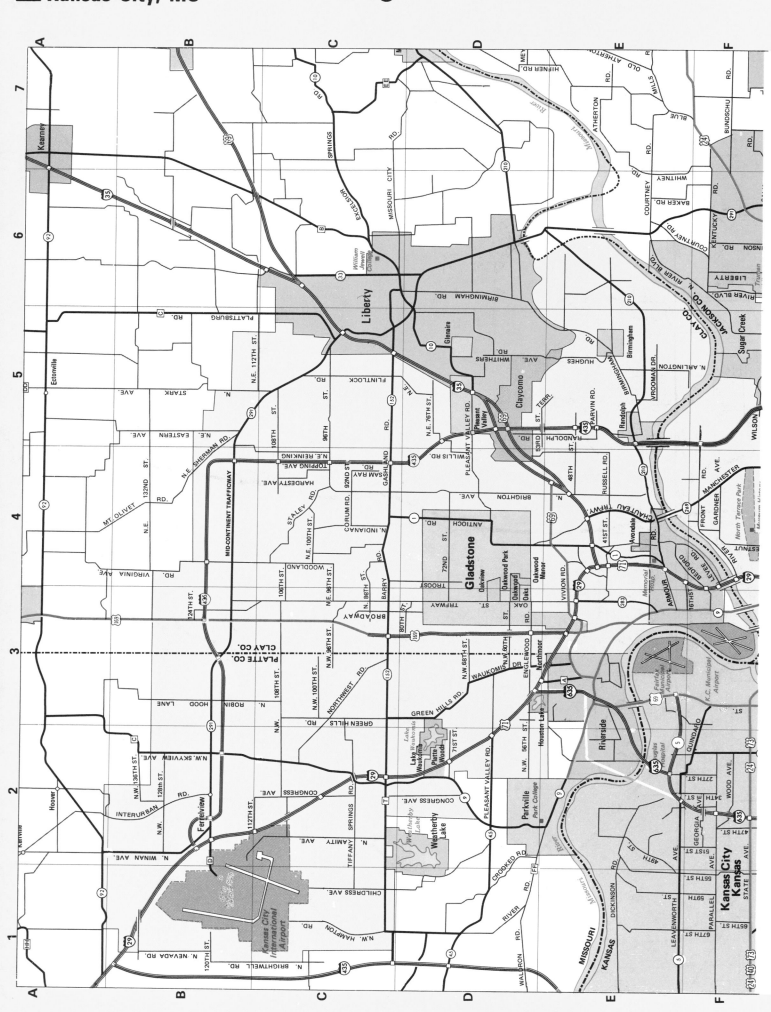

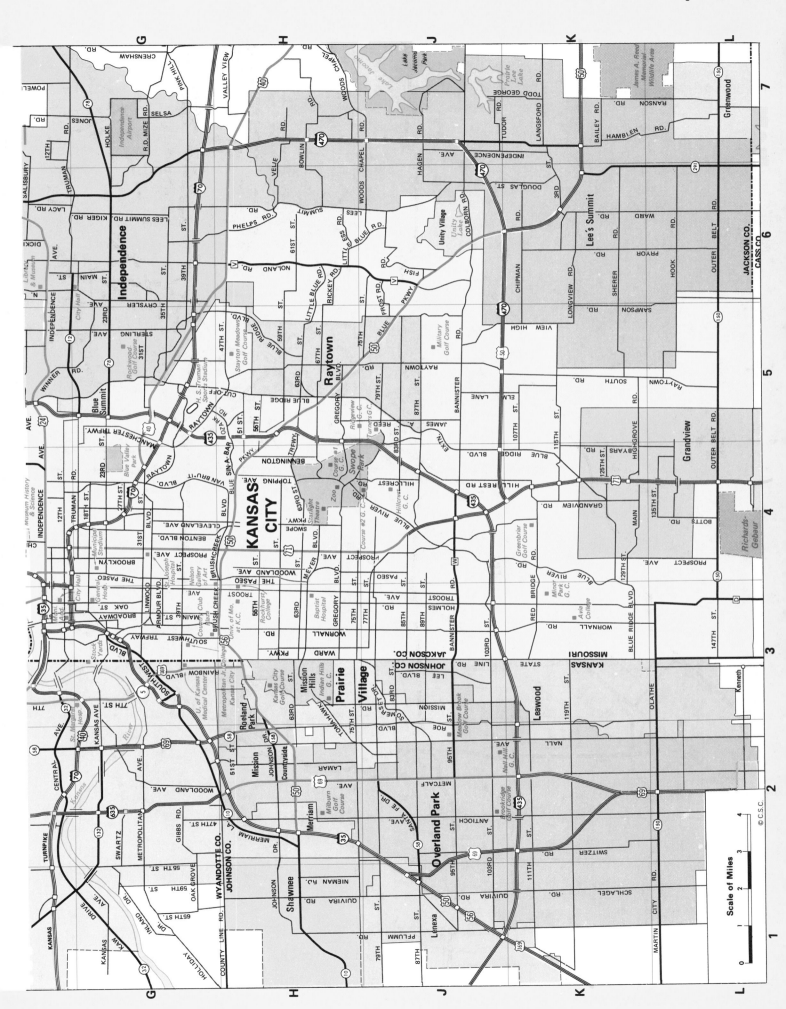

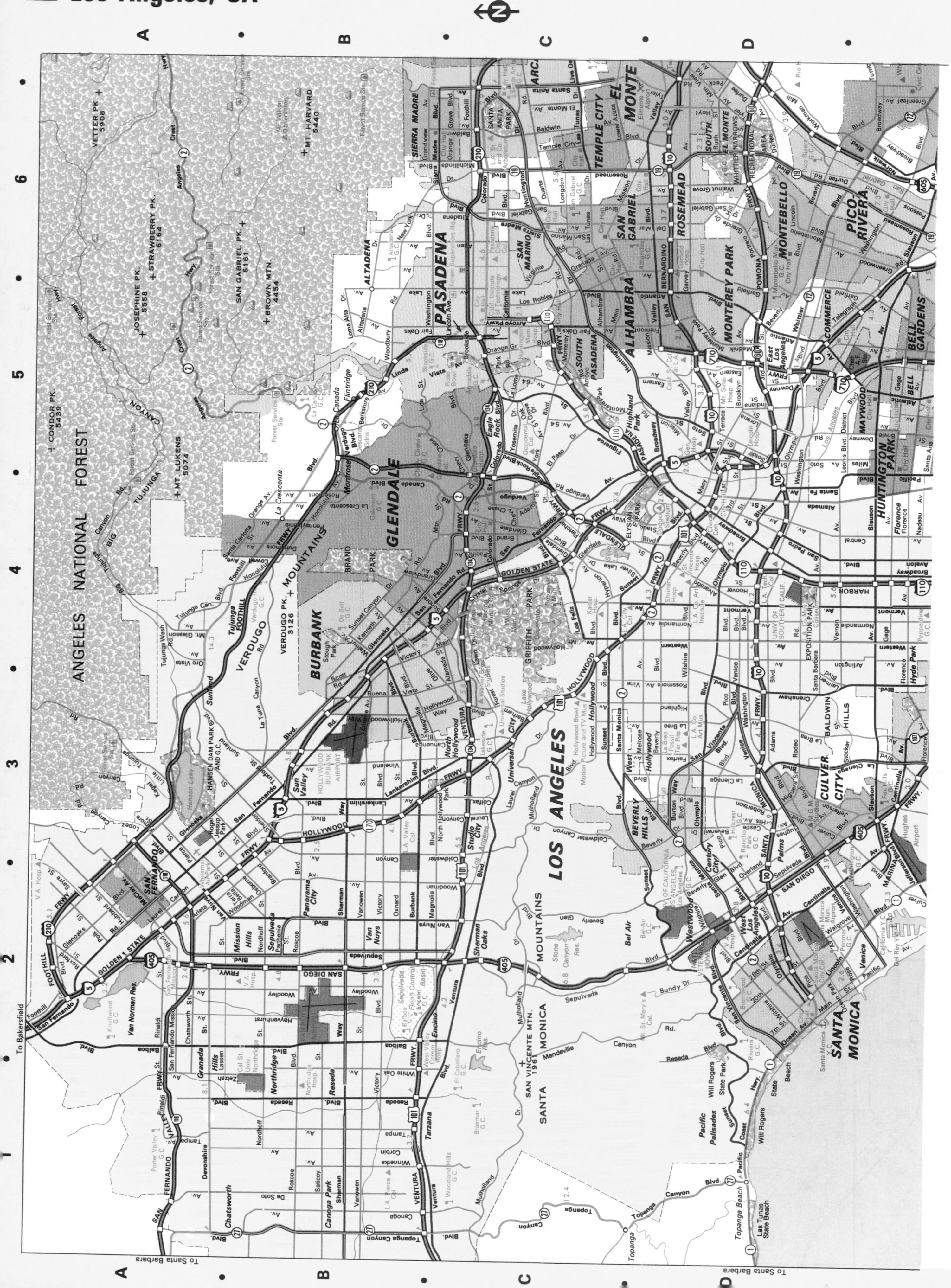

© C.S.C.

ANGELES NATIONAL FOREST

VETTER PK. 5908
JOSEPHINE PK. 5558
STRAWBERRY PK. 6164
SAN GABRIEL PK. 6161
BROWN MTN. 4454
CONDOR PK. 5439
MT. LUKENS 5074

ALTADENA
PASADENA
SAN MARINO
SIERRA MADRE
ARCADIA
TEMPLE CITY
EL MONTE
SOUTH EL MONTE
SAN GABRIEL
ROSEMEAD
ALHAMBRA
MONTEREY PARK
MONTEBELLO
PICO RIVERA
COMMERCE
BELL GARDENS
MAYWOOD
BELL
HUNTINGTON PARK
EAST LOS ANGELES

GLENDALE
BURBANK
VERDUGO MOUNTAINS
VERDUGO PK. 3126
FOOTHILL

LOS ANGELES
GRIFFITH PARK
GOLDEN STATE FRWY
HOLLYWOOD
North Hollywood
Studio City
Universal City

SAN FERNANDO
GOLDEN STATE
Panorama City
Sherman Oaks
Van Nuys
Mission Hills
Sepulveda
Northridge
Reseda
Tarzana
Encino

BEVERLY HILLS
Century City
Westwood
West Los Angeles
CULVER CITY
BALDWIN HILLS
MARINA DEL REY

SANTA MONICA MOUNTAINS
San Vicente Mtn. 1961
Bel Air
Pacific Palisades
Will Rogers State Park

SANTA MONICA
Topanga Beach
Las Tunas State Beach

To Bakersfield
To Santa Barbara

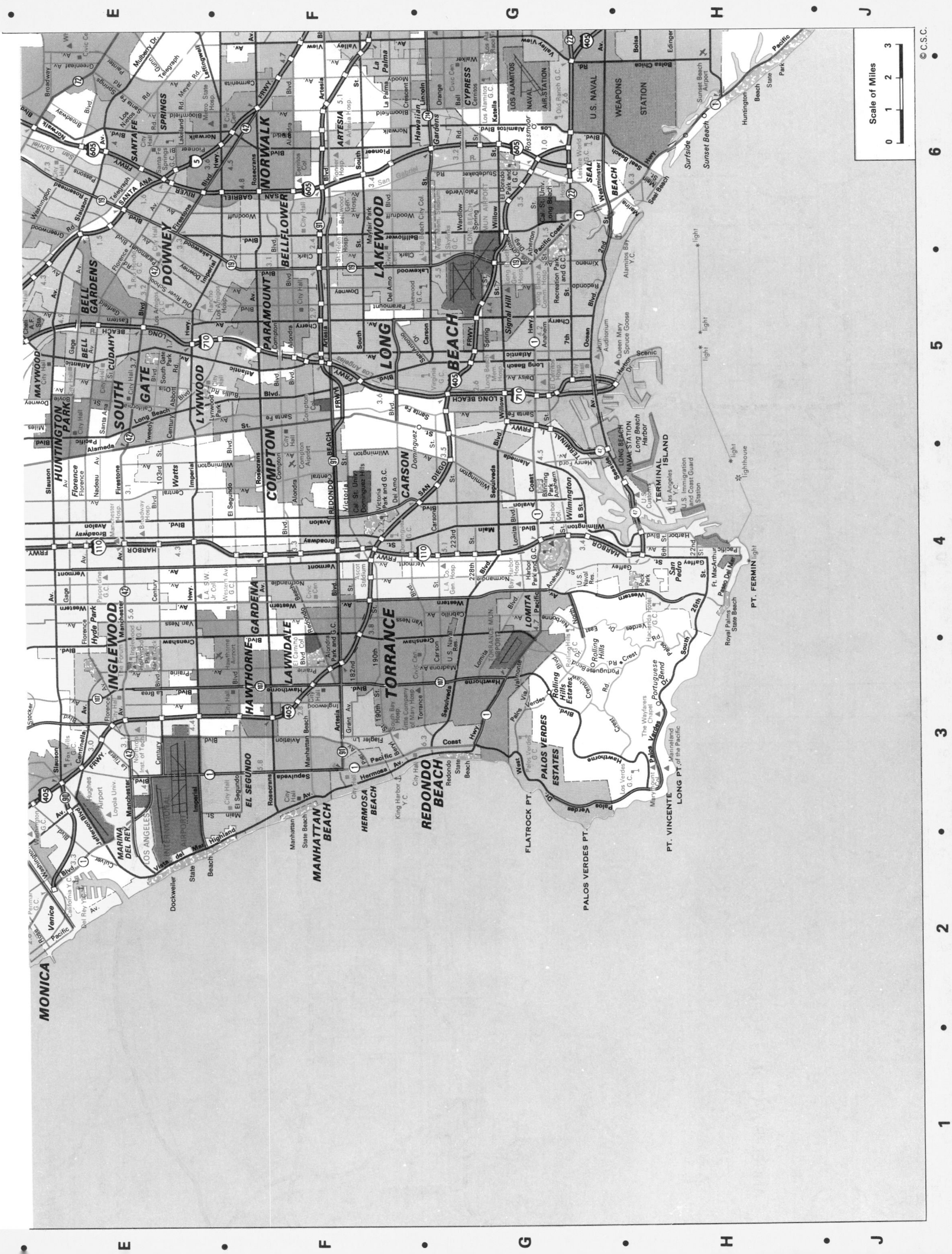

E   F   G   H   J

© C.S.C.

Scale of Miles

0   1   2   3

6

MONICA

SANTA FE SPRINGS

NORWALK

SANTA ANA

BELLFLOWER

DOWNEY

BELL GARDENS

CUDAHY

PARAMOUNT

LAKEWOOD

CYPRESS

LOS ALAMITOS

U.S. NAVAL AIR STATION

U.S. NAVAL WEAPONS STATION

SEAL BEACH

MAYWOOD

BELL

SOUTH GATE

LYNWOOD

LONG BEACH

SIGNAL HILL

HUNTINGTON PARK

COMPTON

WATTS

CARSON

WILMINGTON

TERMINAL ISLAND

LONG BEACH NAVAL STATION

Long Beach Harbor

GARDENA

HAWTHORNE

LAWNDALE

TORRANCE

LOMITA

INGLEWOOD

Hyde Park

EL SEGUNDO

MANHATTAN BEACH

HERMOSA BEACH

REDONDO BEACH

PALOS VERDES ESTATES

Rolling Hills Estates

Rolling Hills

PALOS VERDES

San Pedro

Palos Verdes

PT. FERMIN

PT. VINCENTE

LONG PT.

FLATROCK PT.

PALOS VERDES PT.

MARINA DEL REY

Venice

Dockweiler

LOS ANGELES INTERNATIONAL AIRPORT

Loyola Univ.

Culver

Pacific

Huntington Beach

Sunset Beach

Bolsa Chica

Surfside

2   1

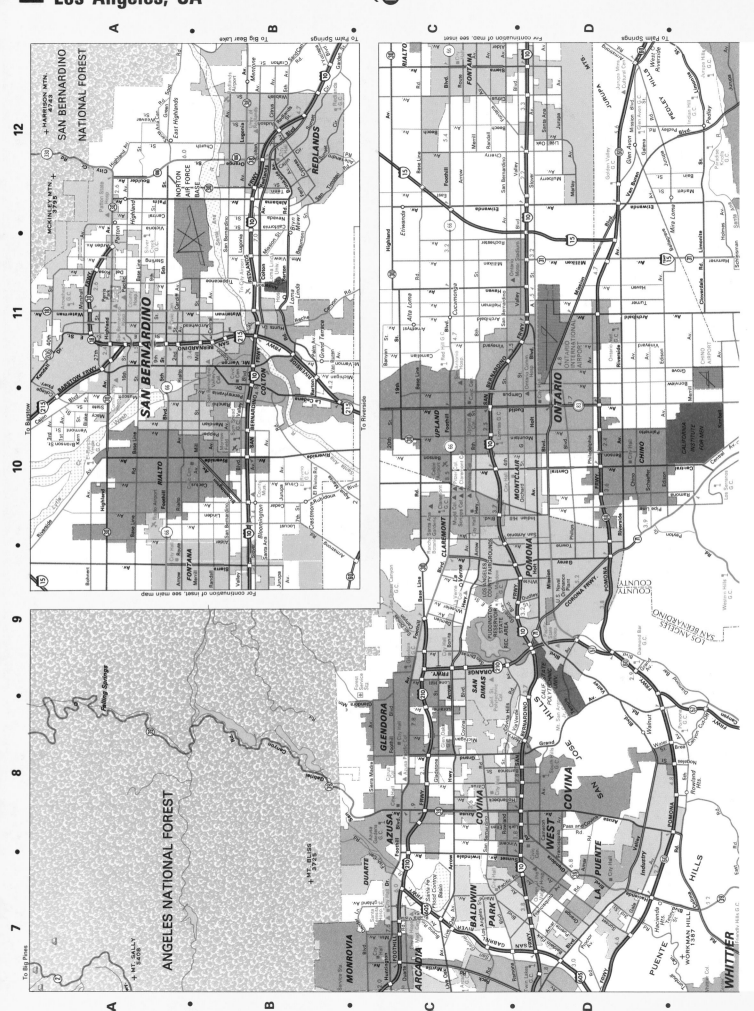

Scale of Miles
0 1 2 3

© C.S.C.

RIVERSIDE
RIVERSIDE MUNICIPAL AIRPORT
HOME GARDENS
NORCO
CORONA
U.S. NAVAL RESERVATION
EL CERRITO
CLEVELAND NATIONAL FOREST
LAKE MATHEWS
CHINO AIRPORT
CALIF. INST. FOR WOMEN
PRADO FLOOD CONTROL BASIN
CALIFORNIA INSTITUTE FOR MEN
SAN BERNARDINO COUNTY
RIVERSIDE COUNTY
CHINO HILLS
SAN BERNARDINO COUNTY
ORANGE COUNTY
LOS ANGELES COUNTY
ORANGE COUNTY
WORKMAN HILL 1387
ROWLAND HTS.
LA HABRA
BREA
FULLERTON
BUENA PARK
LA MIRADA
PLACENTIA
YORBA LINDA
ANAHEIM
GARDEN GROVE
STANTON
WESTMINSTER
HUNTINGTON BEACH
FOUNTAIN VALLEY
ORANGE
SANTA ANA
TUSTIN
COSTA MESA
NEWPORT BEACH
IRVINE
SANTA ANA U.S.M.C. AIR FACILITY
EL TORO U.S.M.C. AIR STATION
UNIV. OF CALIFORNIA IRVINE CAMPUS
LAKE FOREST
Silverado
Trabuco Canyon
Santiago Res.
Villa Park Res.
To San Bernardino
To San Diego
To San Clemente

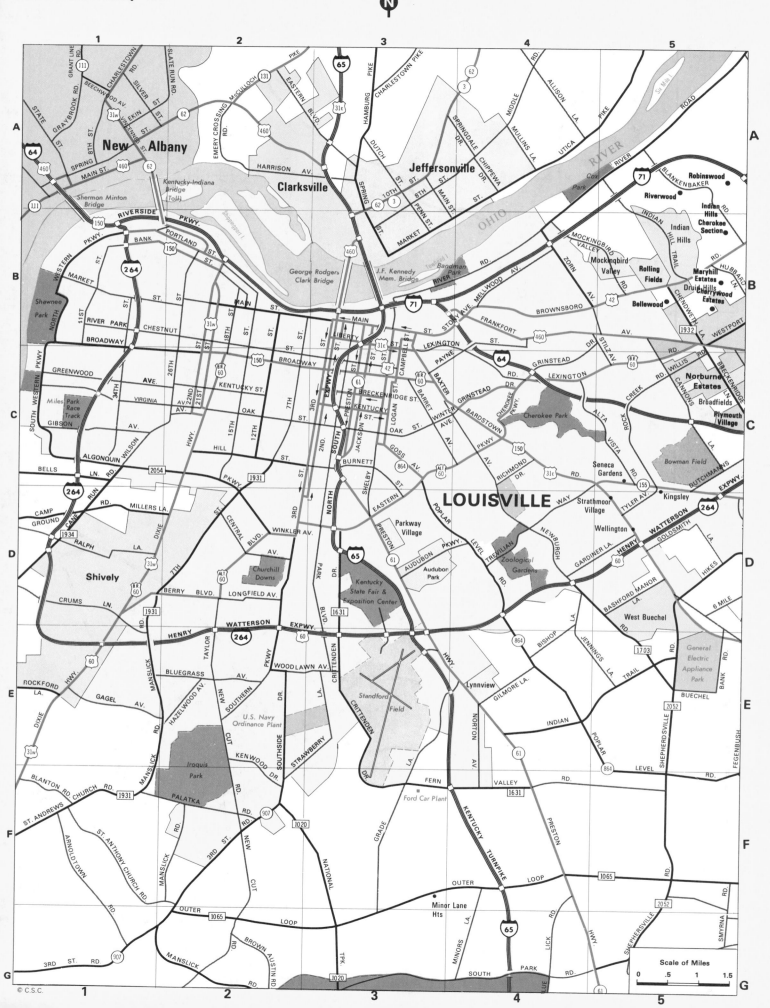

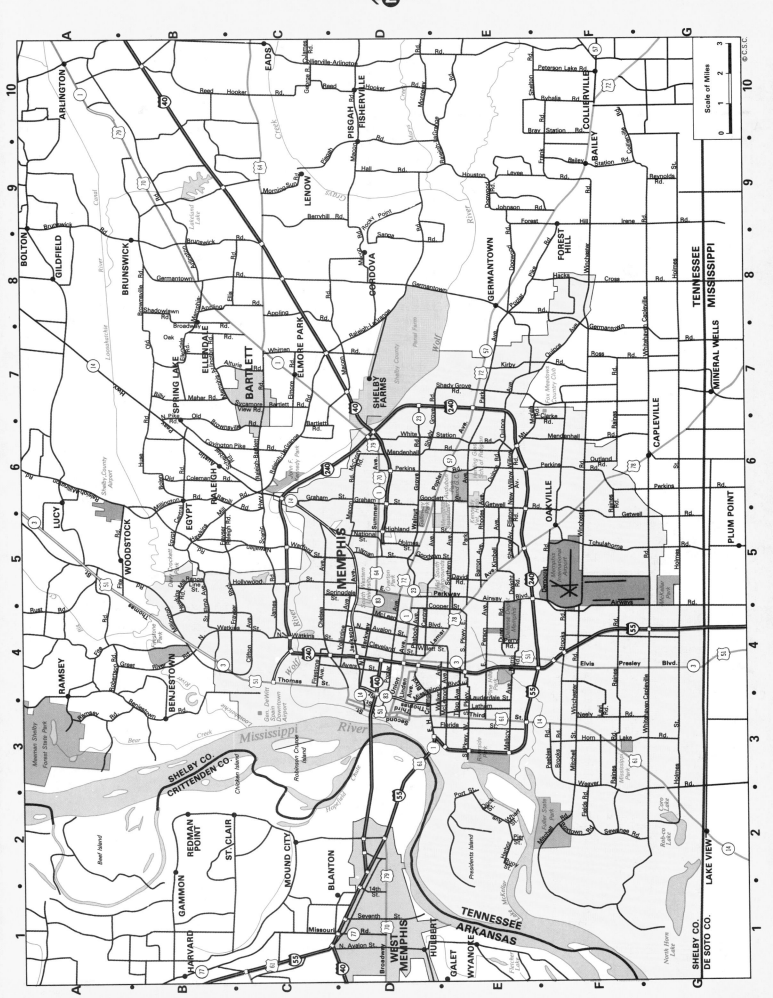

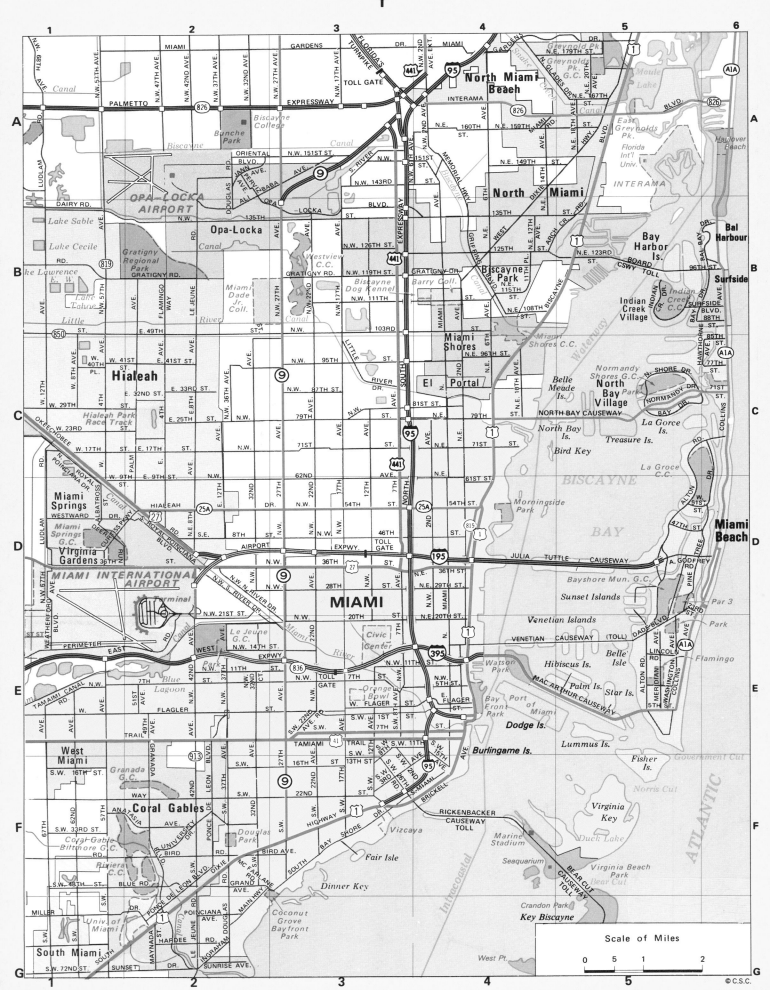

Scale of Miles

© C.S.C.

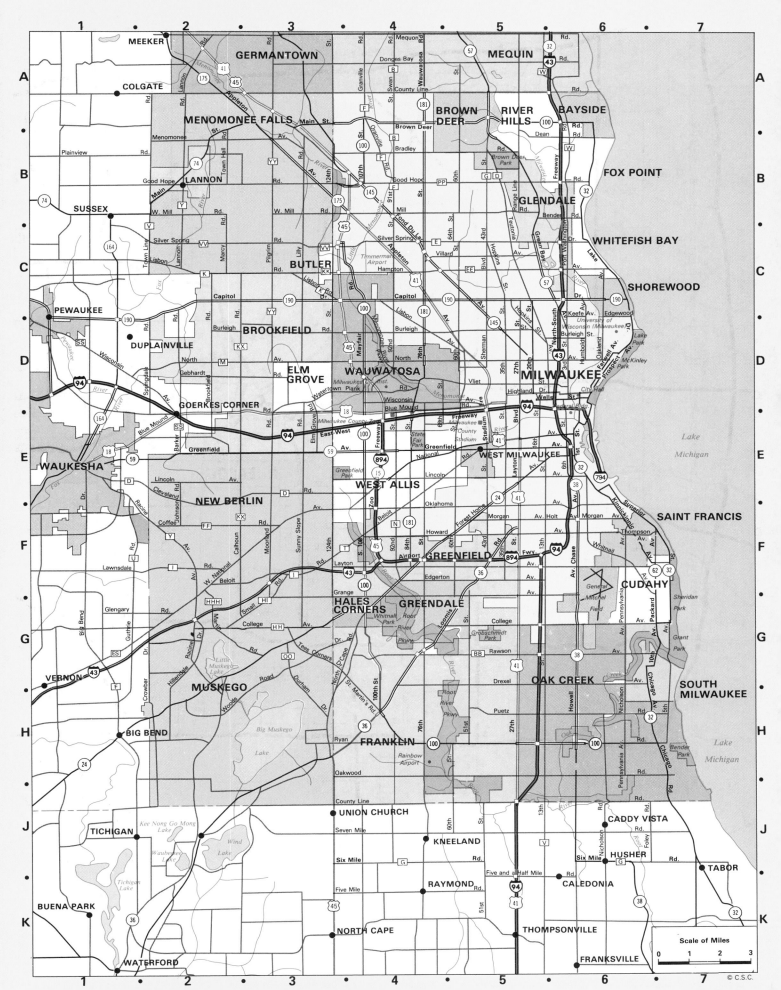

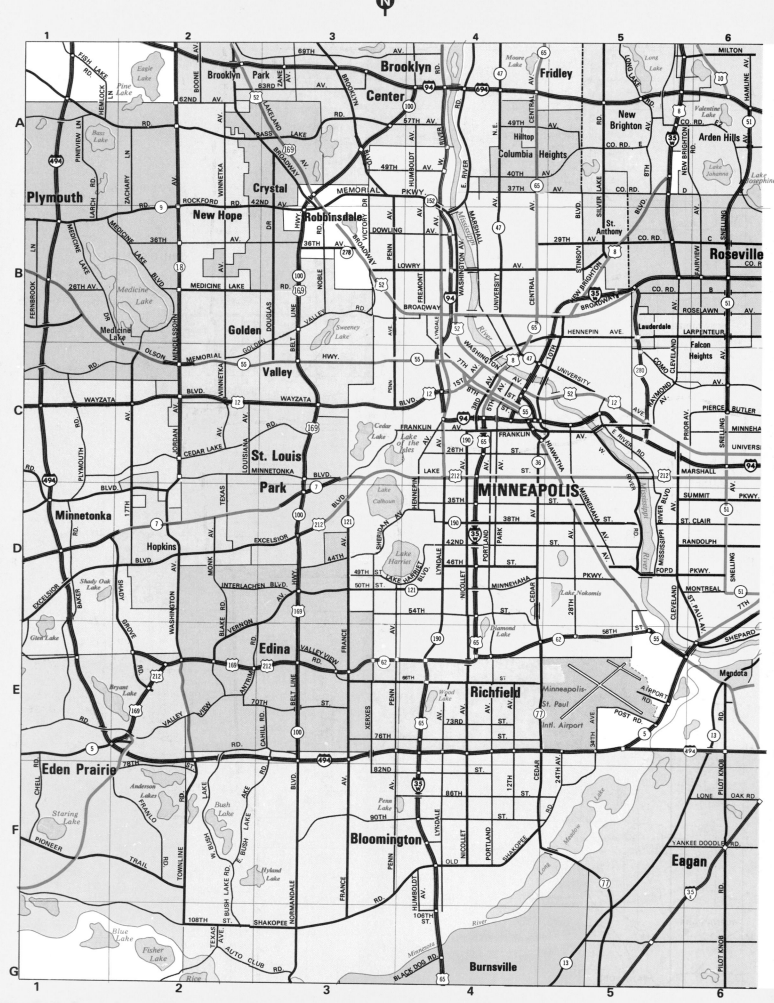

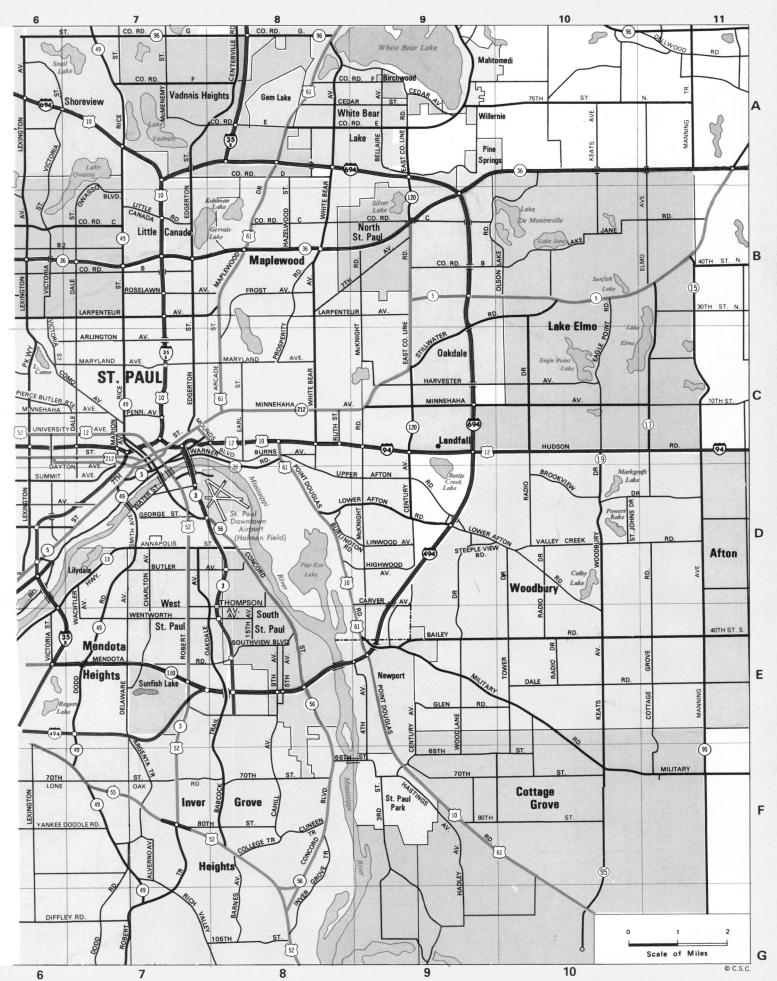

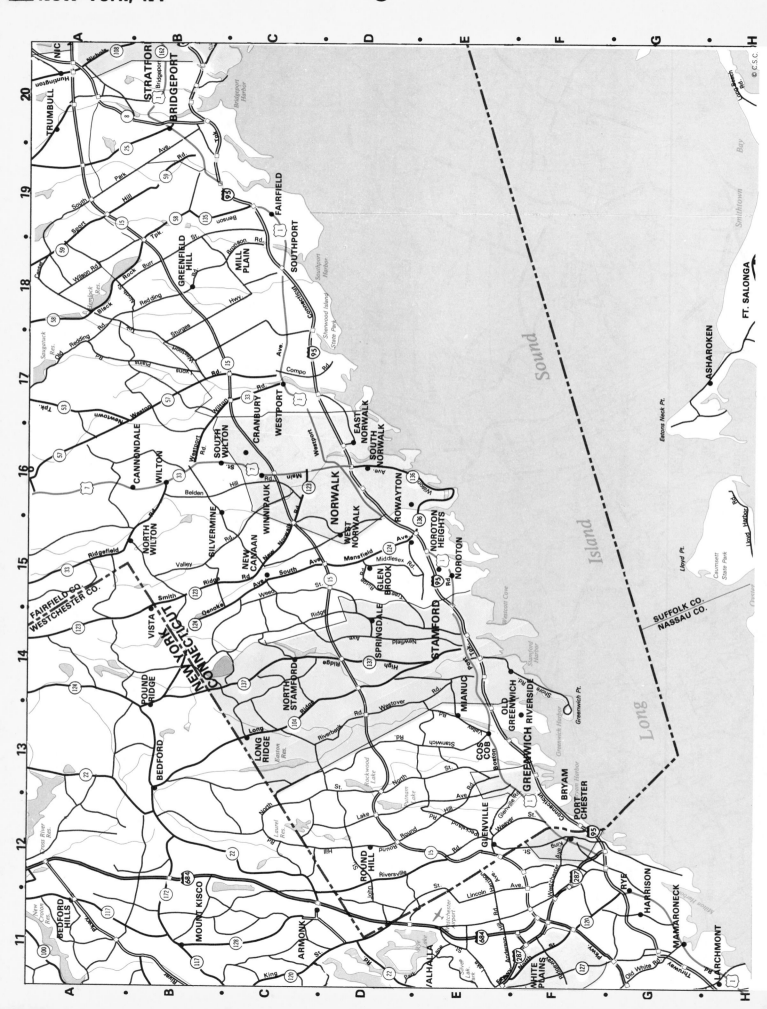

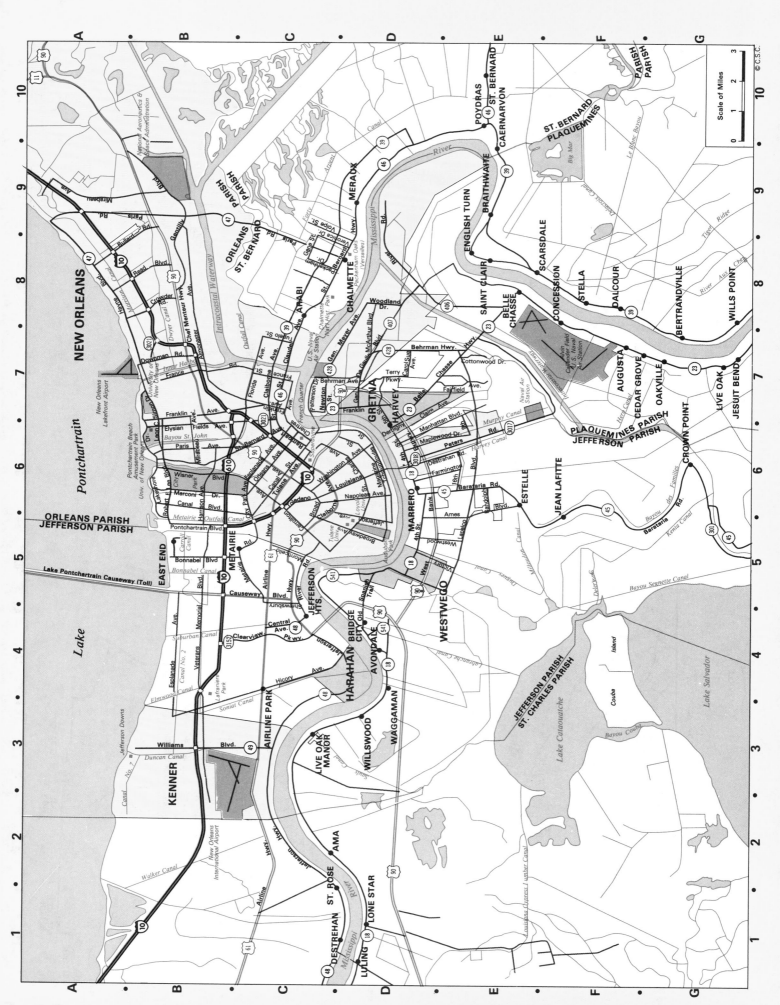

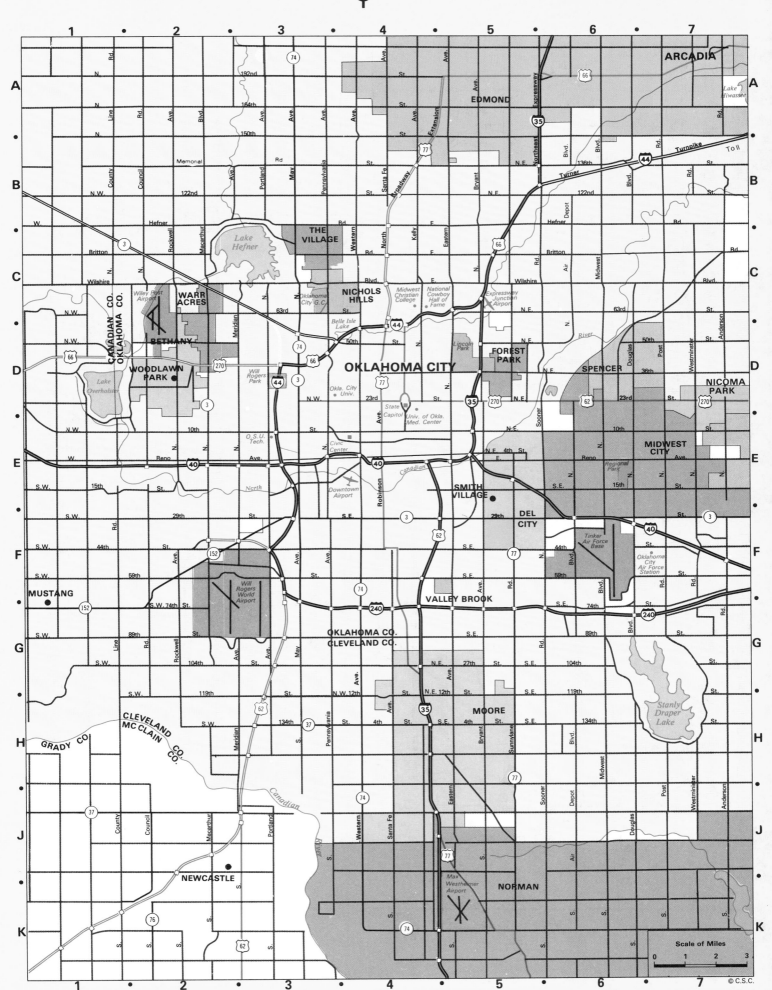

N

A

B

C

D

E

F

G

1 2 3 4 5 6

Beardsley Canal
McMicken Dam Outlet Canal

CAVE CREEK

PINNACLE PEAK RD.

PINNACLE PEAK RD.

Currys Corner

Deer Valley Airport

Adobe
DEER VALLEY RD.

Thunderbird
Regional Park

BEARDSLEY RD.

BEARDSLEY RD.

SCOTTSDALE RD.

89 Beardsley

93

UNION HILLS DR.
W. UNION HILLS RD.

BELL

Paradise
City

Paradise Valley
Park

Scottsdale
Mun. Airport

Surprise

GREENWAY RD.

GREENWAY RD.

Turf Paradise
Race Track

32ND ST.

GREENWAY RD.

El Mirage

American Inst.
for Foreign Trade

Moon
Valley
C. C.

40TH ST.

48TH ST.

56TH ST.

Scottsdale
Century
C. C.

GREENWAY
WADDELL

BULLARD

LITCHFIELD

DYSART RD.

EL MIRAGE RD.

OLIVE

THUNDERBIRD

75TH AVE.

67TH AVE.

59TH AVE.

43RD AVE.

35TH AVE.

BLACK CANYON HWY.

7TH ST.

CAVE CREEK RD.

NORTHERN AVE.

CACTUS RD.

SHEA BLVD.

Sun City

CACTUS

17

North
Mountain
Park

64TH ST.

Scottsdale

Youngstown

PEORIA AVE.

Squaw Peak
Park

Paradise
Valley

Scottsdale
C. C.

Peoria
DUNLAP

Glendale
Com. Col.

NORTHERN

Paradise
Valley
C. C.

MOCKINGBIRD LN.

HAYDEN

Glendale

GLENDALE

19TH AVE.

PHOENIX

LINCOLN

McDONALD DR.

91ST AVE.

BETHANY HOME RD.

BETHANY HOME RD.

7TH ST.

16TH ST.

Arizona
Biltmore

INVERGORDON RD.

CAMELBACK

60

89

Grand
Canyon
Col.

CAMELBACK

24TH ST.

32ND ST.

44TH ST.

56TH ST.

SCOTTSDALE RD.

HAYDEN RD.

CAMELBACK

93

INDIAN SCHOOL RD.

Mun.
G.C.

SCHOOL RD.

Arizona
C.C.

INDIAN SCHOOL

Litchfield Park

Irrigation District

THOMAS RD.

Enchanto
Park

Heard Mus.

Phoenix
C.C.

32ND ST.

56TH ST.

Military Res.

MC DOWELL

123RD AVE.

McDOWELL

Heard Mus.

McDOWELL

Canal

83RD AVE.

75TH

67TH AVE.

51ST AVE.

State Fair
Grounds

State
St. Hosp.

Arizona
C.C.

Papago
Park

Zoological
Park

VAN BUREN ST.

VAN BUREN

Arizona
St. Hosp.

89

Goodyear

WAR ST.

Phoenix
Litchfield
Airfield

Avondale

Cashion

BUCKEYE RD.

State Capitol

Mun.
Bldg.

WASHINGTON ST.

Mun.
Sta.

Temple
Park

Tolleson

85

17

Zoological
Park

University

BULLARD

115TH

107TH

99TH AVE.

91ST AVE.

83RD AVE.

75TH

67TH

LOWER BUCKEYE RD.

BUCKEYE RD.

MARICOPA

10

Ariz. State
Univ.

APACHE

60

Sky Harbor
Mun. Airport

10

85

BROADWAY RD.

BROADWAY RD.

16TH ST.

24TH ST.

32ND ST.

40TH ST.

48TH ST.

PRIEST DR.

RURAL RD.

MIL AVE.

360

Tempe

Salt River (Dry)

SOUTHERN AVE.

SOUTHERN AVE.

Canal

Western

Manzanita
Speedway

35TH AVE.

27TH AVE.

19TH AVE.

7TH AVE.

Central

7TH ST.

GUADALUPE RD.

Guadalupe

ELLIOT RD.

BASE LINE RD.

DOBBINS RD.

43RD AVE.

35TH

27TH

19TH

Thunderbird C.C.

CANYON RD.

49TH ST.

WARNER RD.

Kyrene

KYRENE DR.

Laveen

ELLIOT RD.

59TH AVE.

51ST AVE.

STEPHEN RD.

MATHER RD.

TELEGRAPH PASS

BUENA VISTA RD.

RAY RD.

KYRENE

MC CLINTOCK DR.

WILLIAMS RD.

ESTRELLA DR.

San Juan Rd.

SAN JUAN RD.

Phoenix South
Mountain Park

International Harvester
Proving Ground

56TH

Canal

Chandler

Estrella
Mountain
Regional
Park

Casey Abbott
Semi-Regional
Park

GILA

Gila River

PECOS RD.

Highway

Goodyear
Air Force
Mil. Field

10

PIMA FREEWAY

MARICOPA CO.
PINAL CO.

MARICOPA CO.
PINAL CO.

INDIAN

RIVER

RESERVATION

Scale of Miles
0 1 2 3 4 5

© C.S.C.

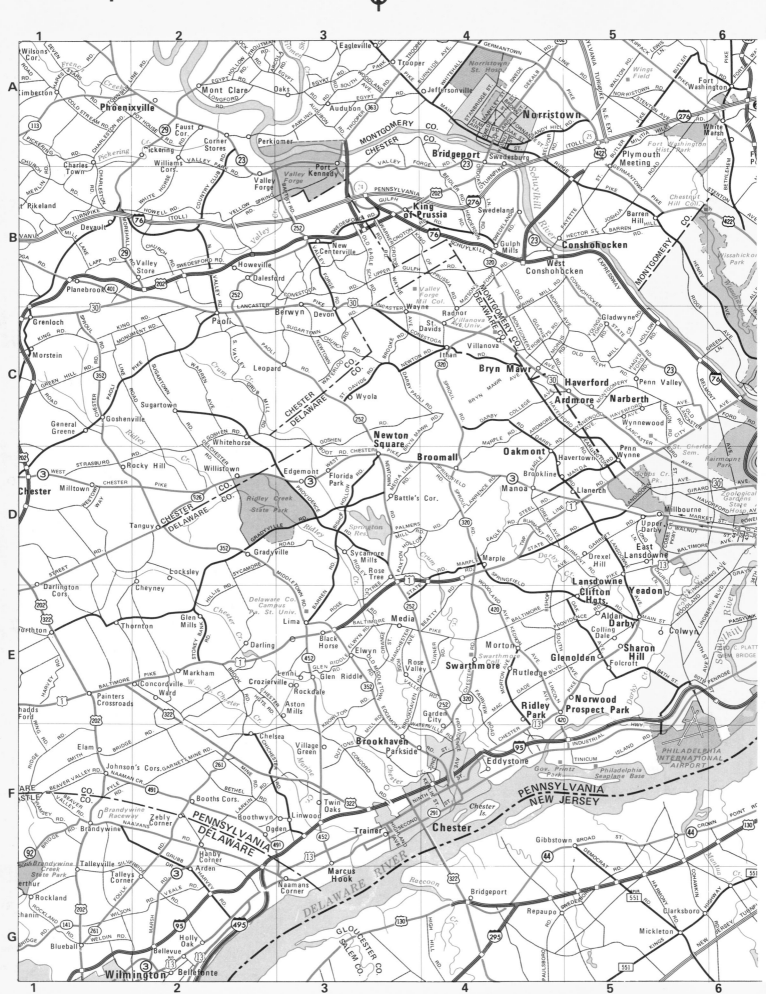

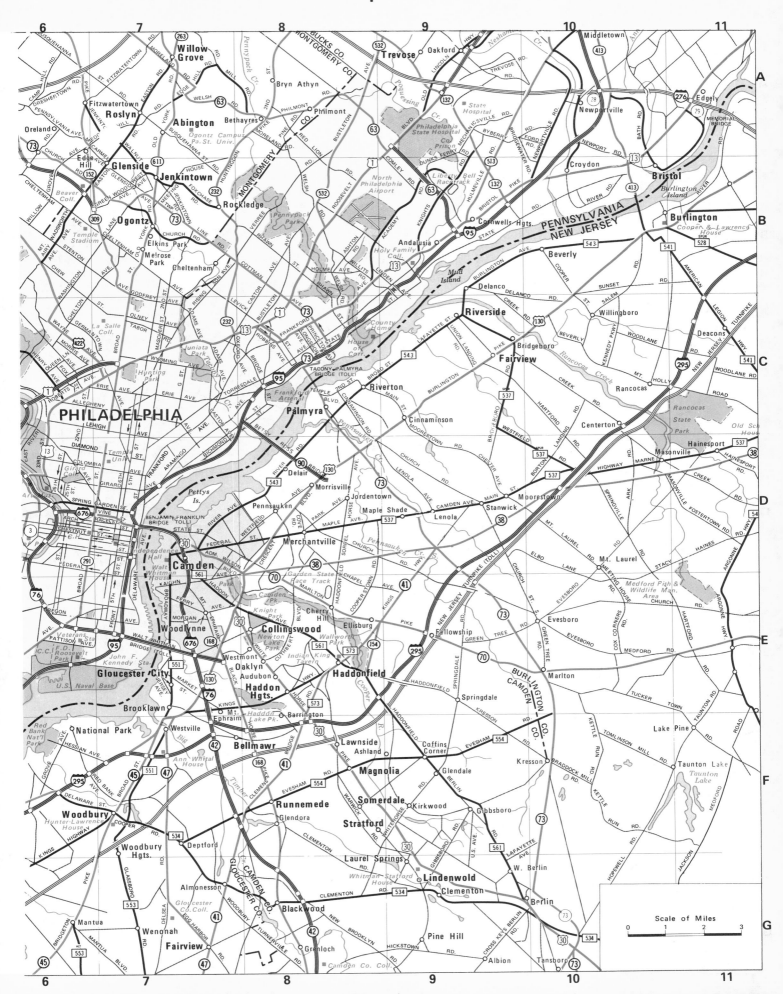

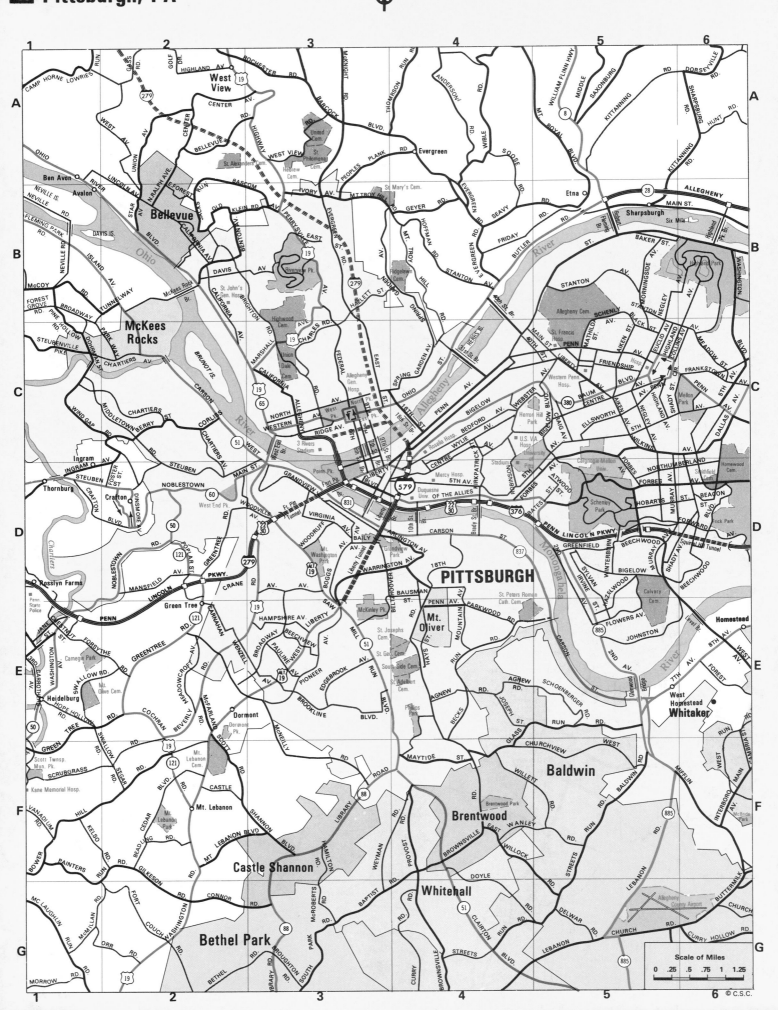

N

Vancouver

WASHINGTON
OREGON

Hayden Island

Columbia River

Smith Lake
Tomahawk Island
Exposition Center

PEARSON FIELD
E. MILL PLAIN BLVD.
500

CLARK COUNTY
MULTNOMAH COUNTY

N. FESSENDEN ST.
N. COLUMBIA BLVD.
N. PORTSMOUTH AVE.
N. LOMBARD ST.
N. WILLAMETTE BLVD.
N. PENINSULAR AVE.
Columbia Park
N. PORTLAND RD.
N. DENVER AVE.
Delta Park
Columbia Edgewater G.C.
Portland G.C.
GERTZ
SUNDERLAND
N.E. MARINE RD.
PORTLAND INT'L AIRPORT
Riverside G.C.
Broadmoor G.C.
Colwood G.C.

Univ. of Portland
N. PORTLAND AVE.
Peninsula Park
Alberta Park
N.E. LOMBARD ST.
N.E. COLUMBIA BLVD.
N.E. KILLINGSWORTH ST.

W. NEWBERRY RD.
N.W. SKYLINE BLVD.

Forest Park
PORTLAND

Swan Island
Willamette River
N. GREELEY AVE.
N. INTERSTATE AVE.
UNION
99E
N.E. FREMONT ST.
N.E. 33RD AVE.
N.E. 39TH AVE.
ALAMEDA
N.E. 42ND AVE.
N.E. CULLY RD.
N.E. 57TH
BR 30
Rose City G.C.

N.W. CORNELL RD.
N.W. SKYLINE BLVD.
MacLeay Park
N.W. VAUGHN ST.
N.W. LOVEJOY ST.
N.W. 23RD AVE.
N.W. 19TH AVE.
405
Memorial Coliseum
N.E. BROADWAY
SANDY
N.E. HALSEY ST.
84
30

Washington Park
N.E. GLISAN ST.
E. BURNSIDE
Laurelhurst Park
S.E. STARK ST.
S.E. MORRISON ST.
S.E. BELMONT ST.

JENKINS RD.
BARNES RD.
CEDAR HILLS BLVD.
S.W. CANYON RD.
26
W. HUMPHREY
Zoological Gardens and Museum
Portland State Univ.
S.W. VISTA AVE.
MARKET
26
405
99E
S.E. HAWTHORNE BLVD.
Mt. Tabor Park

West Slope
8
S.W. FATTON RD.
Council Crest
S.W. BROADWAY DR.
5
S.E. DIVISION ST.
26 POWELL

10
Raleigh Hills
BEAVERTON - HILLSDALE HWY.
S.W. SHATTUCK RD.
10
TERWILLIGER BLVD.
10
43
Ross Island
S.E. HOLGATE BLVD.
Reed College
S.E. 26TH AVE.
S.E. 39TH AVE.
S.E. 52ND AVE.
S.E. FOSTER RD.
S.E. HAROLD ST.

217
FARMINGTON RD.
Hillsdale
S.W. VERMONT ST.
Gabriel Park
Multnomah
S.W. MULTOMAH BLVD.
99
S.W. TERWILLIGER
MACADAM AVE.
Pioneer Park
Bybee Blvd.
S.E. 13TH AVE.
S.E. 17TH AVE.
S.E. TACOMA
S.E. WOODSTOCK BLVD.
Eastmoreland Golf Course
S.E. DUKE ST.
JOHNSON CREEK BLVD.
Kendall

Beaverton
SCHOLLS FERRY RD.
R.H. BALDOCK FWY.
FERRY RD.
S.W. BARBUR BLVD.
S.W. TAYLORS
5
43
STROWBRIDGE
FLAVEL

Metzger
99W
Portland Comm. College
KERR RD.
BOONES FERRY RD.
Lewis & Clark College
Waverly C.C.
99E
Riverside DR.
HARRISON ST.
KING RD.
Milwaukie

Tigard
PACIFIC HWY.
217
Tryon Creek State Park
224

210
BONITA RD.
CARMAN DR.
Lake Grove
Waluga Park
COUNTRY CLUB RD.
Lake Oswego C.C.
Lake Oswego
Oak Grove
N. Clackamas Central Park
OATFIELD RD.
McLOUGHLIN
Kellogg Creek
WEBSTER RD.

King City
WASHINGTON CO.
CLACKAMAS CO.
Durham
99W
Lake Oswego
STAFFORD RD.
ROSEMONT
PORTLAND AVE.
43
Maryhurst College
Willamette River

Tualatin River

Scale of Miles
0   .5   1   1.5

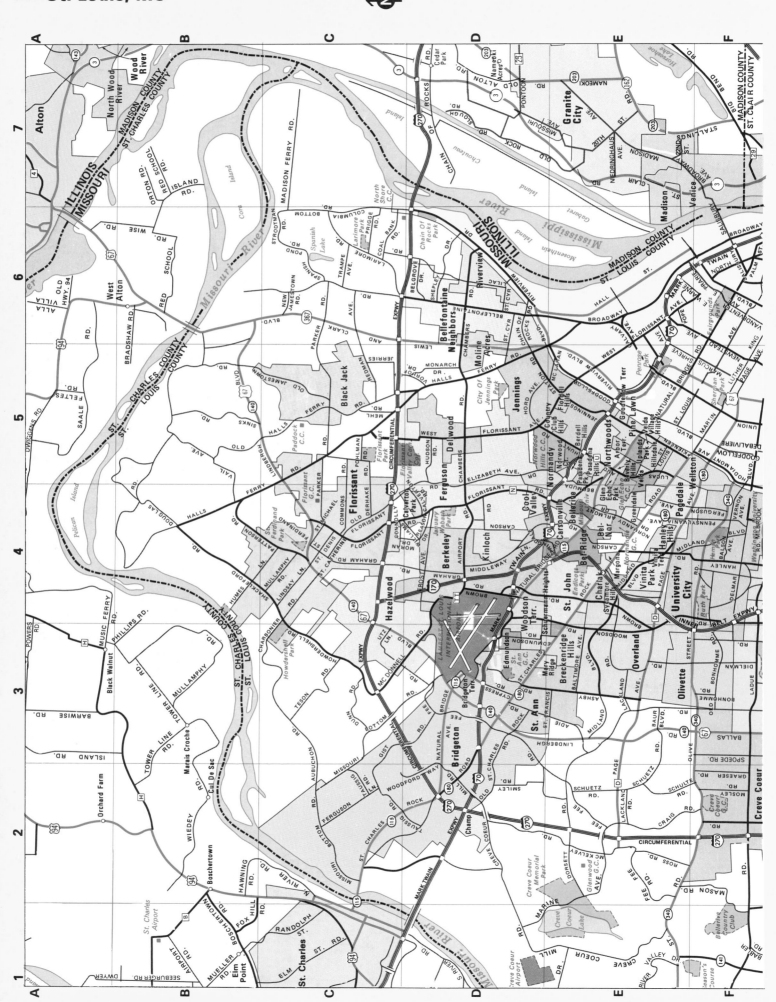

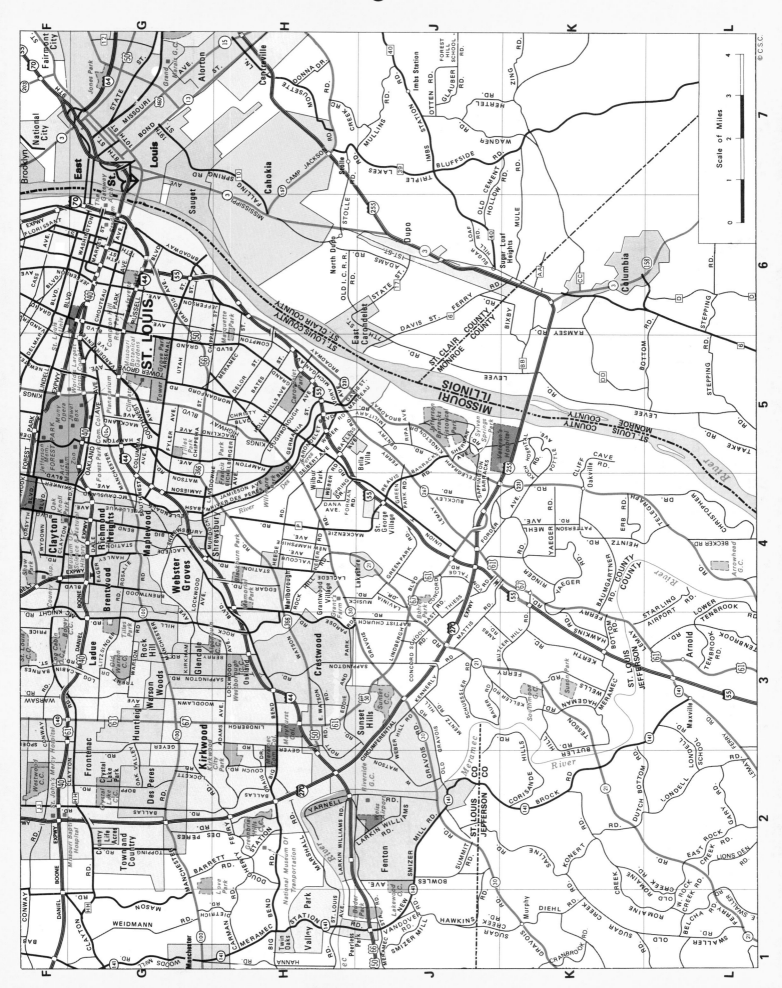

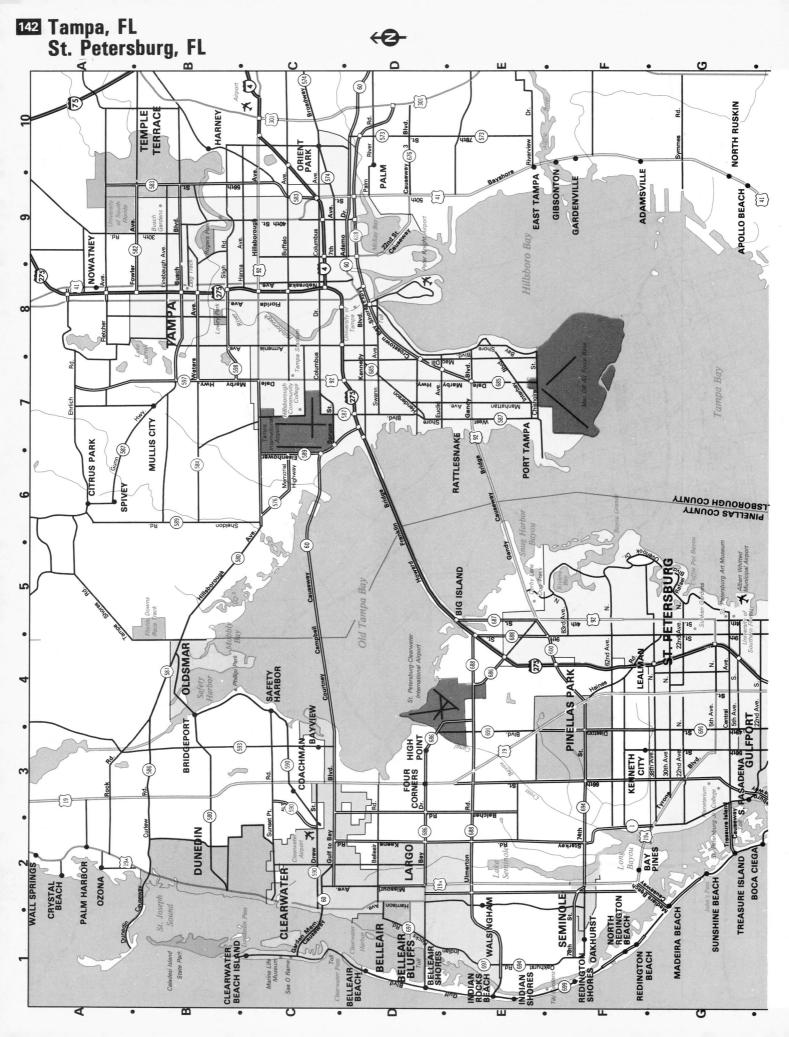

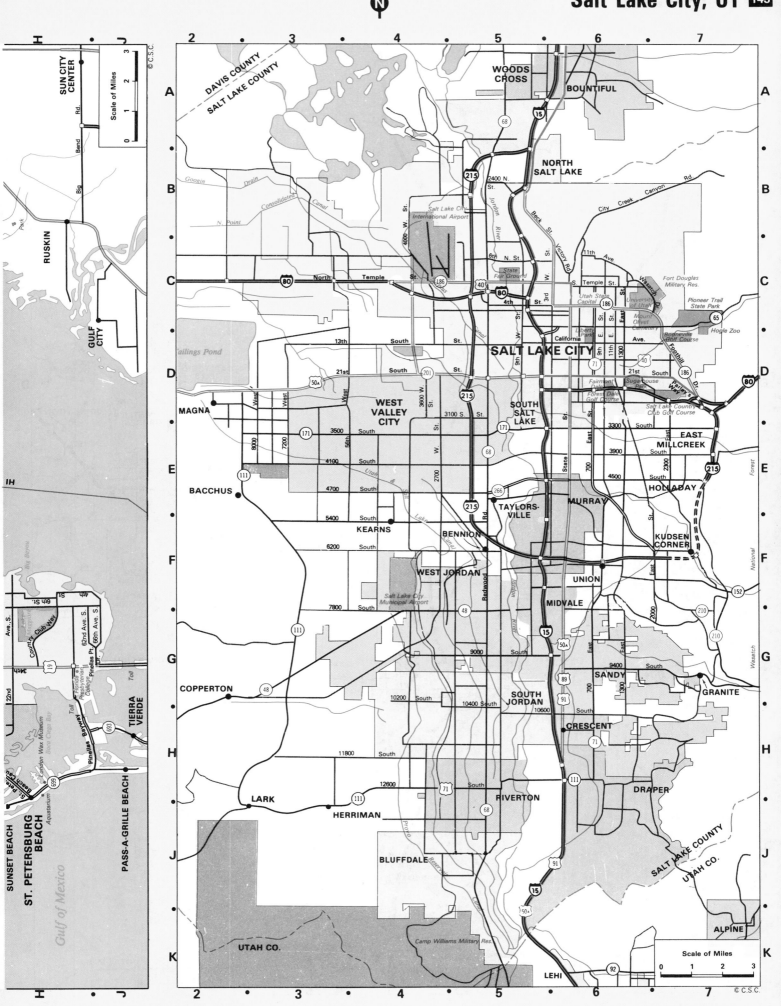

N

DAVIS COUNTY
SALT LAKE COUNTY

WOODS CROSS
BOUNTIFUL

NORTH SALT LAKE

Salt Lake City International Airport

2400 N. St.

6th N. St.

City Creek Canyon Rd.

Fort Douglas Military Res.

Pioneer Trail State Park

North Temple
Utah State Capitol
University of Utah
Mount Olivet Cemetery
Bonneville Golf Course
Hogle Zoo

SALT LAKE CITY

13th South
Liberty Park
California Ave.

21st South
Fairmont Park
Sugar House Park
Forest Dale Golf Course
Salt Lake Country Club Golf Course
Warley's

MAGNA

WEST VALLEY CITY
3600 W. St.
3100 S. St.

3500 South
56th South
4100 South
4700 South
5400 South

SOUTH SALT LAKE

EAST MILLCREEK
3300 South
3900 South
4500 South

HOLLADAY

BACCHUS

KEARNS
BENNION
6200 South

TAYLORS-VILLE
MURRAY

KUDSEN CORNER

WEST JORDAN
Salt Lake City Municipal Airport
7800 South

UNION
MIDVALE

COPPERTON

9000 South

SANDY
9400 South

GRANITE

SOUTH JORDAN
10200 South
10400 South
10600 South

CRESCENT

11800 South

LARK
HERRIMAN
12600 South

RIVERTON

DRAPER

BLUFFDALE

SALT LAKE COUNTY
UTAH CO.

UTAH CO.

Camp Williams Military Res.

ALPINE

LEHI

Scale of Miles
0 1 2 3

© C.S.C.

Scale of Miles
0 1 2 3
© C.S.C.

SUN CITY CENTER
Bend Rd.
Big Bend

RUSKIN

GULF CITY

Tailings Pond

Gulf of Mexico

ST. PETERSBURG BEACH

SUNSET BEACH

Florida Presbyterian College
County Club Way
34th
22nd
Pinellas Bayway
London Wax Museum
Boca Ciega Bay
Aquatarium

TIERRA VERDE

PASS-A-GRILLE BEACH

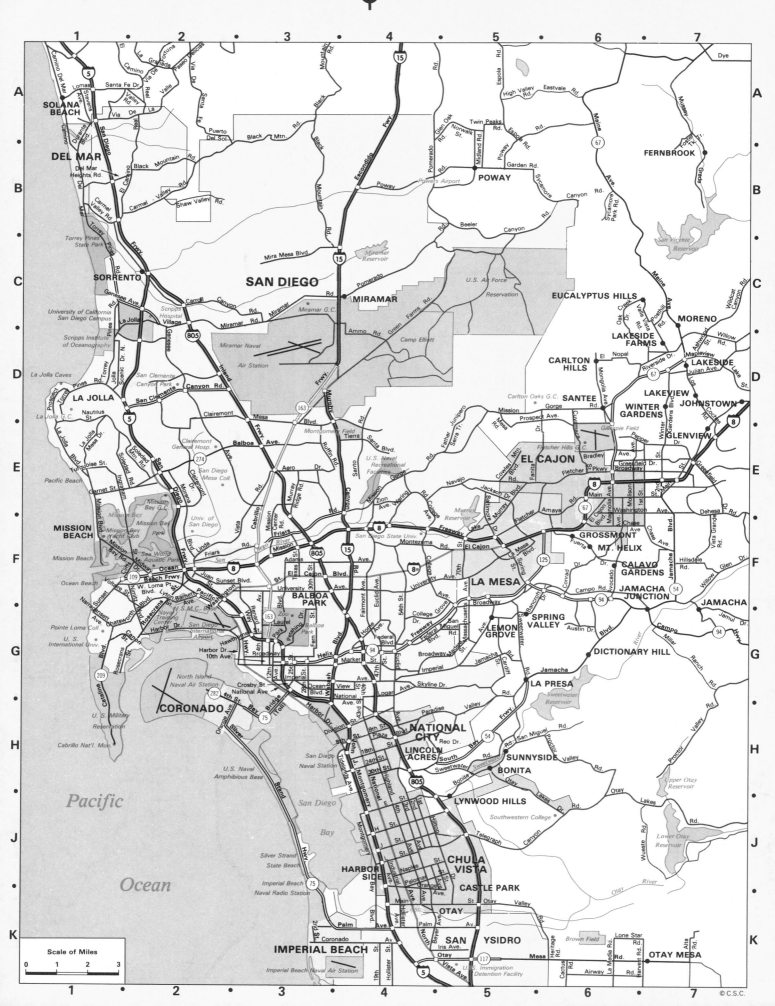

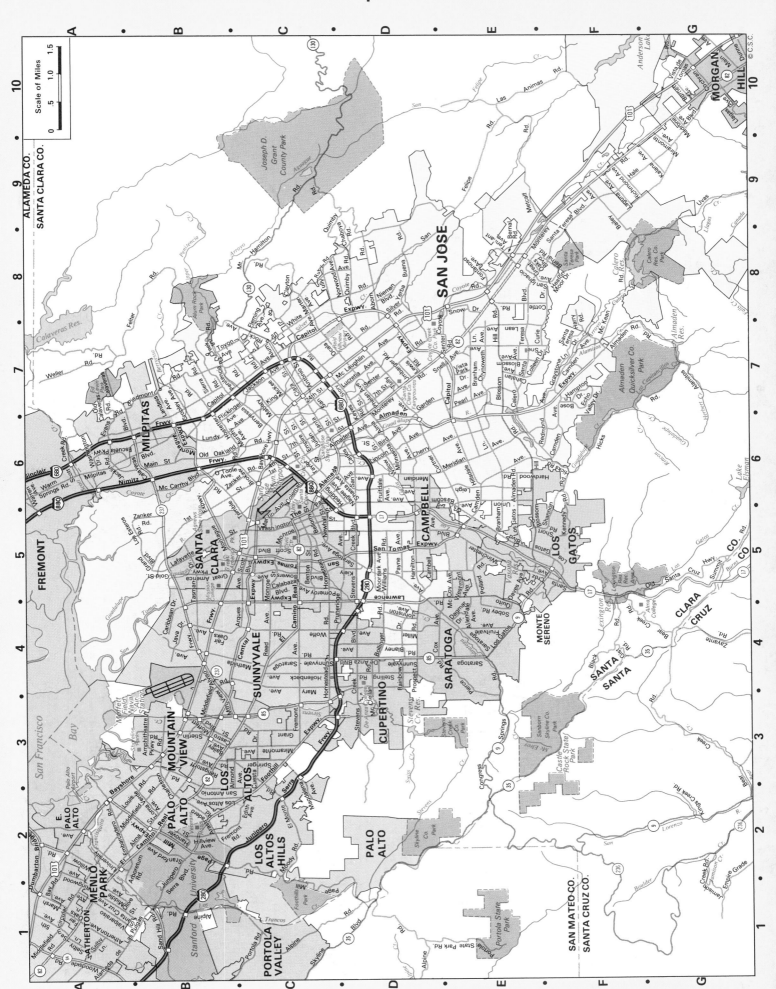

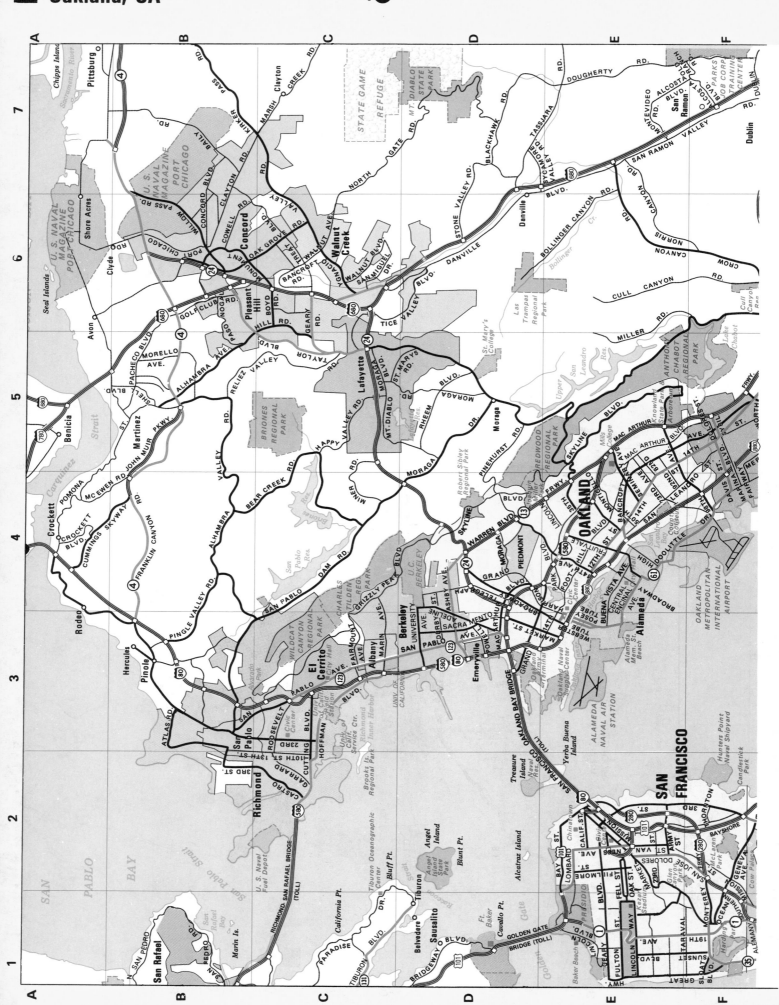

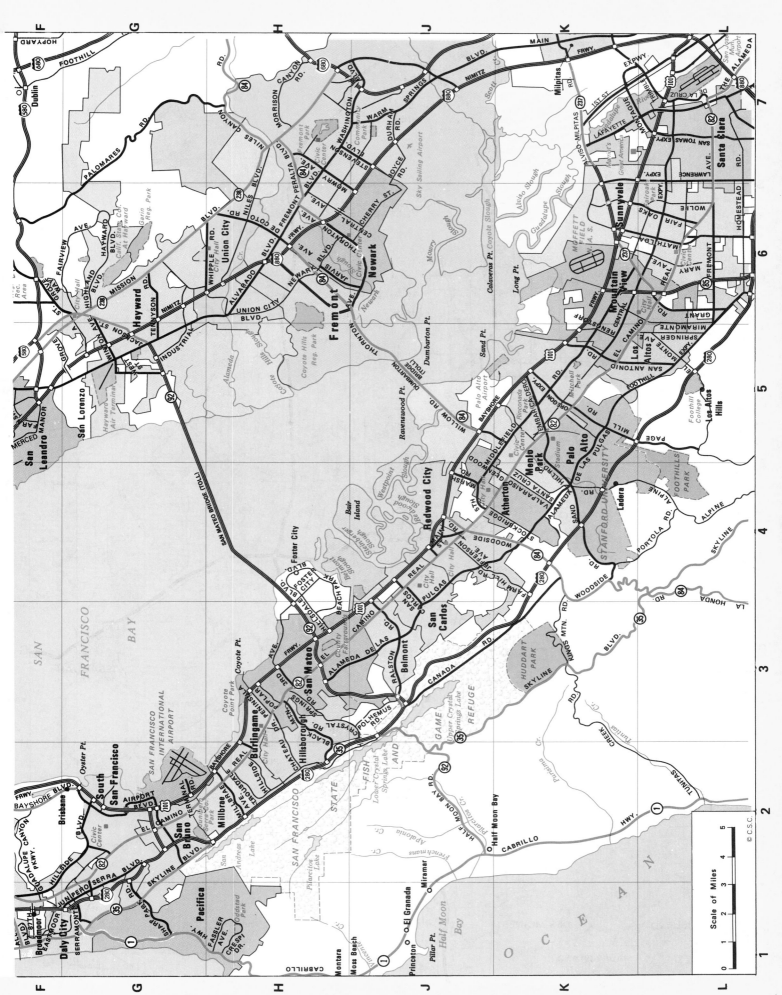

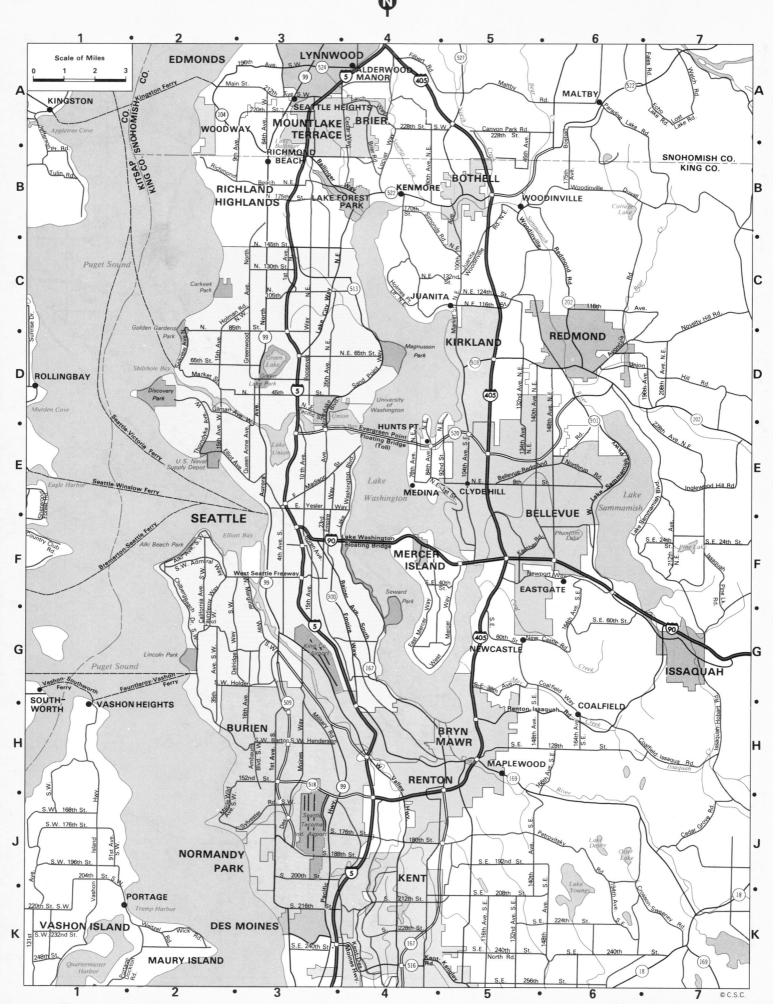

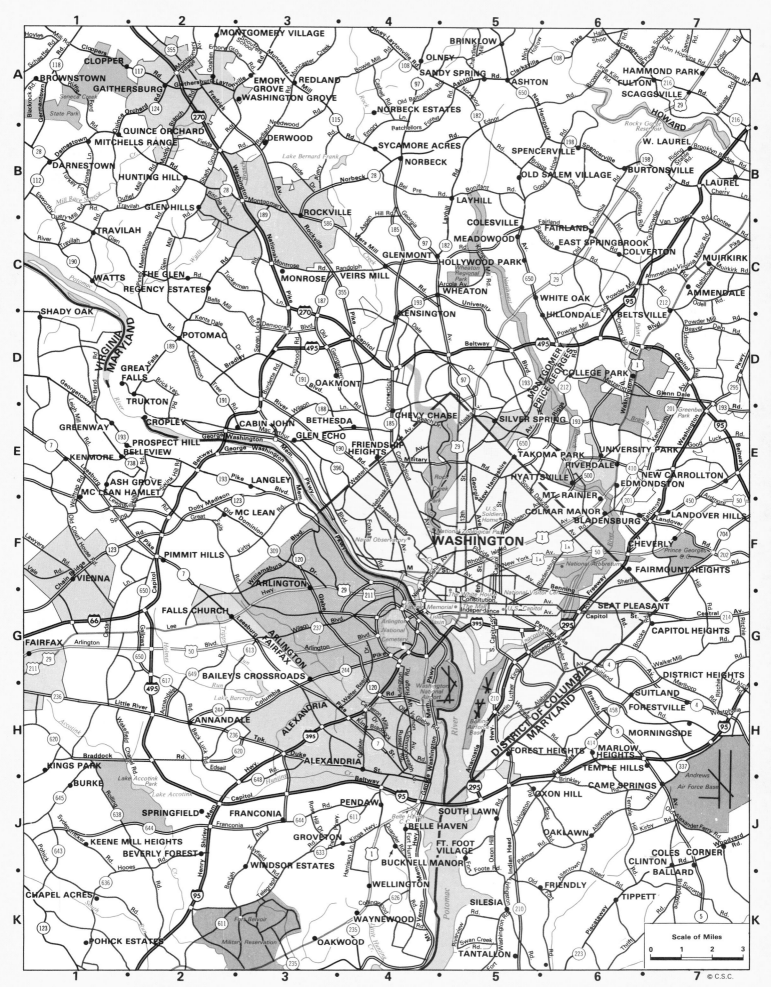

INDEX
To The United States

Index to Canadian Cities and Towns on Pages 12-13.
Index to Mexican Cities and Towns on Page 15.

ALABAMA
Pages 22

Abanda E-6
Abbeville H-6
Ada G-5
Adamsville D-4
Addison B-4
Akron E-3
Alabaster D-4
Albertville B-5
Aldrich D-4
Alexander City E-5
Aliceville E-2
Allgood C-5
Almond E-6
Altoona C-5
Andalusia H-4
Anderson A-4
Anniston C-5
Arab B-5
Arcus H-5
Ardilla H-6
Ardmore A-4
Argo D-5
Arkadelphia C-4
Arley C-4
Ashbury C-6
Ashford H-7
Ashland D-6
Ashville C-5
Athens A-4
Atmore J-3
Attalla C-5
Auburn F-6
Awin G-4
Axis J-2
Babbie H-5
Bailyton B-4
Bakerhill G-6
Ballplay C-6
Bangor C-4
Banks G-5
Bankston C-3
Barfield D-6
Barton A-3
Bass A-6
Batesville G-7
Bay Minette J-3
Bayou LaBatre K-2
Bear Creek B-3
Beatrice G-3
Belgreen B-3
Belk D-3
Bellamy F-2
Belleville H-4
Benton F-4
Berry C-3
Bessemer D-4
Big Cove B-5
Billingsley E-4
Birmingham D-4
Blacksher H-3
Blount Springs C-4
Blountsville C-4
Blue Mountain D-5
Bluff C-3
Boaz C-5
Bolinger G-2
Bon Secour K-3
Borden Springs C-6
Boyd F-2
Braggs G-4
Brantley H-5
Bremen C-4
Brent E-4
Brewton J-4
Bridgeport A-6
Brilliant C-3
Brinn H-4
Brooklyn H-4
Brooksville C-5
Brookwood D-3
Browns F-4
Brundidge G-6
Bryant A-6
Butler G-2

Bynum D-5
Calera E-4
Calvert H-2
Camden G-4
Campbell G-3
Camphill E-6
Carbon Hill C-3
Carrollton D-2
Carrville F-6
Castleberry H-4
Catalpa G-6
Catherine F-3
Cedar Bluff B-6
Center Point C-4
Centre C-6
Centreville E-4
Chase A-4
Chastang J-2
Chatom H-2
Chelsea D-5
Cherokee A-3
Chickasaw J-2
Childersburg D-5
Chrysler H-3
Citronelle H-2
Claiborne H-3
Clanton E-4
Claud F-5
Clayhatchee H-6
Clayton G-6
Cleveland C-5
Clinton E-2
Clio G-6
Clopton H-6
Cloverdale A-3
Cochrane E-2
Coden K-2
Coffeeville G-2
Collinsville B-6
Columbia H-7
Columbiana D-5
Consul F-3
Cooper E-5
Cordova C-4
Cottonwood J-6
Courtland B-4
Coxey A-4
Creola J-2
Crossville B-6
Cuba F-2
Cullman C-4
Cullomburg G-2
Dadeville E-6
Daleville H-6
Damascus H-4
Dancy E-2
Danville B-4
Daphne J-3
Dauphin Island K-2
Daviston E-6
Davisville F-6
Dawes J-2
Dayton F-3
Deans H-4
Deatsville F-5
Decatur B-4
Deer Park H-2
Dees K-2
Delta D-6
Demopolis F-3
Devenport G-5
Dill H-6
Dixie J-4
Dixon Mills G-2
Dixonville J-4
Dora C-4
Dothan H-7
Double Sprs. C-3
Dry Forks G-3
Dublin D-5
Duke C-6
Duncanville E-3
E. Brewton J-4
Easton E-6
Echola D-3
Eclectic F-6
Edwin G-6

Eight Mile J-2
Elamville H-6
Elba H-5
Elberta K-3
Elias E-5
Eliska H-3
Elmore F-5
Elrod D-3
Emelle E-2
Enterprise H-5
Eoline E-4
Epes E-2
Equality E-5
Ethelsville D-2
Eufaula G-7
Eunola J-6
Eutaw E-3
Evergreen H-4
Ewell H-6
Fairfax E-7
Fairfield D-4
Fairhope K-2
Falkville B-4
Fatama G-4
Fayette D-3
Flat Rock A-6
Flomaton J-4
Florala J-5
Florence A-3
Foley K-3
Forkland F-2
Forkville B-3
Fort Mitchell F-7
Fort Payne B-6
Fountain H-3
Francisco A-5
Frisco City H-3
Fruitdale H-2
Fulton G-3
Fultondale D-4
Gadsden C-6
Gainesville E-2
Gallant C-5
Gallion F-3
Gantt H-5
Gardendale C-4
Geiger E-2
Geneva J-5
Georgiana G-4
Geraldine B-5
Glencoe C-6
Goodwater E-5
Gordo D-3
Grand Bay K-2
Graysville C-4
Green Bay H-5
Green Pond D-4
Greensboro E-3
Greenville G-5
Grove Hill G-3
Guin C-2
Gulf Shores K-3
Guntersville B-5
Gurly B-5
Hackleburg B-3
Haleyville B-3
Halsell F-2
Halsos Mill G-5
Hamilton B-2
Hammondville B-6
Hanceville C-4
Harpersville D-5
Harrisburg E-3
Hartford J-6
Hartselle B-4
Hatton B-4
Havana E-3
Hawk D-6
Hayneville F-4
Hazel Green A-5
Hazen H-4
Headland H-7
Helena D-4
Henagar B-6
Hobson City D-6
Hodges B-3

Hokes Bluff C-6
Hollis D-6
Hollywood A-6
Holt D-3
Holy Trinity F-7
Homewood D-4
Hoover D-4
Hueytown D-4
Hulaco B-4
Hunter F-5
Huntsville A-5
Hurtsboro F-6
Huxford H-3
Hybart G-3
Ider B-6
Ino H-5
Intercourse F-2
Irondale D-4
Isney G-2
Jachin F-2
Jack H-5
Jackson H-3
Jacksonville C-6
Jasper C-3
Jefferson F-2
Jemison E-4
Jenkins Crossroads G-4
Jones F-4
Jones Chapel C-4
Kellerman D-3
Kelly's Crossroads E-5
Kellyton E-5
Kennedy C-2
Kimberly D-4
Kimbrough G-3
Kinsey H-6
Kinston H-5
Knoxville E-3
Koenton H-2
Lafayette E-6
Lanett E-7
Langdale E-7
La Place F-6
Lawrenceville H-6
Leeds D-5
Leighton A-3
Letohatchee F-4
Lexington A-4
Lillian K-3
Lincoln D-5
Linden F-3
Lineville D-6
Littleville B-3
Livingston E-2
Lockhart J-5
Lottie H-3
Louisville G-6
Lowndesboro F-5
Loxley K-3
Luverne G-5
Lyeffion H-4
Lynn C-3
Magnolia F-3
Malvern J-6
Maplesville E-4
Marbury E-5
Marion E-3
Marvyn F-6
McDade H-5
McIntosh H-3
McKenzie G-5
Mentone B-6
Meridianville A-5
Midland City H-6
Midway G-4
Millbrook F-5
Millerville D-6
Millport D-2
Millry F-6
Mitchell F-6
Mobile J-2
Monroeville H-4
Monrovia ★
Montevallo E-4
Montgomery F-5
Montrose K-2
Moores Bridge D-3

Mooresville B-4
Morgan City B-5
Morris C-4
Morvin G-3
Moulton B-4
Moundville E-3
Mountain Brook D-4
Mt. Andrew G-6
Mt. Olive E-5
Mount Vernon J-2
Mt. Willing G-4
Mulberry F-4
Munford D-6
Muscle Shoals A-3
Nanafalia G-2
Natural Bridge C-3
New Brockton H-5
Newburg B-3
New Hope B-5
New Lexington D-3
New Market A-5
Newsite E-6
Newton H-6
Newtonville D-3
Newville H-6
Normal A-5
Northport D-3
Notasulga F-6
Oak Grove D-4
Oakland B-4
Oakman C-3
Oakwood F-5
Odenville D-5
Ohatchee C-5
Omaha G-7
Oneonta C-5
Opelika E-6
Opp H-5
Owens Crossroads B-5
Oxford D-6
Ozark H-6
Palmerdale C-4
Parrish C-4
Pelham D-4
Pell City D-5
Pennington F-2
Perdido J-3
Peterman H-3
Peterson D-3
Petersville A-3
Phenix City F-7
Phil Campbell B-3
Pickensville D-2
Piedmont C-6
Pine Apple G-4
Pine Grove C-6
Pine Hill G-3
Pisgah B-6
Plantersville F-4
Pleasant Grove D-2
Pt. Clear K-2
Prattville F-5
Priceville B-4
Prichard J-2
Rabun J-3
Rainbow City C-5
Rainsville B-6
Ranburne D-6
Range H-3
Red Bay B-2
Red Level H-4
Reece City C-5
Reform D-3
Repton H-3
Riley G-4
River Falls H-5
River View J-4
Roanoke D-6
Robertsdale K-3
Rockdale F-6
Rockford E-5
Rogersville A-3
Rome H-5
Romulus D-3
Russellville B-3
Rutledge G-5
Saco G-6

Safford F-3
St. Elmo K-2
Salitpa G-2
Samson H-5
Sanford H-5
Santuck E-5
Saraland J-2
Sardis F-4
Satsuma J-2
Scottsboro B-5
Seale F-7
Section B-6
Selma F-4
Shawmut E-7
Sheffield A-3
Shelby E-5
Shepardville G-4
Shorter F-6
Silver Hill K-3
Simcoe B-4
Sipsey C-4
Sledge F-3
Slocomb J-6
Smiths F-7
Smut Eye G-6
Snead C-5
Snowdoun F-5
Somerville B-4
Southside C-5
Spring Hill H-6
Springville C-5
Sprott E-3
Spruce Pine B-3
Stafford D-2
Steele C-5
Stevenson A-6
Stewartville E-5
Stockton J-3
Sulligent C-2
Sumiton C-4
Summerdale K-3
Sunflower H-2
Suttle E-4
Sweet Water G-3
Sylacauga D-5
Sylvania B-6
Talladega D-5
Tallasee F-6
Tensaw H-3
Terese G-7
Theodore K-2
Thomaston F-3
Thomasville G-3
Thorsby E-4
Three Notch H-4
Tibbie H-2
Town Creek B-4
Toxey G-2
Trinity B-4
Troy G-5
Trussville D-5
Tuscaloosa D-3
Tuscumbia A-3
Tuskegee F-6
Union E-3
Union Grove B-5
Union Springs F-6
Uniontown F-3
Uriah H-3
Valley Head B-6
Vance E-3
Vernon C-2
Vestavia Hills D-4
Victoria G-5
Vidette G-5
Vineland D-3
Vinemont B-4
Wadley E-6
Wagarville H-2
Walnut Grove C-5
Walnut Hill C-6
Warrior C-4
Weaver C-5
Wedowee D-6
West Blockton D-4
West Point B-4

Wetumpka F-5
Whatley G-3
Whiteoak B-2
Whitson D-3
Wicksburg H-6
Wilburn C-4
Wilmer J-2
Wilton E-4
Windham Springs D-3
Winfield C-3
Wing J-4
Winterboro D-5
Woodville B-5
Wren B-3
Yantley F-2
Yarbo H-2
York F-2
Zip City A-3

ALASKA
Page 17

Akhiok F-5
Akolmiut D-4
Akulurak D-4
Akutan F-2
Alakanuk C-4
Alaktak A-7
Aleknagik E-4
Allakaket C-6
Alice Arm H-10
Ambler B-6
Anaktuvuk Pass B-7
Anchorage E-6, B-2
Anderson D-7
Aniak D-4
Angoon G-9
Anvik D-5
Arctic Village C-8
Atka H-8
Atkasuk A-6
Attu H-5
Baranof G-9
Barrow A-7
Beaver C-7
Beechey Pt. A-7
Bethel D-4
Bettles B-7
Big Delta D-7
Big Port Walter G-9
Black D-4
Brevig Mission B-4
Buckland C-5
Candle C-5
Cape Romanzof D-3
Cape Yakataga F-8
Caro C-7
Cantwell D-7
Chalkyitsik C-8
Chatanika D-7
Chevak D-4
Chenik E-5
Chickaloon A-2, E-7
Chicken D-8
Chignik F-4
Chitina E-7
Christian C-8
Christochina E-7
Chugiak B-2
Circle C-8
Circle Hot Springs D-8
Clarks Pt. E-4
College D-7
Copper Center E-7
Cordova E-7
Crooked Creek D-4
Deadhorse B-7
Deering B-5
Delta Junction D-7
Dillingham E-4
Donnelly D-7
Dot Lake D-8
Douglas G-9
Dutch Harbor F-2
Eagle D-8

Eagle River B-2, E-6
Edna Bay H-9
Eek E-4
Egavic C-5
Elfin Cove G-9
Emmonak C-4
English Bay F-6
Eureka D-7
Fairbanks D-7
False Pass F-3
Farewell D-6
Flat D-5
Fort Yukon C-8
Gakona E-7
Galena C-6
Gambell C-3
Glen Alps E-6
Glennallen E-7
Golovin C-5
Goodnews Bay E-4
Gordon B-8
Grayling D-5
Gulkana E-7
Gustavus G-9
Haines F-9
Hawk Inlet G-9
Healy D-7
Homer C-1, E-6
Hoonah G-9
Hooper Bay D-3
Hope B-2, E-6
Houston A-2
Hughes C-6
Huslia C-6
Hydaburg H-9
Itulillic E-5
Juneau G-9
Kachemak E-6
Kake G-9
Kaktovik B-8
Kalla C-6
Kaltag C-5
Karluk F-5
Kashegelok E-5
Kasilof B-1, B-2
Kaskanak E-5
Katalla F-7
Kenai B-1
Ketchikan H-10
Kiana B-5
King Cove F-3
King Salmon F-5
Kipnuk E-4
Kivalina B-5
Klawock H-9
Knik B-2
Kobuk C-6
Kodiak F-6
Kokrines C-6
Koliganek E-5
Kotlick C-4
Kotzebue B-5
Koyuk C-5
Koyukuk C-5
Kuliuk F-5
Kwigillingok E-4
Kwethluk D-4
Larsen Bay F-5
Latouche F-7
Levelock E-5
Lime Village E-5
Livengood C-7
Long D-6
Lower Tonsina E-7
Manley Hot Springs D-6
Manokotak E-4
McCarthy E-8
McGrath D-5
McKinley Park D-7
Medfra D-6
Mekoryuk D-3
Mentasta Lake E-7
Metlakata H-10
Minto D-7
Moose Pass B-2
Mountain Village D-4

★ Denotes City located only on Vicinity map. City not located on State map.

Index 151

ALASKA ARKANSAS

Mount Edgecumber ... G-9
Naknek ... F-5
Napakiak ... D-4
Napaskiak ... E-4
Nenana ... D-7
Newhalen ... E-5
New Stuyahok ... E-5
Nikolai ... D-6
Nikolski ... F-1
Noatak ... B-5
Nogamut ... E-5
Nome ... C-4
Nondalton ... E-5
Noorvik ... B-5
Northeast Cape ... C-4
North Pole ... D-7
Northway ... E-8
Nulato ... C-5
Old Harbor ... F-5
Ooliktok ... A-7
Ouzinkie ... F-6
Palmer ... A-2, E-7
Pedro Bay ... E-5
Pelican ... G-9
Perryville ... F-4
Petersburg ... E-6
Pilgrim Springs ... C-4
Pilot Point ... D-4, F-4
Platinum ... E-4
Point Hope ... A-5
Pt. Lay ... A-5
Porcupine ... F-9
Port Alexander ... G-9
Port Lions ... F-5
Portlock ... F-6
Prudhoe Bay ... B-8
Quinhagak ... E-4
Rampart ... C-7
Rampart House ... C-8
Richardson ... D-7
Ruby ... C-6
Russian Mission ... D-4
Sagwon ... B-7
Salchaket ... D-7
St. George ... E-2
St. Michael ... C-5
St. Marys ... D-4
St. Paul ... E-2
Sand Pt. ... F-4
Saroonga ... C-3
Selawik ... B-5
Seward ... C-2, E-6
Shageluk ... D-5
Shaktoolik ... C-5
Shishmaref ... B-4
Shungnak ... B-6
Sinuk ... C-4
Sitka ... G-9
Skagway ... F-9
Sleetmute ... D-5
Soldotna ... B-1, E-6
Squaw Harbor ... G-3
Stebbins ... C-4
Stevens Village ... C-7
Stony River ... D-5
Stuyahok ... D-5
Susitna ... B-2
Takotna ... D-5
Talkeetna ... A-1, E-7
Tanalian Pt ... E-5
Tanana ... C-6
Tatitlek ... E-7
Teller ... C-4
Tetlin ... E-8
Tetlin Junction ... E-8
Togiak ... E-4
Tok ... D-8
Toksook Bay ... D-4
Tonsina ... E-7
Tuluksak ... D-4
Tununak ... D-4
Tyonek ... B-1, E-6
Ugashik ... F-5
Umiat ... B-7
Unalakleet ... C-5
Unalaska ... F-2
Valdez ... E-7
Venetie ... C-8
Wainwright ... A-6
Wales ... B-4
Wasila ... A-2
Wasilla ... E-7
White Mountain ... C-5
Whittier ... B-2
Willow ... A-1, E-6
Wiseman ... C-7
Wrangell ... G-10
Yakutat ... F-8

ARIZONA
Pages 20-21

Adobe ... ★
Vicinity ... Pg. 135, B-4
Aguila ... J-3
Ajo ... L-4
Ak Chin ... K-5
Alpine ... J-10
Amado ... N-7
Anegam ... L-5
Angell ... F-7
Apache Jct. ... K-6
Arivaca Jct. ... N-7
Arizona City ... L-6
Arlington ... K-4
Artesa ... M-6
Ash Fork ... F-5
Ashurst ... K-9
Avondale ... K-5
Bagdad ... G-4
Bapchule ... K-6
Beaverhead ... J-10
Bellemont ... F-6
Benson ... M-8
Bisbee ... N-9
Black Canyon City ... H-5
Bonita ... L-9
Bouse ... J-2
Bowie ... M-10
Brenda ... J-2
Buckeye ... K-5
Bullhead City ... F-2
Bylas ... K-9
Cameron ... E-7
Camp Verde ... H-6
Carefree ... J-6
Carmen ... N-7
Carrizo ... J-8
Casa Grande ... K-6
Vicinity ... Pg. 135, E-2
Catalina ... L-7
Cave Creek ... J-6
Cedar Creek ... J-9
Cedar Ridge ... D-7
Cedar Springs ... F-8
Central ... L-9
Chambers ... F-9
Chandler ... K-6
Vicinity ... Pg. 135, F-7
Chandler Heights ... K-6
Childs ... L-4
Chilchinbito ... D-9
Chinle ... E-10
Chino Valley ... G-5
Chloride ... F-2
Christmas ... K-8
Christopher Creek ... H-7
Chuichu ... L-6
Circle City ... J-4
Clarkdale ... G-5
Claypool ... K-7
Clay Spring ... H-8
Clifton ... K-10
Cochise ... M-9
Colorado City ... C-4
Concho ... H-9
Congress ... H-4
Continental ... M-7
Coolidge ... K-6
Cordes Jct. ... H-5
Cornville ... G-6
Cortaro ... L-7
Cottonwood ... G-6
Covered Wells ... M-5
Cow Springs ... D-8
Cross Canyon ... E-10
Currys Corner ... ★
Vicinity ... Pg. 135, B-6
Curtis ... M-8
Dateland ... K-3
Desert View ... E-6
Dewey ... H-5
Dilkon ... F-8
Dinnehotso ... D-9
Dolan Springs ... F-2
Dome ... L-2
Dos Cabezas ... M-10
Douglas ... N-10
Dragoon ... M-9
Dudleyville ... L-8
Duncan ... L-10
Eager ... H-10
Eden ... K-9
Ehrenberg ... J-2
Elfrida ... N-9
Eloy ... L-6
Fairbank ... N-8
Flagstaff ... F-6
Florence ... K-7
Florence Jct. ... K-7
Fort Apache ... J-9
Fort Defiance ... E-10
Fort Grant ... L-9
Ft. McDowell ... J-6
Fort Thomas ... K-8
Fountain Hills ... J-6
Francisco Grande ... K-5
Franklin ... L-10
Fredonia ... C-5
Gadsden ... L-1
Ganado ... F-9
Gila Bend ... K-4
Gilbert ... K-6
Glendale ... J-5
Vicinity ... Pg. 135, C-3
Globe ... K-8
Golden Shores ... G-2
Goldroad ... F-2
Goodyear ... K-5
Vicinity ... Pg. 135, E-1
Grand Canyon ... E-5
Gray Mountain ... E-7
Greasewood (Lower) ... F-9
Green Valley ... M-7
Greer ... J-10
Guadalupe ... K-5
Vicinity ... Pg. 135, F-5
Hackberry ... F-3
Hannigan Meadow ... K-10
Harcuvar ... J-3
Hawley Lake ... J-9
Hayden ... K-7
Heber ... H-8
Higley ... K-6
Hillside ... H-4
Holbrook ... G-8
Hope ... J-3
Hotevilla ... E-8
Houck ... F-10
Huachuca City ... N-8
Humboldt ... H-5
Indian Wells ... F-8
Jacob Lake ... D-5
Jerome ... G-5
Johnson ... M-8
Joseph City ... G-8
Katherine ... F-2
Kayenta ... D-8
Keams Canyon ... E-9
Kearney ... K-7
Kelvin ... K-7
Kingman ... F-2
Kirkland ... H-4
Kirkland Jct. ... H-5
Kohls Ranch ... H-7
Kyrene ... ★
Vicinity ... Pg. 135, F-6
Lake Havasu City ... G-2
Lake Montezuma ... G-6
Lakeside ... H-9
La Palma ... L-6
Laveen ... ★
Vicinity ... Pg. 135, E-3
Leupp ... F-7
Litchfield ... J-5
Littlefield ... C-3
Lukeville ... M-4
Lupton ... F-10
Mammoth ... L-8
Many Farms ... D-10
Marana ... L-6
Marble Canyon ... D-6
Maricopa ... K-5
Mayer ... H-5
McGuireville ... G-6
McNeal ... N-9
Meadview ... E-2
Mesa ... J-6
Mexican Water ... C-9
Miami ... K-7
Moenkopi ... E-7
Mountainaire ... G-6
Mt. View ... M-7
Moqui ... E-5
Morenci ... K-10
Morristown ... J-4
Munds Park ... G-6
Naco ... N-9
Navajo ... G-10
Nelson ... F-3
New River ... J-5
Nicksville ... N-8
Nogales ... N-7
North Rim ... E-6
Nutrioso ... J-10
Oatman ... G-2
Ocotillo ... K-6
Olberg ... K-6
Old Oraibi ... E-8
Oracle ... L-7
Oracle Jct. ... L-7
Oraibi ... E-8
Oro Valley ... L-7
Overgaard ... H-8
Page ... C-7
Palm Springs ... K-6
Palominas ... N-8
Palo Verde ... K-4
Pantano ... M-8
Paradise City ... ★
Vicinity ... Pg. 135, B-5
Paradise Valley ... J-5
Parker ... H-2
Parks ... F-6
Patagonia ... N-7
Paulden ... G-5
Paul Spur ... N-9
Payson ... H-6
Peach Springs ... F-4
Peeples Valley ... H-4
Peoria ... J-5
Vicinity ... Pg. 135, C-3
Peridot ... K-8
Phoenix ... J-6
Vicinity ... Pg. 135
Picacho ... L-6
Pima ... L-9
Pine ... H-6
Pinedale ... H-8
Pine Springs ... F-5
Pinetop ... H-9
Pirtleville ... N-9
Polacca ... E-8
Pomerene ... M-8
Prescott ... G-5
Prescott Valley ... G-5
Punkin Center ... J-7
Quartzite ... J-2
Queen Creek ... K-6
Quijotoa ... M-5
Randolph ... K-6
Red Lake ... D-7
Red Rock ... L-7
Rillito ... L-7
Riviera ... F-2
Riverside ... K-7
Rock Point ... D-10
Rock Springs ... J-5
Roosevelt ... J-7
Rough Rock ... D-9
Round Rock ... D-10
Rye ... H-7
Sacaton ... K-6
Safford ... L-9
Sahuarita ... M-7
St. David ... M-9
St. Johns ... H-10
St. Michaels ... F-10
Salome ... J-3
San Carlos ... K-8
Sanders ... F-10
San Luis ... L-1
San Luis ... L-5
San Manuel ... L-8
San Simon ... M-10
Sasabe ... N-6
Scottsdale ... J-6
Vicinity ... Pg. 135, C-6
Seba Dalkai ... F-8
Second Mesa ... E-8
Sedona ... G-6
Seligman ... F-4
Sells ... M-5
Sentinel ... K-3
Shonto ... D-8
Show Low ... H-9
Shumway ... H-9
Sierra Vista ... N-8
Silver Bell ... N-6
Snowflake ... H-9
Solomon ... L-9
Somerton ... L-1
Sonoita ... N-7
South Tuscon ... M-7
Springerville ... H-10
Stanfield ... L-6
Stargo ... K-10
Star Valley ... H-7
Strawberry ... H-6
Sun City ... J-5
Vicinity ... Pg. 135, C-2
Sun City West ... J-5
Sunflower ... J-6
Sunizona ... M-9
Sun Lake ... K-6
Sun Valley ... G-9
Sunrise ... F-7
Sunsites ... M-9
Superior ... K-7
Suprise ... J-5
Tacna ... L-2
Taylor ... H-9
Teec Nos Pas ... D-10
Tempe ... K-6
Vicinity ... Pg. 135, E-6
Tes Nez Iha ... C-9
Thatcher ... L-9
The Gap ... D-7
Three Points (Robles Jct.) ... M-6
Tolleson ... K-5
Vicinity ... Pg. 135, D-2
Tombstone ... N-9
Tonalea ... D-7
Tonopah ... J-4
Topock ... G-2
Tortilla Flat ... J-7
Tracy ... M-5
Truxton ... F-3
Tsaile ... D-10
Tsegi ... D-8
Tuba City ... E-7
Tubac ... N-7
Tucson ... M-7
Tumacacori ... N-7
Tusayan ... E-5
Vail ... M-8
Valentine ... F-3
Valle ... F-5
Vernon ... H-9
Vicksburg ... J-2
Wellton ... L-2
Wenden ... J-3
Whiteriver ... J-9
Why ... M-4
Wickenburg ... J-4
Wikieup ... G-3
Wilhoit ... H-5
Willcox ... M-9
Williams ... F-5
Window Rock ... F-10
Winkelman ... K-8
Winona ... F-7
Winslow ... H-7
Wintersburg ... K-4
Wittmann ... J-4
Yampai ... F-4
Yarnell ... H-4
Young ... H-7
Youngtown ... J-5
Yucca ... G-2
Yuma ... L-1

ARKANSAS
Page 19

Abbott ... D-3
Aberdeen ... D-7
Acorn ... D-3
Alabam ... B-3
Alco ... B-5
Algoa ... C-7
Alicia ... B-7
Allison ... B-6
Allport ... D-6
Alma ... C-2
Alpena ... A-4
Alpine ... E-4
Alread ... C-5
Altheimer ... E-7
Altus ... C-3
Aly ... D-4
Amagon ... C-7
Amity ... E-4
Amy ... F-5
Antioch ... C-6
Antonie ... E-4
Aplin ... D-5
Appleton ... C-5
Arden ... F-2
Arkadelphia ... E-5
Arkansas City ... F-8
Ashdown ... F-3
Ash Flat ... A-7
Athelstan ... B-9
Athens ... E-3
Atkins ... C-4
Atlanta ... G-5
Aubrey ... D-8
Augsburg ... C-4
Augusta ... C-7
Aurora ... B-3
Austin ... D-6
Auvergne ... C-7
Avilla ... D-5
Back Gate ... E-7
Bald Knob ... C-7
Banks ... F-6
Banner ... B-6
Barling ... C-2
Bassett ... C-9
Bates ... D-2
Batesville ... B-7
Bay ... B-8
Bay Village ... C-8
Bear Creek Springs ... A-4
Bearden ... F-5
Beaver ... A-3
Beck ... D-9
Beebe ... C-6
Bee Branch ... C-5
Beedeville ... C-7
Beirne ... F-4
Belfast ... E-5
Bella Vista ... A-2
Belleville ... C-4
Ben Hur ... B-4
Benton ... D-5
Bentonville ... A-2
Bergman ... A-5
Berlin ... G-7
Berryville ... A-4
Best ... A-3
Bethesda ... B-6
Beaverage Town ... C-5
Beulah ... D-7
Bexar ... A-6
Big Flat ... B-5
Big Fork ... E-3
Birdeye ... C-8
Birdtown ... C-5
Birta ... D-4
Biscoe ... D-7
Bismark ... E-4
Black Rock ... B-7
Black Springs ... E-3
Blackton ... D-8
Blackwell ... C-5
Blakely ... D-4
Blakemore ... D-6
Blanton ... ★
Vicinity ... Pg. 123, C-2
Blevins ... F-4
Blue Ball ... D-3
Blue Springs ... D-4
Bluff City ... F-4
Bluffton ... D-4
Blytheville ... B-9
Bodcaw ... F-4
Bogg Springs ... E-2
Boles ... D-3
Bolding ... G-6
Bonanza ... C-2
Bono ... B-8
Booneville ... C-3
Boston ... B-3
Boswell ... B-6
Boughton ... F-4
Boxley ... B-4
Boydsville ... A-8
Bradley ... G-4
Bradford ... C-7
Brasfield ... D-7
Brashears ... B-3
Brentwood ... B-3
Brickeys ... D-8
Briggsville ... D-4
Brinkley ... D-7
Brockwell ... B-7
Brookland ... B-8
Bruins ... D-9
Bruno ... B-5
Bryant ... D-5
Buckeye ... B-9
Buena Vista ... F-5
Bull Shoals ... A-5
Burlington ... A-4
Bunn ... E-5
Burstell ... E-4
Busch ... A-3
Bussey ... G-4
Butlerville ... D-6
Cabot ... D-6
Cades ... E-7
Caddo City ... E-4
Caddo Gap ... E-4
Calamine ... B-7
Cale ... F-4
Calico Rock ... B-6
Calion ... F-5
Calmer ... E-6
Camden ... F-5
Cammack Village ... D-6
Camp ... A-6
Canaan ... B-5
Canfield ... G-4
Capps ... B-4
Caraway ... B-9
Carlisle ... D-7
Carthage ... E-5
Carryville ... A-9
Casa ... D-4
Cash ... B-8
Cass ... B-3
Catalpa ... C-4
Caulksville ... C-3
Cauthorn ... D-2
Cave City ... B-7
Cecil ... C-3
Cedar Creek ... D-3
Cedarville ... C-2
Center ... B-7
Central City ... C-2
Chapel Hill ... E-2
Charleston ... C-3
Charlotte ... B-7
Chatfield ... C-8
Chelford ... B-8
Cherry Hill ... D-3
Cherry Valley ... C-8
Chester ... C-3
Chickalah ... C-4
Chicot Junction ... G-7
Chidester Bragg City ... F-5
Childres ... B-9
Choctaw ... C-5
Chelford ... B-8
Clarendon ... D-7
Clarkedale ... C-9
Clarkridge ... A-5
Clarksville ... C-4
Clay ... C-6
Cleveland ... C-5
Clinton ... C-5
Clow ... F-3
Clyde ... B-2
Coal Hill ... C-3
College City ... B-8
Collins ... F-7
Colt ... C-8
Combs ... B-3
Cornerville ... F-6
Compton ... B-4
Concord ... B-6
Conway ... C-6
Cord ... B-7
Cornerstone ... E-7
Corning ... A-8
Cotter ... A-5
Cotton Plant ... C-7
Cove ... E-2
Cowell ... B-4
Cowlingsville ... F-3
Coy ... D-6
Cozahome ... B-5
Crawfordsville ... C-9
Crocketts Bluff ... E-7
Crosses ... B-3
Crossett ... G-7
Cross Roads ... E-8
Cross Roads ... F-2
Crows ... D-5
Culler ... D-6
Culpepper ... C-5
Curtis ... E-5
Cushman ... B-6
Daisy ... E-3
Dalark ... E-5
Dallas ... E-2
Dalton ... A-7
Danville ... D-4
Dardanelle ... C-4
Datto ... A-8
Deans Market ... C-3
Decatur ... A-2
Deep Elm ... D-7
Deer ... B-4
Delaplaine ... A-8
Delaware ... C-4
Delight ... E-4
Dell ... B-9
Delpro ... C-9
Deluce ... E-7
Denmark ... C-7
DeQueen ... E-2
Dermott ... F-7
Des Arc ... D-7
De Valls Bluff ... D-7
Dewey ... C-6
Dewey ... F-7
DeWitt ... E-7
Diamond City ... A-5
Diaz ... B-7
Dierks ... E-3
Dixie ... B-8
Dixie ... C-7
Doddridge ... G-3
Dogpatch ... B-4
Dolph ... A-6
Donaldson ... E-5
Dover ... C-4
Dowdy ... B-7
Drasco ... B-6
Driver ... B-9
Dumas ... F-7
Durham ... B-3
Dutton ... B-3
Dyer ... C-2
Dyess ... B-9
Eagleton ... D-3
Earle ... C-8
East Camden ... F-5
East End ... D-6
Eaton ... B-7
Elaine ... E-8
El Dorado ... G-5
Elgin ... B-7
Elliot ... F-5
Elkins ... B-3
Elm Sprs. ... B-3
Elm Store ... A-7
Emerson ... G-4
Emmet ... F-4
England ... D-6
Ethel Ratio ... E-8
Eudora ... G-7
Eureka Springs ... A-3
Evening Shade ... B-7
Evening Shade ... F-4
Everton ... B-4
Excelsior ... C-2
Falcon ... F-4
Falls Chapel ... F-2
Fallsville ... B-4
Fargo ... D-7
Farindale ... E-6
Farmington ... B-3
Fayetteville ... B-3
Fifty-Six ... B-6
Figure Five ... C-2
Fisher ... C-8
Fitzhugh ... C-7
Flippin ... A-5
Floral ... C-6
Floyd ... C-6
Fontaine ... B-8
Fordyce ... F-6
Foreman ... F-2
Formosa ... C-5
Forrest City ... C-8
Fort Smith ... C-2
Fouke ... G-3
Fountain ... D-5
Fountain Hill ... G-7
Fourche Junction ... D-4
Franklin ... B-6
Friendship ... E-5
Fresno ... E-7
Furlow ... D-6
Fulton ... F-3
Gainesville ... A-8
Gaither ... B-4
Galet ... ★
Vicinity ... Pg. 123, E-1
Gamaliel ... A-6
Gammon ... ★
Vicinity ... Pg. 123, B-1
Gardner ... G-6
Garfield ... A-3
Garland ... G-3
Garrett Bridge ... F-7
Gassville ... A-5
Gateway ... A-3
Genoa ... G-3
Gentry ... B-2
Georgetown ... C-7
Gethsemane ... E-7
Gibson ... B-8
Gillett ... E-7
Gillham ... E-2
Gilmore ... C-9
Gin City ... G-3
Glendale ... E-8
Glen Rose ... E-5
Glenwood ... E-4
Gold Creek ... D-5
Goodwin ... D-8
Goshen ... B-3
Gosnell ... B-9
Gould ... E-7
Grady ... E-7
Grannis ... E-2
Grandview ... A-4
Grapevine ... E-6
Gravelly ... D-3
Gravette ... A-2
Green Brier ... C-6
Green Forest ... A-4
Greene High ... B-8
Green Hill ... F-6
Greenland ... B-3
Greenway ... A-9
Greenwood ... C-3
Greers Ferry ... C-6
Gregory ... C-7
Grider ... B-9
Grubbs ... B-7
Guernsey ... F-4
Gum Springs ... E-5
Gurdon ... F-4
Guy ... C-6
Hackett ... C-2
Hagarville ... C-4
Halliday ... A-8
Hamburg ... G-6
Hamlet ... D-6
Hardin ... E-6
Hardy ... A-7
Harmon ... B-4
Harmony ... F-7
Harmony ... C-4
Harmony Grove ... F-5
Harrell ... F-6
Harriet ... B-5
Harrisburg ... B-8
Harrison ... A-4
Hartford ... D-2
Hartman ... C-4
Harvard ... ★
Vicinity ... Pg. 123, B-1
Harvey ... D-3
Haskell ... E-5
Hasty ... B-4
Hatton ... E-2
Havana ... C-4
Hazen ... D-7
Heafer ... C-9
Heber Sprs. ... C-6
Helena ... D-8
Hempwallace ... E-5
Hermitage ... F-6
Hervey Patmos ... F-4
Hickory Plaines ... D-7
Hickory Ridge ... C-8
Hicks Station ... C-8
Higginson ... C-7
Highland ... E-3
Hillcrest ... C-4
Hillemann ... C-7
Hollis ... D-4
Holly Grove ... D-7
Holly Island ... A-9
Holly Springs ... F-5
Hollywood ... E-4
Holub ... D-8
Hon ... D-3
Hope ... F-4
Hopper ... E-3
Horatio ... F-2
Horseshoe Bend ... A-6
Hot Springs ... D-4
Hot Springs Village ... D-4
Hoxie ... B-8
Hughes ... D-8
Hulbert ... ★
Vicinity ... Pg. 123, D-1
Humphrey ... E-7
Hunt ... C-3
Huntington ... C-3
Huntsville ... B-3
Huttig ... G-6
Imboden ... A-7
Immanuel ... E-7
Indian ... G-7
Indian Bay ... E-8
Ingalls ... F-6
Ink ... D-3
Ione ... D-3
Ivan ... F-6
Jacksonville ... D-6
Jasper ... B-4
Jennie ... G-7
Jericho ... C-9
Jerome ... F-7
Jersey ... F-6
Jerusalem ... C-5
Jessieville ... D-4
Jesup ... B-7
Johnson ... B-3
Johnsville ... F-6
Joiner ... C-9
Jonesboro ... B-8
Joy ... C-6
Judsonia ... C-7
Julius ... C-9
Junction ... D-5
Junction City ... G-5
Keiser ... B-9
Kelso ... F-7
Kensett ... C-7
Kenwood ... C-5
Keo ... D-6
Kerr ... D-6
Kibler ... C-2
Kilbourne ... G-7
Kingsland ... F-6
Kingston ... B-4
Kirby ... E-4
Knob ... A-8
Knobel ... A-8
Lacey ... F-7
Ladd ... E-6
Ladelle ... F-7
La Grange ... D-8
Lake Dick ... E-7
Lake Nance ... D-5
Lakeview ... A-5
Lake View ... E-8
Lake Village ... C-7
Lakeway ... A-5
Lamar ... C-4
Lamartine ... F-4
Langley ... E-3
Lake City ... B-9
Lavaca ... C-2
Leachville ... B-9
Lead Hill ... A-4
Lee Creek ... B-2
Lehi ... C-9
Leola ... E-5
Lepanto ... B-8
Leslie ... B-5
Levesque ... C-8
Lewisville ... G-4
Lincoln ... B-2
Lisbon ... G-5
Little Dixie ... D-7
Little Flock ... A-3
Little Rock ... D-5
Lockesburg ... F-3
Locust Bayou ... F-5
Lodge Corner ... E-7
Lodi ... E-3
London ... C-4
Lone Pine ... B-5
Lono ... E-5
Lonoke ... D-6
Lorado ... B-8
Louann ... F-5
Lowell ... A-3
Lurton ... B-4
Luxora ... B-9
Madison ... C-8
Magazine ... C-3
Magnolia ... G-4
Malvern ... E-5
Mammoth Springs ... A-7
Mandeville ... F-3
Manila ... B-9
Manning ... E-5
Mansfield ... D-3
Manson ... A-7
Marble ... B-3
Marcella ... B-6
Marcus ... E-4
Marianna ... D-8
Marion ... C-9
Marked Tree ... C-8
Marmaduke ... A-9
Marshall ... B-5
Martinville ... C-5
Marvell ... D-8
Marysville ... G-5
Masonville ... F-7
Maumelle ... D-6
Mayflower ... D-6
Maynard ... A-7
McArthur ... F-7
McCaskil ... F-4
McCrory ... C-7
McDougal ... A-8
McGehee ... F-7
McKamie ... G-4
McNeil ... G-4
McRae ... C-6
Melbourne ... B-6
Mellwood ... E-8
Menifee ... C-5
Mena ... D-3
Meridian ... G-7
Metalton ... B-4
Midland ... D-2
Middlebrook ... A-7
Midway ... C-4
Miller Chapel ... G-7
Minturn ... B-8
Mineral Springs ... F-3
Mist ... G-7
Mitchellville ... E-7
Moark ... A-8
Moko ... A-6
Monette ... B-9
Monroe ... D-8
Monticello ... F-7
Montongo ... F-6
Montrose ... G-7
Moreland ... C-4
Morning Star ... B-5
Morning Star ... E-5
Moro ... D-8
Morrilton ... C-5
Morris ... F-4
Morton ... C-7
Mound City ... ★
Vicinity ... Pg. 123, C-2
Mounds ... A-9
Mountainburg ... C-3
Mountain Home ... A-5
Mountain Pine ... D-4
Mountain View ... B-6
Mount George ... C-4
Mount Holly ... G-5
Mount Ida ... D-4
Mount Judea ... B-4
Mount Olive ... B-6
Mount Pleasant ... B-6
Mount Vernon ... C-6
Mulberry ... C-3
Murfreesboro ... E-3
Myron ... A-7
Nail ... B-4
Nashville ... E-3
Natural Dam ... C-2
Nebo ... E-6
Needmore ... D-3
Newark ... B-7
New Caledonia ... G-5
New Edinburg ... F-6
Newhope ... E-3
New Hope ... F-6
Newell ... G-5
Newport ... B-7
Nimmons ... A-9
Nimrod ... D-4
Noble Lake ... E-7
Nola ... D-3
Norfork ... A-5
Norman ... E-4
Norphlet ... G-5
North Crossett ... G-6
N. Little Rock ... D-6
Nunley ... E-3
Oak Forest ... D-8
Oak Grove ... C-5
Oak Grove ... E-4
Oak Grove ... E-5
Oak Grove ... F-4
Oakhaven ... F-3
Oakland ... A-5
Oark ... C-3
Oden ... D-3
Ogden ... F-3
Oil Trough ... B-7
O'Kean ... A-8
Okolona ... E-4
Ola ... D-4
Old Alabam ... B-3
Old Milo ... G-6
Old Union ... G-5
Omaha ... A-4
Oneida ... D-8
Onia ... B-5

★ Denotes City located only on Vicinity map. City not located on State map.

ARKANSAS

CALIFORNIA

Onyx D-4
Oppelo D-5
Osceola B-9
Ouachita F-5
Overcup C-5
Oxford A-6
Ozan F-4
Ozark C-3
Ozone C-4
Palestine D-8
Pangburn C-6
Pansy F-6
Paragould B-8
Paris C-3
Parkdale G-7
Parkin C-8
Parks D-3
Parnell D-6
Parthenon B-4
Patterson C-7
Peach Orchard A-8
Pearcy E-4
Pea Ridge A-3
Pelsor D-3
Pencil Bluff D-3
Perla C-5
Perrytown F-4
Perryville D-5
Pettus D-6
Pfeiffer B-7
Pickens C-6
Piggott A-9
Pike City E-4
Pindall B-4
Pinebergen E-6
Pine Bluff E-6
Pine Grove F-6
Pine Ridge D-3
Pine Tree C-8
Pineville A-6
Plainview D-4
Pleasant Grove B-6
Pleasant Plains C-7
Pleasant Valley D-5
Plumerville C-5
Pocahontas A-7
Pollard A-9
Ponca B-4
Pontoon C-4
Portland G-7
Potter D-2
Pottsville C-4
Poughkeepsie B-7
Powhatan B-7
Poyen E-5
Prairie Grove B-2
Prairie View C-4
Princeton E-5
Proctor C-9
Provo E-3
Quinn G-5
Quitman C-6
Raggio C-8
Ralph B-5
Ravenden Springs A-7
Rector A-9
Redfield E-6
Redman Pt. ★
  Vicinity Pg. 123, D-8
Reed F-7
Reyno A-8
Rich D-8
Richmond F-2
Rich Mountain D-2
Rison C-6
Rivervale B-8
Rocky D-2
Roe D-7
Rohwer F-7
Roland D-5
Romance F-3
Rondo F-3
Rose Bud C-6
Roseland B-9
Rosie B-7
Ross Van Ness G-7
Rover D-4
Rowell F-6
Rudy C-2
Rule A-4
Rupert C-5
Rushing B-6
Russellville C-4
Rye A-6
Saddle A-6
Saffell B-7
Sage B-6
St. Charles D-7
St. Claire ★
  Vicinity Pg. 123, B-2
St. James D-8
Salem A-8
Salem D-5
Salem B-7
Saltillo D-6
Salus B-3
Saratoga F-3
Savoy A-2
Scotland C-5
Scott Valley C-6
Searcy C-6
Sedgwick B-8
Sellers Store A-7
Selma F-6
Shannon A-8
Sharum A-8
Shearerville C-9
Shell Lake B-8
Sheridan E-5
Sherwood D-6
Shirley B-6
Shoffner C-7
Shuler G-5
Sidney C-6
Sidon C-6
Siloam A-7
Silver Hill B-5
Sitka D-7
Slovak D-7
Smackover G-6
Snowball B-5
Snow Lake E-8
Snyder G-7
Soudan D-8

Sparkman F-5
Spear Lake B-8
Springdale B-3
Springhill C-6
Spring Hill F-3
Stacy B-8
Stamps G-4
Stanford B-8
Star City E-6
Staves E-6
Stephens F-5
Stewart C-8
Stokes A-7
Story D-4
Strawberry B-7
Strickler B-2
Strong G-6
Stuttgart D-7
Subiaco C-3
Success A-8
Summit F-6
Sumpter C-7
Sunnydale C-7
Sunset C-9
Supply A-8
Swain B-4
Swan Lake E-7
Sweet Home D-6
Swifton B-7
Taylor G-4
Temple F-3
Thebes G-7
Thornburg D-5
Thornton F-5
Three Creeks G-5
Three Forks C-9
Tichnor E-7
Tilly B-5
Tilton C-8
Timbo Fox B-5
Tinsman F-5
Tipp C-7
Tollville E-5
Tontitown B-2
Traskwood D-8
Trenton D-8
Trumann B-9
Tuckerman B-7
Tulip D-5
Tupelo C-7
Turrell C-9
Twist C-8
Tyronza C-9
Ulm D-7
Umpire E-3
Union A-6
Union Hill D-3
Uniontown C-2
Valley Gin G-3
Valley View B-8
Van Buren C-2
Vandervoort E-2
Vanndale C-8
Vesta C-3
Vick G-6
Victoria A-6
Vidette A-6
Vilonia C-6
Village F-5
Viola A-6
Wabbaseka E-7
Waldenburg C-8
Waldo G-4
Waldron D-3
Walker F-2
Walkerville G-4
Wallace F-2
Walnut Corner B-9
Walnut Ridge B-8
Waltreak D-6
Ward D-6
Warren C-3
Washburn C-3
Washington F-4
Washita F-4
Waterloo A-7
Watson F-7
Watson Chapel E-6
Waveland D-3
Webb City F-4
Weber E-8
Wedington B-2
Weiner B-8
Weldon C-7
Welsh E-4
West Crossett G-6
Western Grove B-5
West Fork B-3
West Memphis C-9
  Vicinity Pg. 123, D-1
West Ridge D-8
Wheatley D-8
Wheeling J-4
Whelen Springs F-4
White G-7
Whitehall C-8
White Hall E-6
Whiteville A-5
Wickes E-2
Widener C-8
Wild Cherry A-6
Williams D-5
Williford A-7
Willisville F-4
Wilmar F-6
Wilmot G-7
Wilson B-9
Winchester F-7
Winthrop F-2
Wirth A-7
Witter B-3
Witts Springs B-5
Wolf Bayou B-6
Woodberry F-5
Wooster D-5
Wrightsville D-6
Wyanoke ★
  Vicinity Pg. 123, E-1
Wynne C-8
Yarbro B-9
Yardelle B-5
Yellville B-5

Yorktown E-6
Zent D-7
Zion B-6

CALIFORNIA
Pages 24-27

Acton S-10
Adelanto S-11
Adin C-7
Agoura Hills S-9
Ahwahnee L-8
Alameda ★
  Vicinity Pg. 146, C-3
Alamorio V-15
Albany ★
  Vicinity Pg. 146, C-3
Albion G-2
Alder Springs F-3
Alderpoint E-2
Alhambra ★
  Vicinity Pg. 118, C-5
Alleghany F-6
Almanor E-6
Alpaugh P-8
Alpine V-12
Altadena ★
  Vicinity Pg. 118, B-6
Alta Loma ★
  Vicinity Pg. 120, C-11
Altaville J-6
Alton D-1
Alturas C-7
Amboy S-15
Anaheim T-11
  Vicinity Pg. 121, G-7
Anderson D-4
Angels Camp J-7
Angwin H-3
Annapolis H-2
Antioch J-4
Apple Valley S-12
Aptos L-4
Arbuckle G-4
Arcadia ★
  Vicinity Pg. 120, C-7
Arcata C-1
Arnold J-7
Arroyo Grande Q-6
Artesia ★
  Vicinity Pg. 119, F-6
Artois G-4
Arvin Q-9
Atascadero P-6
Atherton ★
  Vicinity Pg. 145, A-1
Atwater L-6
Auberry L-8
Auburn H-6
Avalon U-10
Avenal N-7
Avon ★
  Vicinity Pg. 146, A-6
Azusa ★
  Vicinity Pg. 120, C-8
Badger M-9
Bakersfield Q-9
Balboa Park ★
  Vicinity Pg. 144, F-3
Baldwin Park ★
  Vicinity Pg. 120, C-7
Banning T-13
Barstow R-12
Bartlett Sprs. G-3
Bass Lake L-8
Baxter G-6
Baywood Park Q-5
Bear Valley K-7
Bear Valley J-8
Beaumont T-13
Beckwourth F-7
Bel Air ★
  Vicinity Pg. 118, C-2
Belden E-6
Bell ★
  Vicinity Pg. 119, E-5
Bellflower ★
  Vicinity Pg. 119, F-6
Bella Vista D-4
Bells Station L-5
Belmont ★
  Vicinity Pg. 147, J-3
Belvedere ★
  Vicinity Pg. 146, D-1
Ben Lomond L-4
Bend E-4
Benicia ★
  Vicinity Pg. 146, A-5
Benton K-10
Berenda L-7
Berkeley J-4
  Vicinity Pg. 146, D-3
Berry Creek F-6
Bertsch Terr. A-1
Beverly Hills S-10
  Vicinity Pg. 118, D-3
Bieber C-6
Big Bar D-3
Big Bear Lake S-13
Big Bend C-5
Big Creek L-9
Big Oak Flat K-7
Big Pine L-10
Big Sur N-4
Biggs G-5
Bishop L-10
Blackwells Corner P-7
Blairsden F-7
Bloomington ★
  Vicinity Pg. 120, B-10
Blue Lake C-2
Blythe U-16
Bodega Bay H-2
Bodfish P-10
Bolinas J-3
Bonita ★
  Vicinity Pg. 144, H-5
Boonville G-3
Boron R-11
Borrego Springs U-13
Boulder Creek L-4
Boulevard V-13

Bowman G-6
Bradley P-6
Brawley V-15
Brea ★
  Vicinity Pg. 121, E-8
Brentwood J-5
Bridgeport J-9
Bridgeville D-2
Broderick H-5
Browns Valley G-5
Brownsville G-6
Bryn Mawr ★
  Vicinity Pg. 120, B-12
Buellton R-6
Buena Park ★
  Vicinity Pg. 121, F-7
Burbank ★
  Vicinity Pg. 118, B-4
Burlingame ★
  Vicinity Pg. 147, H-2
Burney D-5
Burnt Ranch C-3
Butte City F-4
Butte Meadows E-5
Buttonwillow Q-8
Cabazon T-13
Cahuilla U-13
Calexico V-15
Caliente Q-10
California City R-11
California Hot Sprs. P-9
Calipatria V-15
Calistoga H-3
Callahan D-4
Calpella G-2
Calpine F-7
Cambria P-5
Cameron Park H-7
Camino H-7
Camp Nelson N-10
Camp Richardson H-8
Campbell ★
  Vicinity Pg. 145, E-6
Campo V-13
Camptonville G-6
Canby C-7
Canoga Park ★
  Vicinity Pg. 118, B-1
Cantil Q-10
Canyon Dam E-6
Capay H-4
Capetown D-1
Capitola L-4
Carlotta D-1
Carlsbad U-11
Carmel Valley M-4
Carnelian Bay G-8
Carpinteria S-8
Carson ★
  Vicinity Pg. 119, F-4
Carson Hill Cartago N-11
Caruthers M-7
Casmalia R-6
Caspar F-1
Castaic S-9
Castella C-4
Castroville M-4
Cayucos P-5
Cedar Grove M-10
Cedarville C-8
Century City ★
  Vicinity Pg. 118, D-3
Ceres K-6
Chalfant L-10
Challenge F-6
Chatsworth ★
  Vicinity Pg. 118, B-1
Chester E-6
Chico E-5
China Lake P-11
Chino ★
  Vicinity Pg. 120, D-10
Cholame P-7
Chowchilla L-7
Chula Vista V-12
Cima Q-15
Claremont ★
  Vicinity Pg. 120, C-10
Clarksburg H-5
Clay J-6
Clayton ★
  Vicinity Pg. 146, C-7
Clearlake G-3
Clovis M-8
Clyde ★
  Vicinity Pg. 146, A-6
Coachella U-14
Coalinga N-6
Coarsegold L-8
Cobb G-3
Colfax G-6
Coloma H-6
Colton ★
  Vicinity Pg. 120, B-11
Colusa G-4
Commerce ★
  Vicinity Pg. 118, D-5
Compton ★
  Vicinity Pg. 119, F-5
Concord J-4
  Vicinity Pg. 146, B-6
Copperopolis J-6
Corcoran N-8
Corning E-4
Corona ★
  Vicinity Pg. 121, F-11
Coronado V-12
Costa Mesa ★
  Vicinity Pg. 121, H-8
Cottonwood D-4
Coulterville K-7
Courtland J-5
Covelo E-3
Covina ★
  Vicinity Pg. 120, C-8
Cranell C-1
Crescent City B-1
Crescent Mills E-6
Crestmore ★
  Vicinity Pg. 120, B-10
Crockett ★
  Vicinity Pg. 146, A-6
Cucamonga ★
  Vicinity Pg. 120, C-11

Cudahy ★
  Vicinity Pg. 119, E-5
Culver City ★
  Vicinity Pg. 118, D-3
Cummings E-2
Cupertino ★
  Vicinity Pg. 145, D-6
Cutler N-9
Cutten C-1
Cuyama R-8
Cypress ★
  Vicinity Pg. 119, G-6
Daggett R-12
Dairyville E-4
Dales E-4
Daly City K-3
Dana Point U-10
Danville ★
  Vicinity Pg. 146, D-6
Darwin N-11
Davis H-5
Davis Creek B-8
Death Valley Jct. N-13
Del Mar V-11
Delano P-9
Desert Center T-15
Desert Hot Sprs. T13
Desert Shores U-14
Devils Den P-7
Diamond Sprs. H-6
Dictionary Hill ★
  Vicinity Pg. 144, G-6
Dillon Beach J-2
Dinkey Creek L-9
Dixon H-5
Dobbins G-6
Dominguez ★
  Vicinity Pg. 119, F-5
Dorris B-5
Dos Palos L-6
Dos Rios F-2
Douglas City D-3
Downey ★
  Vicinity Pg. 119, E-5
Doyle F-8
Duarte ★
  Vicinity Pg. 120, C-7
Ducor P-9
Dulzura V-12
Dunsmuir C-4
Durham F-5
Eagle Rock ★
  Vicinity Pg. 118, C-5
Earlimart P-9
Earp T-17
East Highlands ★
  Vicinity Pg. 120, A-12
East Los Angeles ★
  Vicinity Pg. 118, D-5
East Palo Alto ★
  Vicinity Pg. 145, A-2
Edison Q-9
El Cajon V-13
El Centro V-15
El Cerrito ★
  Vicinity Pg. 146, C-3
El Granada ★
  Vicinity Pg. 147, J-1
El Monte ★
  Vicinity Pg. 118, C-6
El Nido L-7
El Portal K-8
El Segundo ★
  Vicinity Pg. 119, F-3
El Toro T-11
Elk Creek F-4
Elk Grove J-5
Elmira J-4
Emeryville ★
  Vicinity Pg. 146, D-3
Encinitas V-11
Encino ★
  Vicinity Pg. 118, C-2
Enterprise D-4
Escalon K-6
Escondido U-11
Esparto H-4
Essex R-15
Etiwanda ★
  Vicinity Pg. 120, C-12
Etna B-4
Eucalyptus Hills ★
  Vicinity Pg. 144, C-6
Eureka C-1
Exeter N-9
Fairfield J-4
Fall River Mills C-6
Fallbrook U-12
Farmersville N-9
Farmington J-6
Fawnskin S-12
Feather Falls F-6
Fellows Q-8
Fenner R-15
Ferndale D-1
Fields Landing D-1
Fillmore S-9
Firebaugh M-6
Fish Camp L-8
Fish Rock G-2
Five Points N-7
Flint Ridge ★
  Vicinity Pg. 118, B-5
Florence ★
  Vicinity Pg. 119, G-6
Florin H-5
Floriston G-8
Flournoy E-4
Folsom H-6
Fontana ★
  Vicinity Pg. 120, C-11
Forbestown F-5
Ford City Q-8
Forest Glen D-3
Forest Ranch F-5
Foresthill G-6
Forks Of Salmon C-3
Fortuna D-1
Foster City ★
  Vicinity Pg. 147, H-4

Fountain Valley ★
  Vicinity Pg. 121, H-7
Fowler M-8
Frazier Park R-9
Fredericksburg H-8
Fremont K-4
  Vicinity Pg. 145, D-3
Frenbrook ★
  Vicinity Pg. 144, B-7
French Gulch D-4
Fresno M-8
Ft. Bidwell B-8
Ft. Bragg F-1
Ft. Dick A-1
Ft. Jones B-4
Fullerton T-11
  Vicinity Pg. 121, F-7
Furnace Creek Ranch N-13
Galt J-5
Garberville E-2
Gardena ★
  Vicinity Pg. 119, F-4
Garden Grove ★
  Vicinity Pg. 121, G-7
Gasquet A-2
Gaviota R-7
Gazelle B-4
Georgetown H-6
Gerber E-4
Geyserville H-3
Giant Forest Village N-10
Gilroy L-5
Glamis V-16
Glen Avon ★
  Vicinity Pg. 120, D-12
Glen Ellen H-3
Glendale S-10
Glendora S-11
  Vicinity Pg. 120, C-8
Glenn F-4
Glennville P-10
Glenview ★
  Vicinity Pg. 144, F-7
Goffs R-16
Goleta S-7
Gonzales M-5
Gorman R-9
Goshen N-8
Graeagle F-7
Granada Hills ★
  Vicinity Pg. 118, A-1
Grand Terrace ★
  Vicinity Pg. 120, B-11
Graniteville G-7
Grass Valley G-6
Green Acres C-8
Greenfield N-5
Greenview B-4
Greenville E-6
Grenada B-4
Gridley G-5
Grimes G-4
Grossmont ★
  Vicinity Pg. 144, F-5
Groveland K-7
Grover City Q-6
Guadalupe Q-6
Gualala G-2
Guerneville H-2
Guinda H-4
Gustine L-6

Independence M-10
Indio T-14
Industry ★
  Vicinity Pg. 120, D-7
Inglewood ★
  Vicinity Pg. 119, E-3
Ingot D-5
Inverness J-3
Inyokern P-11
Ione J-6
Irvine ★
  Vicinity Pg. 121, H-9
Isleton J-5
Ivanpah Q-15
Jackson J-6
Jamacha ★
  Vicinity Pg. 144, F-7
Jamacha Junction ★
  Vicinity Pg. 144, F-6
Jamestown K-7
Jamul V-12
Janesville E-7
Johannesburg Q-11
Johnstonville E-7
Johnstown ★
  Vicinity Pg. 144, D-7
Jolon N-5
Joshua Tree S-13
Julian V-13
Junction City D-3
June Lake K-9
Keddie E-6
Keeler N-11
Keene Q-10
Kelseyville G-3
Kelso R-15
Kenwood H-3
Kerman M-7
Kernville P-10
Kettleman City N-7
King City N-5
Kings Beach G-8
Kingsburg N-8
Kirkwood H-7
Kit Carson H-7
Klamath Glen B-2
Klamath River B-4
Kneeland D-2
Knights Ferry K-6
Knights Landing H-5
Kyburz H-7
La Canada ★
  Vicinity Pg. 118, B-5
La Crescenta ★
  Vicinity Pg. 118, B-4
Lafayette ★
  Vicinity Pg. 146, C-5
La Grange K-7
La Habra ★
  Vicinity Pg. 121, E-7
La Honda K-4
La Jolla V-11
La Jolla Mesa ★
  Vicinity Pg. 144, D-1
Lakeside ★
  Vicinity Pg. 144, D-7
Lakewood ★
  Vicinity Pg. 119, F-5
La Mesa ★
  Vicinity Pg. 144, F-5
La Mirada ★
  Vicinity Pg. 121, F-7
La Palma ★
  Vicinity Pg. 119, F-6
La Presa ★
  Vicinity Pg. 144, G-6
La Puente ★
  Vicinity Pg. 120, D-7
La Quinta T-13
Laguna Beach U-10
Lake City B-8
Lake Elsinore T-12
Lakehead D-4
Lakeport G-3
Lamont Q-9
Lancaster R-10
Lawndale ★
  Vicinity Pg. 119, F-3
Laytonville F-2
Lee Vining K-9
Leggett E-2
Lemon Grove ★
  Vicinity Pg. 144, G-5
Lemoore N-7
Lenwood R-12
Leucadia V-11
Likely C-7
Lincoln H-5
Lincoln Acres ★
  Vicinity Pg. 144, H-4
Linda G-5
Linden J-6
Lindsay N-9
Litchfield E-7
Little Lake P-11
Live Oak G-5
Livermore K-4
Livingston L-6
Lockeford J-6
Lockwood N-5
Lodi J-5
Loleta D-1
Loma Linda ★
  Vicinity Pg. 120, B-11
Lomita ★
  Vicinity Pg. 119, G-4
Lompoc R-6
Lone Pine N-10
Long Beach T-10
  Vicinity Pg. 119, G-5
Longvale F-2
Lookout C-6
Loomis H-6
Los Alamos R-7
Los Altos ★
  Vicinity Pg. 145, C-3
Los Altos Hills ★
  Vicinity Pg. 145, C-2
Los Angeles T-9
  Vicinity Pgs. 118-121
Los Banos L-6
Los Gatos ★
  Vicinity Pg. 145 F-5

Los Molinos E-4
Los Olivos R-7
Lost Hills P-8
Lower Lake G-3
Loyalton F-7
Lucerne H-3
Lucerne Valley S-12
Ludlow R-14
Lynwood ★
  Vicinity Pg. 119, E-5
Lynwood Hills ★
  Vicinity Pg. 144, A-5
Macdoel B-5
Mad River D-2
Madeline D-7
Madera M-7
Malaga M-8
Malibu S-9
Mammoth Lakes K-9
Manchester G-2
Manhattan Beach ★
  Vicinity Pg. 119, F-3
Manteca K-5
Manton D-5
Maricopa Q-8
Marina M-4
Markleeville H-8
Martinez ★
  Vicinity Pg. 144, D-1
Marysville G-4
Maxwell G-4
Maywood ★
  Vicinity Pg. 118, E-5
McArthur C-6
McCloud C-5
McFarland P-9
McKinleyville C-1
McKittrick Q-8
Meadow Valley F-6
Meadow Vista G-6
Mecca U-14
Meeks Bay H-6
Mendocino F-2
Mendota M-7
Menlo Park ★
  Vicinity Pg. 145, A-2
Merced L-7
Mercy Hot Springs M-6
Meridian G-5
Meyers H-7
Mi-wuk Village K-7
Middletown H-3
Midpines K-8
Milford E-7
Millbrae ★
  Vicinity Pg. 147, H-2
Millville D-4
Milpitas K-4
  Vicinity Pg. 145, A-4
Mineral E-5
Mira Loma ★
  Vicinity Pg. 120, E-12
Miramar ★
  Vicinity Pg. 147, J-1
Miranda E-2
Mission Beach ★
  Vicinity Pg. 144, F-1
Mission Hills ★
  Vicinity Pg. 118, B-2
Modesto K-6
Mojave R-10
Mokelumne Hill J-7
Mono Hot Springs L-9
Mono Lake K-9
Monrovia ★
  Vicinity Pg. 120, C-7
Montague B-4
Montara ★
  Vicinity Pg. 147, H-1
Montclair ★
  Vicinity Pg. 120, D-10
Monte Rio H-2
Montebello ★
  Vicinity Pg. 118, D-6
Montecito S-8
Monterey M-4
Monterey Park ★
  Vicinity Pg. 118, D-5
Monte Serona ★
  Vicinity Pg. 145, E-4
Montgomery Creek C-5
Montrose ★
  Vicinity Pg. 118, B-5
Moorpark S-9
Moraga ★
  Vicinity Pg. 146, D-5
Moreno ★
  Vicinity Pg. 144, C-7
Morgan ★
  Vicinity Pg. 145, G-10
Morgan Hill L-5
Mormon Bar L-8
Morongo Valley T-13
Morro Bay Q-5
Moss Beach ★
  Vicinity Pg. 147, J-1
Moss Landing M-4
Mt. Bullion K-7
Mt. Helix ★
  Vicinity Pg. 144, F-6
Mt. Laguna V-13
Mt. Shasta C-4
Mountain View ★
  Vicinity Pg. 145, B-3
Mugginsville B-4
Murphys J-7
Murrieta U-12
Napa J-4
National City ★
  Vicinity Pg. 144, H-4
Navarro G-2
Needles R-16
Nevada City G-6
Newark ★
  Vicinity Pg. 147, J-6
New Cuyama R-7
New Idria N-6
Newberry Springs R-13
Newell B-7
Newport Beach T-10
  Vicinity Pg. 121, J-8
Nice G-3

★ Denotes City located only on Vicinity map. City not located on State map.

Niland U-15
Nipomo Q-6
Nipton Q-15
Norco ★
Vicinity Pg. 121, E-11
Nord F-5
North Edwards R-11
North Fork L-8
North Hollywood ★
Vicinity Pg. 118, C-3
Northridge ★
Vicinity Pg. 118, B-1
Norwalk ★
Vicinity Pg. 119, F-6
Novato J-3
Noyo F-2
Nubieber C-6
O'Neals L-8
Oakdale K-6
Oakhurst L-8
Oakland K-4
Vicinity Pg. 146, E-4
Oasis L-11
Oceano Q-6
Oceanside U-11
Ocotillo V-14
Ocotillo Wells V-14
Ojai S-8
Olancha N-11
Old Station D-6
Olema J-3
Olinda D-4
Ono D-4
Ontario T-11
Vicinity Pg. 120, D-10
Onyx P-10
Orange ★
Vicinity Pg. 121, G-8
Orange Cove M-9
Orcutt R-6
Ordbend F-5
Orick B-1
Orland F-4
Orleans C-2
Oro Grande R-12
Orosi N-9
Oroville F-5
Otay ★
Vicinity Pg. 144, K-5
Otay Mesa ★
Vicinity Pg. 144, K-7
Oxnard S-8
Pacific Grove M-4
Pacific Palisades ★
Vicinity Pg. 118, D-1
Pacifica K-3
Vicinity Pg. 147, G-1
Paicines M-5
Pala U-12
Palermo G-5
Palms ★
Vicinity Pg. 118, D-3
Palm Desert T-13
Palm Springs T-13
Palmdale S-10
Palo Alto K-4
Vicinity Pg. 145, B-2
Vicinity Pg. 147, K-5
Palo Verde U-16
Palomar Mountain U-13
Palos Verdes Estates ★
Vicinity Pg. 119, G-3
Panamint Sprs. N-12
Panoche M-6
Panorama City ★
Vicinity Pg. 118, B-2
Paradise F-5
Paramount ★
Vicinity Pg. 119, F-5
Parker Dam S-17
Parkfield P-6
Pasadena S-10
Vicinity Pg. 118, C-5
Paskenta E-4
Paso Robles P-6
Patterson K-6
Patton ★
Vicinity Pg. 120, A-12
Pauma Valley U-12
Pearblossom S-11
Pebble Beach M-4
Pedley ★
Vicinity Pg. 121, E-12
Penn Valley G-6
Perris T-12
Petrolia D-1
Phelan S-11
Philo G-2
Pico Rivera ★
Vicinity Pg. 118, E-6
Piedmont ★
Vicinity Pg. 146, D-4
Piercy E-2
Pine Grove J-6
Pine Valley V-13
Pinecrest J-8
Pinedale M-8
Pinole ★
Vicinity Pg. 146, B-3
Pioneer Station J-7
Pismo Beach Q-6
Pittsburg ★
Vicinity Pg. 146, A-7
Pixley P-9
Placentia ★
Vicinity Pg. 121, F-8
Placerville H-6
Planada L-7
Pleasant Hill ★
Vicinity Pg. 146, B-6
Plymouth H-6
Pollock Pines H-7
Pomona T-11
Vicinity Pg. 120, D-9
Pondosa C-5
Poplar P-9
Port Hueneme S-8
Portola F-7
Portola Valley ★
Vicinity Pg. 145, C-1
Portuguese Bend ★
Vicinity Pg. 119, H-3
Potter Valley F-3
Poway V-12
Vicinity Pg. 144, B-5
Pozo Q-6
Princeton G-4
Vicinity Pg. 147, J-1
Project City D-4
Point Arena G-2
Pulga F-6
Pumpkin Center Q-9
Quincy F-6
Raisin M-7
Ramona V-12
Rancho Palos VerdesT-9
Randsburg Q-11
Ravendale D-7
Red Bluff E-4
Red Mountain Q-11
Redcrest D-1
Redding D-4
Redlands ★
Vicinity Pg. 120, B-12
Redondo Beach T-9
Vicinity Pg. 119, G-3
Redway E-2
Redwood City ★
Vicinity Pg. 147, J-4
Redwood Valley G-2
Requa B-1
Reseda ★
Vicinity Pg. 118, B-1
Rialto ★
Vicinity Pg. 120, A-10
Richmond ★
Vicinity Pg. 146, C-2
Richvale F-5
Ridgecrest Q-11
Rio Dell D-1
Rio Linda H-5
Rio Oso G-5
Rio Vista J-5
Ripley U-16
Ripon K-5
River Pines H-6
Riverdale N-7
Riverside T-12
Vicinity Pg. 121, E-12
Rocklin H-5
Rockport F-2
Rodeo ★
Vicinity Pg. 146, A-3
Rohnert Park J-3
Rolling Hills ★
Vicinity Pg. 119, G-3
Rolling Hills Estates ★
Vicinity Pg. 119, G-3
Romoland T-12
Rosamond R-10
Rosemead ★
Vicinity Pg. 118, D-6
Roseville H-5
Rough And Ready G-6
Round Mountain D-5
Rovana L-10
Rowland Heights ★
Vicinity Pg. 120, E-8
Rutherford H-4
South Lake Tahoe H-8
Sacramento H-5
Salinas M-5
Salton City U-14
Salyer C-2
San Andreas J-7
San Ardo N-6
San Bernardino T-12
Vicinity Pg. 120, A-11
San Carlos ★
Vicinity Pg. 147, J-3
San Clemente U-11
San Diego V-12
Vicinity Pg. 144
San Dimas ★
Vicinity Pg. 120, C-9
San Fernando S-10
Vicinity Pg. 118, A-2
San Francisco K-3
Vicinity Pg. 147
San Gabriel ★
Vicinity Pg. 118, C-6
San Jacinto T-12
San Joaquin M-7
San Jose L-5
Vicinity Pg. 145, D-8
San Juan Bautista M-5
San Juan CapistranoU-11
San Leandro ★
Vicinity Pg. 147, F-5
San Lorenzo ★
Vicinity Pg. 147, G-5
San Lucas N-5
San Luis Obispo Q-6
San Marcos U-12
San Marino ★
Vicinity Pg. 118, C-6
San Martin L-5
San Mateo K-3
Vicinity Pg. 147, H-3
San Miguel P-6
San Pablo ★
Vicinity Pg. 146, B-3
San Pedro ★
Vicinity Pg. 119, H-4
San Rafael J-3
Vicinity Pg. 146, B-1
San Simeon P-5
Sanger M-8
Santa Ana T-11
Vicinity Pg. 121, H-8
Santa Barbara S-7
Santa Clara L-4
Vicinity Pg. 145, C-5
Vicinity Pg. 147, L-3
Santa Cruz L-4
Santa Fe Springs ★
Vicinity Pg. 119, E-6
Santa Margarita Q-6
Santa Maria Q-6
Santa Monica T-9
Vicinity Pg. 118, E-2
Santa Nella L-6
Santa Paula S-8
Santa Rosa H-3
Santa Ysabel V-13
Santee V12
Vicinity Pg. 144, D-6
Saratoga L-4
Vicinity Pg. 145, E-4
Saticoy S-9
Sattley F-7
Sausalito J-3
Vicinity Pg. 146, D-1
Sawyers Bar C-3
Scotia D-1
Scotts Valley L-4
Scottys Castle M-12
Seeley V-14
Selma M-8
Shafter Q-8
Shandon P-6
Shasta D-4
Shaver Lake L-8
Shelter Cove E-1
Shingletown D-5
Short Acres ★
Vicinity Pg. 146, A-6
Shoshone P-14
Sierra City F-7
Sierraville F-7
Signal Hill ★
Vicinity Pg. 119, G-5
Simi Valley S-9
Simmler Q-7
Sisquoc R-7
Sites G-4
Sloat F-7
Smith River A-2
Snelling K-7
Soda Springs G-7
Solana Beach ★
Vicinity Pg. 144, A-1
Soledad N-5
Solvang R-7
Somerset H-7
Somes Bar B-3
Sonoma J-3
Sonora K-7
Soquel L-4
Sorrento ★
Vicinity Pg. 144, C-1
South El Monte ★
Vicinity Pg. 118, D-6
South Gate ★
Vicinity Pg. 119, E-5
South Pasadena ★
Vicinity Pg. 118, C-5
S. San Francisco ★
Vicinity Pg. 147, G-2
Spring Valley ★
Vicinity Pg. 144, G-6
Springville N-9
Squaw Valley G-7
St. Helena H-3
Standish E-7
Stanton ★
Vicinity Pg. 121, G-7
Stewarts Point H-2
Stirling City F-5
Stockton K-5
Stonyford G-4
Stovepipe Wells N-12
Stratford N-8
Strathmore N-9
Studio City ★
Vicinity Pg. 118, C-3
Suisun City J-4
Summerland S-8
Summit City D-4
Sun City T-12
Sunland ★
Vicinity Pg. 118, A-3
Sunnyside ★
Vicinity Pg. 144, H-6
Sunnyvale K-5
Vicinity Pg. 145, C-4
Vicinity Pg. 147, K-6
Sun Valley ★
Vicinity Pg. 118, B-3
Surf R-6
Susanville E-7
Sutter G-5
Sutter Creek J-6
Taft Q-8
Tahoe City G-7
Talmage G-3
Tarzana ★
Vicinity Pg. 118, C-1
Taylorsville E-7
Tecopa P-14
Tehachapi Q-10
Temecula U-12
Temple City ★
Vicinity Pg. 118, C-6
Templeton P-6
Terminous J-5
Termo D-7
Terra Bella P-8
Thornton J-5
Thousand Oaks S-9
Thousand Palms T-13
Three Rivers N-9
Tiburon ★
Vicinity Pg. 146, D-1
Tipton N-8
Tomales J-3
Toms Place L-10
Topanga ★
Vicinity Pg. 118, D-1
Topanga Beach ★
Vicinity Pg. 118, D-1
Topaz H-8
Torrance ★
Vicinity Pg. 119, F-3
Tracy K-5
Tranquillity M-7
Tres Pinos M-5
Trinidad C-1
Trinity Center C-3
Trona P-12
Truckee G-7
Tujunga ★
Vicinity Pg. 118, A-4
Tulelake B-6
Tuolumne K-7
Turlock K-6
Tustin ★
Vicinity Pg. 121, G-9
Tuttletown K-7
Twain F-6
Twain Harte K-7
Twentynine Palms S-14
Ukiah G-2
Union City ★
Vicinity Pg. 147, H-6
Universal City ★
Vicinity Pg. 118, C-3
Upland ★
Vicinity Pg. 120, C-10
Upper Lake G-3
Vacaville H-4
Valencia S-9
Vallejo J-4
Valley Springs J-6
Van Nuys ★
Vicinity Pg. 118, B-2
Venice ★
Vicinity Pg. 118, E-2
Ventura S-8
Victorville S-12
Vidal T-17
Vina E-4
Vinton F-8
Visalia N-8
Vista U-12
Walker J-8
Walnut ★
Vicinity Pg. 120, D-8
Walnut Creek K-4
Vicinity Pg. 146, C-6
Walnut Grove J-5
Wasco P-8
Washington G-7
Waterford K-6
Waterloo J-6
Watsonville L-4
Watts ★
Vicinity Pg. 119, E-4
Wawona K-8
Weaverville D-3
Weed C-4
Weitchpec C-2
Wendel E-8
Weott D-1
West Covina ★
Vicinity Pg. 120, D-8
West Hollywood ★
Vicinity Pg. 118, D-3
West Los Angeles ★
Vicinity Pg. 118, D-2
West Point J-7
West Riverside ★
Vicinity Pg. 120, D-12
Westley K-6
Westminister ★
Vicinity Pg. 121, G-7
Westmorland V-14
Westport F-2
Westwood E-6
Wheatland G-5
Whitmore D-5
Whittier ★
Vicinity Pg. 121, E-7
Williams G-4
Willits F-2
Willow Creek C-2
Willow Ranch B-8
Willow Springs R-10
Willows F-4
Wilmington ★
Vicinity Pg. 119, G-4
Winter Gardens ★
Vicinity Pg. 144, D-6
Winterhaven V-16
Winters H-4
Winton L-6
Wofford Heights P-10
Woodfords H-8
Woodlake N-9
Woodland H-4
Woodville N-9
Woody P-9
Wrightwood S-11
Yermo R-12
Yolo H-5
Yorba Linda ★
Vicinity Pg. 121, F-9
Yorkville G-2
Yosemite Village K-8
Yreka B-4
Yuba City G-5
Yucaipa T-12
Yucca Valley S-13

COLORADO
Pages 28-29

Abarr C-12
Agate D-10
Aguilar J-9
Akron C-11
Alamosa J-7
Allison J-5
Alma E-7
Almont F-6
Amherst B-12
Antero Jct. F-7
Anton D-11
Antonito J-7
Arapahoe E-12
Arboles J-5
Arlington G-11
Aroya F-11
Arriba E-11
Arriola J-3
Arvada D-8
Vicinity Pg. 112, C-3
Aspen E-6
Atwood B-11
Ault B-9
Aurora D-9
Vicinity Pg. 112, E-7
Austin F-4
Avon D-6
Avondale G-9
Bailey E-8
Barnesville B-9
Bartlett H-12
Basalt E-6
Baxter G-9
Bayfield J-5
Bedrock G-3
Bellvue B-8
Bennett D-9
Berthoud C-8
Beshoar Jct. J-10
Bethune D-12
Beulah G-8
Black Hawk D-8
Blanca H-8
Blue Mountain C-3
Blue River E-7
Bond D-6
Boone G-10
Boulder C-8
Bowie F-5
Boyero E-11
Brandon F-12
Branson J-10
Breckenridge D-7
Breen J-4
Briggsdale B-9
Brighton C-8
Bristol G-12
Broomfield D-8
Vicinity Pg. 112, A-3
Brush C-10
Buckingham B-9
Buena Vista F-7
Buford D-5
Burlington D-12
Byers D-9
Cahone H-3
Calhan E-10
Cameo E-4
Campo J-12
Canon City F-8
Capulin J-7
Carbondale E-5
Carlton G-12
Castle Rock E-9
Cedaredge F-4
Center H-7
Chama J-8
Chatfield Acres ★
Vicinity Pg. 112, K-4
Cheney Center G-12
Cheraw G-11
Cherry Hill Village ★
Vicinity Pg. 112, G-5
Cheyenne Wells F-12
Chimney Rock J-5
Chipita Park F-9
Chivington F-12
Chromo J-6
Cimarron G-5
Clarkville B-12
Clifton E-4
Climax E-7
Coaldale G-7
Coalmont C-6
Collbran E-4
Colorado City G-8
Colorado Springs F-9
Columbine ★
Vicinity Pg. 112, J-3
Commerce City D-9
Vicinity Pg. 112, D-6
Como E-7
Conejos J-7
Conifer D-8
Cope D-11
Cortez J-3
Cotopaxi G-8
Cowdrey B-6
Craig B-5
Crawford F-5
Creede H-6
Crested Butte F-6
Cripple Creek F-8
Critchell ★
Vicinity Pg. 112, K-1
Crook B-11
Crowley G-10
Cuchara J-9
Dailey B-11
DeBeque E-4
Deckers E-8
Deermont ★
Vicinity Pg. 112, K-2
Deer Ridge C-7
Deer Trail D-10
Delhi H-10
Del Norte H-7
Delta F-4
Denver D-9
Vicinity Pg. 112
Dillon D-7
Dinosaur C-3
Divide F-8
Dolores H-4
Dotsero D-5
Dove Creek H-3
Dowd D-6
Doyleville F-6
Drake C-8
Dunton H-4
Dupont ★
Vicinity Pg. 112, C-6
Durango J-5
Eads F-12
Eagle D-6
Eastlake ★
Vicinity Pg. 112, A-5
Eaton B-9
Echo Lake D-7
Eckley C-12
Edgewater ★
Vicinity Pg. 112, E-3
Edwards D-6
Egnar H-3
El Dorado Estates ★
Vicinity Pg. 112, C-1
Eldorado Springs D-8
Elizabeth E-9
Elk Springs C-4
Ellicott F-9
Empire D-7
Englewood D-8
Vicinity Pg. 112, G-5
Estes Park C-7
Evans C-9
Fairplay E-7
Falcon E-9
Farisita H-8
Federal Heights ★
Vicinity Pg. 112, B-4
Fenders ★
Vicinity Pg. 112, J-1
Firstview F-12
Flagler E-11
Fleming B-11
Florence G-8
Florissant F-8
Ford G-10
Fort Collins B-8
Fort Garland J-8
Fort Lewis J-4
Fort Lupton C-9
Fort Lyon G-11
Fort Morgan C-10
Fountain F-9
Fowler G-10
Franktown E-9
Fraser D-7
Frederick C-8
Frisco D-7
Fruita E-3
Galatea F-11
Garcia J-8
Gardner H-8
Garfield F-7
Garo Park E-7
Gateway F-3
Genoa E-11
Georgetown D-7
Gilcrest C-9
Gilman D-7
Gladstone H-5
Glendale ★
Vicinity Pg. 112, F-5
Glenwood Springs D-5
Golden D-8
Vicinity Pg. 112, E-1
Goodrich C-10
Gould B-7
Granada G-12
Granby C-7
Grand Junction E-3
Grand Lake C-7
Grandview Estates ★
Vicinity Pg. 112, J-7
Granite E-7
Grant E-7
Greeley C-9
Greenwood Village D-8
Vicinity Pg. 112, H-5
Guffey F-8
Gunnison F-6
Gypsum D-6
Hamilton C-5
Hartsel E-7
Hasty G-11
Haswell F-11
Hawley G-10
Haxtun B-12
Hayden C-5
Henderson ★
Vicinity Pg. 112, A-6
Hermosa J-5
Hesperus J-4
Hiawatha B-3
Hillrose C-10
Hillside G-8
Hoehne J-9
Holly G-12
Holyoke B-12
Homelake H-7
Homewood Park ★
Vicinity Pg. 112, K-1
Hooper H-7
Hotchkiss F-5
Hot Sulphur Springs C-7
Howard F-7
Hudson C-9
Hugo E-11
Idaho Sprs. D-8
Idalia D-12
Idledale ★
Vicinity Pg. 112, G-1
Ignacio J-5
Iliff B-11
Indian Hills ★
Vicinity Pg. 112, H-1
Jansen J-9
Jaroso J-8
Jefferson E-7
Joes D-12
Johnson Village F-7
Julesburg A-12
Keenesburg C-9
Kersey C-9
Kim J-11
Kiowa E-9
Kit Carson F-12
Kline J-4
Komman G-12
Kremmling C-7
Lafayette C-8
Laird C-12
La Jara J-7
La Junta G-11
Lake City G-5
Lake George F-8
Lakewood D-8
Vicinity Pg. 112, F-2
Lamar G-12
Larkspur E-9
La Salle C-9
Las Animas G-11
Last Chance D-10
La Veta H-9
Lay C-4
Lazear F-5
Leadville E-7
Lebanon J-4
Lewis J-3
Limon E-10
Lindon D-11
Littleton D-8
Vicinity Pg. 112, H-4
Lochbule C-9
Log Lane Village C-10
Loma E-3
Longmont C-8
Loveland C-8
Lucerne C-9
Lycan H-12
Lyons C-8
Mack E-3
Maher F-5
Malta E-6
Manassa J-7
Mancos J-4
Manitou Springs F-8
Manzanola G-10
Marble E-5
Marshall ★
Vicinity Pg. 112, A-1
Marvel J-4
Masters C-9
Matheson E-10
Maybell B-4
McClave G-11
McCoy D-6
Meeker C-4
Meeker Park C-8
Merino B-10
Mesa E-4
Mesita J-8
Milliken C-8
Milner B-5
Mineral Hot Springs G-7
Minturn D-6
Model J-10
Moffat G-7
Molina E-4
Monte Vista H-7
Montrose G-4
Monument E-9
Monument Park J-9
Morrison ★
Vicinity Pg. 112, G-2
Mosca H-7
Mount Princeton
Hot Springs F-7
Mountain View ★
Vicinity Pg. 112, D-3
Nathrop F-7
Naturita G-3
Nederland D-7
New Castle D-5
Northglenn D-8
Vicinity Pg. 112, B-5
Norwood G-4
Nucla G-4
Nunn B-9
Oak Creek C-6
Olathe F-4
Olney Springs G-10
Ophir H-4
Orchard C-9
Orchard City F-4
Ordway G-10
Otis C-11
Ouray G-5
Ovid A-12
Oxford J-5
Pagosa Sprs. J-6
Palisade E-4
Palmer Lake E-8
Paoli B-12
Paonia F-5
Parachute E-4
Paradox G-3
Park F-9
Parkdale F-8
Parker D-9
Parlin F-6
Parshall C-7
Peetz A-11
Penrose F-9
Peyton E-9
Phillipsburg ★
Vicinity Pg. 112, K-2
Phippsburg C-6
Pierce B-9
Pinewood Springs C-8
Placerville G-4
Platoro J-6
Platteville C-9
Pleasant View H-3
Vicinity Pg. 112, E-1
Poncha Springs F-7
Poudre Park B-8
Powderhorn G-6
Pritchett H-12
Proctor B-11
Prospect Valley C-9
Pryor H-9
Pueblo G-9
Pueblo West G-9
Punkin Center F-10
Ramah E-9
Rand C-7
Rangely C-3
Raymer B-10
Raymond C-8
Redcliff E-6
Redmesa J-4
Red Mountain H-5
Redstone E-5
Redvale G-4
Rico H-4
Ridgway G-4
Rifle D-4
Rio Blanco D-4
Riverside ★
Vicinity Pg. 112, K-3
Rockport B-9
Rockwood H-5
Rocky G-10
Roggen C-9
Rollinsville D-7
Romeo J-7
Royal Gorge G-8
Rush F-10
Rustic B-7
Rye H-8
Saguache G-7
Salida F-7
San Acacio J-8
Sanford J-7
San Francisco J-8
San Luis J-8
San Pablo J-8
Sapinero G-5
Sargents G-7
Saw Pit G-4
Security-Widefield F-9
Sedalia E-8
Sedgewick A-11
Segundo J-9
Seibert E-11
Shawnee E-8
Sheridan F-12
Vicinity Pg. 112, G-4
Sheridan Lake F-12
Silt D-5
Silver Cliff G-8
Silverthorne D-7
Silverton H-5
Simla E-10
Skyway E-4
Slick Rock G-3
Snow Mass E-6
Snow Mass Village E-6
Snyder C-10
Somerset F-5
South Fork H-6
Spar City H-6
Springfield H-12
Starkville J-9
State Bridge D-6
Steamboat Springs B-6
Sterling B-11
Stoneham B-10
Stoner H-4
Stonewall J-9
Strasburg D-9
Stratton D-12
Sugar City G-11
Summitville H-6
Sunbeam B-4
Superior ★
Vicinity Pg. 112, A-2
Swink G-10
Tabernash D-7
Telluride H-5
Texas Creek G-8
Thatcher H-10
The Forks B-8
Thronton D-8
Vicinity Pg. 112, B-5
Timpas H-10
Tiny Town ★
Vicinity Pg. 112, H-1
Tobe J-10
Toonerville H-11
Toponas C-6
Towaoc J-3
Towner F-12
Trinidad J-9
Twin Forks ★
Vicinity Pg. 112, H-1
Twinlakes E-6
Two Buttes H-12
Tyrone H-10
Uravan G-3
Utleyville J-11
Vail D-7
Valdez J-9
Vancorum G-3
Victor F-8
Vilas H-12
Villa Grove G-7
Vineland G-9
Virginia Dale B-8
Vona E-11
Wagon Wheel Gap H-6
Walden B-6
Walsenburg H-9
Walsh H-12
Watkins D-9
Welby ★
Vicinity Pg. 112, C-5
Weldona C-10
Westcliffe G-8
Westcreek E-8
Westminster D-8
Vicinity Pg. 112, B-3
Weston J-9
Wetmore G-8
Wheat Ridge D-8
Vicinity Pg. 112, E-3
Whitewater F-4
Wiggins C-10
Wild Horse F-11
Wiley G-12
Windsor B-8
Winter Park D-7
Wolcott D-6
Woodland Park E-8
Woodrow C-10
Woody Creek E-6
Wray C-12
Yampa C-6
Yellow Jacket H-3
Yoder F-9
Yuma C-12

CONNECTICUT
Page 30

Almyville D-6
Andover D-4
Ansonia F-2
Attawaugan D-5
Avon D-2
Bakersville D-2
Bantam D-2
Barkhamsted D-3
Bashan E-4
Belltown F-1
Berlin E-3
Bethel F-2
Black Point F-5
Bloomfield D-3
Blue Hills D-3
Bozrah Street E-5
Branford F-3
Bridgeport F-2
Vicinity Pg. 131, B-20
Bristol E-3
Broad Brook C-3
Brookfield E-2
Brookfield Ctr. E-2
Brooklyn D-5
Bryam ★
Vicinity Pg. 131, F-12
Bulls Bridge E-1
Canaan C-2
Cannondale E-1
Vicinity Pg. 131, B-16
Cheshire E-3
Chester F-4
Chesterfield E-5
Clarks Corner D-5
Clarks Falls E-5
Clinton F-4
Colchester E-4

★ Denotes City located only on Vicinity map. City not located on State map.

CONNECTICUT   FLORIDA

CONNECTICUT

Cornwall D-2
Cornwall Bridge ... D-2
Cornwall Center ... D-2
Cos Cob G-1
  Vicinity ...... Pg. 131, E-13
Coventry D-5
Cranbury .... ★
  Vicinity ...... Pg. 131, C-16
Cromwell D-3
Crystal Lake D-4
Danbury F-2
Danielson D-5
Darien G-1
Dayville D-5
Deep River F-4
Derby E-2
Dobsonville D-4
Dodgingtown F-2
Durham E-3
Durham Center E-3
Eagleville D-5
E. Brooklyn D-5
East Haddam E-4
East Hampton E-4
E. Hartford D-3
E. Haven E-3
East Lyme F-5
East Morris E-2
East Norwalk ...... ★
  Vicinity ...... Pg. 131, D-16
East Plymouth D-3
East Putnam D-5
Ekonk E-6
Ellington D-4
Ellsworth D-1
Enfield D-3
Essex F-4
Exeter E-4
Fairfield F-2
  Vicinity ...... Pg. 131, C-19
Fitchville D-1
Flanders D-1
Foxon F-3
Franklin E-5
Georgetown F-2
Germantown E-2
Glasgo E-6
Glastonbury D-3
Glen Brook .... ★
  Vicinity ...... Pg. 131, D-15
Glenville .... ★
  Vicinity ...... Pg. 131, E-12
Granby D-3
Greenfield Hill ... ★
  Vicinity ...... Pg. 131, B-18
Greenwich G-1
  Vicinity ...... Pg. 131, F-13
Groton E-4
Guilford E-3
Haddam E-4
Hamden F-3
Hartford D-3
Hartland D-2
Harwinton D-2
Hazardville D-4
Hopeville E-6
Huntsville D-2
Ivoryton F-4
Jewett City E-5
Joyceville C-2
Kensington D-3
Kent D-1
Killingly Center .. D-6
Lakeville D-1
Laysville F-4
Lebanon E-4
Ledyard Center E-5
Litchfield D-2
Long Ridge .... ★
  Vicinity ...... Pg. 131, C-13
Lower City .... ★
Manchester D-4
Mansfield D-4
Marion E-3
Meriden E-3
Mianuc .... ★
  Vicinity ...... Pg. 131, E-16
Middlebury E-2
Middlefield E-3
Middletown E-4
Milford E-2
Millington E-4
Mill Plain .... ★
  Vicinity ...... Pg. 131, C-18
Mohegan E-5
Montville E-5
Moosup D-5
Mystic F-5
Naugatuck E-2
New Britain E-3
New Canaan F-1
  Vicinity ...... Pg. 131, C-15
New Fairfield E-1
New Hartford D-2
New Haven F-3
New London F-5
New Milford E-2
New Preston D-1
Newington F-5
Niantic F-5
Norfolk D-2
Noroton .... ★
  Vicinity ...... Pg. 131, E-15
Noroton Hts. ...... ★
  Vicinity ...... Pg. 131, E-15
North Branford E-3
North Granby C-3
N. Grosvenordale .. D-6
North Guilford F-3
North Haven F-3
North Madison F-4
North Norfolk C-2
North Somers C-4
North Stamford .... ★
  Vicinity ...... Pg. 131, C-14
  Vicinity ...... Pg. 131, D-16
Norwalk F-2
  Vicinity ...... Pg. 131, D-16
Norwich E-5
Norwichtown E-5
Oakville E-2
Old Greenwich ..... ★
  Vicinity ...... Pg. 131, F-13
Old Lyme F-4

Old Saybrook F-4
Oneco E-6
Orange F-2
Pachaug E-5
Packerville E-5
Park Lane E-1
Pawcatuck F-5
Phoenixville D-5
Plainfield E-5
Poquetaruck D-3
Poquonock D-3
Port Chester ...... ★
  Vicinity ...... Pg. 131, F-12
Portland E-4
Prospect E-2
Putnam D-5
Putnam Heights D-6
Quaker Farms E-2
Quinebaug C-5
Redding F-1
Ridgefield F-1
Riverside .... ★
  Vicinity ...... Pg. 131, F-13
Riverton D-3
Robertsville D-2
Rockfall E-4
Rockland E-4
Rocky Hill D-3
Round Hill F-1
  Vicinity ...... Pg. 131, D-12
Rowayton .... ★
  Vicinity ...... Pg. 131, E-16
Roxbury E-2
Salisbury D-1
Sandy Hook E-1
Sharon D-1
Shelton F-2
Sherman E-1
Silvermine .... ★
  Vicinity ...... Pg. 131, B-16
Simsbury D-3
Somers C-4
Southbury E-2
Southington E-3
South Killingly ... D-6
South Norwalk ..... ★
  Vicinity ...... Pg. 131, D-16
Southport .... ★
  Vicinity ...... Pg. 131, C-18
South Wilton ...... ★
  Vicinity ...... Pg. 131, B-16
South Windsor D-4
Springdale .... ★
  Vicinity ...... Pg. 131, D-14
Stafford Springs .. C-4
Staffordville C-4
Stamford G-1
  Vicinity ...... Pg. 131, E-14
Sterling D-6
Sterling Hill E-6
Stonington F-6
Storrs D-5
Stratford F-2
  Vicinity ...... Pg. 130, B-20
Suffield C-3
Tariffville C-3
Taftville E-5
Terryville D-2
Thomaston E-2
Thompsonville C-3
Titicus D-1
Tolland D-4
Torrington D-2
Trumbull F-2
  Vicinity ...... Pg. 130, A-20
Tylerville E-4
Union D-5
Unionville D-3
Vernon D-4
Voluntown E-6
Wallingford E-3
Warren D-1
Warrenville D-5
Washington E-2
Waterbury E-2
Watertown E-2
Wauregan D-5
Westchester E-4
Westford D-5
West Hartford D-3
West Haven F-3
West Norwalk ...... ★
  Vicinity ...... Pg. 131, D-15
Weston F-1
Westport F-2
  Vicinity ...... Pg. 131, C-16
Wethersfield D-3
Willimantic D-5
Willington D-5
Wilsonville D-6
Wilton F-1
  Vicinity ...... Pg. 131, B-16
Windsor D-3
Windsor Locks D-3
Winnipauk .... ★
  Vicinity ...... Pg. 131, C-16
Winsted D-2
Winthrop E-4
Wolcott E-2
Woodbridge F-2
Woodmont F-3
Woodstock D-5
Woodville D-2
Yantic E-5

DELAWARE
Page 31

Adamsville J-2
Anderson Crossroads J-3
Angola K-3
Arden .... ★
  Vicinity ...... Pg. 136, G-2
Argos Corner J-3
Armstrong G-2
Bacons K-2
Barratts Chapel ... H-2
Bayville K-3
Bear F-2
Bellefonte .... ★
  Vicinity ...... Pg. 136, G-2
Bellevue .... ★
  Vicinity ...... Pg. 136, G-2
Bennetts Pier H-3
Bethany Beach K-3
Bethesda H-2
Big Oak Corners ... H-2
Big Stone Beach ... J-3
Bishops Corner H-2
Blackbird G-2
Blackiston H-2
Blades K-2
Blueball .... ★
  Vicinity ...... Pg. 136, G-1
Bowers Beach H-3
Brandywine .... ★
  Vicinity ...... Pg. 136, F-1
Bridgeville J-2
Broad Creek K-2
Broadkill Beach ... J-3
Burrsville J-2
Camden H-2
Centerville H-2
Cheswold H-2
Clarksville K-3
Clayton H-2
Cokesbury Church .. K-2
Coldwell Corners .. G-1
Collins Beach G-2
Columbia K-2
Concord K-2
Coverdale Crossroads H-3
Cowgills Corner ... H-3
Delaware City G-2
Delmar K-2
Dewey Beach K-3
Downs Chapel H-1
Dublin Hill J-2
Ellendale J-2
Elsmere F-2
Everetts Corner ... H-1
Fairmount K-3
Farmington J-2
Felton H-2
Fenwick Island K-3
Fieldsboro G-2
Flemings Landing .. H-2
Flower Station K-2
Frankford K-3
Frederica J-2
Fowler Beach J-3
Georgetown K-3
Glasgow G-2
Gravel Hill J-2
Greenwood J-2
Gumboro K-2
Hanby Cor. .... ★
  Vicinity ...... Pg. 136, F-2
H & H Corner G-2
Harbeson K-3
Hartly H-2
Harrington J-2
Henlopen Acres J-3
Hickman J-2
Hollandsville H-2
Holletts Corners .. G-1
Holly Oak .... ★
  Vicinity ...... Pg. 136, G-2
Horseys Grove Church K-2
Hughes Crossroads . J-1
Kenton H-2
Kitts Hummock H-3
Laurel K-2
Leipsic H-2
Lewes J-3
Little Creek H-2
Little Heaven H-2
Lynch Hts. H-2
Marshtown K-3
Marydel H-1
Masseys Landing ... K-3
McDonough G-2
Middletown G-2
Midway K-3
Milford J-2
Milford Crossroads K-3
Millsboro K-3
Millville K-3
Milton J-3
Mispillion Light .. K-2
Mission K-2
Mt. Pleasant G-2
Mt. Pleasant Church K-1
Naamans Cor. ...... ★
  Vicinity ...... Pg. 136, G-3
Newark F-2
New Castle F-2
Newport F-2
Oakley J-2
Oak Orchard K-3
Odessa G-2
Ogletown G-2
Owens J-2
Packing House Corner K-1
Pepper J-2
Pepperbox K-2
Phillips Hill K-2
Pickering Beach ... H-3
Port Mahon H-3
Port Penn G-2
Postels Corner H-3
Prices Corners G-1
Primehook Beach ... J-3
Redden J-2
Redden Crossroads . J-2
Reliance K-2
Rehoboth Beach K-3
Reynolds Mill J-2
Rockland .... ★
  Vicinity ...... Pg. 136, G-1
Roxana K-3
Sandtown H-1
Sandy Landing K-2
Seaford K-2
Selbyville K-3
Shawnee J-2
Slaughter Beach ... J-3
Smyrna H-2
South Bethany K-3
South Bowers H-3
Southwood F-2
Stanton F-2
Staytonville J-2
Summit Bridge G-1
Talleys Cor. ...... ★
  Vicinity ...... Pg. 136, G-1
Talleyville F-2
  Vicinity ...... Pg. 136, F-1
Taylors Bridge G-2
Thomas Corners H-2
Townsend G-2
VanDyke G-2
Vernon J-2
Warwick K-3
Whiteleysburg J-1
Willow Grove H-2
Wilmington F-2
  Vicinity ...... Pg. 136, G-2
Woodenhawk J-1
Woodland Beach H-2
Woodside H-2
Wrights Crossroads H-1
Wyoming H-2

DIST. OF COLUMBIA
Pg. 47, D-10

  Vicinity ...... Pg. 149, F-5

FLORIDA
Pages 32-33

Adams Beach D-3
Adamsville .... ★
  Vicinity ...... Pg. 142, F-9
Alachua E-5
Alford M-4
Allandale F-8
Altamonte Sprs. ... A-6
Altha N-5
Altoona E-6
Alva K-7
Andalusia E-7
Andytown L-8
Apalachicola P-5
Apopka A-3
Appollo Beach H-5
  Vicinity ...... Pg. 142, G-9
Arcadia J-6
Archer M-3
Argyle M-3
Ariel F-8
Aripeka G-5
Arran D-2
Arredondo E-5
Ashton G-7
Astor M-9
Astor Park F-6
Atlantic Beach C-7
Auburndale H-7
Aucilla C-3
Avon Park H-7
Bagdad N-2
Baker M-2
Baldwin D-6
Barberville F-7
Bartow H-6
Bascom M-5
Bay Harbor Is. .... ★
  Vicinity ...... Pg. 124, B-5
Bay Lake G-6
Bayonet Point G-4
Bay Pines .... ★
  Vicinity ...... Pg. 142, F-2
Bayview .... ★
  Vicinity ...... Pg. 142, G-4
Belle Glade K-8
Belle Isle G-7
Belleair H-4
  Vicinity ...... Pg. 142, D-1
Belleair Beach .... ★
  Vicinity ...... Pg. 142, D-1
Belleair Bluffs ... ★
  Vicinity ...... Pg. 142, D-1
Belleair Shores ... ★
  Vicinity ...... Pg. 142, D-1
Belleview F-6
Bevilles Corner ... G-6
Big Island .... ★
  Vicinity ...... Pg. 142, E-5
Biscayne Park ..... ★
  Vicinity ...... Pg. 124, B-4
Bithlo G-7
Blackman M-2
Blountstown N-5
Blue Springs C-3
Bluff Springs M-1
Boca Ciega .... ★
  Vicinity ...... Pg. 142, G-2
Boca Grande K-5
Boca Raton L-10
Bogia M-1
Bonifay M-4
Bonita Springs L-6
Bostwick E-6
Bowling Green H-6
Boynton Beach K-10
Bradenton Beach ... J-5
Bradfordville F-5
Brandenton J-5
Branton M-1
Branford D-4
Bridgeport E-6
Bridgeport .... ★
  Vicinity ...... Pg. 142, B-3
Brighton J-8
Bristol C-1
Broad Branch N-5
Bronson M-3
Brooksville G-5
Brownsville N-1
Bruce M-3
Bryceville C-6
Buckingham K-7
Bunnell E-7
Bushnell G-5
Callahan C-6
Cambellton M-5
Canal Point K-8
Cantonment N-1
Cape Canaveral G-8
Cape Coral K-6
Cape Haze K-5
Capps C-3
Carnestown L-7
Carol City L-9
Carrabelle E-1
Carrabelle Beach .. E-1
Caryville M-4
Casselberry G-7
Cedar Key F-4
Center Hill G-8
Century M-1
Chason N-5
Chattahoochee M-5
Chiefland E-4
Chipley M-4
Christmas G-7
Chuluota G-7
Chumuckla M-1
Citrus Park H-5
  Vicinity ...... Pg. 142, A-6
Citrus Springs F-5
Clarcona .... ★
  Vicinity ...... Pg. 134, C-3
Clarksville N-5
Clearwater H-4
  Vicinity ...... Pg. 142, C-2
Clearwater Beach Island ★
  Vicinity ...... Pg. 142, B-2
Clermont G-6
Cleveland K-6
Clewiston K-8
Coachman .... ★
  Vicinity ...... Pg. 142, C-3
Cocoa G-8
Cocoa Beach G-8
Coleman G-6
College Station ... N-4
Compass Lake N-4
Concord C-2
Conway .... ★
  Vicinity ...... Pg. 134, F-7
Cooks Hammock D-3
Cooper City L-9
Coral Gables M-9
  Vicinity ...... Pg. 124, C-3
Coral Springs L-9
Cottondale M-5
Crawford C-6
Crawfordville D-2
Crescent Beach E-7
Crescent City E-7
Crestview M-3
Cross City E-4
Crystal Beach ..... ★
  Vicinity ...... Pg. 142, A-2
Crystal Lake N-4
Crystal River F-5
Cutler Ridge M-9
Dade City G-6
Dallas F-6
Dania L-9
Davenport G-7
Day D-3
Daytona Beach E-8
Daytona Beach Shores E-8
De Bary F-7
Deerfield Beach ... L-9
De Funiak Springs . M-3
Deland F-7
De Leon Springs ... F-7
Dellwood M-5
Delray Beach K-10
Deltona F-7
Destin N-3
Dover H-5
Dowling Park D-4
Dundee H-7
Dunedin H-4
  Vicinity ...... Pg. 142, B-2
Dunnellon F-5
Dupont E-7
East Palatka E-7
Eastpoint E-1
East Tampa .... ★
  Vicinity ...... Pg. 142, E-9
Eatonville .... ★
  Vicinity ...... Pg. 134, C-4
Ebro N-4
Edgewater F-8
Edgewood G-7
El Portal .... ★
  Vicinity ...... Pg. 124, C-4
Elfers G-5
Ellaville D-4
Ellenton J-5
Ellinton D-5
Ellisville D-5
Eloise H-6
Englewood K-5
Estero L-6
Esto M-4
Eugene E-4
Eustis F-6
Everglades City ... M-7
Fairvilla .... ★
  Vicinity ...... Pg. 134, D-5
Falmouth D-4
Fanlew D-2
Fanning Springs ... E-4
Fedhaven H-7
Fellowship F-5
Felsmere H-8
Fernandina Beach .. C-7
Ferndale G-6
Flagler Beach E-7
Flamingo N-8
Flamingo Bay K-5
Flemington E-5
Florahome E-6
Floral City F-5
Florida City M-8
Foley D-3
Forest City .... ★
  Vicinity ...... Pg. 134, B-5
Fort Basinger J-7
Fort Green Springs H-6
Fort Lauderdale ... L-10
Fort Lonesome H-6
Fort McCoy E-6
Fort Meade H-6
Fort Meyers K-6
Fort Ogden J-6
Fort Pierce J-9
Fort Walton Beach . N-2
Fort White D-5
Four Corners ...... ★
  Vicinity ...... Pg. 142, D-3
Freeport N-3
Frostproof H-7
Fruitland Park F-6
Fountain N-4
Gainesville E-5
Garden City M-2
Gardenville .... ★
  Vicinity ...... Pg. 142, F-9
Gardner M-3
Gaskin M-3
Geneva F-7
Gibsonia H-6
Gibsonton H-5
  Vicinity ...... Pg. 142, F-10
Gifford H-9
Glen St. Mary D-5
Glendale M-3
Godley's Bluff E-1
Golden Beach L-9
Golden Gate L-7
Gonzalez N-1
Gotha .... ★
  Vicinity ...... Pg. 134, C-4
Goulds M-9
Gowers Corners G-5
Graceville M-4
Grand Island F-6
Grand Ridge M-5
Grant C-1
Grayton Beach N-3
Greenacres City ... N-9
Green Cove Springs D-6
Greenhead N-4
Green Point C-1
Greensboro C-1
Greenville C-3
Greenwood M-5
Grove Park E-5
Groveland G-6
Gulf Breeze N-2
Gulf Cove K-5
Gulf Hammock F-4
Gulfport .... ★
  Vicinity ...... Pg. 142, G-4
Gulfstream K-9
Haines City G-7
Hallandale L-9
Hampton E-6
Hampton Springs ... D-3
Harlem K-8
Harney .... ★
  Vicinity ...... Pg. 142, B-10
Harold M-2
Hastings E-7
Havana C-2
Hawthorne E-5
Hernando F-5
Hialeah M-9
  Vicinity ...... Pg. 124, C-2
Hialeah Gardens ... L-9
Highland D-6
Highland Beach K-10
Highland View P-4
High Springs E-5
Hilliard C-6
Hobe Sound J-9
Holder H-5
Holley N-2
Holly Hill E-7
Hollywood L-10
Holmes Beach J-4
Holopaw H-8
Holt M-2
Homestead M-8
Homosassa F-5
Homosassa Sprs. ... F-5
Hosford D-1
Hudson H-4
Immokalee K-7
Indialantic H-9
Indian Creek Village ★
  Vicinity ...... Pg. 124, B-5
Indian Harbour Beach H-9
Indian Lakes Estates H-7
Indian Rocks Beach H-4
  Vicinity ...... Pg. 142, E-1
Indian Shores ..... ★
  Vicinity ...... Pg. 142, E-1
Indiantown J-9
Inglis F-5
Interlachen E-6
Inverness F-5
Islamorada N-9
Ivan D-2
Jacksonville C-7
Jacksonville Beach C-7
Jasper C-4
Jay M-2
Jennings C-4
Jensen Beach J-9
Johnson E-6
June Park H-9
Juno Beach K-9
Jupiter K-9
Jupiter Island J-9
Kenansville H-8
Kendall M-9
Kendrick E-5
Kenneth City ...... ★
  Vicinity ...... Pg. 142, F-4
Key Biscayne N-9
  Vicinity ...... Pg. 124, C-5
Key Colony Beach .. N-9
Key Largo M-9
Key West P-9
Kinard N-5
Kingsley D-6
Kissimmee G-7
Korona E-7
LaBelle K-7
Lacoochee G-5
Lake Alfred G-6
Lake Buena Vista .. ★
  Vicinity ...... Pg. 134, J-3
Lake Butler D-5
Lake City D-4
Lake Hamilton G-7
Lake Harbor K-8
Lake Helen F-7
Lake Mary F-7
Lake Park K-10
Lake Placid J-7
Lake Wales H-6
Lake Worth K-10
Lakeland H-6
Lakewood M-3
Lamont D-2
Land O' Lakes G-5
Lantana K-9
Largo H-4
  Vicinity ...... Pg. 142, D-2
Lauderdale-by-the-Sea L-9
Laurel J-5
Laurel Hill M-3
Lawtey D-5
Lealman .... ★
  Vicinity ...... Pg. 142, F-4
Lebanon Station ... F-5
Lecanto F-5
Lee D-3
Leesburg F-6
Leisure City M-9
Leonia M-4
Lighthouse Point .. L-9
Live Oak D-4
Lloyd C-2
Lockhart .... ★
  Vicinity ...... Pg. 142, C-9
Longboat Key J-5
Longwood G-7
Lower Matecumbe Beach N-9
Loxahatchee K-9
Lulu D-5
Lynn Haven N-4
Lynne F-6
Macclenny D-5
Madeira Beach H-4
  Vicinity ...... Pg. 142, G-1
Madison B-6
Maitland F-7
Malabar H-9
Malone M-5
Marathon P-8
Marco Island L-6
Margate L-9
Marianna M-5
Marineland E-7
Mary Esther N-2
Mascotte G-6
Mayo D-4
McDavid M-1
McIntosh E-5
McNeils P-5
Medart D-1
Medley .... ★
  Vicinity ...... Pg. 124, B-2
Melbourne G-9
Melbourne Beach ... H-9
Melbourne Village . G-8
Memphis J-5
Merritt Island G-8
Miami L-9
  Vicinity ...... Pg. 124
Miami Beach M-9
  Vicinity ...... Pg. 124, D-6
Miami Shores L-9
Miami Shores ...... ★
  Vicinity ...... Pg. 124, D-1
Miami Springs L-9
Miami Springs ..... ★
  Vicinity ...... Pg. 124, D-1
Micco H-9
Midway .... ★
Mikesville D-5
Miles City L-7
Miller Crossroads . M-4
Milligan M-2
Milton M-2
Mims F-8
Minneola G-6
Miramar .... ★
Monroes Corner F-6
Monroe Station L-8
Monticello C-2
Moore Haven K-8
Morriston E-5
Mossy Head M-3
Moultrie E-7
Mt. Carrie D-5
Mt. Dora F-6
Mt. Plymouth F-7
Mulberry H-6
Mullis City .... ★
  Vicinity ...... Pg. 142, B-7
Munson M-2
Murdock K-5
Myakka City J-6
Naples L-6
Naples Park L-6
Naranja M-9
National Gardens .. E-7
Navarre N-2
Neptune Beach D-7
Newberry E-5
New Harmony M-3
New Hope N-4
New Point Comfort . K-5
Newport D-2
New Port Richey ... G-4
New River D-5
New Smyrna Beach .. F-8
Niceville N-3
Nocatee J-6
Nokomis J-5
North Bay Village . ★
  Vicinity ...... Pg. 124, C-5
North Lauderdale .. L-9
North Miami L-10
  Vicinity ...... Pg. 124, B-4
North Miami Beach . L-9
North Miami Beach . ★
  Vicinity ...... Pg. 124, A-4
North Palm Beach .. K-9
North Port J-5
North Redington Beach ★
  Vicinity ...... Pg. 142, F-2
North Ruskin ...... ★
  Vicinity ...... Pg. 142, G-10
Nowatney .... ★
  Vicinity ...... Pg. 142, A-8
Oak Hill F-8
Oakland Park L-10
Oakhurst .... ★
  Vicinity ...... Pg. 142, F-1
Ocala F-5
Ochopee M-7
Ocoee .... ★
  Vicinity ...... Pg. 134, D-3
Odessa G-5
Okahumpka F-6
Okeechobee J-8
Okeelanta K-8
Old Town E-4
Oldsmar H-5
  Vicinity ...... Pg. 142, B-4
Olustee D-5
Ona J-6
Oneco J-5
O'Neil L-9
Opa-Locka L-9
  Vicinity ...... Pg. 124, B-2
Orange City F-7
Orange Heights E-5
Orange Lake E-5
Orange Park D-6
Orange Springs E-6
Orient Park .... ★
  Vicinity ...... Pg. 142, C-9
Orlando G-7
  Vicinity ...... Pg. 134, E-6
Orlouista .... ★
  Vicinity ...... Pg. 134, D-4
Ormond Beach E-8
Ormond-By-The Sea . E-8
Ortona K-7
Oslo H-9
Osprey J-5
Osteen F-7
Otter Creek E-4
Overstreet P-5
Oviedo F-7
Owl's Head N-3
Oxford F-6
Ozona .... ★
  Vicinity ...... Pg. 142, A-2
Pace N-2
Pahokee K-8
Palatka E-6
Palm .... ★
  Vicinity ...... Pg. 142, D-9
Palm Bay H-8
Palm Beach K-10
Palm Beach Gardens K-9
Palm Shores G-8
Palm Springs K-9
Palmdale K-7
Palmetto J-5
Palm Harbor H-4
  Vicinity ...... Pg. 142, A-2
Panacea D-2
Panama City N-4
Panama City Beach . N-4
Parker N-4
Parrish J-5
Pasco G-5
Paxton M-3
Pembroke L-9
Pembroke Pines L-9
Pensacola N-2
Pensacola Beach ... N-2
Penney Farms D-6
Perrine M-9
Perry D-3
Peters M-3
Pierce H-6
Pierson F-7
Pine Castle .... ★
  Vicinity ...... Pg. 134, G-6
Pine Hills D-4
Pinellas Park H-4
  Vicinity ...... Pg. 142, F-4
Pinetta C-3
Piney Point H-5
Pinland D-3
Placida K-5
Plantation L-9
Plant City H-5
Plymouth .... ★
  Vicinity ...... Pg. 134, A-2
Polk City G-6
Pomona Park E-6
Pompano Beach L-10
Ponte Vedra Beach . D-7
Port Charlotte K-5
Portland N-3
Port Manatee H-4
Port Mayaca K-9
Port Orange F-8
Port Richey G-4
Port St. Lucie J-9
Port St. Joe P-5
Port Salerno J-9
Port Sewall J-9
Port Tampa .... ★
  Vicinity ...... Pg. 142, F-7
Providence G-6
Punta Gorda K-6
Punta Russa K-6
Quincy C-1
Raleigh E-5
Rattlesnake .... ★
  Vicinity ...... Pg. 142, E-7
Red Bay N-3
Redington Beach ... H-4
Redington Shores .. ★
  Vicinity ...... Pg. 142, F-1
Red Level F-5
Richmond Heights .. M-8
Ridge Manor G-5
Riverview H-5
Riviera Beach K-10
Robertsdale N-1
Rockledge G-8
Roseland H-8
Rotonda K-5
Round Lake M-5
Royal Palm Beach .. K-9
Ruskin H-5
Safety Harbor ..... ★
  Vicinity ...... Pg. 142, C-4
St. Augustine D-7
St. Augustine Beach D-7
St. Augustine Shores D-7
St. Cahterine G-6
St. Cloud G-7

FLORIDA (continued)

St. James City K-6
St. Leo G-5
St. Lucie H-9
St. Marks D-2
St. Petersburg H-4
  Vicinity .... Pg. 142,G-4
St. Petersburg Beach H-4
St. Teresa E-2
Salem E-3
Salt Springs E-6
Samoset J-5
San Antonio G-5
San Carlos Park K-6
San Mateo E-7
Sanford F-7
Sanibel L-6
Sanlando Springs .... ★
  Vicinity Pg.134,A-5
Sarasota J-5
Satellite Beach G-8
Satsuma E-6
Scottsmoor F-8
Seagrove Beach N-3
Sebastian H-9
Sebring J-7
Seminole .... ★
  Vicinity .... Pg. 142,F-2
Shady Grove D-3
Shalimar N-3
Shamrock E-3
Sharpes G-8
Silver Springs F-6
Sneads C-1
Sopchoppy D-1
South Bay K-8
South Daytona F-8
South Miami M-9
South Miami .... ★
  Vicinity .... Pg. 124, F-2
South Pasadena .... ★
  Vicinity Pg.142,G-3
Southport N-4
Spanish Harbor P-7
Spivey .... ★
  Vicinity Pg.142,A-6
Spring Lake G-5
Springfield N-4
Starke D-5
Steel City M-4
Steinhatchee E-3
Stuart J-9
Sumatra D-1
Summerfield F-6
Sun City Center J-5
Sunland Gardens J-8
Sunrise .... ★
Sunshine Beach .... ★
  Vicinity .... Pg. 142,G-2
Suwannee F-4
Sweetwater M-9
Taft G-7
Tallahassee D-1
Tampa H-5
  Vicinity .... Pg. 142,B-8
Tamarac L-9
Tarpon Springs H-4
Tavares F-6
Tavernier N-9
Telogia D-1
Temple Terrace .... ★
  Vicinity .... Pg. 142,B-10
Tennille E-4
Thomas City D-2
Tice K-6
Titusville G-8
Treasure Island .... ★
  Vicinity .... Pg. 142,G-2
Trenton E-4
Trilby G-5
Vaklaria H-8
Valparaiso N-2
Venice J-5
Venus J-7
Virginia Gardens .... ★
  Vicinity .... Pg. 124, D-1
Vernon M-4
Vero Beach H-9
Wabasso H-9
Waldo E-6
Wall Springs .... ★
  Vicinity .... Pg. 142,A-4
Walnut Hill M-1
Warrington N-1
Walsingham .... ★
  Vicinity .... Pg. 142,E-2
Watertown D-5
Wauchula J-6
Waucissa D-2
Waukeenah D-2
Wausau N-4
Waverly H-7
Weeki Wachee G-4
Wellborn D-4
Wesley Chapel G-5
Westbay N-3
West Melbourne H-8
West Miami .... ★
  Vicinity .... Pg. 124, F-1
West Palm Beach K-10
Westwood Lake M-8
Wewahitchka P-4
White City J-9
White City P-5
White Springs D-5
Wilbur-By-The-Sea F-6
Wildwood F-5
Williston E-5
Wilton Manors L-10
Wimauma H-5
Winderemere G-7
Winter Garden G-7
Winter Beach H-9
Winter Haven H-7
Winter Park .... ★
  Vicinity Pg.134,C-6
Winter Springs F-7
Woodville D-1
Worthington Sprs. D-5
Wynnhaven N-2
Yalaha F-6
Yankeetown F-4
Yehawe Junction H-7
Yulee C-6
Zephyrhills G-5
Zolfo Springs J-6

GEORGIA
Page 23

Abbeville G-10
Adairsville B-7
Adel H-9
Adrian F-11
Alamo G-10
Alapaha H-9
Albany H-8
Aldora E-8
Allentown F-10
Alpharetta H-11
Alston G-11
Alto B-9
Alvaton E-8
Ambrose H-10
Americus G-8
Amsterdam J-8
Andersonville F-8
Apalachee C-9
Appling D-11
Arabi G-9
Aragon B-6
Arlington H-7
Arnoldsville C-10
Ashburn G-9
Athens C-9
Atlanta C-8
  Vicinity .... Pg. 98 & 99
Attapulgus J-8
Augusta D-11
Austell C-7
Avalon B-10
Avondale Estates .... ★
  Vicinity .... Pg. 98, F-7
Axson H-11
Baconton H-8
Baldwin B-9
Ball Ground B-7
Barnesville E-8
Barney J-9
Barwick J-9
Baxley G-11
Bellville H-7
Belvidere .... ★
  Vicinity .... Pg. 99, G-7
Benevolence G-8
Berlin H-9
Berryton B-6
Between C-9
Bishop C-9
Blackshear H-11
Blairsville A-8
Blakely H-7
Blue Ridge A-8
Bluffton F-9
Bonaire F-9
Boston J-9
Bowdon C-6
Bowersville B-10
Bowman C-10
Bremen C-6
Bridgeboro H-9
Bristol H-11
Brookhaven .... ★
  Vicinity .... Pg. 98, D-6
Bronwood G-8
Broxton H-10
Brunswick H-13
Buchanan C-6
Buena Vista F-8
Buford C-9
Byron E-8
Cadwell F-10
Cairo J-8
Calhoun B-7
Calvary J-8
Camilla H-8
Candler B-9
Canon C-10
Canoochee F-11
Canton C-7
Carrollton C-7
Cartersville C-7
Cave Spring B-6
Cecil J-10
Cedar Springs H-7
Cedartown C-6
Center C-9
Centerville .... ★
  Vicinity .... Pg. 98, C-6
Chatsworth B-7
Chatahoochee Plantation .... ★
  Vicinity .... Pg. 98, B-3
Chauncey G-10
Chester F-10
Chickamauga A-6
Cisco A-7
Clarkston .... ★
  Vicinity .... Pg. 98, E-7
Clarkesville B-9
Claxton F-11
Cleveland B-9
Climax J-8
Clyattville J-10
Clyo F-12
Cobbtown F-11
Cogdell H-10
Cohutta A-7
Coleman's Lake E-11
College Park .... ★
  Vicinity .... Pg. 99, J-3
Colquitt H-7
Columbus F-7
Comer C-10
Commerce C-9
Conley .... ★
  Vicinity .... Pg. 99, J-5
Conyers C-9
Cochran F-9
Coolidge J-9
Copperhill A-8
Cordele G-9
Cornelia B-9
Cotton H-8
Covena F-11
Covington D-9
Crandall A-7
Crawford C-10
Crawfordville D-10
Culloden E-8
Cumming C-8
Cusseta F-7
Cuthbert G-8
Dacula C-8
Dahlonega B-9
Dallas C-7
Dalton A-7
Damascus H-7
Danielsville C-10
Darien H-12
Dasher J-10
Davisboro E-11
Dawson G-8
Dawsonville B-9
Decatur .... ★
  Vicinity .... Pg. 98, F-6
Demorest B-9
Dexter F-9
Doerun H-9
Donalsonville J-7
Doraville .... ★
  Vicinity .... Pg. 98, C-6
Douglas H-10
Douglasville C-7
Druid Hills .... ★
  Vicinity .... Pg. 98, F-6
Dublin F-10
Duluth C-8
Dunwoody .... ★
  Vicinity .... Pg. 98, B-5
DuPont J-10
East Dublin F-10
Eastman F-10
East Point C-8
  Vicinity .... Pg. 99, J-3
Eatonton D-10
Edgehill E-11
Edison G-7
Elberton C-10
Elko F-9
Ellaville F-8
Ellenton H-9
Ellijay A-8
Emerson C-7
Empire F-10
Enigma H-10
Evans C-11
Faceville J-7
Fairmount B-8
Fair Oaks .... ★
  Vicinity .... Pg. 98, B-1
Fargo J-11
Fayetteville D-8
Finleyson G-9
Fitzgerald G-9
Flemington G-12
Folkston J-11
Forest Beach F-13
Forest Park .... ★
  Vicinity .... Pg. 99, K-5
Forsyth E-8
Fort Gaines H-7
Fort Oglethorp A-6
Fort Valley F-8
Franklin D-7
Franklin Springs C-10
Gainsville C-8
  Vicinity .... Pg. 98, D-6
Garden City G-13
Georgetown G-7
Gibson D-11
Gillsville B-9
Gilmore .... ★
  Vicinity .... Pg. 98, D-2
Girard E-12
Glennville G-12
Glenwood G-11
Gordon E-10
Gracewood D-11
Grantville D-7
Gray E-9
Greensboro D-10
Greenville E-8
Griffin D-8
Grovania F-9
Grovetown D-11
Haddock E-10
Hagan F-11
Hahira J-10
Hamilton E-7
Hampton D-8
Hapeville .... ★
  Vicinity .... Pg. 99, J-4
Haralson E-8
Harlem D-11
Harrison E-11
Hartwell B-10
Hawkinsville F-9
Hazlehurst G-10
Helen B-9
Helena G-10
Hephzibah D-11
Herod G-8
Hiawassee A-9
Hickox H-12
Hilton H-7
Hiltonia E-12
Hinesville G-12
Hoboken H-11
Hogansville D-7
Holly Springs B-8
Homeland J-11
Homer B-9
Homerville J-11
Ideal F-8
Irwinton E-10
Irwinville G-9
Ivey E-10
Jackson D-9
Jacksonville G-10
Jasper B-8
Jefferson B-9
Jeffersonville F-9
Jesup G-11
Jonesboro D-8
Juno E-12
Kingsland J-12
Kingston B-7
Kite E-11
Knoxville E-9
LaFayette B-7
La Grange E-7
Lakeland J-10
Lakemont A-9
Lavonia B-10
Lawrenceville C-8
Leary H-8
Leesburg H-9
Lenox H-9
Leslie G-8
Lexington C-10
Lilburn C-8
Lilly F-9
Lincolnton C-10
Lithonia C-8
Locust Grove C-9
Loganville C-9
Louisville E-11
Lovett E-10
Ludowici G-12
Lula B-9
Lumber City G-10
Lumpkin G-7
Lyons F-11
Macon E-9
Madison D-9
Manassas E-11
Manchester E-8
Manor H-11
Margret A-8
Marietta C-7
  Vicinity .... Pg. 98, A-1
Marshallville F-8
Maxeys D-10
Maysville B-9
McCaysville A-8
McIntyre E-10
McRae G-10
Meigs J-8
Merrillville B-8
Metcalf F-11
Metter F-11
Middleton E-9
Midville E-11
Milan G-10
Milledgeville E-10
Millen E-11
Mineral Bluff A-8
Molena E-8
Moncrief H-7
Monroe C-9
Montezuma F-9
Monticello D-10
Morgan H-7
Morganton A-8
Morris G-7
Morven H-9
Moultrie H-9
Mountain City A-9
Mount Airy A-9
Mount Bethel .... ★
  Vicinity .... Pg. 98, A-4
Mount Vernon F-11
Mt. Zion D-7
Nahunta H-12
Nashville H-10
Naylor J-10
Nelson B-8
Newington F-12
Newnan D-7
Newton H-8
Nicholls H-11
Norman Park H-9
Norristown F-11
North Decatur .... ★
  Vicinity .... Pg. 98, E-6
Oakdale .... ★
  Vicinity .... Pg. 98, E-2
Oakfield H-9
Oakman B-7
Oak Park F-11
Oakwood B-9
Ochlocknee J-8
Ocilla H-10
Odum G-12
Offerman H-11
Oglethorpe F-8
Ohoopee F-11
Oliver F-12
Omega H-9
Oxford D-9
Palmetto D-7
Panthersville .... ★
  Vicinity .... Pg. 99, H-7
Parrott G-8
Patterson H-11
Pavo H-8
Peachtree City D-7
Pearson H-10
Pelham H-8
Pembroke F-12
Percale E-9
Perry F-9
Philomath C-10
Pinehurst F-9
Pine Park J-8
Pineview F-10
Pitts G-8
Plains F-8
Pooler F-12
Portal F-12
Port Wentworth F-13
Poulan H-9
Pulaski F-12
Quitman J-9
Radium Springs H-8
Raleigh E-7
Ray City J-10
Rayle C-10
Rebecca G-9
Red Oak .... ★
  Vicinity .... Pg. 99, K-2
Rehobeth .... ★
  Vicinity .... Pg. 98, E-6
Reidsville G-11
Resaca A-7
Reynolds E-8
Rhine G-10
Richland G-7
Richmond Hill F-12
Rincon F-12
Riverdale .... ★
  Vicinity .... Pg. 99, L-4
Roberta E-8
Rochelle G-9
Rockmart C-7
Rocky Ford F-12
Rome B-6
Roopville D-7
Rossville A-6
Roswell C-7
Royston B-10
Saffold J-7
St. Marys J-12
St. Simons Island H-13
Sale City H-9
Sandersville E-10
Sandy Springs .... ★
  Vicinity .... Pg. 98, B-4
Sardis E-12
Sasser G-8
Savannah F-13
Scottsdale .... ★
  Vicinity .... Pg. 98, F-7
Screven H-11
Senoia D-7
Seville G-9
Shady Dale D-9
Shellman G-8
Shiloh E-7
Smithville G-8
Smyrna .... ★
  Vicinity .... Pg. 98, C-2
Social Circle D-9
Soperton F-11
Sparks H-9
Sparta D-10
Springfield F-12
Statenville J-10
Statesboro F-12
Stillmore F-11
Stockbridge D-8
Stockton J-10
Stone Mtn. C-8
Stonewall .... ★
  Vicinity .... Pg. 99, L-1
Suches B-8
Sugar Hill B-8
Summertown E-11
Summerville B-7
Sumner H-9
Surrency G-11
Suwanee C-8
Swainsboro F-11
Sycamore G-9
Sylvania E-12
Sylvester H-9
Tallapoosa C-6
Tallulah Falls B-9
Talmo H-11
Tarrytown F-10
Tarver J-10
Tarversville F-9
Tazewell F-8
Tennga A-7
Thunderbolt G-13
Thomasville J-8
Thomaston E-8
Thomson D-11
Tifton H-9
Tignall C-10
Toco Hills .... ★
  Vicinity .... Pg. 98, E-6
Toccoa B-9
Toombsboro E-9
Trenton A-6
Trion A-6
Twin City F-11
Ty Ty H-9
Unadilla F-9
Union City .... ★
  Vicinity .... Pg. 99, L-1
Union Point C-9
Upatoi F-7
Uvalda G-11
Valdosta J-10
Vanna B-10
Van Wert C-7
Vidalia F-11
Vienna F-9
Villa Rica C-7
Wadley E-11
Waleska B-8
Waresboro H-11
Warner Robins F-9
Warrenton D-10
Warthen E-10
Warwick G-8
Washington C-10
Watkinsville C-9
Waycross H-11
Waynesboro E-11
West Point E-7
Whigham J-8
White B-8
Whitesburg D-7
Willacoochee H-10
Winder C-9
Woodbine J-12
Woodbury E-8
Woodland E-7
Wrens D-11
Wrightsville E-10
Zebulon E-8

HAWAII
Page 18

Aiea G-6
Anahola D-3
Apole Pt. C-9
Apua Pt. G-10
Barbers Pt. G-4
Cape Halawa A-6
Cape Hanamanioa C-8
Captain Cook F-9
Eleele F-2
Ewa G-5
Glenwood F-9
Haena D-2
Haena Pt. D-2
Haiku A-9
Hakalau F-10
Halawa A-6
Haleiwa F-5
Hana B-10
Hanamalo Pt. G-8
Hanamaulu E-3
Hanapepe F-1
Hauula F-6
Hawaii Kai G-6
Hekili Pt. B-7
Hilo F-10
Honaunau F-8
Honohina E-10
Honokaa E-9
Honokahau F-8
Honokahua A-7
Honolulu G-5
Honuapo G-9
Hookena F-8
Kaalualu G-9
Kaena Pt. G-10
Kahakuloa A-8
Kahana F-6
Kahe Pt. A-4
Kahuku E-5
Kahuku Pt. E-5
Kahului A-8
Kailua (Hawaii) F-9
Kailua (Oahu) G-7
Kaimu F-10
Kainaliu F-8
Ka Lae G-9
Kalaheo F-2
Kalahu Pt. B-10
Kalalau D-2
Kalaoa F-8
Kalapana F-10
Kamalo A-5
Kamiloloa A-5
Kaneohe G-6
Kapaa E-3
Kaumakani F-1
Kaunakakai A-5
Kauna Pt. G-8
Kaupo C-10
Kawaihae E-8
Keaau F-10
Keahole Pt. F-8
Keauhou F-8
Kekaha E-1
Keokea B-8
Keomuku B-7
Kepuhi Pt. F-4
Kihei B-8
Kikipua Pt. A-5
Kilauea D-2
Kipahulu B-10
Koele D-6
Koheo Pt. F-2
Koko Head G-7
Kolo A-4
Koloa F-2
Kualapapa A-4
Kualoa Pt. F-6
Kukaiau E-9
Kukuihaele E-9
Kurtistown F-10
Laau Pt. A-4
Lahaina A-7
Laie F-6
Lamaloa Head A-6
Lanai City C-7
Lawai F-2
Lihue E-3
Lilo Pt. A-3
Maalaea B-8
Mahukona E-8
Maili F-4
Makaha F-4
Makaha Pt. D-1
Makahuena Pt. F-3
Makakilo City G-4
Makalanua Pen A-5
Makapuu Pt. G-7
Makawao A-9
Makena B-8
Mana E-1
Mauna Loa F-9
Mililani Town F-5
Mililii F-2
Mokapuu Pt. F-7
Moloaa D-3
Mopua B-7
Mountain View F-9
Muolea Pt. B-10
Naalehu G-9
Nakalele Pt. A-7
Nanakuli F-4
Napoopoo F-8
Nawiliwili E-3
Ninini Pt. E-3
Niulii E-8
Nukuele Pt. C-8
Olowalu A-7
Ookala E-9
Opihikao F-10
Pahala F-9
Pahoa F-10
Paia A-9
Papa F-9
Papaaloa F-10
Papaikou F-10
Paukaa F-10
Pauwalu A-6
Pauwela A-9
Pauwela Pt. A-9
Pearl City G-5
Pearl Harbor G-4
Pepeekeo F-10
Pepeekeo Pt. F-10
Pohoiki F-10
Port Allen F-2
Puako E-8
Puhi E-2
Pukaula Pt. A-5
Pukoo A-5
Puunene A-8
Puuwai B-1
Ulapue A-5
Ulupalakua B-9
Upola Pt. E-8
Wahiawa F-5
Waiahukini G-9
Waiaka E-8
Waiakea F-10
Waialua (Molokai) A-6
Waialua (Ohau) F-5
Waianae F-4
Waihee Pt. A-8
Waikiki G-6
Wailea B-8
Wailua (Kauai) E-3
Wailua (Maui) A-9
Wailuku A-8
Waimea (Hawaii) E-9
Waimea (Kauai) E-1
Waimanalo G-7
Waiohinu G-9
Walpahu G-5

IDAHO
Pages 34-35

Aberdeen M-7
Acequia N-6
Albion N-6
American Falls M-7
Ammon L-9
Arco L-7
Arimo N-8
Ahsahka F-3
Ashton K-9
Athol C-3
Atlanta L-4
Atomic City L-7
Avery D-4
Avon E-3
Baker E-3
Bancroft N-9
Banida N-9
Banks K-3
Basalt M-8
Bayview C-3
Bellevue L-6
Bennington N-9
Bern N-9
Blackfoot M-8
Blanchard C-2
Bliss M-4
Bloomington P-9
Blue Dome K-8
Boise L-3
Bonners Ferry B-3
Bovill E-3
Bowmont L-2
Bruneau M-3
Buhl N-5
Burley N-6
Butte City L-7
Calder D-3
Caldwell L-2
Cambridge J-2
Cape Horn J-4
Carey M-6
Cascade J-3
Castleford N-4
Cataldo D-3
Centerville K-3
Challis J-5
Chatcolet D-2
Chester K-9
Chubbuck M-8
Clark Fork C-3
Clarkia E-3
Clayton K-5
Clifton P-8
Coeur D'Alene D-2
Colburn B-3
Conda N-9
Coolin B-3
Corral L-4
Cottonwood F-3
Council J-3
Craigmont F-3
Crouch K-3
Culdesac F-2
Dalton Gardens C-3
Darlington L-7
Dayton P-8
Deary E-3
Declo N-6
Dietrich M-5
Dingle P-10
Donnelly J-3
Dover C-3
Downey N-8
Driggs K-8
Dubois K-8
Eagle L-3
East Hope C-3
Eastport A-3
Eden N-6
Elk City G-4
Elk River E-3
Emida E-3
Emmett K-2
Enaville D-3
Fairfield L-5
Ferdinand F-3
Fernan Lake D-3
Fernwood E-3
Filer N-5
Firth M-8
Fish Haven P-9
Fort Hall M-8
Franklin P-9
Fruitland K-2
Fruitvale J-2
Gannett L-5
Gardena K-3
Garden City L-3
Garden Valley K-3
Garfield Bay C-3
Garwood C-3
Gem E-2
Genesee E-2
Georgetown N-9
Gibbonsville H-5
Givens Hot Springs L-2
Glens Ferry M-4
Golden G-4
Gooding M-5
Grace N-9
Grand View M-2
Grangeville G-3
Granite C-3
Greer F-3
Greencreek F-3
Hagerman M-4
Hailey L-5
Hammett M-3
Hansen N-5
Harpster G-3
Harrison D-3
Harvard E-3
Hauser C-2
Hayden C-3
Hazelton N-6
Headquarters E-4
Heise L-9
Heyburn N-6
Hill City L-4
Holbrook P-8
Hollister N-5
Homedale L-2
Hope C-3
Horseshoe Bend K-3
Howe L-7
Idaho City K-3
Idaho Falls L-9
Inkom N-8
Iona L-9
Irwin L-10
Island Park K-10
Jerome M-5
Joel E-2
Juliaetta F-3
Kamiah F-3
Kellogg D-3
Kendrick E-3
Ketchum L-5
Kimama M-6
Kimberly N-5
King Hill M-4
Kooskia F-3
Kootenai B-3
Kuna L-2
Laclede C-3
Lamb Creek B-2
Lane D-3
Lapwai F-2
Lava Hot Springs N-9
Leadore J-7
Lemhi J-6
Lewiston F-2
Lewisville L-8
Lorenzo L-9
Lowell F-4
Lowman K-4
Mackay K-7
Malad City N-8
Marsing L-2
May J-6
McCall H-3
McCammon N-8
Meadows H-3
Melba L-2
Menan L-9
Meridian L-2
Middleton K-2
Midvale J-2
Minidoka M-7
Mink Creek N-9
Montour K-3
Montpelier N-10
Moreland M-8
Morre E-2
Moscow E-2
Mountain Home M-3
Mount Idaho G-3
Moyie Springs B-4
Mud Lake K-8
Mullan D-4
Murphy L-2
Murray D-4
Murtaugh N-5
Myrtle F-2
Nampa L-2
Naples B-3
Newdale K-9
New Meadows H-3
New Plymouth K-2
Nezperce F-3
Nordman B-2
North Fork H-6
Notus K-2
Oakley N-6
Onaway E-2
Orofino F-3

★ Denotes City located only on Vicinity map. City not located on State map.

IDAHO   ILLINOIS

Osburn D-3
Ovid N-9
Paris N-9
Parker K-9
Parma K-2
Patterson J-6
Paul N-6
Pauline N-8
Payette K-2
Peck F-3
Picabo M-6
Pierce F-4
Pinehurst D-3
Pingree M-8
Pioneerville K-3
Plummer D-2
Pocatello M-8
Pollock H-3
Ponderay B-3
Porthill A-3
Portneuf M-8
Post Falls C-2
Potlatch E-2
Preston P-9
Priest River C-2
Raft River N-7
Rathdrum C-2
Reubens F-3
Rexburg L-9
Richfield M-6
Rigby L-9
Riggins H-3
Ririe L-9
Roberts L-8
Robin N-8
Rock Creek N-5
Rockford M-8
Rockland N-7
Rogerson N-5
Rupert N-6
Sagle C-3
St. Anthony L-9
St. Charles P-9
St. Joe D-3
St. Maries D-3
Salmon H-6
Samaria P-8
Sandpoint B-3
Santa D-3
Shelley M-9
Shoshone M-5
Silver City M-2
Silverton D-4
Smelterville D-3
Soda Springs N-9
Spirit Lake C-2
Springfield M-7
Springston D-3
Stanley K-5
Starkey H-2
Sterling M-7
Stites F-3
Sugar City L-9
Sunbeam K-5
Sun Valley L-5
Swanlake N-9
Swan Valley L-10
Sweet K-2
Tamarack H-2
Tendoy J-6
Tensed D-2
Teton L-9
Tetonia L-10
Thatcher N-9
Thornton L-9
Troy E-2
Twin Falls N-5
Ucon L-9
Victor L-10
Viola E-2
Virginia N-8
Wallace D-3
Warm Lake J-4
Warm River K-10
Weippe F-4
Weiser J-2
Wendell M-5
White Bird G-3
Whitney P-9
Wilder L-2
Winchester F-2
Woodland F-3
Worley D-2

ILLINOIS
Pages 36-37

Abingdon G-4
Addison .... ★
  Vicinity .... Pg. 104, G-4
Albany E-5
Albers L-6
Albion M-8
Aledo F-4
Alexis G-4
Algonquin .... ★
  Vicinity .... Pg. 102, C-1
Alhambra L-6
Allendale L-8
Allerton J-7
Alma L-7
Alorton .... ★
  Vicinity .... Pg. 141, G-7
Alpha G-4
Alpine .... ★
  Vicinity .... Pg. 105, J-4
Alsip .... ★
  Vicinity .... Pg. 105, H-5
Altamont K-7
Alton L-5
Altona G-5
Alto Pass N-6
Alvin H-9
Amboy F-6
Andalusia F-4
Andover F-5
Anna N-6
Annawan F-5
Antioch D-8
Apple River D-5
Aptakisik .... ★
  Vicinity .... Pg. 102, C-4

Arcola J-7
Arenzville J-5
Argenta J-7
Arlington Heights .... ★
  Vicinity .... Pg. 102, D-3
Aroma Park G-8
Arthur J-7
Ashkum G-8
Ashland J-5
Ashmore J-8
Ashton E-6
Assumption J-7
Astoria H-4
Athens J-5
Atkinson F-5
Atlanta H-6
  Vicinity .... Pg. 141, H-7
Atwood J-7
Auburn J-5
Aubury Hills .... ★
  Vicinity .... Pg. 105, K-5
Augusta J-4
Aurora E-7
  Vicinity .... Pg. 104, G-1
Ava M-6
Avon G-4
Baldwin M-5
Bannockburn .... ★
  Vicinity .... Pg. 103, C-4
Barrington .... ★
  Vicinity .... Pg. 102, C-2
Barrington Hills .... ★
  Vicinity .... Pg. 102, D-2
Barry J-4
Bartlett .... ★
  Vicinity .... Pg. 102, E-4
Bartonville H-5
Batavia .... ★
  Vicinity .... Pg. 104, F-1
Bath H-5
Beardstown J-5
Beaverville G-8
Beckemeyer L-6
Belle River M-7
Belleville L-5
Bellflower H-7
Bellevue G-5
Bellwood .... ★
  Vicinity .... Pg. 105, F-4
Belvidere D-7
Bement J-7
Benld K-5
Bensenville .... ★
  Vicinity .... Pg. 105, E-4
Benton M-7
Berkeley .... ★
  Vicinity .... Pg. 105, F-4
Berwyn .... ★
  Vicinity .... Pg. 105, F-5
Bethalto L-5
Bethany J-7
Blandinsville H-4
Bloomingdale .... ★
  Vicinity .... Pg. 102, E-3
Bloomington H-7
Blue Island .... ★
  Vicinity .... Pg. 105, H-6
Blue Mound J-6
Bluford M-7
Bolingbrook .... ★
  Vicinity .... Pg. 104, H-3
Bone Gap L-8
Bourbonnais F-8
Bowen H-3
Bradford G-6
Bradley F-8
Braidwood F-8
Breese L-6
Bridgeport L-9
Bridgeview .... ★
  Vicinity .... Pg. 105, G-5
Brighton K-5
Brimfield G-5
Broadview .... ★
  Vicinity .... Pg. 105, F-4
Brocton J-8
Brookfield .... ★
  Vicinity .... Pg. 105, G-5
Brookport P-7
Browning H-5
Brownstown K-7
Buckley H-6
Buckner N-7
Buda F-6
Buffalo E-5
Buffalo Grove .... ★
  Vicinity .... Pg. 102, C-4
Bunker Hill K-5
Burbank .... ★
  Vicinity .... Pg. 105, H-5
Burnham .... ★
  Vicinity .... Pg. 105, J-7
Burr Ridge .... ★
  Vicinity .... Pg. 104, G-4
Burtons Bridge .... ★
  Vicinity .... Pg. 102, B-2
Bushnell H-4
Byron E-6
Cahokia .... ★
  Vicinity .... Pg. 141, H-7
Cairo P-6
Calumet City .... ★
  Vicinity .... Pg. 105, J-7
Calumet Park .... ★
  Vicinity .... Pg. 105, H-6
Cambridge F-5
Campbell Hill M-6
Camp Point J-3
Canton H-5
Capron D-7
Carbondale N-6
Carlinville K-5
Carlock H-6
Carlyle L-6
Carmi M-8
Carol Stream .... ★
  Vicinity .... Pg. 104, F-2
Carpentersville .... ★
  Vicinity .... Pg. 102, D-1
Carrier Mills N-7

Carrollton K-5
Carthage H-3
Cary .... ★
  Vicinity .... Pg. 102, C-1
Casey K-8
Catlin H-8
Caton Farm .... ★
  Vicinity .... Pg. 104, J-2
Cave in Rock N-8
Cedar Pk .... ★
  Vicinity .... Pg. 140, H-7
Cedarville D-6
Central City L-7
Centralia L-7
Central Park H-9
Centreville L-5
  Vicinity .... Pg. 141, H-7
Cerro Gordo J-7
Chadwick E-5
Champaign H-7
Chandlerville J-5
Channahon F-8
Chapin J-5
Charleston J-8
Chatham J-5
Chatsworth G-8
Chebanse G-7
Chenoa G-7
Cherry F-6
Chester M-5
Chicago E-9
  Vicinity .... Pgs. 102-105
Chicago Heights F-9
  Vicinity .... Pg. 105, K-6
Chicago Ridge .... ★
  Vicinity .... Pg. 105, H-5
Chillicothe G-6
Chrisman J-8
Christopher N-7
Cicero .... ★
  Vicinity .... Pg. 105, F-5
Cisco J-7
Cisne L-7
Cissna Park H-8
Clarendon Hills .... ★
  Vicinity .... Pg. 104, G-4
Clay City L-8
Clayton J-3
Clifton G-8
Clinton H-7
Cloverdale .... ★
  Vicinity .... Pg. 104, F-2
Coal City F-8
Coal Valley F-4
Cobden N-6
Coffeen K-6
Colchester H-4
Colfax H-7
Collinsville L-5
Colona F-4
Columbia L-5
  Vicinity .... Pg. 141, J-6
Compton F-7
Cordova F-5
Cornell G-7
Cortland E-7
Coulterville M-6
Country Club Hills .... ★
  Vicinity .... Pg. 105, J-5
Countryside .... ★
  Vicinity .... Pg. 105, G-4
Cowden K-7
Creal Springs N-7
Crescent City G-8
Crest Hill .... ★
  Vicinity .... Pg. 104, J-2
Creston E-7
Crestwood .... ★
  Vicinity .... Pg. 105, J-5
Crete F-9
Creve Coeur G-6
Crossville M-8
Crystal City M-5
Crystal Lake .... ★
  Vicinity .... Pg. 102, B-1
Cuba H-5
Cullom G-8
Cutler M-6
Dahlgren M-7
Dakota D-6
Dallas City H-3
Dalton City J-7
Danforth G-8
Danville H-9
Danvers H-6
Darien .... ★
  Vicinity .... Pg. 104, G-4
Davis D-6
Decatur J-7
Deerfield .... ★
  Vicinity .... Pg. 103, C-4
Deer Park .... ★
  Vicinity .... Pg. 102, C-3
DeKalb E-7
De Land H-7
Delavan H-6
De Pue F-6
De Soto M-6
Des Plaines .... ★
  Vicinity .... Pg. 103, D-4
Diamond F-8
Diamond Lake .... ★
  Vicinity .... Pg. 102, B-3
Dieterich K-8
Dix L-7
Dixmoor .... ★
  Vicinity .... Pg. 105, J-6
Dixon E-6
Divernon J-5
Dolton .... ★
  Vicinity .... Pg. 105, J-6
Donovan G-8
Downers Grove .... ★
  Vicinity .... Pg. 104, G-4
Dunlap G-5
Dupo .... ★
  Vicinity .... Pg. 141, J-6
Du Quoin M-6
Durand D-6
Dwight F-7
Earlville E-7
East Alton L-5

East Cape Girardeau L-5
East Carondelet .... ★
  Vicinity .... Pg. 141, J-5
East Chicago Heights .... ★
  Vicinity .... Pg. 105, K-7
East Dubuque D-4
East Dundee .... ★
  Vicinity .... Pg. 102, D-1
East Galesburg G-5
East Hazelcrest .... ★
  Vicinity .... Pg. 105, J-6
East Moline F-4
Easton H-5
East Peoria G-6
East St. Louis L-5
  Vicinity .... Pg. 141, G-7
Edgewood L-7
Edinburg J-6
Edwardsville L-5
Effingham K-7
Elburn E-7
Eldorado N-8
Eldred K-4
Elgin E-8
  Vicinity .... Pg. 102, E-1
Elizabeth D-5
Elizabethtown N-7
Elk Grove Village .... ★
  Vicinity .... Pg. 102, E-3
Elkhart J-6
Elkville M-6
Ellis Grove M-5
Elmhurst .... ★
  Vicinity .... Pg. 104, F-4
Elmwood G-5
Elmwood Park .... ★
  Vicinity .... Pg. 105, F-5
El Paso G-7
Elsah K-4
Energy N-7
Enfield M-8
Eola .... ★
  Vicinity .... Pg. 104, G-2
Erie F-5
Eureka G-6
Evanston E-9
  Vicinity .... Pg. 103, E-6
Evansville M-5
Evergreen Park .... ★
  Vicinity .... Pg. 105, H-6
Fairbury G-7
Fairfield L-8
Fairmont .... ★
  Vicinity .... Pg. 104, J-3
Fairmont City .... ★
  Vicinity .... Pg. 141, H-7
Fairview G-5
Farina L-7
Farmer City H-7
Farmersville K-5
Farmington G-5
Findlay J-7
Fisher H-8
Fithian H-8
Flanagan G-7
Flat Rock L-9
Flora L-7
Flossmoor .... ★
  Vicinity .... Pg. 105, K-6
Flower Field .... ★
  Vicinity .... Pg. 104, F-3
Forest City H-5
Forest Park .... ★
  Vicinity .... Pg. 105, F-5
Forest View .... ★
  Vicinity .... Pg. 105, G-5
Forrest G-8
Forreston E-6
Fox Lake .... ★
  Vicinity .... Pg. 102, A-2
Fox Lake Hills .... ★
  Vicinity .... Pg. 102, A-2
Fox River Grove .... ★
  Vicinity .... Pg. 102, C-2
Fox River Valley Gardens .... ★
  Vicinity .... Pg. 102, B-2
Frankfort .... ★
  Vicinity .... Pg. 105, K-4
Franklin J-5
Franklin Grove E-6
Franklin Park .... ★
  Vicinity .... Pg. 105, F-4
Freeburg M-5
Freeport D-6
Fremont Center .... ★
  Vicinity .... Pg. 102, B-3
Frontenac .... ★
  Vicinity .... Pg. 104, H-2
Fulton E-5
Galatia M-7
Galena D-5
Galesburg G-4
Galva G-5
Gardner F-8
Geneseo F-5
Geneva .... ★
  Vicinity .... Pg. 104, F-1
Genoa E-7
Georgetown J-8
Germantown L-6
German Valley E-6
Gibson City H-8
Gifford H-8
Gilberts .... ★
  Vicinity .... Pg. 102, D-1
Gillespie K-5
Gilman G-8
Gilmer .... ★
  Vicinity .... Pg. 102, B-3
Girard J-5
Glasford H-5
Glen Carbon L-5
Glencoe .... ★
  Vicinity .... Pg. 103, C-5
Glendale Heights .... ★
  Vicinity .... Pg. 104, F-3
Glen Ellyn .... ★
  Vicinity .... Pg. 104, F-3
Glenview .... ★
  Vicinity .... Pg. 103, D-5
Glenwood .... ★
  Vicinity .... Pg. 105, K-6
Godfrey K-5
Golconda N-7

Golden H-4
Goodfield G-6
Good Hope H-4
Goodings Grove .... ★
  Vicinity .... Pg. 104, J-4
Goreville N-7
Grafton K-4
Grand Ridge F-7
Grand Tower N-6
Granite City .... ★
  Vicinity .... Pg. 140, H-7
Grant Park F-8
Granville F-6
Grayslake .... ★
  Vicinity .... Pg. 102, A-3
Grayville M-8
Green Oaks .... ★
  Vicinity .... Pg. 102, B-4
Green Rock F-4
Greenup K-8
Green Valley H-6
Greenville L-6
Gridley G-7
Griggsville J-4
Gurnee .... ★
  Vicinity .... Pg. 103, A-4
Hainesville .... ★
  Vicinity .... Pg. 102, A-3
Half Day .... ★
  Vicinity .... Pg. 103, C-4
Hamel L-5
Hamilton H-3
Hampshire E-7
Hanaford (Logan) M-7
Hanover E-5
Hanover Park .... ★
  Vicinity .... Pg. 102, E-3
Hardin K-4
Harrisburg N-7
Harristown J-6
Hartford L-5
Hartsburg H-6
Harvard D-7
Harvey .... ★
  Vicinity .... Pg. 105, J-6
Harwood Heights .... ★
  Vicinity .... Pg. 103, E-5
Hastings .... ★
  Vicinity .... Pg. 104, H-4
Havana H-5
Hawthorn Woods .... ★
  Vicinity .... Pg. 102, C-3
Hazel Crest .... ★
  Vicinity .... Pg. 105, J-6
Hebron D-7
Hecker M-5
Hennepin F-6
Henning H-9
Henry G-6
Herrin N-6
Herscher G-8
Heyworth H-7
Hickory Hills .... ★
  Vicinity .... Pg. 105, H-5
Highland L-6
Highland Hills .... ★
  Vicinity .... Pg. 104, F-4
Highland Park E-9
  Vicinity .... Pg. 103, C-5
Highwood .... ★
  Vicinity .... Pg. 103, C-5
Hillcrest E-6
Hillsboro K-6
Hillside .... ★
  Vicinity .... Pg. 105, F-4
Hillview E-7
Hindsboro J-8
Hinsdale .... ★
  Vicinity .... Pg. 104, G-4
Hodgkins .... ★
  Vicinity .... Pg. 105, G-4
Hoffman Estates .... ★
  Vicinity .... Pg. 102, D-2
Holiday Hills .... ★
  Vicinity .... Pg. 102, B-2
Homer J-8
Hometown .... ★
  Vicinity .... Pg. 105, H-5
Homewood .... ★
  Vicinity .... Pg. 105, J-6
Hoopeston H-8
Hopedale H-6
Hopkins Park G-9
Hudson H-7
Hull J-3
Humboldt J-7
Huntley E-7
Hurst N-6
Hutsonville K-8
Illiopolis J-6
Imbs Station .... ★
  Vicinity .... Pg. 141, J-7
Ina M-7
Indian Creek .... ★
  Vicinity .... Pg. 102, C-4
Indian Head Park .... ★
  Vicinity .... Pg. 105, G-4
Industry H-4
Ingalls Park .... ★
  Vicinity .... Pg. 104, K-3
Inverness .... ★
  Vicinity .... Pg. 102, D-3
Ipava H-5
Irving K-6
Irvington L-6
Island Lake .... ★
  Vicinity .... Pg. 102, B-2
Itasca .... ★
  Vicinity .... Pg. 102, E-3
Iuka L-7
Ivanhoe .... ★
  Vicinity .... Pg. 102, B-3
Jacksonville J-4
Jeffersonville L-7
Jerome J-5
Jerseyville K-4
Johnsburg .... ★
  Vicinity .... Pg. 102, A-1
Joliet F-8
  Vicinity .... Pg. 104, K-2
Jonesboro N-6
Joppa P-7
Joy G-4

Justice .... ★
  Vicinity .... Pg. 105, G-5
Kampsville K-4
Kankakee G-8
Kansas J-8
Karnak N-7
Keeneyville .... ★
  Vicinity .... Pg. 102, E-2
Keithsburg G-4
Kenilworth .... ★
  Vicinity .... Pg. 103, D-6
Kenney H-7
Kewanee F-5
Keyesport L-6
Kilbourne H-5
Kildeer .... ★
  Vicinity .... Pg. 102, C-3
Kincaid J-6
Kinderhook J-3
Kingston E-7
Kinmundy L-7
Kirkland E-7
Kirkwood G-4
Knollwood .... ★
  Vicinity .... Pg. 103, B-4
Knoxville G-4
La Grange .... ★
  Vicinity .... Pg. 105, G-4
La Grange Park .... ★
  Vicinity .... Pg. 105, G-4
La Harpe H-3
Lake Barrington .... ★
  Vicinity .... Pg. 102, C-2
Lake Bluff .... ★
  Vicinity .... Pg. 103, B-5
Lake Forest .... ★
  Vicinity .... Pg. 103, B-5
Lake in the Hills .... ★
  Vicinity .... Pg. 102, C-1
Lakemoor .... ★
  Vicinity .... Pg. 102, B-2
Lakewood .... ★
  Vicinity .... Pg. 102, C-1
Lake Villa .... ★
  Vicinity .... Pg. 102, A-3
Lake Zurich .... ★
  Vicinity .... Pg. 102, C-3
La Moille F-6
Lanark E-5
Lansing .... ★
  Vicinity .... Pg. 105, J-7
La Salle F-6
Lawrenceville L-9
Latham J-6
Leaf River E-6
Lebanon L-5
Leland F-7
Lemont .... ★
  Vicinity .... Pg. 104, H-3
Lena D-5
Lerna K-8
Le Roy H-7
Lewistown H-5
Lexington H-7
Liberty J-3
Libertyville .... ★
  Vicinity .... Pg. 102, B-4
Lilymoor .... ★
  Vicinity .... Pg. 102, A-2
Lincoln H-6
Lincoln Estates .... ★
  Vicinity .... Pg. 105, K-6
Lincolnshire .... ★
  Vicinity .... Pg. 103, C-4
Lincolnwood .... ★
  Vicinity .... Pg. 103, E-5
Lindenhurst .... ★
  Vicinity .... Pg. 102, A-3
Lisle .... ★
  Vicinity .... Pg. 104, G-3
Litchfield K-5
Livingston K-5
Loami J-5
Lockport .... ★
  Vicinity .... Pg. 104, J-3
Loda H-8
Lomax G-4
Lombard .... ★
  Vicinity .... Pg. 104, F-3
London Mills G-4
Long Creek J-7
Long Grove .... ★
  Vicinity .... Pg. 102, C-3
Long Lake .... ★
  Vicinity .... Pg. 102, A-3
Loraine H-3
Lostant G-6
Louisville L-7
Loves Park D-6
Lovington J-7
Ludlow H-8
Lyndon F-5
Lynwood .... ★
  Vicinity .... Pg. 105, K-7
Lyons .... ★
  Vicinity .... Pg. 105, G-5
Machesney Park D-6
Mackinaw H-6
Macomb H-4
Macon J-7
Madison .... ★
  Vicinity .... Pg. 140, H-7
Mahomet H-7
Malta E-7
Manchester J-5
Manhattan .... ★
  Vicinity .... Pg. 104, K-4
Manito H-5
Manlius F-5
Manteno F-8
Maquon G-5
Marengo D-7
Marine L-5
Marion N-7
Marissa M-5
Markham .... ★
  Vicinity .... Pg. 105, J-6
Marley .... ★
  Vicinity .... Pg. 104, K-4
Maroa H-7
Marquette Hts. .... ★
  Vicinity .... Pg. 105, F-5
Marseilles F-7
Marshall K-9

Martinsville K-9
Martinton G-9
Marywood .... ★
  Vicinity .... Pg. 104, G-1
Mascoutah L-5
Mason K-7
Mason City H-6
Matteson .... ★
  Vicinity .... Pg. 105, K-5
Mattoon K-7
Maunie M-8
Maywood .... ★
  Vicinity .... Pg. 105, F-5
Mazon F-7
McCook .... ★
  Vicinity .... Pg. 105, G-4
McCullom Lake .... ★
  Vicinity .... Pg. 102, A-1
McHenry D-8
McHenry Shores .... ★
  Vicinity .... Pg. 102, B-1
McLean H-6
McLeansboro M-8
Medinah .... ★
  Vicinity .... Pg. 102, E-3
Medora K-5
Melrose Park .... ★
  Vicinity .... Pg. 105, F-4
Melvin H-7
Mendon H-3
Mendota F-7
Meredosia J-4
Merrionette Park .... ★
  Vicinity .... Pg. 105, H-6
Metamora G-6
Metropolis P-7
Mettawa .... ★
  Vicinity .... Pg. 103, B-4
Middletown J-5
Midlothianb .... ★
  Vicinity .... Pg. 105, J-5
Milford G-8
Milledgeville E-6
Mill Shoals M-8
Millstadt L-5
Milton J-4
Minier H-6
Minonk G-6
Minooka .... ★
  Vicinity .... Pg. 104, K-1
Mokena .... ★
  Vicinity .... Pg. 105, K-4
Moline F-4
Momence G-9
Monmouth G-4
Montgomery .... ★
  Vicinity .... Pg. 104, H-1
Monticello J-7
Mooseheart .... ★
  Vicinity .... Pg. 104, G-1
Morris F-8
Morrison E-5
Morrisonville K-6
Morristown E-6
Morton H-6
Morton Grove .... ★
  Vicinity .... Pg. 103, D-5
Mound City P-7
Mounds P-6
Mount Carmel L-8
Mount Carroll E-5
Mt. Morris E-6
Mt. Olive K-6
Mount Prospect .... ★
  Vicinity .... Pg. 102, D-4
Mount Pulaski J-6
Mount Sterling J-4
Mount Vernon M-7
Mt. Zion J-7
Moweaqua J-6
Mulberry Grove L-6
Mundelein .... ★
  Vicinity .... Pg. 102, B-3
Murphysboro N-6
Murrayville J-4
Nameoki Acres .... ★
  Vicinity .... Pg. 140, D-7
Naplate .... ★
Naperville .... ★
  Vicinity .... Pg. 104, G-2
Nashville M-6
National City .... ★
  Vicinity .... Pg. 141, F-7
Nauvoo H-3
Neoga K-7
Neponset F-5
Newark F-7
New Athens M-5
New Berlin J-5
New Boston F-4
New Canton J-3
New Holland H-6
New Lenox .... ★
  Vicinity .... Pg. 104, K-4
Newman J-8
New Windsor G-4
Niantic J-6
Niles .... ★
  Vicinity .... Pg. 103, E-5
Noble L-8
Nokomis K-6
Normal H-7
Normantown .... ★
  Vicinity .... Pg. 104, H-1
Norridge .... ★
  Vicinity .... Pg. 103, D-5
Norris City M-8
North Aurora .... ★
  Vicinity .... Pg. 104, G-1
North Barrington .... ★
  Vicinity .... Pg. 102, C-2
Northbrook .... ★
  Vicinity .... Pg. 103, D-4
North Chicago .... ★
  Vicinity .... Pg. 103, A-4
North Dupo .... ★
  Vicinity .... Pg. 141, H-6
Northfield .... ★
  Vicinity .... Pg. 103, D-5
Northlake .... ★
  Vicinity .... Pg. 105, F-4
North Riverside .... ★
  Vicinity .... Pg. 105, F-5

North Wood River .... ★
  Vicinity .... Pg. 140, A-7
Oak Brook .... ★
  Vicinity .... Pg. 104, G-4
Oakbrook Terrace .... ★
  Vicinity .... Pg. 104, F-4
Oak Forest .... ★
  Vicinity .... Pg. 105, J-5
Oakland J-8
Oak Lawn .... ★
  Vicinity .... Pg. 105, H-5
Oak Park .... ★
  Vicinity .... Pg. 105, F-5
Oakwood H-8
Oakwood Hills .... ★
  Vicinity .... Pg. 102, B-2
Oblong K-8
Odell G-7
Odin L-7
Ogden H-8
Ohio F-6
Okawville L-6
Old Mill Creek .... ★
  Vicinity .... Pg. 102, A-4
Olmsted P-7
Olney L-8
Olympia Fields .... ★
  Vicinity .... Pg. 105, K-6
Omaha M-8
Onarga G-8
Oneida G-4
Oquawka G-4
Orangeville D-6
Oregon E-6
Orion F-4
Orland Hills .... ★
  Vicinity .... Pg. 105, J-4
Orland Park .... ★
  Vicinity .... Pg. 105, J-4
Oswego F-8
Ottawa F-7
Owaneco K-6
Palatine E-8
  Vicinity .... Pg. 102, D-3
Palestine K-9
Palmyra K-5
Palos Heights .... ★
  Vicinity .... Pg. 105, H-5
Palos Hills .... ★
  Vicinity .... Pg. 105, H-5
Palos Park .... ★
  Vicinity .... Pg. 105, H-5
Pana K-6
Panama K-6
Paris J-8
Park City .... ★
  Vicinity .... Pg. 103, A-4
Park Forest F-9
  Vicinity .... Pg. 105, K-6
Park Ridge .... ★
  Vicinity .... Pg. 103, E-5
Patoka L-7
Pawnee J-6
Paw Paw F-6
Paxton H-8
Payson J-3
Pearl J-4
Pearl City E-5
Pecatonica D-6
Pekin H-6
Peoria G-6
Peoria Hts. G-6
Peotone F-8
Percy M-6
Peru F-6
Perry J-4
Pesotum J-8
Petersburg J-5
Philo H-8
Phoenix .... ★
  Vicinity .... Pg. 105, J-6
Pinckneyville M-6
Pierron L-6
Piper City G-8
Pittsfield J-4
Plainfield .... ★
  Vicinity .... Pg. 104, J-2
Plainview J-3
Plano F-7
Pleasant Hill K-4
Pleasant Plains J-5
Plymouth H-3
Polo E-6
Pontiac G-7
Poplar Grove D-7
Port Byron F-4
Posen .... ★
  Vicinity .... Pg. 105, H-6
Potomac H-8
Prairie City H-4
Prairie Du Rocher M-5
Prairie Grove .... ★
  Vicinity .... Pg. 102, B-1
Prairie View .... ★
  Vicinity .... Pg. 102, C-4
Preston Heights .... ★
  Vicinity .... Pg. 104, K-3
Prince Crossing .... ★
  Vicinity .... Pg. 104, F-2
Princeton F-6
Princeville G-5
Prophetstown F-5
Prospect Heights .... ★
  Vicinity .... Pg. 102, C-4
Pulaski P-6
Quincy J-3
Raleigh M-7
Ramsey K-6
Rankin H-8
Rantoul H-8
Rapids City F-4
Raymond K-6
Red Bud M-5
Reynolds F-4
Richton Park .... ★
  Vicinity .... Pg. 105, K-5
Ridge Farm J-8
Ridgefield .... ★
  Vicinity .... Pg. 102, B-1
Ridgewood .... ★
  Vicinity .... Pg. 104, K-3
Ringwood .... ★
  Vicinity .... Pg. 102, A-1

★ Denotes City located only on Vicinity map. City not located on State map.

ILLINOIS

Riverdale ... ★
 Vicinity ... Pg. 105, J-6
River Forest ... ★
 Vicinity ... Pg. 105, F-5
River Grove ... ★
 Vicinity ... Pg. 105, F-5
Riverside ... ★
 Vicinity ... Pg. 105, G-5
Riverton ... J-6
Riverwoods ... ★
 Vicinity ... Pg. 103, C-4
Roanoke ... G-6
Robbins ... ★
 Vicinity ... Pg. 105, J-6
Roberts ... H-8
Robinson ... K-9
Rochelle ... E-6
Rochester ... J-6
Rockdale ... ★
 Vicinity ... Pg. 104, K-2
Rock Falls ... F-6
Rockford ... E-6
Rock Island ... F-4
Rockton ... D-7
Rolling Meadows ... ★
 Vicinity ... Pg. 102, D-3
Rome ... G-6
Romeoville ... F-8
 Vicinity ... Pg. 104, J-3
Rondout ... ★
 Vicinity ... Pg. 103, C-4
Roodhouse ... K-4
Roscoe ... D-6
Roselle ... ★
 Vicinity ... Pg. 102, E-3
Rosemont ... ★
 Vicinity ... Pg. 103, E-4
Roseville ... G-4
Rosiclare ... N-7
Rossville ... H-8
Round Lake ... ★
 Vicinity ... Pg. 102, A-2
Round Lake Beach ... ★
 Vicinity ... Pg. 102, A-3
Round Lake Heights ... ★
 Vicinity ... Pg. 102, A-3
Round Lake Park ... ★
 Vicinity ... Pg. 102, A-3
Roxana ... L-5
Royalton ... M-6
Rushville ... H-4
Sadorus ... J-7
Sag Bridge ... ★
 Vicinity ... Pg. 104, H-4
St. Anne ... G-9
St. Charles ... ★
 Vicinity ... Pg. 104, F-1
St. David ... H-5
St. Elmo ... K-7
St. Francisville ... L-8
St. Joseph ... H-8
St. Libory ... M-6
St. Peter ... L-7
Salem ... L-7
Sandoval ... L-6
Sandwich ... F-7
San Jose ... H-6
Sauget ... ★
 Vicinity ... Pg. 141, G-6
Sauk Village ... ★
 Vicinity ... Pg. 105, K-7
Saunemin ... G-8
Savanna ... E-5
Savoy ... H-8
Saybrook ... H-7
Schaumburg ... ★
 Vicinity ... Pg. 102, E-3
Schiller Park ... ★
 Vicinity ... Pg. 103, E-4
Schram City ... K-6
Seneca ... F-7
Sesser ... M-7
Shabbona ... E-7
Shannon ... E-5
Shawneetown ... N-8
Sheffield ... F-5
Shelbyville ... K-7
Sheldon ... G-9
Sheridan ... F-7
Shipman ... K-5
Shorewood ... ★
 Vicinity ... Pg. 104, K-2
Sidell ... J-8
Sims ... L-7
Skokie ... ★
 Vicinity ... Pg. 103, E-5
Sleepy Hollow ... ★
 Vicinity ... Pg. 102, D-1
Smithton ... M-5
Somonauk ... F-7
Sorento ... L-5
South Barrington ... ★
 Vicinity ... Pg. 102, B-2
South Beloit ... D-6
South Chicago Heights ... ★
 Vicinity ... Pg. 105, K-6
South Elgin ... ★
 Vicinity ... Pg. 102, E-1
South Holland ... ★
 Vicinity ... Pg. 105, J-6
South Jacksonville ... ★
South Lockport ... ★
 Vicinity ... Pg. 104, G-7
South Pekin ... H-6
South Streator ... G-7
South Wilmington ... G-8
Sparland ... G-6
Sparta ... M-5
Spencer ... ★
 Vicinity ... Pg. 104, K-4
Springfield ... J-5
Spring Grove ... D-8
Spring Valley ... F-6
Stanford ... K-5
Staunton ... K-5
Steeleville ... M-5
Steger ... F-9
 Vicinity ... Pg. 105, K-6
Sterling ... E-5
Steward ... E-7
Stickney ... ★
 Vicinity ... Pg. 105, G-5
Stillman Valley ... E-6
Stockton ... D-5

Stolle ... ★
 Vicinity ... Pg. 141, H-7
Stonefort ... N-7
Stone Park ... ★
 Vicinity ... Pg. 105, F-4
Stonington ... J-7
Streamwood ... ★
 Vicinity ... Pg. 102, E-2
Streator ... G-7
Stronghurst ... G-4
Sublette ... F-6
Sugarloaf Hts. ... ★
 Vicinity ... Pg. 141, H-7
Sullivan ... J-7
Summit ... ★
 Vicinity ... Pg. 105, G-5
Sumner ... L-8
Sunnyside ... ★
 Vicinity ... Pg. 102, A-2
Sunrise Ridge ... ★
 Vicinity ... Pg. 102, A-1
Sycamore ... E-7
Table Grove ... H-4
Tallula ... J-5
Tamaroa ... M-6
Tamms ... P-6
Taylor Springs ... K-6
Taylorville ... J-6
Teutopolis ... K-8
Thayer ... J-5
Thebes ... P-6
Third Lake ... ★
 Vicinity ... Pg. 102, A-3
Thomasboro ... H-8
Thompsonville ... M-7
Thomson ... E-5
Thornton ... ★
 Vicinity ... Pg. 105, J-6
Tilden ... M-6
Tilton ... H-9
Tinley Park ... ★
 Vicinity ... Pg. 105, J-5
Toledo ... K-8
Tolono ... J-8
Toluca ... G-6
Tonica ... G-6
Toulon ... G-5
Tovey ... J-6
Towanda ... H-7
Tower Hill ... K-7
Tower Lakes ... ★
 Vicinity ... Pg. 102, C-2
Tremont ... H-6
Troy ... L-5
Troy Grove ... F-7
Tuscola ... J-7
Union ... E-7
University Park ... ★
 Vicinity ... Pg.105, K-5
Urbana ... H-8
Ursa ... J-3
Valier ... N-7
Valley View ... ★
 Vicinity ... Pg. 102, E-1
Valmeyer ... M-5
Vandalia ... L-7
Venetian Village ... ★
 Vicinity ... Pg. 102, A-3
Venice ... ★
 Vicinity ... Pg. 140, F-7
Vermont ... H-4
Vernon Hills ... ★
 Vicinity ... Pg. 102, B-4
Versailles ... J-4
Vienna ... N-7
Villa Grove ... J-8
Villa Park ... ★
 Vicinity ... Pg. 104, F-3
Viola ... F-4
Virden ... K-5
Virginia ... J-5
Volo ... ★
 Vicinity ... Pg. 102, B-2
Wadsworth ... ★
 Vicinity ... Pg. 102, A-4
Walnut ... F-5
Waltonville ... M-6
Wamac ... L-6
Wapella ... H-6
Warren ... D-5
Warrensburg ... J-6
Warrenville ... ★
 Vicinity ... Pg. 104, F-2
Warsaw ... H-3
Washington ... G-6
Wataga ... G-5
Waterloo ... M-5
Waterman ... E-7
Watseka ... G-8
Watson ... K-7
Wauconda ... ★
 Vicinity ... Pg. 102, B-2
Waukegan ... D-8
 Vicinity ... Pg. 102, A-4
Waverly ... J-5
Wayne ... ★
 Vicinity ... Pg. 102, E-1
Wayne City ... M-7
Waynesville ... H-7
Weldon ... H-7
Wenona ... G-7
Westchester ... ★
 Vicinity ... Pg. 104, F-4
West Chicago ... ★
 Vicinity ... Pg. 104, F-2
West City ... M-7
West Dundee ... ★
 Vicinity ... Pg. 102, D-1
Western Springs ... ★
 Vicinity ... Pg. 104, G-4
Westfield ... K-8
West Frankfort ... N-7
West Miltmore ... ★
 Vicinity ... Pg. 102, A-3
Westmont ... ★
 Vicinity ... Pg. 104, G-3
West Salem ... L-8
Westville ... J-8
Wheaton ... ★
 Vicinity ... Pg. 104, F-3
Wheeling ... ★
 Vicinity ... Pg. 102, D-4
White Hall ... K-4
Wildwood ... ★
 Vicinity ... Pg. 102. A-4

Williamsfield ... G-5
Williamsville ... J-6
Willisville ... M-6
Willowbrook ... ★
 Vicinity ... Pg. 104, G-4
Willow Hill ... K-8
Willow Springs ... ★
 Vicinity ... Pg. 104, H-4
Wilmette ... ★
 Vicinity ... Pg. 103, D-6
Wilmington ... F-8
Winchester ... J-4
Windsor ... K-7
Winfield ... ★
 Vicinity ... Pg. 104, F-2
Winnebago ... E-6
Winnetka ... ★
 Vicinity ... Pg. 103, D-5
Winslow ... D-5
Winthrop Harbor ... D-8
Witt ... K-6
Wolfs ... ★
 Vicinity ... Pg. 104, H-1
Wonder Lake ... ★
 Vicinity ... Pg. 102, A-1
Wood Dale ... ★
 Vicinity ... Pg. 104, F-3
Woodhull ... G-5
Woodridge ... ★
 Vicinity ... Pg. 104, H-3
Wood River ... ★
 Vicinity ... Pg. 140, D-7
Woodson ... J-4
Woodstock ... D-7
Worden ... L-5
Worth ... ★
 Vicinity ... Pg. 105, H-5
Wyanet ... F-6
Wyoming ... G-6
Xenia ... L-7
York Center ... ★
 Vicinity ... Pg. 105, J-5
Yorkville ... F-7
Zeigler ... M-7
Zion ... D-8

INDIANA
Page 38

Advance ... E-4
Adyeville ... J-3
 Vicinity ... Pg 103, C-7
Aix ... C-3
Akron ... C-4
Albany ... E-5
Albion ... C-5
Alexandria ... E-4
Alfordsville ... H-3
Allendale ... E-5
Alliance ... E-5
Allisonville ... ★
 Vicinity ... Pg. 113, B-5
Alton ... J-3
Amboy ... D-4
Amity ... F-4
Anderson ... E-5
Andersonville ... F-5
Andrews ... D-5
Angola ... B-6
Anoka ... D-4
Antioch ... E-4
Antrim ... ★
Arcadia ... E-4
Arcana ... D-5
Argos ... C-4
Arlington ... F-5
Asherville ... F-3
Atlanta ... E-4
Attica ... D-3
Auburn ... C-6
Augusta ... ★
 Vicinity ... Pg. 113, B-3
Aurora ... G-6
Austin ... H-5
Avoca ... G-3
Bainbridge ... F-3
Bargersville ... F-4
Bass Lake ... C-4
Batesville ... G-5
Battle Ground ... D-3
Beardstown ... C-4
Bedford ... H-4
Beech Grove ... ★
 Vicinity ... Pg. 113, B-5
Belleview ... F-4
Bellmore ... F-3
Belmont ... G-4
Ben Davis ... ★
 Vicinity ... Pg. 113, B-5
Benton ... B-5
Berne ... D-6
Bicknell ... G-2
Bippus ... C-5
Birdseye ... H-3
Bloomfield ... G-3
Bloomington ... G-4
Bluffton ... D-6
Bobo ... J-2
Boonville ... J-2
Boston ... F-6
Boswell ... D-2
Boundary City ... E-6
Bourbon ... C-4
Brazil ... F-3
Bremen ... C-4
Bridgeport ... ★
 Vicinity ... Pg. 113, E-1
Brook ... C-3
Brooklyn ... F-4
Brooksburg ... H-5
Brookston ... D-3
Brookville ... F-6
Brownsburg ... E-4
Brownstown ... G-4
Bruceville ... G-2
Bryan ... D-6
Bud ... ★
Bunker Hill ... D-4
Burlington ... D-4
Butler ... C-6
Cairo ... D-3

Cambridge City ... F-5
Camby ... ★
 Vicinity ... Pg. 113, C-4
Camden ... D-4
Campbellsburg ... H-4
Cannelton ... J-3
Carefree ... J-4
Carmel ... E-4
Carthage ... F-5
Castleton ... ★
 Vicinity ... Pg. 113, B-5
Cayuga ... E-2
Cedar Grove ... G-6
Cedar Lake ... B-3
Central ... J-4
Chalmers ... D-3
Chandler ... J-2
Charleston ... H-5
Chesterton ... B-3
Chile ... D-4
China ... H-5
Chrisney ... J-3
Churubusco ... C-5
Cicero ... E-4
Cincinnati ... G-3
Clarks Hill ... E-3
Clarksville ... ★
 Vicinity ... Pg. 122, A-2
Clarmont ... E-4
Clay City ... G-3
Clayton ... F-4
Clear Lake ... B-6
Clermont ... ★
 Vicinity ... Pg. 113, C-1
Clifford ... F-5
Clinton ... F-3
Cloverdale ... F-3
Clymers ... D-4
Colfax ... E-3
Collegeville ... C-3
Columbia City ... C-5
Columbus ... G-4
Connersville ... F-6
Converse ... E-5
Cope ... F-4
Correct ... G-5
Corydon ... J-4
Covington ... E-2
Crandall ... J-4
Crawfordsville ... E-3
Cromwell ... C-5
Cross Plains ... G-5
Crothersville ... H-4
Crown Point ... B-3
Crystal ... H-3
Culver ... C-4
Cumberland ... F-4
Cynthiana ... J-1
Dale ... J-3
Dana ... E-2
Danville ... F-4
Darlington ... E-3
Darmstadt ... J-2
Dayton ... D-3
Dayville ... J-2
Decatur ... D-6
Decker ... H-2
Deerfield ... E-6
Delphi ... D-3
Demotte ... C-3
Denver ... D-5
Dexter ... J-3
Dillsboro ... G-5
Doans ... G-3
Dogwood ... J-3
Doolittle Mills ... J-3
Dover Hill ... H-3
Dublin ... F-5
Dudleytown ... C-5
Dugger ... G-2
Dunkirk ... E-5
Dupont ... G-5
Dyer ... ★
 Vicinity ... Pg. 103, B-5
 Vicinity ... Pg. 105, K-7
Earl Park ... D-2
East Chicago ... ★
 Vicinity ... Pg. 103, A-6
 Vicinity ... Pg. 105, J-7
East Gary ... ★
 Vicinity ... Pg. 103, B-7
Eaton ... E-5
Eby ... ★
Economy ... E-5
Edgewood ... E-5
 Vicinity ... Pg. 113, F-4
Edinburgh ... G-4
Edwardsport ... G-2
Effner ... E-2
Elberfeld ... J-2
Elizabeth ... J-4
Elizabethtown ... G-5
Elkhart ... B-5
Elletsville ... G-3
Ellis ... B-3
Elnora ... G-3
Elrod ... G-5
Elwood ... E-5
Eminence ... F-3
English ... H-3
Enos ... C-2
Erie ... D-4
Etna Green ... C-4
Evansville ... J-2
Everton ... F-6
Fairfield ... F-5
Fairmount ... E-5
Fairview Park ... F-2
Farmersburg ... G-2
Farmland ... E-5
Fayette ... E-4
Ferdinand ... J-3
Fiat ... D-6
Fishers ... E-4
Five Points ... ★
 Vicinity ... Pg.113, E-5
Flint ... B-6
Flora ... D-4
Foresman ... C-3
Foresman ... C-3
Fort Branch ... H-2
Fortville ... E-4
Fort Wayne ... C-6
Foster ... E-2

Fountain City ... E-6
Fountain Park ... C-3
Fowler ... D-3
Francesville ... C-3
Francisco ... H-2
Frankfort ... E-4
Franklin ... F-4
Freeman ... G-3
Fremont ... B-6
French Lick ... H-3
Friendship ... G-5
Galveston ... D-4
Garden City ... ★
Garfield ... E-3
Garrett ... B-6
Gary ... B-3
 Vicinity ... Pg. 103, B-6
Gasburg ... F-4
Gas City ... D-5
Gaston ... E-5
Geneva ... D-6
Gentryville ... J-3
Georgetown ... D-3
Georgetown ... H-4
Gilead ... C-4
Glenns Valley ... ★
 Vicinity ... Pg. 113, E-3
Glenwood ... F-5
Goodland ... C-3
Goshen ... B-5
Gosport ... F-3
Grandview ... J-3
Gravel Hill ... D-2
Graysville ... G-2
Greencastle ... F-3
Greendale ... G-6
Greenfield ... F-5
Green Oak ... C-4
Greensburg ... G-5
Greentown ... D-5
 Vicinity ... Pg. 103, B-6
Greenville ... H-4
Greenwood ... F-4
Griffith ... ★
 Vicinity ... Pg. 103, B-6
Guilford ... G-6
Hagerstown ... E-5
Hamilton ... B-6
Hamilton ... B-6
Hamlet ... C-4
Hammond ... B-3
 Vicinity ... Pg. 103, A-5
 Vicinity ... Pg. 105, J-7
Hanover ... H-5
Hardinsburg ... H-4
Harrisburg ... F-5
Hartford City ... D-5
Haubstadt ... H-2
Haysville ... H-3
Hazleton ... H-2
Hebron ... C-3
Heilman ... J-2
Henryville ... H-5
Heusler ... J-2
Highland ... ★
 Vicinity ... Pg. 103, B-6
 Vicinity ... Pg. 105, K-7
Highwoods ... ★
 Vicinity ... Pg. 113, C-3
Hillham ... H-3
Hillsboro ... E-3
Hillsdale ... F-2
Hobart ... B-3
 Vicinity ... Pg. 103, B-7
Holland ... H-2
Holton ... G-5
Homecroft ... ★
 Vicinity ... Pg. 113, C-4
Hope ... G-4
Hovey ... J-1
Huntertown ... C-5
Huntingburg ... H-3
Huntington ... D-5
Huron ... H-3
Hymera ... G-2
Indianapolis ... F-4
 Vicinity ... Pg. 113
Ingalls ... E-4
Ireland ... H-3
Iva ... H-3
Jamestown ... E-3
Jasonville ... G-2
Jasper ... H-3
Jeffersonville ... J-5
 Vicinity ... Pg. 122, A-3
Johnsburg ... J-3
Kempton ... E-4
Kendallville ... B-6
Kent ... G-5
Kewanna ... C-4
Kingman ... E-2
Kingsbury ... B-4
Kingsford Hts. ... B-4
Kirby ... G-3
Kirklin ... E-3
Knightstown ... F-5
Knightsville ... F-3
Knox ... C-4
Kokomo ... D-4
Koutz ... B-3
Kyle ... G-6
Laconia ... J-4
Ladoga ... E-3
Lafayette ... D-3
La Fontaine ... D-5
La Grange ... B-5
Lagro ... D-5
Lake Station ... see E. Gary
Lakeville ... B-4
Lanesville ... J-4
Lapaz ... B-4
Lapel ... E-4
LaPorte ... B-4
Laurel ... F-5
Lawrenceburg ... G-6
Leavenworth ... J-3
Lebanon ... E-4
Leo ... C-6
Leroy ... B-3
Lewis ... F-2
Lewisville ... F-5
Liberty ... F-6
Ligonier ... C-5
Linden ... E-3
Linton ... G-3

Little York ... H-4
Livonia ... H-4
Logansport ... D-4
Long Beach ... B-3
Loogootee ... H-3
Losantville ... E-6
Lowell ... B-3
Lynn ... E-6
Lynnville ... J-2
Lyons ... G-2
Mace ... E-3
Mackey ... H-2
Madison ... H-5
Mahon ... C-5
Majenica ... D-5
Marco ... G-2
Marengo ... H-4
Marion ... D-5
Markle ... D-5
Marshall ... E-3
Marshfield ... D-2
Mars Hill ... ★
 Vicinity ... Pg. 113, E-2
Martinsville ... F-3
Matthews ... D-5
Mauckport ... J-4
Maywood ... ★
Medaryville ... C-3
Medora ... H-4
Mellott ... E-3
Mentone ... C-4
Meridian Hills ... ★
 Vicinity ... Pg. 113, B-5
Merom ... G-2
Merriam ... C-5
Merrillville ... B-3
 Vicinity ... Pg. 103, B-6
Mexico ... D-4
Michigan City ... B-3
Mickeyville ... ★
 Vicinity ... Pg. 113, C-3
Middlebury ... B-5
Middletown ... E-5
Milan ... G-5
Milford ... B-5
Millersburg ... B-5
Milligan ... E-3
Milltown ... H-4
Milport ... H-4
Milroy ... F-5
Milton ... F-5
Mishawaka ... B-4
Mitchell ... H-4
Modoc ... E-6
Monon ... C-3
Monroe ... D-6
Monroe City ... H-2
Monroeville ... C-6
Monterey ... C-4
Montezuma ... E-2
Monticello ... D-3
Montmorenci ... D-3
Montpelier ... D-5
Moores Hill ... G-5
Mooresville ... F-4
Morgantown ... F-4
Morocco ... C-2
Morristown ... F-5
Morton ... E-3
Mt. Ayr ... C-3
Mt. Carmel ... G-6
Mt. Etna ... D-5
Mt. Healthy ... G-4
Mt. Sterling ... H-6
Mt. Summit ... E-5
Mt. Vernon ... J-1
Mulberry ... F-5
Muncie ... E-5
Munster ... B-2
 Vicinity ... Pg. 103, A-5
 Vicinity ... Pg. 105, K-7
Nappanee ... B-4
Nashville ... G-4
Natchez ... H-3
Needmore ... G-3
New Albany ... J-5
 Vicinity ... Pg. 122, A-1
New Augusta ... ★
 Vicinity ... Pg. 113, B-2
New Bethel ... ★
 Vicinity ... Pg. 113, B-5
New Carlisle ... B-4
New Castle ... E-5
New Haven ... C-6
New Liberty ... H-5
New Palestine ... F-5
New Pekin ... H-4
New Providence ... H-5
New Salem ... F-5
Newtown ... E-3
New Whiteland ... F-4
Nobelsville ... E-4
Nora ... ★
 Vicinity ... Pg. 113, B-4
North Crows Nest ... ★
 Vicinity ... Pg.113, B-3
N. Liberty ... B-4
N. Manchester ... C-5
N. Salem ... E-3
N. Vernon ... G-5
N. Webster ... C-5
Oakland City ... H-2
Oaktown ... G-2
Odell ... D-3
Odon ... G-3
Oldenburg ... G-5
Onward ... D-4
Oolitic ... G-3
Orestes ... E-5
Orland ... B-6
Orleans ... H-4
Osgood ... G-5
Ossian ... D-6
Otisco ... H-5
Otterbein ... D-3
Otwell ... H-3
Owensburg ... H-1

Oxford ... D-3
Palmyra ... H-4
Paoli ... H-4
Paragon ... F-3
Patoka ... H-2
Patricksburg ... F-3
Patriot ... G-6
Pembroke ... C-3
Pendleton ... E-4
Pennville ... D-6
Perrysville ... E-2
Peru ... D-4
Petersburg ... H-2
Pierceton ... C-5
Pike ... F-2
Pimento ... F-2
Pittsboro ... E-4
Plainfield ... F-4
Plainville ... H-3
Plattsburg ... H-4
Pleasant ... ★
Pleasant Mills ... D-6
Plum Tree ... D-6
Plymouth ... C-4
Poneto ... D-5
Portage ... B-3
Porter ... B-3
Portland ... D-6
Poseyville ... J Pleasant
Pleasant Mills ... D-6
Plum Tree ... D-6
Plymouth ... C-4
Poneto ... D-5
Portage ... B-3
Porter ... B-3
Portland ... D-6
Poseyville ... J-1
Prairie City ... F-3
Prairie Creek ... F-2
Prairieton ... F-2
Prince's Lake ... F-4
Princeton ... H-2
Prospect ... H-3
Quaker ... E-2
Ravenswood ... ★
 Vicinity ... Pg. 113, B-4
Redkey ... E-6
Reiffsburg ... D-6
Remington ... C-3
Rensselaer ... C-3
Reynolds ... D-3
Richmond ... F-6
Riddle ... J-3
Riley ... F-2
Rising Sun ... G-6
Roachdale ... E-3
Roann ... C-4
Roanoke ... D-5
Rochester ... C-4
Rockfield ... D-4
Rockport ... J-2
Rockville ... F-2
Rocky Ripple ... ★
 Vicinity ... Pg. 113, C-3
Roland ... H-3
Roll ... D-5
Rome City ... B-5
Romney ... E-3
Roseburg ... F-5
Rosedale ... F-2
Roselawn ... C-3
Rossville ... D-4
Royal Center ... C-4
Rushville ... F-5
St. Joe ... C-6
St. John ... ★
 Vicinity ... Pg. 105, K-7
St. Leon ... G-6
Salem ... H-4
Saltillo ... H-4
Sandborn ... G-2
Sandford ... F-2
Saratoga ... E-6
Savah ... J-1
Schererville ... B-2
Schneider ... C-2
Scottsburg ... H-5
Seelyville ... F-2
Sellersburg ... H-5
Selma ... E-5
Servia ... C-5
Sexton ... F-5
Seymour ... G-4
Shadeland ... D-2
Shelburn ... G-2
Shelbyville ... F-5
Sheridan ... E-4
Shipshewana ... B-5
Shirley ... F-5
Shoals ... H-3
Shore Acres ... ★
 Vicinity ... Pg. 113, B-4
Sidney ... C-5
Silver Lake ... C-5
Sleeth ... D-3
Sloan ... D-2
Smyrna ... G-5
Solsberry ... G-3
South Bend ... B-4
Southport ... F-3
Southport ... ★
 Vicinity ... Pg. 113, G-4
South Whitley ... C-5
Speedway ... E-4
Spencer ... F-3
Spencerville ... C-6
Spring Hills ... ★
 Vicinity ... Pg. 113, C-3
Springport ... E-5
Springville ... G-3
Spurgeon ... H-2
Star City ... C-4
Staser ... J-2
State Line City ... D-2
Stearleyville ... F-3
Stewart ... D-2
Stilesville ... F-3
Stinesville ... G-3
Story ... H-4
Sullivan ... G-2
Sulphur ... J-3
Summit ... B-6

Summitville ... E-5
Sunman ... G-5
Swalls ... F-2
Swanington ... D-2
Swayzee ... D-5
Switz City ... G-3
Syracuse ... B-5
Tab ... D-2
Tampico ... H-4
Tell City ... J-3
Tennyson ... J-2
Terre Huate ... F-2
Thornhope ... C-3
Thorntown ... E-4
Tiosa ... C-4
Tipton ... E-4
Tocsin ... D-6
Topeka ... B-5
Townley ... C-6
Traders Point ... ★
 Vicinity ... Pg. 113, B-1
Trafalgar ... F-4
Treaty ... D-5
Tremont ... D-6
 Vicinity ... Pg. 113, D-1
Trinity ... D-6
Trinity Springs ... G-3
Troy ... J-3
Union City ... E-6
Unindale ... D-6
Upland ... D-5
Valentine ... B-5
Vallonia ... G-4
Valparaiso ... B-3
Van Buren ... D-5
Veedersburg ... E-2
Vera Cruz ... D-6
Vernon ... G-5
Versailles ... G-5
Vevay ... H-6
Vincennes ... H-2
Vienna ... H-5
Wabash ... D-5
Wadesville ... J-1
Wakarusa ... B-4
Walkerton ... B-4
Wallace ... E-3
Walnut ... D-4
Walton ... D-4
Wanatah ... B-3
Warren ... D-5
Warsaw ... C-5
Washington ... H-2
Waterloo ... B-6
Waveland ... E-3
Webster ... E-6
W. Baden Springs ... H-3
W. College Corner ... F-6
Westfield ... E-4
West Lafayette ... D-3
W. Lebanon ... E-2
West Newton ... ★
 Vicinity ... Pg. 113, C-3
Westphalia ... G-2
Westport ... G-5
W. Terre Haute ... F-2
Westville ... B-3
Wheatfield ... C-3
Whiteland ... F-4
Whiting ... ★
 Vicinity ... Pg. 103, A-6
 Vicinity ... Pg. 105, H-7
Williams Creek ... ★
 Vicinity ... Pg. 113, B-4
Williamsport ... D-2
Wilson ... H-4
Winamac ... C-4
Winchester ... E-6
Windfall ... E-4
Wingate ... E-3
Winona Lake ... C-5
Winslow ... H-2
Wirt ... H-5
Wolcott ... C-3
Wolcottville ... B-5
Woodland ... B-4
Woodstock ... ★
 Vicinity ... Pg. 113, C-3
Worthington ... G-3
Wynnedale ... ★
 Vicinity ... Pg. 113, C-3
Yorktown ... E-5
Zanesville ... D-6
Zionsville ... E-4
 Vicinity ... Pg. 113, A-2
Zulu ... C-6

IOWA
Page 39

Abingdon ... F-7
Ackley ... C-6
Adair ... E-4
Adel ... E-4
Afton ... F-4
Agency ... E-6
Ainsworth ... E-8
Akron ... C-1
Albia ... F-6
Alden ... C-5
Algona ... B-4
Allendorf ... B-2
Allerton ... G-5
Allison ... C-6
Alta ... C-3
Alta Vista ... A-7
Alton ... B-2
Altoona ... E-5
Amana ... D-7
Ames ... D-5
Anamosa ... D-8
Anderson ... F-2
Anita ... E-3
Anthon ... C-2
Aplington ... C-6
Arcadia ... D-3
Archer ... B-2
Aredale ... C-6
Arispe ... F-4
Arlington ... C-8
Armstrong ... B-4
Arnolds Park ... A-3

IOWA

Arthur D-3
Asbury C-9
Ashton B-2
Athelstan G-4
Atkins D-7
Atlantic E-3
Attica E-6
Auburn D-3
Audubon E-3
Aurelia C-3
Avery F-7
Avoca E-3
Badger C-4
Baldwin D-9
Bancroft B-4
Barnes City E-7
Barnum C-4
Bartlett F-2
Battle Creek D-2
Baxter D-6
Bayard D-4
Beacon E-6
Beaver D-5
Bedford F-4
Beech E-6
Belle Plaine D-7
Bellevue D-9
Belmond C-5
Bennett E-9
Bettendorf E-10
Bevington E-5
Blakesburg F-7
Blencoe D-2
Bloomfield F-7
Bonaparte F-8
Boone D-5
Boyden B-2
Bradford C-6
Brayton E-3
Breda C-3
Bridgewater F-4
Brighton F-8
Britt B-5
Brooklyn D-6
Brooks F-3
Brunsville C-1
Bryant D-10
Buffalo Center B-5
Burlington F-8
Burr Oak A-8
Burt B-4
Bussey E-6
Calmar B-8
Calumet B-2
Camanche D-9
Cantril G-7
Carbon F-3
Carpenter B-6
Carroll D-3
Carson F-3
Carter Lake E-1
Carter Lake ★
Cascade D-9
Casey E-4
Cedar F-7
Cedar Falls C-7
Cedar Rapids D-8
Center Point D-8
Centerville F-6
Central City D-8
Chapin C-6
Chariton F-6
Charles City B-6
Charlotte D-10
Charter Oak D-2
Chatsworth B-1
Chelsea D-7
Cherokee C-2
Churdan D-4
Cincinnati G-6
Clare C-4
Clarence D-9
Clarinda F-3
Clarion C-5
Clarksville C-7
Clear Lake B-6
Clearfield F-4
Cleghorn C-2
Clinton D-10
Clutier D-7
Coggon D-8
Colesburg C-9
Colfax E-6
Collins D-6
Colo D-5
Columbus Jct. E-8
Columbia F-6
Colwell B-7
Conesville E-8
Conrad D-6
Conroy E-8
Conway F-4
Coon Rapids D-3
Cooper D-4
Coppock F-8
Coralville E-8
Corley E-2
Corning F-3
Correctionville C-2
Corwith B-5
Corydon F-6
Council Bluffs E-1
Crescent E-1
Cresco B-7
Creston F-4
Curlew C-4
Dakota City C-4
Dallas Center D-4
Dana D-4
Danbury D-2
Danville F-8
Davenport E-9
Davis City G-5
Dayton C-4
Decorah B-8
Dedham D-3
Defiance D-3
Delhi D-8
Delmar D-9
Deloit D-3
Delta E-7
Denison D-3
Denver C-7

Derby F-5
Des Moines E-5
De Soto E-5
De Witt D-9
Dexter E-4
Dickens B-3
Dike C-7
Dolliver B-4
Donnellson F-8
Doon B-1
Dorchester B-8
Dow City D-2
Dows C-5
Drakesville F-7
Dubuque C-9
Dumont C-6
Duncombe C-4
Dunkerton C-7
Dunlap D-2
Durant D-9
Dyersville C-9
Dysart D-7
Eagle Grove C-5
Earlham E-4
Earling E-3
Early C-3
Eddyville E-6
Edgewood B-8
Eldon F-7
Eldora C-6
Eldridge E-10
Elgin B-7
Elkader C-8
Elkhart D-5
Elk Horn F-3
Elliott F-3
Ellsworth D-5
Emerson F-2
Emmetsburg B-4
Epworth C-9
Essex F-3
Estherville B-4
Evansdale C-7
Everly B-3
Exira E-3
Fairbank C-8
Fairfield F-8
Farley C-9
Farnhamville D-4
Farragut F-2
Fayette C-8
Fenton B-4
Floyd B-6
Fonda C-3
Fontanelle E-3
Forest City B-5
Fort Dodge C-4
Fort Madison F-9
Fostoria B-3
Frankville B-8
Fraser D-5
Fredericksburg B-7
Fremont E-7
Froelich B-8
Frytown E-8
Fulton D-9
Garden City D-6
Garnavillo C-9
Garner C-5
Garwin D-7
George B-2
Gibson E-7
Gilbert D-5
Gilman D-6
Gilmore City C-4
Gladbrook C-6
Glenwood E-2
Glidden D-4
Goldfield C-5
Goodell B-5
Gowrie C-4
Graettinger B-4
Grafton B-6
Grand Jct. D-4
Grand River F-4
Grandview E-9
Grant E-3
Greene C-6
Greenfield E-4
Green Mountain D-6
Grimes E-5
Grinnell D-7
Griswold F-3
Grundy Center C-6
Guernsey E-7
Gunder B-8
Guthrie Center E-4
Guttenberg B-8
Hamburg G-2
Hamilton E-6
Hamlin E-3
Hampton C-6
Hancock E-2
Hanlontown B-5
Harcourt C-4
Harlan E-3
Harper's Ferry B-9
Hartley B-3
Hartwick D-7
Harvey E-6
Havelock C-4
Haven E-6
Hawarden B-1
Hawkeye B-7
Hazleton C-8
Hedrick E-7
Henderson F-2
Hepburn F-3
Hesper B-8
Hiawatha D-7
Hillsboro F-8
Hinton C-1
Holmes C-5
Holstein C-2
Hopeville F-5
Hopkinton C-8
Hospers B-1
Hubbard C-5
Hudson C-7
Hull B-1
Humboldt C-4
Humeston F-5
Huxley D-5
Iconium F-6

Ida Grove C-2
Imogene F-3
Independence C-8
Indianola E-5
Inwood B-1
Iowa Center D-6
Iowa City E-8
Iowa Falls C-6
Ireton B-1
Irwin E-3
Jackson Jct. B-7
Jamaica D-4
Janesville C-7
Jefferson D-4
Jerico B-7
Jesup C-7
Jewell D-5
Joice B-5
Kalona E-8
Kamrar C-5
Kanawha B-4
Kellogg D-6
Kendallville B-8
Kent F-4
Keokuk G-8
Keosauqua F-8
Keota E-7
Keswick E-7
Keystone D-7
Killduff E-6
Kingsley C-2
Kirkman E-3
Kiron D-3
Klemme B-5
Knoxville E-6
Lake City C-3
Lake Mills B-5
Lake Park B-3
Lake View D-3
Lamoni G-4
Lamont C-8
Lansing B-8
La Porte City D-7
Larchwood A-1
Larrabee C-2
Laurel D-6
Laurens C-3
Le Claire D-9
Ledyard B-4
Lehigh D-5
Leighton E-7
Le Mars C-2
Lenox F-4
Leon F-5
Lester B-1
Lewis E-3
Liberty Center F-6
Lime Springs B-7
Linderwood E-2
Lineville G-5
Lisbon D-8
Little Rock B-2
Little Sioux D-2
Livermore C-5
Logan E-2
Lohrville C-4
Long Grove D-9
Long Tree E-8
Lorimor F-5
Lost Nation D-9
Lovilia F-6
Lowden D-9
Lucas F-5
Luther D-5
Luxemburg C-9
Lyman F-3
Lytton C-3
Macedonia F-2
Macksburg E-4
Madrid D-5
Malcom E-7
Mallard B-4
Malvern F-2
Manchester C-8
Manilla D-3
Manly B-6
Manning D-3
Manson C-4
Mapleton D-2
Maquoketa D-10
Marathon C-3
Marcus C-2
Marengo E-7
Marne E-3
Marquette B-9
Marshalltown D-6
Mason City B-6
Matlock B-2
Maxwell D-5
May City B-3
Maynard C-8
McGregor B-9
McIntire B-7
Mechanicsville D-8
Mediapolis F-9
Melbourne D-6
Melcher E-6
Melrose F-5
Melvin B-2
Meriden C-2
Merrill C-1
Milford B-3
Millerton F-6
Millville C-9
Milo E-5
Milton G-7
Minburn E-5
Minden E-2
Mingo E-6
Missouri Valley E-2
Mitchellville D-5
Mondamin E-2
Moneta B-3
Monona B-8
Monroe E-6
Monticello D-8
Montour D-6
Montrose G-8
Moorhead D-2
Moravia F-6
Morning Sun F-8
Morse E-8
Moulton G-7

Mount Ayr F-4
Mount Etna F-4
Mt. Pleasant F-8
Mount Sterling G-8
Mount Union F-8
Mt. Vernon D-8
Moville C-1
Murray F-4
Muscatine E-9
Nashua B-7
Neola E-2
Nevada D-5
Nevinville F-4
New Albin A-8
Newburg D-7
New Hampton B-7
New Haven B-7
Newkirk B-2
New London F-9
New Market F-3
New Providence D-6
New Sharon E-7
Newton E-6
New Virginia F-5
Nichols E-9
Nora Springs B-6
Northboro G-3
North English E-7
Northwood B-6
Norwalk D-7
Norway E-7
Oakland E-3
Ocheyedan A-2
Odebolt D-3
Oelwein C-8
Ogden D-4
Olin D-9
Onawa D-2
Onslow D-9
Orange City B-2
Orient F-4
Osage B-6
Osceola F-5
Oskaloosa E-6
Otho D-5
Oto D-2
Ottumwa F-7
Oxford E-8
Oyens C-2
Panama E-3
Panora E-4
Park View E-10
Parnell E-7
Paullina B-2
Pella E-6
Percival F-2
Perkins B-1
Perry D-5
Persia E-2
Pierson C-2
Pilot Mound D-5
Pisgah D-2
Plainfield C-7
Plain View E-9
Plano F-6
Pleasant Plain F-8
Pleasantville E-5
Pocahontas C-3
Polk City E-5
Pomeroy C-4
Popejoy C-6
Portsmouth E-3
Postville B-8
Prairie City E-6
Preston D-10
Primghar B-2
Prole E-5
Pulaski F-7
Quasqueton C-8
Quimby C-2
Rathbun F-6
Reasnor E-6
Redding G-4
Redfield E-4
Red Oak F-3
Reinbeck D-7
Rembrandt C-3
Remsen C-2
Renwick C-5
Rhodes D-6
Riceville B-6
Richland F-8
Rickardsville C-9
Ridgeway B-7
Ringsted B-4
Rippey D-4
Rochester E-9
Rock Rapids B-2
Rock Valley B-1
Rockford B-6
Rockwell B-6
Rockwell City C-4
Roland D-5
Rolfe C-4
Roselle D-3
Rossie B-3
Rossville B-8
Royal B-3
Russell F-6
Ruthven B-3
Ryan C-8
Sabula C-10
Sac City C-3
St. Ansgar B-6
St. Charles E-5
St. Joseph B-4
St. Marys E-5
Sanborn B-3
Saratoga B-7
Schaller C-3
Schleswig D-2
Schley B-7
Scranton D-4
Searsboro E-7
Selma F-7
Sergeant Bluff C-2
Seymour G-5
Shambaugh G-3
Sharpsburg F-3
Sheffield B-6
Sheldon B-2
Shell Rock B-6
Shellsburg D-8

Shenandoah F-3
Sibley A-2
Sidney F-2
Sigourney E-7
Sioux Center C-1
Sioux City C-1
Sioux Rapids B-3
Sloan D-1
Smithland D-2
Soldier D-2
Solon D-8
Somers C-4
Spencer B-3
Spirit Lake B-3
Springville D-8
Stacyville A-6
Stanton E-2
Stanwood D-8
State Center D-6
Steamboat Rock C-6
Storm Lake C-3
Story City D-5
Stout C-7
Strawberry Point C-8
Stuart E-4
Sully D-6
Sulphur Springs C-3
Sumner C-7
Superior B-3
Sutherland B-2
Swea City B-4
Tabor F-2
Tama D-7
Templeton D-3
Terril B-3
Thompson B-5
Thornton B-5
Thurman F-2
Tiffin E-8
Tipton D-9
Titonka B-5
Toledo D-6
Traer D-7
Trenton F-8
Tripoli C-7
Truesdale C-3
Truro F-5
Turin D-2
Ulmer D-3
Underwood E-2
Union C-5
Urbana D-7
Urbandale E-5
Ute D-2
Vail D-3
Valeria E-6
Van Horne D-7
Van Wert F-5
Ventura B-5
Victor E-7
Villisca F-3
Vining D-7
Vinton D-7
Walker D-8
Walford D-8
Wall Lake C-2
Wallingford B-4
Wapello F-9
Washington E-8
Washta C-2
Waterloo C-7
Waukee D-4
Waukon B-8
Waverly C-7
Webster E-7
Webster City C-5
Weldon F-5
Wellman E-8
Welton D-9
Wesley B-5
West Bend B-4
W. Burlington F-8
West Des Moines D-4
Westfield C-1
West Grove F-7
West Liberty E-9
West Point F-9
West Union B-8
Wheatland D-9
Whiting D-2
Whittemore B-4
Wick E-5
Willey D-3
Williams D-5
Williamsburg E-8
Williamson F-6
Wilton D-9
Winfield F-8
Winterset E-4
Wiota E-3
Woodbine E-2
Woodburn F-5
Woodward D-4
Woolstock C-5
Wyoming D-9
Yale E-4
Zwingle C-9

KANSAS

KANSAS
Pages 40-41

Abbeyville F-7
Abilene E-10
Admire E-10
Agenda C-9
Agra C-6
Alexander E-6
Alden F-7
Allen E-10
Alma E-10
Almena B-6

Alta Vista E-10
Altamont H-12
Alton D-7
Altoona G-11
Americus F-11
Andale F-8
Andover G-9
Anthony H-8
Arcadia G-12
Argonia H-8
Arkansas City H-9
Arlington G-8
Arma G-13
Ashland H-5
Atchison C-12
Athol C-7
Atlanta H-9
Attica H-7
Atwood C-4
Augusta G-9
Axtell C-10
Baldwin City D-8
Barnard D-8
Barnes C-10
Bartlett H-12
Basehor D-12
Bassett G-11
Baxter Springs H-12
Bazine E-5
Beattie C-10
Belle Plaine H-9
Belleville C-8
Beloit D-8
Belpre F-6
Belvue D-11
Benedict G-11
Bennington D-9
Bentley G-8
Benton G-9
Bern C-11
Beverly D-8
Bird City C-3
Bison E-6
Blue Mound F-12
Blue Rapids C-10
Bluff City H-8
Bogue D-6
Brewster C-3
Bronson G-12
Brookville E-8
Brownell E-6
Bucklin G-5
Buffalo G-11
Buhler F-8
Bunker Hill E-7
Burden H-9
Burdett F-6
Burlingame E-11
Burlington F-11
Burns G-9
Burr Oak C-8
Burrton F-8
Bushong E-10
Bushton E-7
Byers G-7
Caldwell H-8
Cambridge H-10
Caney H-11
Canton F-9
Carbondale E-11
Carlton E-9
Cassoday G-10
Cawker City C-7
Cedar C-7
Cedar Point F-10
Cedar Vale H-10
Centralia C-11
Chanute G-12
Chapman E-10
Chase F-8
Chautauqua H-10
Cheney G-8
Cherokee G-12
Cherryvale H-11
Chetopa H-12
Cimarron G-5
Circleville D-11
Claflin E-7
Clay Center D-9
Clayton C-5
Clearwater G-8
Clifton C-9
Climax G-10
Clyde C-9
Coats G-7
Coffeyville H-11
Colby D-4
Coldwater H-6
Collyer D-5
Colony F-11
Columbus H-12
Concordia C-9
Conway Springs H-8
Coolidge G-2
Copeland G-4
Cottonwood Falls F-10
Council Grove E-10
Countryside ★
  Vicinity Pg. 117, H-2
Courtland C-9
Cuba C-9
Cullison G-7
Culver E-8
Cunningham G-7
Damar D-6
Danville H-8
Dearing H-11
Deerfield F-3
Delphos D-8
Denison D-11
Denton C-12
Derby G-9
De Soto E-12
Dexter H-10
Dighton E-5
Dodge City G-4
Douglass G-9
Downs C-7
Dresden C-5
Dunlap E-10
Dwight E-10

Edmond C-5
Edna H-11
Effingham D-12
Elbing F-9
El Dorado G-10
Elgin H-10
Elk City H-11
Elk Falls H-10
Elkhart H-2
Ellinwood F-7
Ellis E-7
Ellsworth E-7
Elmdale F-10
Emmett D-11
Emporia F-11
Englewood H-5
Ensign G-5
Enterprise E-9
Erie G-12
Esbon C-7
Eskridge E-11
Eudora D-12
Eureka G-10
Fairview C-11
Florence F-9
Fontana F-12
Ford G-5
Formoso C-8
Fort Scott G-13
Fowler G-5
Frankfort C-10
Fredonia G-11
Freeport H-8
Frontenac H-12
Fulton F-12
Galatia E-7
Galena H-13
Galesburg G-11
Galva F-9
Garden City F-4
Garden Plain G-8
Gardner E-12
Garfield F-6
Garnett F-12
Gaylord C-7
Gem D-4
Geneseo E-8
Girard G-12
Glade C-6
Glasco C-8
Glen Elder C-8
Goessel F-9
Goodland D-3
Gorham E-7
Gove D-5
Grainfield D-4
Grandview Plaza E-10
Great Bend F-7
Greenleaf C-9
Greensburg G-6
Gridley F-11
Grinnell D-5
Halstead F-9
Hamilton G-11
Hamlin C-11
Hanover C-10
Hanston F-6
Hardtner H-7
Harper H-8
Harris F-12
Hartford F-11
Harveyville E-11
Havana H-11
Haven G-8
Havensville C-10
Haviland G-6
Hays E-6
Haysville G-9
Hazelton H-7
Healy E-4
Herington E-10
Herndon B-4
Hesston F-9
Hiawatha C-11
Highland C-11
Hill City D-5
Hillsboro F-9
Hoisington E-7
Holcomb F-4
Hollenberg C-9
Holton D-11
Holyrood E-7
Hope E-9
Horton C-11
Howard G-10
Hoxie D-5
Hoyt D-11
Hudson F-7
Hugoton H-3
Humboldt G-12
Hunter D-7
Huron C-11
Hutchinson F-8
Independence H-11
Ingalls G-4
Inman F-8
Iola G-12
Isabel G-7
Jamestown C-9
Jennings C-5
Jetmore F-5
Jewell C-8
Johnson City G-3
Junction City D-9
Kanapolis E-8
Kansas City E-13
  Vicinity Pg. 116, F-2
Kenneth ★
  Vicinity Pg. 117, L-3
Kensington C-6
Kincaid F-12
Kingman G-7
Kinsley G-6
Kiowa H-7
Kirwin C-7
Kismet H-4
La Crosse E-6
La Cygne F-13
La Harpe G-12
Lakin F-3
Lane F-12
Langdon G-8
Lansing D-12
Larned F-7

Latham H-10
Latimer E-9
Lawrence D-12
Leavenworth D-12
Leawood ★
  Vicinity Pg. 117, K-3
Lebanon C-7
Lebo F-11
Lecompton D-12
Lehigh F-9
Lenexa ★
  Vicinity Pg. 117, J-1
Lenora C-5
Leon G-10
Leona C-12
Leonardville D-10
Leoti E-3
Le Roy F-11
Lewis F-6
Liberal H-4
Liberty H-11
Lincoln D-8
Lincolnville E-10
Lindsborg E-8
Linn C-9
Linwood D-12
Little River E-8
Logan C-6
Lone Elm F-12
Long Island B-6
Longford D-9
Longton H-11
Lorraine E-8
Lost Springs E-9
Louisburg E-13
Louisville D-10
Lucas D-7
Luka G-7
Luray D-7
Lyndon E-11
Lyons F-8
Macksville F-6
Madison F-10
Mahaska C-9
Maize G-9
Manchester D-9
Manhattan D-10
Mankato C-8
Manter G-3
Maple Hill D-11
Mapleton G-12
Marion F-9
Marquette E-8
Marysville C-10
Matfield Green F-10
Mayetta D-11
Mayfield H-9
McCracken E-6
McCune H-12
McDonald C-3
McFarland D-10
McPherson F-8
Meade H-4
Medicine Lodge H-7
Melvern E-11
Menlo D-4
Meriden D-11
Merriam ★
  Vicinity Pg. 117, H-2
Milan H-8
Mildred G-12
Milford D-10
Miltonvale D-9
Minneapolis D-8
Minneola G-5
Mission ★
  Vicinity Pg. 117, H-2
Mission Hills ★
  Vicinity Pg. 117, H-3
Moline H-10
Montezuma G-4
Moran G-12
Morganville C-9
Morland D-5
Morrill C-11
Morrowville C-9
Moscow G-3
Mound City F-12
Moundridge F-8
Mound Valley H-11
Mount Hope G-8
Mulberry G-13
Mullinville G-6
Mulvane G-9
Munden C-9
Muscotah C-11
Narka C-9
Nashville G-7
Natoma D-6
Neodesha H-11
Neosho Falls F-11
Ness City E-5
Netawaka C-11
New Cambria E-9
New Strawn F-11
Newton G-9
Nickerson F-8
Norcatur C-5
North Newton F-9
Nortonville D-12
Norton C-5
Norwich G-8
Oak Hill D-9
Oakley D-4
Oberlin C-5
Offerle F-6
Ogden D-10
Oketo C-10
Olathe D-12
Olmitz E-7
Olpe F-11
Olsburg D-10
Onaga D-10
Oneida C-11
Osage City E-11
Osawatomie F-12
Osborne D-7
Oskaloosa D-12
Oswego H-12
Otis E-6
Ottawa E-12
Overbrook E-11
Overland Park ★
  Vicinity Pg. 117, J-2

★ Denotes City located only on Vicinity map. City not located on State map.

KANSAS (continued)

City	Ref		City	Ref
Oxford	H-9		Wathena	C-12
Palco	D-6		Waverly	F-11
Palmer	C-9		Webber	C-8
Paola	E-12		Weir Scammon	H-12
Paradise	D-7		Wellington	H-8
Park	D-5		Wellsville	E-12
Park City	F-9		Westmoreland	D-10
Parker	F-12		Westphalia	F-11
Parsons	H-11		Wetmore	C-11
Pawnee Rock	F-7		Wheaton	C-10
Paxico	D-10		White City	E-9
Peabody	F-9		Whitewater	G-9
Perry	D-11		Whiting	C-11
Peru	H-10		Wichita	G-9
Phillipsburg	C-6		Williamsburg	F-12
Pittsburg	H-13		Willowbrook	F-8
Plains	H-4		Wilmore	G-6
Plainville	D-6		Wilsey	E-9
Pleasanton	F-13		Wilson	D-8
Plevna	F-7		Winchester	D-12
Pomona	E-11		Windom	F-8
Portis	C-7		Winfield	H-9
Potwin	G-9		Winona	D-3
Powhattan	C-11		Woodbine	E-9
Prairie View	C-6		Woodston	C-6
Prairie Village	★		Yates Center	G-11
Vicinity	Pg. 117, H-3		Zenda	G-7
Pratt	G-7		Zurich	D-6
Prescott	F-13			
Preston	G-7			
Pretty Prairie	G-8			
Princeton	F-12			
Protection	H-6			
Quenemo	E-11			
Quinter	D-5			
Ramona	E-9			
Randall	C-8			
Randolph	D-10			
Ransom	E-5			
Rantoul	E-12			
Raymond	F-7			
Reading	F-11			
Redfield	G-12			
Republic	C-8			
Rexford	C-4			
Richfield	G-3			
Richmond	F-12			
Riley	D-9			
Robinson	C-12			
Roeland Park	★			
Vicinity	Pg. 117, H-2			
Rolla	G-2			
Rose Hill	G-9			
Roseland	H-12			
Rossville	D-11			
Rozel	F-6			
Rush Center	E-6			
Russel	E-7			
Russel Springs	D-4			
Sabetha	C-11			
St. Francis	C-3			
St. John	F-7			
St. Marys	D-11			
St. Paul	H-12			
Salina	E-8			
Satanta	G-3			
Sawyer	G-7			
Scandia	C-8			
Schoenchen	E-6			
Scott City	E-4			
Scottsville	C-8			
Sedan	H-10			
Sedgwick	G-9			
Selden	C-4			
Seneca	C-11			
Severy	G-10			
Seward	F-6			
Sharon	H-7			
Sharon Springs	D-3			
Shawnee	★			
Vicinity	Pg. 117, H-1			
Silver Lake	D-11			
Simpson	C-7			
Smith Center	C-7			
Smolan	E-8			
Solomon	E-9			
South Haven	H-9			
Spearville	G-5			
Spivey	H-8			
Spring Hill	E-12			
Stafford	G-7			
Stark	G-12			
Sterling	F-8			
Stockton	D-6			
Strong City	F-10			
Sublette	G-4			
Summerfield	C-10			
Sun City	H-7			
Sylvan Grove	D-7			
Sylvia	F-7			
Syracuse	F-3			
Tampa	E-9			
Tescott	D-8			
Thayer	G-12			
Timken	E-6			
Tipton	D-7			
Tonganoxie	D-12			
Topeka	E-11			
Toronto	G-11			
Vicinity	Pg. 106, G-8			
Towanda	G-9			
Tribune	E-3			
Troy	C-12			
Turon	G-7			
Tyro	H-11			
Udall	H-9			
Ulysses	G-3			
Uniontown	G-12			
Utica	E-5			
Valley Center	G-9			
Valley Falls	D-11			
Vermillion	C-10			
Victoria	E-6			
Vining	C-9			
Virgil	G-11			
Wakefield	D-9			
WaKeeney	D-5			
Waldo	D-7			
Waldron	H-8			
Wallace	D-3			
Walnut	G-12			
Walton	F-9			
Wamego	D-10			
Washington	C-9			
Waterville	C-10			

KENTUCKY
Pages 42-43

City	Ref		City	Ref
Adairville	F-6		Dover	B-11
Albany	F-9		Drakesboro	E-6
Alexandria	B-10		Druid Hills	★
Allen	D-12		Vicinity	Pg. 122, B-5
Allensville	F-6		Dry Ridge	B-10
Arlington	F-3		Dunmor	E-6
Ashland	C-12		Dycusburg	E-4
Auburn	F-7		Earlington	E-5
Audobon Park	★		Eddyville	E-4
Vicinity	Pg. 122, D-3		Edmonton	E-7
Augusta	B-10		Ekron	D-7
Barbourville	F-11		Elizabethtown	D-7
Bardstown	D-8		Elkhorn City	E-13
Bardwell	F-3		Elkton	F-6
Barlow	E-3		Elsmere	B-10
Beattyville	D-11		Eminence	C-9
Beaver Dam	E-6		Erlanger	★
Bedford	B-8		Vicinity	Pg. 106, G-2
Beech Grove	D-6		Eubank	E-9
Bellefonte	C-12		Evarts	F-12
Bellevue	★		Fairdale	★
Vicinity	Pg. 106, G-2		Vicinity	Pg. 122, F-3
Benham	E-12		Fairfield	D-9
Benton	F-4		Fairview	F-6
Berea	D-10		Falmouth	B-10
Berry	B-10		Ferguson	E-9
Bloomfield	D-9		Flatwoods	C-12
Bonnieville	E-7		Fleming	C-12
Boone	D-10		Flemingsburg	C-11
Boonesboro	D-10		Florence	B-10
Booneville	E-11		Fordsville	D-7
Bowling Green	E-6		Ft. Mitchell	★
Bradfordsville	D-9		Vicinity	Pg. 106, G-3
Brandenburg	D-7		Ft. Thomas	★
Breckinridge	C-10		Vicinity	Pg. 106, F-5
Bremen	E-6		Ft. Wright	★
Broadfields	★		Vicinity	Pg. 106, F-3
Vicinity	Pg. 122, C-5		Foster	B-10
Brodhead	E-10		Fountain Run	F-8
Bromley	A-10		Frankfort	C-9
Brooksville	B-10		Franklin	F-6
Brownsville	E-7		Fredonia	E-5
Buckhorn	E-11		Freeburn	D-13
Buffalo	D-8		Frenchburg	D-11
Burgin	D-9		Frogue	F-9
Burkesville	F-8		Fulton	F-3
Burnside	F-10		Gamaliel	F-8
Burton	B-10		Georgetown	C-10
Butler	B-10		Ghent	B-9
Cadiz	F-5		Glasgow	E-7
Calhoun	D-6		Golden Pond	F-5
Callaway	F-11		Goshen	C-8
Calvert City	E-4		Gratz	C-9
Campbellsville	E-8		Grayson	C-12
Campbellsburg	B-8		Greensburg	E-8
Campton	D-11		Greenup	B-12
Caneyville	E-7		Greenville	E-6
Carlisle	C-10		Guthrie	F-5
Carrollton	B-9		Hanson	E-5
Carrsville	E-4		Hardin	F-4
Carter	C-11		Hardinsburg	D-7
Caseyville	D-4		Harlan	F-11
Catlettsburg	C-12		Harned	D-7
Cave City	E-8		Harrodsburg	D-9
Centertown	E-6		Hartford	E-6
Central City	E-6		Hawesville	D-6
Cerulean	F-5		Hazard	E-12
Clairfield	F-10		Hazel	F-4
Clay	D-6		Hazel Green	D-11
Clay City	D-10		Hellier	E-12
Clearfield	C-11		Henderson	D-5
Clinton	F-3		Hickman	F-3
Cloverport	D-7		Highland Hts.	★
Cold Spring	★		Vicinity	Pg. 106, G-5
Vicinity	Pg. 106, G-1		Hindman	E-12
Columbia	E-8		Hodgenville	D-8
Columbus	F-3		Hopkinsville	F-5
Concord	B-11		Horse Cave	E-8
Constance	★		Hustonville	E-9
Vicinity	Pg. 106, F-1		Hyden	E-11
Corbin	F-10		Independence	B-10
Corydon	D-5		Indian Hills	★
Covington	B-9		Vicinity	Pg. 122, B-5
Vicinity	Pg. 106, F-2		Inez	D-12
Crab Orchard	E-9		Irvine	D-10
Crescent Springs	★		Irvington	D-7
Vicinity	Pg. 106, F-2		Island	E-6
Crestview Hills	★		Jackson	D-11
Vicinity	Pg. 106, F-2		Jamestown	E-8
Crestwood	C-8		Jeffersontown	C-8
Crittenden	B-10		Jeffersonville	D-10
Crofton	E-5		Jenkins	E-12
Cumberland	E-12		Jimtown	★
Cynthiana	C-10		Junction City	D-9
Danville	D-9		Kenton Vale	★
Dawson Sprgs.	E-5		Vicinity	Pg. 106, G-3
Dayton	★		Kingsley	★
Vicinity	Pg. 106, F-1		Vicinity	Pg. 122, D-5
Dexter	F-4		Kirkmansville	E-6
Dixon	D-5		Knob Creek	D-8
			Kuttawa	E-4
			LaCenter	E-3
			Lackey	E-12
			La Fayette	F-5
			La Grange	C-8
			Lakeside Park	★
			Vicinity	Pg. 106, G-3
			Lakeview	★
			Vicinity	Pg. 106, G-3
			Lancaster	D-10
			Lawrenceburg	C-9
			Leatherwood	E-11
			Lebanon	D-9
			Lebanon Jct.	D-8
			Lee City	D-11
			Leitchfield	D-7
			Lewisburg	E-6
			Lewisport	D-6
			Lexington	D-9
			Liberty	E-9
			Livermore	D-6
			Livingston	E-9
			Lockport	C-9
			London	E-10
			Loretto	D-8
			Lothair	E-12
			Louisa	C-12
			Louisville	Pg. 122
			Vicinity	Pg. 122
			Loyall	F-11
			Lucas	F-8
			Ludlow	★
			Vicinity	Pg. 106, F-1
			Lynch	F-12

City	Ref		City	Ref
Lynnview	★		Valley Station	C-8
Vicinity	Pg. 122, E-4		Vanceburg	B-11
Madisonville	E-5		Vancleve	D-11
Manchester	E-10		Van Lear	D-12
Martin	D-12		Versailles	C-9
Mason	B-10		Villa Hills	★
Mayfield	F-4		Vicinity	Pg. 106, F-2
Maysville	B-11		Vine Grove	D-7
McCreary	D-10		Wallins Creek	F-11
McHenry	E-6		Walton	B-10
McKee	E-10		Warfield	D-13
McRoberts	E-12		Warsaw	B-9
Middlesboro	F-10		Washington	B-11
Middletown	C-8		Water Valley	F-4
Midway	C-9		Waverly	D-5
Midway	F-4		Wayland	D-12
Millersburg	C-10		Webster	D-7
Milton	B-8		Wellington	D-11
Minor Lake Hts.	★		West Beauchel	★
Vicinity	Pg. 122, F-3		Vicinity	Pg. 122, D-5
Mockingbird Valley	★		West Liberty	D-11
Vicinity	Pg. 122, B-5		West Point	D-8
Monkeys Eyebrow	E-3		Wheatley	B-9
Monroe	E-8		Wheelwright	E-12
Monterey	C-9		Whitesburg	E-12
Monticello	F-9		Whitesville	D-6
Morehead	C-11		Whitley City	F-10
Moreland	D-9		Wickliffe	E-3
Morgan	B-10		Wilders	★
Morganfield	D-5		Vicinity	Pg. 106, F-4
Morgantown	E-7		Willard	C-12
Mortons Gap	E-5		Williamsburg	F-10
Mount Olivet	B-10		Williamstown	B-10
Mount Sterling	C-11		Wilmore	D-9
Mount Vernon	E-10		Winchester	D-10
Mt. Washington	C-8		Wingo	F-4
Muldraugh	D-8		Woodburn	F-7
Munfordville	E-8			
Murray	F-4			
Nebo	E-5			
Neon	E-12			
New Castle	C-9			
New Haven	D-8			
Newport	A-10			
Vicinity	Pg. 106, F-4			
Nicholasville	D-9			
N. Middletown	C-10			
Nortonville	E-6			
Oak Grove	F-5			
Oakland	E-7			
Olive Hill	C-11			
Owensboro	D-6			
Owenton	C-9			
Owingsville	C-10			
Paducah	E-4			
Paintsville	D-12			
Paris	C-10			
Park City	E-8			
Park Hills	★			
Vicinity	Pg. 106, F-3			
Parker Lake	F-10			
Parkway Village	★			
Vicinity	Pg. 122, D-3			
Pembroke	F-5			
Perryville	D-9			
Phelps	D-13			
Pikeville	E-13			
Pine Knot	F-10			
Pineville	F-11			
Pleasureville	C-9			
Prestonsburg	D-12			
Prestonville	B-8			
Princeton	E-5			
Prospect	C-8			
Providence	E-5			
Raceland	B-12			
Radcliff	D-7			
Ravenna	D-11			
Reidland	E-4			
Renfro Valley	E-10			
Richmond	D-10			
Rockcastle	F-5			
Rockport	E-6			
Rolling Fields	★			
Vicinity	Pg. 122, B-5			
Russell	B-12			
Russell Springs	E-9			
Russellville	F-6			
Sadieville	C-10			
Salt Lick	C-11			
Salyersville	D-12			
Sanders	B-9			
Sandy Hook	C-12			
Sardis	C-11			
Science Hill	E-9			
Scottsville	F-7			
Sebree	D-5			
Seco	E-12			
Sedalia	F-4			
Seneca Gardens	★			
Vicinity	Pg. 122, C-5			
Sharpsburg	C-11			
Shelby City	D-9			
Shelbyville	C-9			
Shepardsville	C-8			
Shively	★			
Vicinity	Pg. 122, D-1			
Silver Grove	★			
Vicinity	Pg. 106, G-6			
Simpsonville	C-8			
Slaughters	D-5			
Smithland	E-4			
Smiths Grove	E-7			
Somerset	E-10			
Sonora	D-8			
S. Shore	B-12			
S. Williamson	D-13			
Springfield	D-9			
Stamping Ground	C-9			
Stanford	D-10			
Stanton	D-11			
Stearns	F-9			
Strathmoor Village	★			
Vicinity	Pg. 122, D-5			
Sturgis	D-5			
Taylorsville	D-9			
Tompkinsville	F-8			
Trenton	F-6			
Trimble	F-10			
Tyner	E-11			
Union	B-9			
Uniontown	D-5			
Upton	E-8			

LOUISIANA
Page 44

City	Ref		City	Ref		City	Ref		City	Ref
Abbeville	F-5		Boyce	D-4		Felps	D-7		Lafitte	F-8
Abington	B-2		Braithwaite	★		Fenton	E-3		Lafourche	F-7
Abita Springs	E-8		Vicinity	Pg. 132, E-9		Ferriday	C-6		Lake Arthur	F-4
Acme	D-5		Branch	E-5		Fields	E-2		Lake Charles	F-3
Acy	E-6		Breaux Bridge	E-5		Fillmore	A-2		Lake End	C-3
Ada	B-3		Bridge City	★		Fisher	C-3		Lakeland	E-6
Addis	E-6		Vicinity	Pg. 132, D-4		Flatwoods	D-4		Lake Providence	A-6
Adeline	F-6		Broussard	E-5		Flora	D-4		La Place	F-8
Afton	D-3		Brusly	E-6		Florien	D-2		Larose	G-8
Aimwell	C-5		Bryceland	B-3		Folsom	E-8		Larto	D-5
Airline Park	★		Bunkie	D-5		Fondale	E-5		Latanier	D-4
Vicinity	Pg. 132, C-4		Buras	G-9		Fordoche	E-6		Lawtell	E-5
Ajax	C-3		Burnside	F-7		Forest	A-6		Leander	D-4
Albany	E-7		Burr Ferry	D-2		Forest Hill	D-4		Lebeau	E-5
Alco	D-3		Bush	E-9		Fort Jesup	D-3		Le Blanc	E-3
Alden Bridge	A-2		Cade	F-5		Fort Necessity	B-5		Lecompte	D-4
Alexandria	D-4		Caernarvon	★		Foules	C-5		Leesville	D-3
Alsatia	A-6		Vicinity	Pg. 132, D-8		Franklin	F-6		Leeville	G-8
Alto	B-5		Calhoun	B-4		Franklinton	D-8		LeMoyen	E-5
Ama	★		Calvin	C-3		French Settlement	E-7		Leonville	E-5
Vicinity	Pg. 132, C-2		Cameron	F-3		Frenier	★		Libuse	D-5
Amelia	G-6		Campti	C-3		Vicinity	Pg. 132, C-2		Lillie	A-4
Amite	E-8		Caney	D-2		Frogmore	C-5		Linville	A-5
Anacoco	D-2		Carencro	E-5		Frost	D-7		Lisbon	A-3
Angie	D-9		Carlisle	F-9		Galbraith	C-4		Lismore	C-5
Angola	B-6		Carville	E-7		Galion	D-3		Live Oak	★
Ansley	B-4		Caspiana	B-2		Galliano	G-8		Vicinity	Pg. 132, G-7
Antioch	B-3		Castor	B-3		Gandy	D-2		Live Oak Manor	★
Arabi	★		Cedar Grove	★		Gardner	D-4		Vicinity	Pg. 132, C-3
Vicinity	Pg. 132, C-5		Vicinity	Pg. 132, F-7		Georgetown	C-4		Livingston	E-7
Archibald	B-5		Center Point	D-5		Gheens	F-8		Livonia	E-6
Archie	D-5		Centerville	F-6		Gibsland	B-3		Lockport	F-8
Arcola	D-8		Chacahoula	F-7		Gibson	G-7		Logansport	C-2
Armistead	B-3		Chackbay	F-7		Gilbert	B-6		Lone Star	★
Arnaudville	E-5		Chalmette	★		Gillis	E-3		Vicinity	Pg. 132, D-1
Ashland	B-3		Vicinity	Pg. 132, D-8		Glade	C-5		Longstreet	B-2
Athens	A-3		Chase	B-5		Glenmora	D-4		Longville	E-3
Atlanta	C-4		Chataignier	E-4		Gloster	B-2		Lottie	E-6
Augusta	★		Chatham	B-4		Golden Meadow	G-8		Louisa	F-6
Vicinity	Pg. 132, F-7		Chauvin	G-8		Goldonna	B-4		Lucas	B-3
Avery Island	F-5		Cheneyville	D-5		Gonzales	E-7		Lucky	B-3
Avondale	★		Chestnut	B-3		Gordon	A-3		Luling	★
Vicinity	Pg. 132, C-3		Chipola	C-7		Gorum	C-3		Vicinity	Pg. 132, D-1
Aycock	A-3		Chopin	C-4		Goudeau	D-5		Luna	B-5
Bains	D-6		Choudrant	B-4		Grambling	B-4		Lutcher	F-7
Baker	E-6		Church Point	E-5		Grand Bayou	B-3		Madisonville	E-8
Baldwin	F-6		Clarence	C-3		Grand Cane	B-2		Mamou	E-4
Ball	D-4		Clarks	B-5		Grand Chenier	F-4		Manchac	F-8
Bancroft	E-2		Clay	B-4		Grand Coteau	E-5		Mandeville	E-8
Barataria	G-7		Clayton	C-6		Grand Isle	G-9		Mangham	B-5
Basile	E-4		Clifton	D-8		Grand Lake	F-3		Manifest	C-5
Baskin	B-5		Clinton	D-7		Grangeville	E-7		Mansfield	C-2
Bastrop	A-5		Cloutierville	C-4		Grayson	B-5		Many	C-3
Batchelor	D-6		Cocodrie	G-7		Greensburg	D-7		Maringouin	E-6
Baton Rouge	E-7		Colfax	C-4		Gretna	★		Marion	A-5
Bayou Cane	G-7		Collinston	A-5		Vicinity	Pg. 132, D-7		Marksville	D-5
Bayou Chicot	D-5		Colquitt	A-3		Grosse Tete	E-6		Marrero	★
Bayou Sorrel	E-6		Columbia	B-5		Gueydan	F-4		Vicinity	Pg. 132, D-6
Bayou Vista	G-6		Concession	★		Hackberry	F-3		Marsalis	A-3
Baywood	E-7		Vicinity	Pg. 132, F-8		Haile	A-5		Martin	B-3
Beaver	E-4		Convent	F-7		Hall Summit	B-3		Mathews	F-8
Beekman	A-5		Converse	C-2		Hamburg	D-5		Maurepas	E-7
Beggs	E-5		Cottonport	D-5		Hammond	E-8		Maxie	E-4
Bell City	F-3		Cotton Valley	A-3		Hanna	C-3		Mayna	B-5
Belle Chasse	★		Coushatta	B-3		Harahan	★		McDade	B-2
Vicinity	Pg. 132, C-5		Covington	E-8		Vicinity	Pg. 132, D-4		McNary	D-4
Belle Rose	F-7		Cravens	D-3		Harrisonburg	C-5		Melder	C-3
Bellwood	C-2		Creole	F-3		Harvey	★		Melrose	C-3
Belmont	C-2		Creston	B-3		Vicinity	Pg. 132, D-6		Melville	E-5
Benson	A-2		Crowley	F-4		Hathaway	E-4		Meraux	★
Bentley	D-4		Crown Point	★		Hayes	F-4		Vicinity	Pg. 132, D-9
Benton	A-2		Vicinity	Pg. 132, G-6		Haynesville	A-3		Mermentau	F-4
Bernice	A-4		Crowville	B-6		Hebert	B-5		Mer Rouge	A-5
Bertrandville	★		Cullen	A-3		Henderson	E-6		Merryville	E-2
Vicinity	Pg. 132, G-7		Cut Off	G-8		Hermitage	F-7		Metairie	★
Berwick	F-6		Cypress	C-3		Hickory	E-9		Vicinity	Pg. 132, C-5
Bienville	B-3		Dalcour	★		Hico	D-3		Midland	F-4
Blackburn	D-7		Vicinity	Pg. 132, F-8		Highland	A-6		Midway	B-4
Blanchard	A-2		Darlington	D-7		Hodge	B-4		Mill	A-3
Bogalusa	D-9		Darnell	A-6		Holden	E-7		Minden	A-3
Bohemia	★		Davant	F-9		Holloway	D-5		Mira	A-2
Vicinity	Pg. 132, D-5		Deer Park	C-6		Holly Beach	F-2		Mitchel	E-4
Bolinger	A-2		Delacroix	★		Hollywood	★		Mittie	E-4
Bonita	A-5		Vicinity	Pg. 132, F-9		Vicinity	Pg. 132, F-3		Mix	D-7
Bordelonville	D-5		Delcambre	F-5		Holton	A-5		Monroe	A-5
Bosco	D-5		Delhi	A-6		Holum	B-5		Montegut	G-8
Bossier City	A-2		De Quincy	E-3		Hornbeck	D-3		Monterey	C-6
Bourg	G-8		De Ridder	E-3		Hosston	A-2		Montgomery	C-3
			Derry	C-4		Hot Wells	D-4		Montpelier	E-7
			Des Allemands	F-8		Houma	G-7		Mooringsport	A-2
			Destrehan	★		Humphreys	G-7		Moreauville	D-5
			Vicinity	Pg. 132, D-1		Hutton	D-3		Morgan City	F-7
			Deville	D-5		Ida	A-2		Morganza	E-6
			Diamond	G-9		Independence	E-8		Morrow	E-5
			Dixie Inn	A-3		Indian Mound	D-7		Mount Carmel	D-7
			Dodson	B-4		Indian Village	F-6		Myrtle Grove	G-8
			Donaldsonville	F-7		Iota	E-4		Nairn	G-9
			Downsville	A-4		Iowa	F-3		Naomi	F-8
			Dry Creek	E-3		Ivan	A-2		Napoleonville	F-7
			Dry Prong	C-4		Jackson	D-6		Natchez	C-3
			Dubach	A-4		Jamestown	B-3		Natchitoches	C-3
			Dubuisson	E-5		Jeanerette	F-5		Negreet	C-2
			Dulac	G-7		Hessmer	D-5		Newellton	B-6
			Dupont	F-5		Jean Lafitte	★		New Era	D-5
			Duson	E-5		Vicinity	Pg. 132, F-5		New Iberia	F-6
			Duty	C-5		Jefferson Hts.	★		Newlight	B-6
			Easlyville	D-7		Vicinity	Pg. 132, C-4		New Llano	D-3
			East End	★		Jena	C-5		New Orleans	★
			Vicinity	Pg. 132, B-5		Jennings	F-4		Vicinity	Pg. 132, A-7
			Eastwood	B-2		Jesuit Bend	★		New Roads	D-6
			Echo	D-5		Vicinity	Pg. 132, G-7		Noble	C-2
			Edgerly	F-2		Jigger	B-5		North Hodge	B-4
			Effie	D-5		Johnsons Bayou	F-2		Norwood	D-6
			Elizabeth	E-4		Jones	A-5		Oakdale	E-4
			Elm Grove	B-2		Jonesboro	B-4		Oak Grove	A-6
			Elton	E-4		Jonesville	C-5		Oak Grove	F-5
			Empire	G-9		Kaplan	F-4		Oakland	C-3
			Englishturn	★		Keatchie	B-2		Oak Ridge	A-5
			Vicinity	Pg. 132, E-8		Keithville	B-2		Oakvale	★
			Enon	D-8		Kenner	★		Vicinity	Pg. 132, G-7
			Enterprise	C-5		Vicinity	Pg. 132, B-3		Oberlin	E-4
			Erath	F-5		Kentwood	D-8		Oil City	A-2
			Estelle	★		Kilbourne	A-6		Olivier	F-6
			Vicinity	Pg. 132, E-6		Killian	E-7		Olla	C-4
			Esther	F-5		Kinder	E-4		Opelousas	E-5
			Esto	D-2		Kingston	A-3		Oscar	E-6
			Ethel	D-7		Kisatchie	D-3		Oxford	C-4
			Eunice	E-4		Krotz Springs	E-5		Packton	C-4
			Eva	D-3		Kurthwood	D-3		Paincourtville	F-7
			Evangeline	E-4		Labadieville	F-7		Parks	F-5
			Evans	D-3		Labarre	D-6		Patterson	F-6
			Evelyn	B-2		Lacamp	C-3		Pearl River	E-9
			Evergreen	D-5		Lacombe	E-8		Pecan Island	G-4
			Farmerville	A-4		Lafayette	E-5		Perryville	A-5

★ Denotes City located only on Vicinity map. City not located on State map.

LOUISIANA

MARYLAND

Phoenix F-9
Pickering D-3
Pierre Part F-6
Pigeon F-6
Pilot Town G-10
Pine D-8
Pine Grove E-7
Pine Prairie E-4
Pineville D-4
Pitkin D-3
Plain Dealing A-2
Plaquemine E-6
Pleasant Hill C-2
Pointe a La Hache ... G-9
Pollock C-4
Ponchatoula E-8
Port Allen E-6
Port Barre E-5
Port Sulphur G-9
Port Vincent E-7
Powhatan C-3
Poydras .... ★
　Vicinity ... Pg. 132, E-10
Provencal C-3
Quitman B-4
Raceland F-7
Ragley E-3
Ramah E-6
Rayne F-5
Rayville A-5
Readhimer B-3
Reddell E-4
Reeves E-3
Reggio F-9
Reserve F-8
Richmond B-6
Richwood B-5
Ridgecrest C-6
Ringgold B-3
Riverton B-5
Robeline C-3
Robert E-8
Rocky Mountain A-2
Ruddock F-8
Rodessa A-1
Rosedale E-6
Rosefield C-5
Rosepine D-3
Ruston B-3
St. Bernard .... ★
　Vicinity ... Pg. 132, E-10
Saint Clair .... ★
　Vicinity ... Pg. 132, E-8
St. Francisville E-6
Saint Gabriel E-7
Saint Joseph B-6
Saint Martinville ... F-6
St. Rose .... ★
　Vicinity ... Pg. 132, D-1
Sailes B-3
Saline B-3
Sarepta A-3
Scarsdale .... ★
　Vicinity ... Pg. 132, F-8
Scott F-5
Shaw D-6
Sheridan D-9
Shongaloo A-3
Shreveport B-2
Sibley A-3
Sicily Island C-5
Sieper D-4
Sikes B-4
Simmesport D-5
Simpson D-3
Singer E-3
Slagle D-3
Slaughter E-6
Slidell E-9
Slocum D-6
Somerset A-6
Sondheimer A-6
Sorrento F-7
Spearsville A-4
Spencer A-5
Spokane C-6
Springfield E-7
Springhill A-2
Spring Ridge B-2
Stanley C-2
Starks F-2
Start B-5
State Line D-9
Stella .... ★
　Vicinity ... Pg. 132, F-8
Sterlington A-5
Stonewall B-2
Sugartown E-3
Sulphur F-3
Summerfield A-3
Sun E-9
Sunset E-5
Supreme F-7
Talisheek E-9
Tallulah B-6
Tangipahoa D-8
Tannehill B-4
Taylortown B-2
Terry A-6
Tete E-6
Thibodaux F-7
Tidewater G-9
Tioga D-4
Toomey F-2
Toro D-2
Transylvania A-6
Triumph G-10
Truxno A-4
Tullos C-5
Turkey Creek D-4
Urania C-4
Union F-7
Union Hill D-4
Vacherie F-7
Varnado D-8
Venice G-10
Verda C-4
Vidalia C-6
Vienna A-4
Ville Platte E-4
Vinton F-2
Violet F-9
Vivian A-2
Vixen B-5

Waggaman .... ★
　Vicinity ... Pg. 132, D-3
Wakefield D-6
Waldheim E-8
Walters C-5
Warnerton D-8
Washington E-5
Waterproof C-6
Watson E-7
Waverly B-6
Weeks F-6
Welsh F-4
Weston B-4
Westwego F-8
　Vicinity ... Pg. 132, D-3
Weyanoke D-6
White Castle F-7
Whitehall C-5
Williana C-4
Wills Point .... ★
　Vicinity ... Pg. 132, G-8
Wilmer D-8
Wilson D-6
Winnfield C-4
Winnsboro B-5
Woodworth D-4
Worden D-6
Wye E-3
Youngsville F-5
Yscloskey F-9
Zenoria C-4
Zwolle C-2

MAINE
Page 45

Acton J-2
Addison J-6
Albion G-3
Alexander E-6
Allagash B-4
Andover G-2
Anson F-3
Appleton G-4
Ashland C-4
Athens F-3
Auburn H-3
Augusta G-3
Avon F-2
Bailey Island J-3
Balleyville E-6
Bangor F-4
Bar Harbor G-5
Baring E-6
Bath H-3
Beddington F-6
Belfast G-4
Belgrade G-3
Benton F-3
Berwick J-2
Bethel G-2
Biddeford J-3
Bigelow F-2
Bingham F-3
Blaine C-5
Blanchard E-3
Blue Hill G-5
Boothbay H-4
Boothbay Harbor H-4
Bradford F-4
Bradley F-5
Brewer F-4
Bridgewater C-5
Bridgton H-2
Brighton F-3
Bristol H-4
Brooks G-4
Brownfield H-2
Brownville E-4
Brownville Jct. E-4
Brunswick H-3
Buckfield G-3
Bucks Harbor F-6
Bucksport G-5
Burlington E-5
Burnham F-4
Buxton J-2
Bryon G-2
Calais E-6
Cambridge F-3
Camden G-4
Canton G-2
Cape Elizabeth J-3
Cape Porpoise J-2
Caratunk E-3
Caribou B-5
Carmel F-4
Carrabassett F-3
Carroll E-4
Cary D-5
Casco H-2
Castine G-4
Charleston F-4
Cherryfield G-6
Chester E-5
Chisolm G-2
Clayton Lake C-3
Clinton F-3
Coburn Gore E-2
Colby E-6
Columbia G-6
Conner B-5
Cooper E-6
Corea G-6
Corinna F-3
Cornish H-1
Costigan F-4
Crawford F-6
Crouseville B-5
Cumberland Center ... H-2
Cutler F-7
Daigle B-4
Dallas E-2
Damariscotta H-3
Danforth D-5
Deblois F-5
Dedham G-4
Deer Isle G-5
Dennistown E-2
Dennysville F-6

Dickey B-4
Dixfield G-2
Dixmont G-4
Dover Foxcroft F-4
Durham H-3
Eagle Lake B-4
Eastbrook F-5
E. Hampden F-4
E. Machias F-6
E. Millinocket E-4
Easton C-5
Eastport F-7
E. Sumner G-3
E. Wilton G-3
Eaton E-6
Eddington F-4
Ellsworth G-5
Estcourt Sta. A-4
Eustis F-2
Fairfield F-3
Farmingdale G-3
Farmington G-2
Fort Fairfield B-5
Fort Kent A-4
Franklin G-5
Freeport H-3
Frenchville A-4
Friendship H-3
Frye G-2
Fryeburg H-2
Gardiner G-3
Gilead G-1
Glenburn Center F-4
Gorham J-2
Gouldsboro G-5
Grand Isle A-5
Grand Lake Stream ... E-5
Gray H-3
Greenville E-3
Grindstone D-4
Grove F-6
Guerette B-4
Guilford F-3
Hallowell G-3
Hamlin B-5
Hampden F-4
Hancock G-5
Hanover G-2
Hartford Center G-2
Hartland F-3
Harrington F-5
Harrison H-2
Haven D-4
Haynesville D-5
Hersey D-5
Hinckley G-3
Hiram H-2
Hogdon C-5
Houghton F-2
Howland E-4
Houlton D-5
Island Falls D-5
Jackman E-2
Jay H-2
Jefferson H-4
Jemtland B-5
Jonesboro F-6
Jonesport G-6
Keegan B-5
Kenduskeag F-4
Kennebago Lake F-2
Kennebunk J-2
Kennebunkport J-3
Kingfield F-2
Kingman E-4
Kingsbury F-3
Kittery K-2
Kittery Point K-2
Knowles Corner C-4
Kokadjo E-4
Lagrange F-4
Lake Parlin E-2
Lambert Lake E-6
Larone G-3
Lee E-5
Lewiston H-2
Liberty G-4
Lille A-5
Lily Bay E-3
Limerick J-2
Limestone B-5
Limington J-2
Lincoln E-5
Lincolnville G-4
Linneus D-5
Lisbon H-3
Lisbon Falls H-2
Littleton C-5
Livermore Falls G-3
Long Pond E-3
Lovell H-2
Lubec F-7
Ludlow D-5
Lynchville G-2
Machias F-6
Macwahoc D-5
Madawaska A-4
Madison F-3
Madrid F-2
Manchester G-3
Mapleton B-5
Marion F-6
Mars Hill C-5
Masardis C-4
Mattawamkeag D-4
Mayfield F-3
Medway E-4
Medford F-4
Mexico G-2
Milbridge G-5
Milford F-5
Millinocket D-4
Milo E-4
Molunkus D-5
Monarda B-4
Monroe G-4
Monson E-3
Monticello C-5
Moose River E-2
Morrill G-4
Moscow F-3
Naples H-2
Newcastle H-3
Newfield J-2

New Gloucester H-2
Newport F-4
Newry G-2
Nobleboro G-3
Norcross E-4
Norridgewock G-3
N. Amity D-6
N. Anson F-3
N. Berwick J-2
Northeast Carry D-3
Northeast Harbor G-5
Northfield F-6
N. Lovell H-2
N. Lyndon B-5
Norway H-2
Oakland G-3
Ogunquit J-2
Old Orchard Beach ... J-2
Old Town F-5
Oquossoc F-2
Orient D-5
Orland G-4
Orono F-5
Orrington F-4
Ouelette B-4
Oxbow C-4
Oxford G-2
Palmyra F-3
Parkman F-3
Paris G-2
Patten D-4
Perham B-5
Perry F-7
Peru G-2
Phippsburg H-3
Pittsfield F-4
Pittston H-3
Poland H-2
Portage C-4
Portland J-3
Presque Isle B-5
Princeton E-6
Quimby B-4
Randolph F-2
Rangeley F-2
Raymond H-2
Readfield G-3
Red Beach E-6
Reed D-5
Richmond G-3
Ripley F-3
Ripogenus Dam D-4
Robbinston E-7
Robinsons C-5
Rockland H-4
Rockport H-4
Rockwood E-3
Rome G-3
Roxbury G-2
Rumford G-2
Sabattus H-3
Saco J-3
St. Albans F-3
St. David A-5
St. Francis B-4
St. George G-4
St. John B-4
Salem F-2
Sanford J-2
Sangerville F-3
Saponac E-5
Scarborough J-2
Sebago H-2
Sebec Lake F-3
Seboeis E-4
Seboomook D-3
Sedgewick G-5
Seal Harbor G-5
Searsmont G-4
Searsport G-4
Selden D-5
Sheridan B-4
Sherman D-4
Sherman Mills D-5
Sherman Sta. D-4
Shin Pond D-4
Shirley Mills E-3
Sidney G-3
Skowhegan F-3
Smyrna Mills C-5
Soldier Pond A-4
Solon F-3
S. Arm F-2
S. Bancroft D-5
S. China G-4
S. Hope G-4
S. Paris H-2
S. Portland J-3
Southwest Harbor E-5
Springfield E-5
Springvale J-2
Squa Pan C-4
Stacyville D-4
Standish J-2
Starboard F-6
Stetson F-4
Steuben G-6
Stockton Springs G-4
Stonington G-5
Stow H-2
Stratton F-2
Strong F-2
Sweden H-2
The Forks E-3
Thomaston H-4
Topsfield E-6
Topsham H-3
Tremont H-5
Turner H-2
Union G-3
Unity G-3
Upton F-1
Van Buren B-5
Vanceboro E-6
Vienna G-3
Vinalhaven H-4
Waite E-6
Waldoboro H-4
Waltham F-5
Warren H-4
Washburn B-5
Washington G-4
Waterboro J-2
Waterford H-2

Waterville G-4
Wells J-2
Wesley F-6
Westbrook J-3
W. Enfield E-5
W. Forks E-3
W. Newfield J-2
Weston D-5
Whiting F-6
Williamantic E-3
Wilsons Mills F-1
Wilton F-2
Winn E-5
Winslow G-3
Winter Harbor G-5
Winterville B-4
Winterport G-4
Winthrop G-3
Wiscasset H-3
Woodland E-6
Woolwich H-3
Wytopitlock D-5
Yarmouth H-3
York Beach K-2
York Village K-2

MARYLAND
Pages 46-47

Aberdeen C-12
Accident C-7
Alexandria E-11
Allanwood .... ★
Allview Estates ..... ★
　Vicinity ... Pg. 148, B-5
　Vicinity ... Pg. 97, G-1
Altamont C-7
Ammendale .... ★
　Vicinity ... Pg. 148, C-4
Andersontown D-12
Annapolis D-11
Annapolis Junction .. ★
　Vicinity ... Pg. 97, H-2
Arbutus .... ★
　Vicinity ... Pg. 97, E-3
Armiger .... ★
　Vicinity ... Pg. 97, H-6
Arnold D-11
　Vicinity ... Pg. 97, K-6
Ashton .... ★
　Vicinity ... Pg. 148, A-3
Aspen Hill D-10
Ballard .... ★
　Vicinity ... Pg. 148, J-7
Baltimore D-11
　Vicinity ... Pg. 97, D-4
Bayside Beach .... ★
　Vicinity ... Pg. 97, H-7
Beauvue F-11
Beaverdam F-13
Bel Air C-11
Belltown .... ★
　Vicinity ... Pg. 97, A-2
Beltsville D-11
　Vicinity ... Pg. 148, C-6
Benevola C-9
Benfield .... ★
　Vicinity ... Pg. 97, J-4
Berlin E-13
Bethesda .... ★
　Vicinity ... Pg. 148, E-4
Birdtown F-12
Bishop E-13
Bivalve F-12
Bladensburg .... ★
　Vicinity ... Pg. 148, F-6
Bloomsbury .... ★
　Vicinity ... Pg. 97, E-3
Bowie D-11
Braddock Hts. C-9
Brandywine E-11
Brinklow .... ★
　Vicinity ... Pg. 148, A-5
Bristol E-11
Brookландville ...... ★
　Vicinity ... Pg. 97, B-4
Brooklyn .... ★
　Vicinity ... Pg. 97, E-3
Brownstown .... ★
　Vicinity ... Pg. 148, A-1
Brunswick D-10
Burleytown .... ★
　Vicinity ... Pg. 97, G-4
Burtonville .... ★
　Vicinity ... Pg. 148, B-6
Cabin John .... ★
　Vicinity ... Pg. 148, H-7
California E-11
Cambridge E-11
Camp Springs .... ★
　Vicinity ... Pg. 148, J-7
Cape St. Claire D-11
Capitol Hts. .... ★
　Vicinity ... Pg. 148, G-7
Carderock .... ★
　Vicinity ... Pg. 148, E-2
Carney .... ★
　Vicinity ... Pg. 97, B-6
Catonsville .... ★
　Vicinity ... Pg. 97, E-3
Cayots C-12
Cearfoss C-9
Centreville D-11
Champ F-12
Chance F-12
Chaptico E-11
Charlotte Hall E-11
Chattolanee .... ★
　Vicinity ... Pg. 97, B-3
Chesapeake City C-12
Chestertown D-12
Cheverly .... ★
　Vicinity ... Pg. 148, F-6
Chevy Chase .... ★
　Vicinity ... Pg. 148, E-4
Church Creek E-12
Church Hill D-12
Clements F-11
Clinton .... ★
　Vicinity ... Pg. 148, J-7
Clinton Hts. .... ★
　Vicinity ... Pg. 148, J-7

Cockeysville C-11
Colesville .... ★
　Vicinity ... Pg. 148, C-4
College Park .... ★
　Vicinity ... Pg. 148, D-6
Colmar Manor .... ★
　Vicinity ... Pg. 97, C-1
Columbia D-10
Columbus Hills ...... ★
　Vicinity ... Pg. 97, E-1
Colverton .... ★
　Vicinity ... Pg. 148, C-6
Conways .... ★
　Vicinity ... Pg. 97, K-2
Cooksville D-10
Corriganville C-8
Creagerstown C-10
Cresaptown C-8
Crisfield F-12
Crofton D-11
Cronhardt .... ★
　Vicinity ... Pg. 97, A-3
Cropley .... ★
　Vicinity ... Pg. 148, D-2
Crownsville .... ★
　Vicinity ... Pg. 97, K-5
Cub Hill .... ★
　Vicinity ... Pg. 97, B-6
Cumberland C-8
Damascus D-10
Daniels .... ★
　Vicinity ... Pg. 97, D-1
Darnestown .... ★
　Vicinity ... Pg. 148, B-1
Deale E-11
Deer Park .... ★
　Vicinity ... Pg. 97, B-1
Delight .... ★
　Vicinity ... Pg. 97, A-2
Delmar E-12
Denton D-12
Derwood .... ★
　Vicinity ... Pg. 148, B-3
District Heights .... ★
　Vicinity ... Pg. 148, G-6
Dorrs Corner .... ★
　Vicinity ... Pg. 97, B-1
Dorsey .... ★
　Vicinity ... Pg. 97, G-2
Dublin C-11
Dundalk E-11
Easton E-11
East Springbrook .... ★
　Vicinity ... Pg. 148, C-6
Eden F-12
Edgemere .... ★
　Vicinity ... Pg. 97, D-7
Edgewood C-12
Edmonston .... ★
　Vicinity ... Pg.148, E-6
Elder C-12
Eldorado E-12
Elkridge .... ★
　Vicinity ... Pg. 97, F-3
Elkton C-12
Ellicott City D-11
　Vicinity ... Pg. 97, E-2
Ellwood E-12
Emmitsburg C-10
Emory Grove .... ★
　Vicinity ... Pg. 148, A-2
Essex .... ★
　Vicinity ... Pg. 97, D-7
Exline .... ★
　Vicinity ... Pg. 97, G-2
Fairknoll .... ★
　Vicinity ... Pg. 97, F-1
Fairland .... ★
　Vicinity ... Pg. 148, C-6
Fairlee D-11
Fairmount Heights ... ★
　Vicinity ... Pg. 148, F-6
Fairview .... ★
　Vicinity ... Pg. 97, G-6
Fairview Beach E-10
Fallston C-11
Federalsburg E-12
Ferndale .... ★
　Vicinity ... Pg. 97, G-4
Finksburg C-10
Fishtown E-12
Flintstone C-8
Foremans Corner ..... ★
　Vicinity ... Pg. 97, E-4
Forest Hts .... ★
　Vicinity ... Pg. 148, H-5
Forestville .... ★
　Vicinity ... Pg. 148, H-7
Fort Foot Village ... ★
　Vicinity ... Pg. 148, J-5
Fort Howard .... ★
　Vicinity ... Pg. 97, D-7
Fountain Head C-10
Frederick C-10
Fredericktown C-12
Friendly .... ★
　Vicinity ... Pg. 97, F-4
Friendship Heights .. ★
　Vicinity ... Pg. 148, C-7
Friendsville C-7
Frostburg C-8
Fruitland F-12
Fullerton .... ★
　Vicinity ... Pg. 97, C-6
Fulton .... ★
　Vicinity ... Pg. 148, A-7
Galena D-12
Gambrills .... ★
　Vicinity ... Pg. 97, J-4
Gapland C-9
Garrison .... ★
　Vicinity ... Pg. 97, J-6
Georgetown C-12
Germantown .... ★
　Vicinity ... Pg. 148, H-6
Girdletree F-13
Glen Burnie .... ★
　Vicinity ... Pg. 97, G-4
Glen Echo .... ★
　Vicinity ... Pg. 148, E-3
Glen Hills .... ★
　Vicinity ... Pg. 148, B-2

Glenmont .... ★
　Vicinity ... Pg. 148, C-4
Gorman D-7
Gransonville D-11
Grantie .... ★
　Vicinity ... Pg. 97, C-1
Gray Manor .... ★
　Vicinity ... Pg. 97, E-6
Grayton E-10
Greenbelt D-11
Green Haven .... ★
　Vicinity ... Pg. 97, H-6
Greensboro D-12
Guilford .... ★
　Vicinity ... Pg. 97, G-1
Hagerstown C-10
Halethorpe C-9
Halfway C-9
Hammond Park .... ★
　Vicinity ... Pg. 148, A-7
Hampstead C-11
Hampton .... ★
　Vicinity ... Pg. 97, A-5
Hancock C-9
Hanover .... ★
　Vicinity ... Pg. 97, G-3
Harmans .... ★
　Vicinity ... Pg. 97, G-3
Harmony E-12
Harrisonville .... ★
　Vicinity ... Pg. 97, B-1
Harwood .... ★
　Vicinity ... Pg. 97, G-3
Harundale .... ★
　Vicinity ... Pg. 97, H-5
Havre de Grace C-12
Hebbville .... ★
　Vicinity ... Pg. 97, C-2
Hebron E-12
Hermanville F-11
Hernwood .... ★
　Vicinity ... Pg. 97, C-1
Herwood Heights ..... ★
　Vicinity ... Pg. 97, B-1
Hillondale .... ★
　Vicinity ... Pg. 148, D-5
Holbrook .... ★
　Vicinity ... Pg. 97, B-5
Hollywood E-11
Holywood Park .... ★
　Vicinity ... Pg. 148, C-6
Honga E-11
Hopewell F-12
Hughesville E-11
Huntersville E-11
Hunting Hill .... ★
　Vicinity ... Pg. 148, B-2
Huntingtown E-11
Hurlock E-12
Hyattsville .... ★
　Vicinity ... Pg. 148, E-6
Ilchester .... ★
　Vicinity ... Pg. 97, E-2
Ingleside D-12
Inverness .... ★
　Vicinity ... Pg. 97, E-7
Ironshire F-13
Jacobsville .... ★
　Vicinity ... Pg. 97, H-6
Jarrettsville C-11
Jefferson C-9
Jessup .... ★
　Vicinity ... Pg. 97, G-2
Jonestown .... ★
　Vicinity ... Pg. 97, F-1
Josenhans Corner .... ★
　Vicinity ... Pg. 97, D-7
Joyce .... ★
　Vicinity ... Pg. 97, K-6
Kennedyville D-11
Kensington .... ★
　Vicinity ... Pg. 148, D-4
Kettering C-10
Keymar C-10
Keysers Ridge C-7
Kingsville .... ★
　Vicinity ... Pg. 97, B-5
Lake Shore .... ★
　Vicinity ... Pg. 97, J-6
Lakesville E-12
Landover Hills ...... ★
　Vicinity ... Pg. 148, F-7
Lansdowne .... ★
　Vicinity ... Pg. 97, E-4
Lappans C-9
La Plata E-10
Laurel .... ★
　Vicinity ... Pg. 97, J-1
LaVale .... ★
　Vicinity ... Pg. 148, B-7
Layhill .... ★
　Vicinity ... Pg. 97, J-3
Leonardtown E-11
Lexington Park F-11
Linthicum .... ★
　Vicinity ... Pg. 97, F-4
Lipins Corner .... ★
　Vicinity ... Pg. 97, H-5
Lochearn .... ★
　Vicinity ... Pg. 97, C-3
Lonaconing C-8
Londontowne D-11
Longridge F-12
Lusby E-11
Lutherville .... ★
　Vicinity ... Pg. 97, B-5
Macey's Cor. .... ★
　Vicinity ... Pg. 97, J-6
Madison E-11
Mago Mista Beach .... ★
　Vicinity ... Pg. 148, H-6
Manchester C-11
Marbury E-10
Mardela Springs E-12
Marion F-12
Marlow Hts. .... ★
　Vicinity ... Pg. 148, H-6
Maryland City .... ★
　Vicinity ... Pg. 148, H-6
Massey D-12
Maugansville C-9
Mayfield .... ★

McPherson .... ★
　Vicinity ... Pg. 97, G-4
Meadowood .... ★
　Vicinity ... Pg. 148, C-5
Middle River .... ★
　Vicinity ... Pg. 97, D-7
Milford .... ★
　Vicinity ... Pg. 97, C-3
Millersville .... ★
　Vicinity ... Pg. 97, K-4
Millington D-12
Mitchells Range ..... ★
　Vicinity ... Pg. 148, B-1
Monie F-12
Monrose .... ★
　Vicinity ... Pg. 148, C-3
Monrovia C-10
Montgomery Village .. ★
　Vicinity ... Pg. 148, A-2
Morningside .... ★
　Vicinity ... Pg. 148, H-7
Mountain Lake Park .. C-7
Mt. Airy C-10
Mt. Rainier .... ★
　Vicinity ... Pg. 148, F-6
Mt. Wilson .... ★
　Vicinity ... Pg. 97, B-2
Mulkirk .... ★
　Vicinity ... Pg. 148, C-7
Necker .... ★
　Vicinity ... Pg. 97, B-7
Newburg E-11
New Carrollton ...... ★
　Vicinity ... Pg. 148, E-7
Norbeck Estates ..... ★
　Vicinity ... Pg. 148, A-4
N. Beach E-11
North Point E-7
Norrisville C-11
Oakland D-12
Oakland Mills .... ★
　Vicinity ... Pg. 97, F-1
Oaklawn .... ★
　Vicinity ... Pg. 148, J-6
Oakmont .... ★
　Vicinity ... Pg. 148, D-3
Oakwood C-12
Ocean City E-13
Odenton .... ★
　Vicinity ... Pg. 97, J-3
Old Salem Village ... ★
　Vicinity ... Pg. 148, B-5
Oldtown C-8
Orchard Beach .... ★
　Vicinity ... Pg. 97, G-6
Overlea .... ★
　Vicinity ... Pg. 97, C-6
Owings Mills .... ★
　Vicinity ... Pg. 97, B-2
Oxon Hill .... ★
　Vicinity ... Pg. 148, J-5
Parkville .... ★
　Vicinity ... Pg. 97, C-6
Pasadena .... ★
　Vicinity ... Pg. 97, H-5
Perry Hall .... ★
　Vicinity ... Pg. 97, B-7
Pikesville .... ★
　Vicinity ... Pg. 97, C-3
Pinehurst E-11
Pocomoke City F-12
Poolesville D-10
Pt. Lookout F-11
Potomac .... ★
　Vicinity ... Pg. 148, D-2
Potomac Heights E-10
Pot Spring .... ★
　Vicinity ... Pg. 97, A-5
Pratt C-8
Presto .... ★
　Vicinity ... Pg. 97, A-5
Price D-12
Prince Frederick E-11
Princess Anne F-12
Providence .... ★
　Vicinity ... Pg. 97, B-5
Pumphrey .... ★
　Vicinity ... Pg. 97, F-4
Putty Hill .... ★
　Vicinity ... Pg. 97, C-6
Queen Anne D-12
Quince Orchard ...... ★
　Vicinity ... Pg.148, B-1
Randallstown .... ★
　Vicinity ... Pg. 97, C-2
Redhouse D-7
Redland .... ★
　Vicinity ... Pg. 148, A-3
Regency Estates ..... ★
　Vicinity ... Pg. 148, C-2
Reisterstown C-11
Relay .... ★
　Vicinity ... Pg. 97, F-3
Riderwood .... ★
　Vicinity ... Pg. 97, B-4
Ridge F-11
Ridgely D-12
Ridgeway .... ★
　Vicinity ... Pg. 97, H-3
Rising Sun C-12
Rison E-10
Riverdale .... ★
　Vicinity ... Pg. 148, E-6
Riviera Beach .... ★
　Vicinity ... Pg. 97, G-6
Robinson .... ★
　Vicinity ... Pg. 97, J-5
Rockdale .... ★
　Vicinity ... Pg. 97, C-2
Rock Hall D-11
Rockland .... ★
　Vicinity ... Pg. 97, A-4
Rockville D-10
　Vicinity ... Pg. 148, C-3
Rodgers Forge .... ★
Roe D-12
Rosedale .... ★
　Vicinity ... Pg. 97, D-6
Rush C-8
Ruxton .... ★
　Vicinity ... Pg. 97, B-4

★ Denotes City located only on Vicinity map. City not located on State map.

MARYLAND

St. Charles E-11
St. Michaels E-11
Salisbury E-12
Sandy Spring .... ★
  Vicinity .... Pg. 148, A-5
Scaggsville .... ★
  Vicinity .... Pg. 148, A-7
Scotland F-11
Seat Pleasant .... ★
  Vicinity .... Pg. 148, G-7
Secretary E-12
Severn .... ★
  Vicinity .... Pg. 97, H-3
Severna Park .... ★
  Vicinity .... Pg. 97, J-6
Sharpsburg C-9
Sharptown E-12
Shore Acres .... ★
  Vicinity .... Pg. 97, K-7
Showell E-13
Silesia .... ★
  Vicinity .... Pg. 148, K-5
Silver Spring .... ★
  Vicinity .... Pg. 148, E-5
Snow Hill F-13
Solley .... ★
  Vicinity .... Pg. 97, G-6
Southgate D-11
South Lawn .... ★
  Vicinity .... Pg. 148, J-5
Sparrows Point .... ★
  Vicinity .... Pg. 97, F-7
Spencerville .... ★
  Vicinity .... Pg. 148, B-6
Stevenson .... ★
  Vicinity .... Pg. 97, B-3
Stockton F-13
Sudlersville D-12
Suitland .... ★
  Vicinity .... Pg. 148, H-6
Sunshine D-10
Swanton C-7
Sycamore Acres .... ★
  Vicinity .... Pg. 148, B-4
Takomd Park .... ★
  Vicinity .... Pg. 148, E-5
Taneytown C-10
Tantallon .... ★
  Vicinity .... Pg. 148, K-5
Taylors Island E-11
Temple Hills .... ★
  Vicinity .... Pg. 148, H-6
Thayerville C-7
The Glen .... ★
  Vicinity .... Pg. 148, C-2
Timanium .... ★
  Vicinity .... Pg. 97, A-4
Tippett .... ★
  Vicinity .... Pg. 148, K-6
Towson .... ★
  Vicinity E-11
Trappe E-11
Travilan .... ★
  Vicinity .... Pg. 148, C-1
Truxton .... ★
  Vicinity .... Pg. 148, D-1
Turner E-11
  Vicinity .... Pg. 97, E-6
Unionville C-10
University Park .... ★
  Vicinity .... Pg. 148, E-6
Urbana C-10
Valley Lee F-11
Veirs Mills .... ★
  Vicinity .... Pg. 148, C-4
Waldorf E-11
Walkersville C-10
Washington Grove .... ★
  Vicinity .... Pg. 148, A-3
Waterloo .... ★
  Vicinity .... Pg. 97, G-2
Waterview F-12
Webster C-11
Wenona F-12
Wesley F-13
Westernport C-7
West Laurel .... ★
  Vicinity .... Pg. 148, B-7
Westminster C-10
Westover F-12
Wheaton .... ★
  Vicinity .... Pg. 148, C-5
White Marsh .... ★
  Vicinity .... Pg. 97, C-7
Whiteford C-11
White Oak .... ★
  Vicinity .... Pg. 148, C-5
Whiton F-12
Williamsport C-9
Willards E-12
Woodlawn .... ★
  Vicinity .... Pg. 97, D-3
Woodsboro C-10

MASSACHUSETTS
Page 30

Abington C-8
Acushnet D-8
Adams B-2
Agawam C-3
Amesbury A-8
Amherst B-4
Andover A-7
Acton B-6
Arlington .... ★
  Vicinity .... Pg. 100, E-3
Ashburnham A-5
Ashby A-5
Ashfield B-3
Ashland B-6
Ashley Falls C-2
Assinippi C-8
Assonet D-8
Athol B-5
Attleboro C-7
Auburn B-6
Ayer B-6
Ayersville A-7
Baldwinville A-4
Bancroft C-3
Barnstable D-10
Barre B-5
Barre Plains B-5
Becket C-2
Bedford B-7
  Vicinity .... Pg. 100, C-1
Belchertown C-4
Bellingham C-7
Belmont B-7
  Vicinity .... Pg. 100, D-2
Bernardston A-3
Beverly B-8
  Vicinity .... Pg. 100, A-7
Billerica B-7
  Vicinity .... Pg. 100, A-1
Blackstone C-6
Blandford C-3
Bolton B-6
Bondsville B-4
Boston B-7
  Vicinity .... Pgs. 100-101
Boxford A-7
Braintree C-8
  Vicinity .... Pg. 101, H-4
Brant Rock C-8
Brewster D-10
Bridgewater D-7
Briggs Corner C-7
Briggsville A-2
Brockton C-7
Brookfield C-5
Brookline .... ★
  Vicinity .... Pg. 101, G-3
Buckland B-3
Burlington B-7
  Vicinity .... Pg. 100, C-2
Buzzard Bay D-8
Cambridge .... ★
  Vicinity .... Pg. 100, C-7
Canton C-7
  Vicinity .... Pg. 101, L-3
Carver D-8
Cedarville D-9
Centerville E-9
Charlemont B-3
Charlton C-5
Chartley D-7
Chatham D-10
Chelmsford B-6
Chelsea .... ★
  Vicinity .... Pg. 100, B-2
Cheshire B-2
Chester C-2
Clarksburg A-2
Cliftondale .... ★
  Vicinity .... Pg. 100, D-5
Clinton B-6
Cohasset B-8
Coldbrook Springs B-5
Colrain A-3
Concord B-7
Conway B-3
Cotuit D-9
Crossing C-5
Dalton B-2
Danvers B-7
  Vicinity .... Pg. 100, A-6
Dedham C-7
  Vicinity .... Pg. 101, J-3
Dell A-3
Dennis D-10
Dennis Port E-10
Douglas C-6
Dover C-7
  Vicinity .... Pg. 101, J-1
Dracut A-7
Dunstable A-6
Duxbury C-9
Eastham D-10
Easthampton C-3
Edgarton F-9
Egremont C-1
Ellisville D-9
Elmwood C-8
Erving B-4
Essex A-8
Everett B-8
  Vicinity .... Pg. 100, E-5
Fairhaven E-8
Fall River D-8
Falmouth E-9
Farley B-4
Feeding Hills C-3
Fiskdale C-5
Fitchburg B-6
Florida A-3
Forestdale D-9
Framingham C-7
Franklin C-7
Gardner B-5
Georgetown A-7
Gibbs C-4
Gilbertville C-4
Gloucester B-9
Goshen B-3
Granville C-3
Great Barrington C-2
Greenfield B-3
Greenharbor C-9
Groton B-6
Groveland A-8
Hadley B-4
Halifax D-8
Hancock B-2
Hanover C-8
Hartsville C-2
Harwich D-10
Harvard B-6
Haverhill A-8
Haydenville B-3
Heath A-3
Hingham C-8
Hinsdale B-2
Hockanum C-4
Holbrook .... ★
  Vicinity .... Pg. 101, K-5
Holden B-6
Holliston C-6
Holyoke C-3
Housatonic C-2
Hubbardston B-5
Hudson B-6
Hull B-8
Huntington C-3
Hyannis D-9
Ingleside C-4
Ipswich A-8
  Vicinity .... Pg. 100, C-1
Kingston C-8
Lakeville D-8
Lanesborough B-2
Lanesville A-8
Lawrence A-7
Lee C-2
Leeds B-3
Leicester B-5
Lenox B-2
Leominster B-6
Lexington B-7
  Vicinity .... Pg. 100, D-2
Lincoln B-7
Littleton B-6
Longmeadow C-3
Loudville C-3
Lowell A-7
Ludlow C-4
Ludlow Center C-4
Lunenburg A-5
Lynn B-8
  Vicinity .... Pg. 100, D-6
Lynnfield .... ★
  Vicinity .... Pg. 100, B-5
Malden .... ★
  Vicinity .... Pg. 100, D-4
Manchester B-8
Manomet D-9
Marblehead B-8
Marion D-8
Marlborough B-7
Marshfield C-9
Marstons Mills E-9
Mashpee E-9
Mattapoisett E-8
Maynard B-6
Medfield C-7
Medford .... ★
  Vicinity .... Pg. 100, D-5
Melrose .... ★
  Vicinity .... Pg. 100, D-5
Merrimac A-7
Methuen A-7
Middleborough D-8
Middlefield B-2
Middleton B-7
Milford C-6
Millbury B-6
Millers Falls B-4
Millis C-7
Millville C-6
Milton C-7
  Vicinity .... Pg. 101, J-4
Montague B-4
Mount Herman B-4
Nahant .... ★
  Vicinity .... Pg. 100, E-7
Nantucket F-9
Natick C-7
Needham .... ★
  Vicinity .... Pg. 101, H-2
New Ashford B-2
New Bedford E-8
New Boston C-2
New Braintree B-5
Newbury A-8
Newburyport A-8
New Salem B-4
Newton C-7
  Vicinity .... Pg. 100, F-2
Norfolk C-7
North Adams A-2
North Amherst B-4
North Attleborough C-7
North Bernardston A-3
North Eastham D-10
North Egremont C-1
North Field A-4
North Reading A-8
  Vicinity .... Pg. 100, A-4
North Truro C-10
Northbridge C-6
Northborough B-6
Northampton C-3
Norton C-7
Norwell C-8
Norwood C-7
  Vicinity .... Pg. 101, K-2
Oak Bluffs F-9
Onset D-8
Orange B-4
Orleans D-10
Otis C-2
Oxford C-6
Palmer C-4
Paxton C-5
Peabody .... ★
  Vicinity .... Pg. 100, C-6
Pelham B-4
Pembroke C-8
Pepperell A-6
Peru B-2
Petersham B-5
Phillipston B-4
Pigeon Cove B-4
Pinehurst .... ★
  Vicinity .... Pg. 100, B-1
Pittsfield B-2
Plainfield B-2
Plainville C-7
Plymouth C-9
Plympton C-8
Pocasset E-9
Princeton B-5
Provincetown C-10
Quinapoxet B-5
Quincy C-8
  Vicinity .... Pg. 101, J-5
Randolph C-8
Reading B-8
  Vicinity .... Pg. 100, B-4
Revere .... ★
  Vicinity .... Pg. 100, D-6
Richmond B-2
Rochdale C-6
Rockland C-8
Rockport A-9
Rowley A-8
Royalston A-5
Russell C-3
Rutland B-5
Sagamore D-9
Salem B-8
  Vicinity .... Pg. 100, B-7
Salisbury A-8
Sandwich D-9
Saugus .... ★
  Vicinity .... Pg. 100, D-5
Savoy B-3
Scituate C-8
Seekonk D-7
Sharon C-7
Sherborn C-7
Sheffield C-2
Shelburne Falls B-3
Sheldonville C-6
Shirley A-5
Shrewsbury C-6
Somerset D-7
Somerville .... ★
  Vicinity .... Pg. 100, E-4
Southbridge C-5
South Deerfield B-3
S. Hadley C-4
Southwick C-3
South Williamstown A-2
Spencer C-5
Springfield C-4
Spruce Corner B-3
State Line .... ★
Sterling B-6
Still River B-6
Stoneham B-7
  Vicinity .... Pg. 100, C-4
Stoughton C-7
Sturbridge C-5
Sudbury B-7
Sunderland B-4
Swampscott B-8
  Vicinity .... Pg. 100, D-7
Swansea D-7
Taunton D-8
Teaticket E-9
Templeton B-5
Tewksbury B-7
Three Rivers C-4
Truro C-10
Topsfield A-7
Townsend A-6
Tyngsborough A-7
Upton C-6
Uxbridge C-6
Van Duesenville C-1
Vineyard Haven E-8
Wakefield B-7
  Vicinity .... Pg. 100, C-4
Wales C-4
Walpole C-7
Waltham B-7
  Vicinity .... Pg. 100, F-1
Waquoit Village E-9
Ware B-4
Wareham D-8
Warren C-4
Warwick A-4
Washington B-2
Watertown B-7
  Vicinity .... Pg. 100, F-2
Waterville B-5
Wayland B-7
Webster C-5
Wellesley B-7
Wellfleet D-10
West Becket C-2
West Boylston B-5
West Bridgewater C-7
Westhampton C-3
West Otis C-2
West Stockbridge C-1
West Worthington B-3
Westborough C-6
Westfield C-3
Westminister B-6
Weston B-7
Westville D-7
Westwood .... ★
  Vicinity .... Pg. 101, K-2
Wellesley .... ★
  Vicinity .... Pg. 101, G-1,H-1
Weymouth C-8
  Vicinity .... Pg. 101, K-6
Whitinsville C-6
Whitman C-8
Wilbraham C-4
Williamstown A-2
Wilbram C-4
Wilmington B-7
  Vicinity .... Pg. 100, B-3
Winchendon A-5
Winchedon Springs A-5
Winchester .... ★
  Vicinity .... Pg. 100, D-3
Windsor B-2
Winthrop .... ★
  Vicinity .... Pg. 100, F-6
Woburn B-7
  Vicinity .... Pg. 100, C-3
Woods Hole E-9
Worcester C-5
Woronoco C-3
Worthington Corner B-2
Wrentham C-7
Yarmouth D-10
Zoar A-3

MICHIGAN
Pages 48-49

Acme F-6
Adamsville L-4
Adrian L-8
Afton E-7
Agnew J-4
Ahmeek B-2
Akron H-8
Alabaster G-8
Alanson E-7
Alaska J-5
Alba F-6
Alberta B-2
Albion L-6
Alcona F-9
Algonac K-10
Alger G-8
Allegan K-4
Allen L-6
Allendale J-5
Allen Park .... ★
  Vicinity .... Pg. 111, J-8
Allenville D-7
Allouez B-2
Alma J-7
Almena K-4
Almont K-9
Aloha E-7
Alpena F-9
Alpha D-2
Alston B-2
Alverno E-7
Amador J-10
Amasa C-2
Amble H-6
Anchorville K-10
Ann Arbor L-8
  Vicinity .... Pg. 110, H-1
Antrim F-6
Applegate J-10
Arbutus Beach F-7
Arcadia G-4
Argentine K-7
Argyle H-9
Arnheim B-2
Arnold D-3
Arthur J-8
Ashley J-7
Ashton H-6
Assinins B-2
Athens L-6
Atlanta F-8
Atlantic Mine B-2
Attica J-9
Atwood E-6
Auburn H-8
Au Gres H-8
Au Sable G-9
Austin C-3
Au Train H-7
Averill H-7
Bad Axe H-9
Bailey J-5
Baldwin H-5
Banat D-3
Bancorft K-8
Banfield K-6
Bangor K-4
Baraga B-2
Bark River D-3
Barryton H-6
Barton City F-8
Bates F-6
Bath K-7
Battle Creek K-6
Bay City H-8
Bay Mills C-7
Bay Park H-8
Bay Port H-8
Bay View E-7
Beacon C-2
Bear Lake G-5
Beaverton H-7
Beechwood C-2
Belding J-6
Belleville .... ★
  Vicinity .... Pg. 110, H-4
Bellevue K-6
Bendon G-5
Bennington J-7
Benton Harbor L-4
Benzonia F-5
Bergland A-10
Berkley .... ★
  Vicinity .... Pg. 111, F-8
Berrien Sp. L-4
Bessemer B-9
Bethel L-6
Beulah F-5
Beverly Hills .... ★
  Vicinity .... Pg. 110, F-8
Big Bay B-3
Big Rapids H-6
Birch Creek E-3
Birch Run J-8
Birmingham .... ★
  Vicinity .... Pg. 111, E-8
Bitley H-5
Black River F-9
Blanchard H-6
Blaney Park D-5
Blissfield L-8
Bloomfield Hills .... ★
  Vicinity .... Pg. 110, B-7
Boyne City E-6
Boyne Falls E-7
Brampton D-3
Branch H-5
Bravo K-4
Breckenridge J-7
Brethren G-5
Brevort D-6
Bridgeport J-8
Bridgeton J-5
Bridgman L-4
Brighton K-8
Brimley C-7
Britton L-8
Brockway J-9
Brohman H-5
Bronson L-6
Brooklyn L-7
Brown City J-9
Bruce Crossing C-1
Brunswick H-5
Brutus E-7
Buchanan L-4
Buckley G-5
Burlington L-6
Burnside J-9
Burt J-8
Burton J-8
Cadillac G-6
Byron Center J-5
Cadmus L-7
Caledonia K-6
California A-2
Calumet A-2
Camden L-7
Capac J-9
Carbondale E-3
Carleton L-8
Carney D-3
Caro J-8
Carp Lake E-7
Carsonville J-9
Caseville H-9
Casnovia J-5
Cass City H-9
Cassopolis L-4
Cathro F-8
Cedar F-5
Cedar River E-3
Cedar Springs J-6
Cedarville D-8
Center Line .... ★
  Vicinity .... Pg. 111, D-11
Centreville L-5
Ceresco L-6
Channing C-2
Charleviox Bay E-6
Charlotte K-6
Chase H-6
Chatham C-4
Chelsea L-7
Chesaning J-7
Christmas C-4
Cilo J-8
Clare H-6
Clarendon L-6
Clarion E-7
Clarksville K-6
Clawson .... ★
  Vicinity .... Pg. 111, C-9
Clayton J-8
Cliff A-2
Climax K-5
Clinton L-8
Cloverdole K-6
Coldwater L-6
Coleman H-7
Coloma L-4
Colon L-5
Comins F-8
Concord L-7
Constantine L-5
Conway E-7
Coopersville J-5
Copemish G-5
Copper Harbor A-3
Cornell D-3
Corunna J-7
Covert K-4
Covington C-2
Cross Village E-7
Croswell J-10
Croton H-5
Crump H-7
Crystal Falls C-2
Cunard D-3
Curran F-8
Curtis D-6
Custer H-4
Dafter C-7
Daggett E-3
Dansville K-7
Darragh F-6
Davison J-8
Dayton L-4
Dearborn L-9
Dearborn Hts. .... ★
  Vicinity .... Pg. 110, G-7
Decatur L-5
Deer Park C-6
Deerton C-4
Deford J-9
Delano K-8
Delton K-5
Denton L-8
Detour Village D-8
Detroit .... ★
  Vicinity .... Pg. 111, F-8
Devereaux J-7
Dewitt K-7
Dexter L-8
Dighton G-6
Disco K-9
Dixboro .... ★
Donaldson C-7
Donken B-1
Dorr K-5
Douglas K-4
Dowagiac L-4
Dowling K-6
Drayton Plains .... ★
Drummond D-8
Dublin G-5
Dundee L-8
Dunningville K-5
Durand K-8
Eagle Habor A-2
Eagle River A-2
East Detroit .... ★
  Vicinity .... Pg. 111, E-12
East Gilead L-6
E. Grand Rapids J-5
E. Jordan E-6
East Lake G-5
East Leroy K-6
Eastport F-6
East Tawas G-8
Eaton Rapids K-7
Eckerman C-7
Ecorse .... ★
  Vicinity .... Pg. 111, J-9
Edenville H-7
Edmore J-6
Edwardsburg L-5
Elba J-8
Elberta F-5
Elk Rapids F-6
Elkton H-9
Elmer J-9
Elmira F-7
Elwell J-8
Emmett K-9
Empire F-5
Engadine D-6
Ensign D-4
Epoufette D-6
Erie M-8
Escanaba H-8
Essexville H-8
Estey H-7
Evans J-6
Evart H-6
Ewen C-1
Fair Haven K-9
Fairfield J-8
Fairgrove J-8
Fairview F-8
Farmington .... ★
  Vicinity .... Pg. 110, D-5
Farmington Hills .... ★
  Vicinity .... Pg. 110, D-5
Farwell H-6
Fawn River L-6
Fayette D-3
Felch D-3
Felch Mountain D-3
Fennville K-5
Fenton K-8
Ferndale .... ★
  Vicinity .... Pg. 111, F-9
Ferry H-5
Ferrysburg J-4
Fibre D-7
Fife Lake G-6
Filer City G-4
Filion H-9
Filmore K-5
Flat Rock L-9
Flint J-8
Flushing J-8
Ford River D-4
Forest Hill J-7
Forest Lake C-4
Forester J-10
Forestville H-10
Foster City D-3
Fouch F-5
Fountain G-5
Fowler J-7
Fowlerville K-8
Frankenmuth J-8
Frankfort F-5
Franklin .... ★
  Vicinity .... Pg. 110, C-7
Fraser .... ★
  Vicinity .... Pg. 111, C-12
Frederick F-7
Freeland H-7
Freesoil G-5
Fremont H-5
Frontier L-7
Fruitport J-5
Galesburg K-5
Galien L-4
Ganges K-4
Garden City .... ★
  Vicinity .... Pg. 110, G-6
Garden Corners D-5
Gardendale J-10
Garfield J-7
Garnet D-6
Gaylord F-7
Gera J-8
Germfask C-5
Gibraltar L-9
Girard L-6
Gladstone D-4
Gladwin H-7
Glen Arbor F-5
Glen Haven F-5
Glennie F-8
Glenwood L-5
Gobles K-5
Goetzville D-8
Good Hart E-6
Goodrich J-8
Gould City D-5
Gourley D-3
Grand Beach L-4
Grand Blanc J-8
Grand Haven J-4
Grand Ledge K-6
Grand Marais C-5
Grand Rapids J-5
Grange L-5
Granite Bluff D-2
Grant J-5
Grawn F-6
Grayling F-7
Greenbush G-9
Greenland B-1
Greenville J-6
Gregory K-7
Grindstone City H-9
Gros Cap D-7
Grosse Pointe .... ★
  Vicinity .... Pg. 111, F-13
Grosse Pointe Farms .... ★
  Vicinity .... Pg. 111, F-13
Grosse Pointe Park .... ★
  Vicinity .... Pg. 111, G-12
Grosse Pointe Shores .... ★
  Vicinity .... Pg. 111, E-13
Grosse Pointe Woods .... ★
  Vicinity .... Pg. 111, E-13
Gulliver D-5
Gwinn C-3
Hancock B-2
Harbert L-4
Harbor Beach H-10
Harbor Springs E-6
Harper Woods .... ★
  Vicinity .... Pg. 111, E-12
Harris D-3
Harrison H-7
Harrisville F-9
Hart H-4
Hartford L-4
Hartland K-8
Harvey C-3
Hardwood D-3
Hastings K-6
Hawks E-8
Hazel Park .... ★
  Vicinity .... Pg. 111, E-10
Helmer C-5
Hemans J-9
Henderson J-7
Hermansville D-3
Heron H-8
Hesperia H-5
Hessel D-7
Higgans Lake G-7
Highland Park .... ★
  Vicinity .... Pg. 111, F-9
Hilliards K-5
Hillman F-8
Hillsdale L-7
Hockaday G-7
Hodunk L-6
Holland K-4
Holly K-8
Holt K-7
Holton H-5
Homer L-6
Honor F-5
Horton L-7
Houghton B-1
Houghton Heights G-7
Houghton Lake G-7
Howard City J-6
Howell K-8
Hubbard Lake F-8
Hubbell B-2
Hudson L-7
Hudsonville J-5
Huntington Woods .... ★
  Vicinity .... Pg. 111, D-8
Huron Beach E-8
Huron City H-9
Hyde D-4
Idlewild H-5
Imlay City J-9
Ingalls E-3
Inkster .... ★
  Vicinity .... Pg. 110, H-6
Interlochen F-5
Ionia J-6
Iron Mountain D-2
Iron River C-1
Irons G-5
Ironwood B-9
Isabella D-4
Ishpeming C-3
Ithaca J-7
Jackson L-7
Jamestown J-5
Jasper M-8
Jeddo J-10
Johannesburg F-7
Jones L-5
Jonesville L-7
Joppa L-6
Kalamazoo K-5
Kaleva G-5
Kalkaska F-6
Karlin F-5
Kawkawlin H-8
Kearsarge A-2
Kendall K-5
Kent City J-5
Kenton C-1
Kentwood J-5
Keweenaw Bay B-2
Kewadin F-6
Kibbie K-4
Kinde H-9
Kinderhook L-6
Kingsford D-2
Kingston J-9
Kinross C-7
Kiva C-4
LaBranche D-3
Lachine F-8
Lacota K-5
Laingsburg K-7
Lake City G-6
Lake George H-6
Lakeland K-8
Lake Leelanau F-5
Lake Linden A-2
Lake Odessa K-6
Lake Orion K-8
Lake Roland B-1
Lakeport J-10
Lakeview J-6
Lambertville M-8
Langston J-6
L'Anse B-2
Lansing K-7
Lapeer J-8
Laporte J-7
La Salle L-8
Lathrop C-4
Lathrup Village .... ★
  Vicinity .... Pg. 111, D-8
Laurium A-2
Lawton L-5
Leland F-5
Leoni L-7
Leonidas L-5
Le Roy G-6
Leslie K-7
Levering E-7
Lexington J-10
Liberty L-7
Lilley H-5
Lincoln F-9
Lincoln Park .... ★
  Vicinity .... Pg. 111, J-8
Linden K-8

★ Denotes City located only on Vicinity map. City not located on State map.

MICHIGAN

Linwood H-8
Litchfield L-7
Little Lake C-4
Livonia K-8
  Vicinity ..... Pg. 110, F-6
Long Lake G-8
Long Rapids F-8
Longrie E-3
Loomis H-7
Loretto D-3
Lowell J-6
Lucas G-6
Ludington G-4
Luna Pier M-8
Lupton G-7
Luzerne G-7
Mackinaw City D-7
Madison Heights ..... ★
  Vicinity ..... Pg. 111, D-9
Mancelona F-6
Mandan A-3
Manistee G-4
Manistique D-5
Manton F-6
Maple Rapids J-7
Marcellus L-5
Marenisco B-10
Marlette J-9
Marine City K-10
Marion G-6
Marne J-6
Marquette C-3
Marshall L-6
Martin K-5
Mayville J-9
Marysville K-10
Mashek D-3
Mason K-7
Masonville D-4
Mass City B-1
Matchwood B-1
Matherton J-7
Mattawan L-5
McBain G-6
McFarland C-3
McMillan C-5
Meade K-9
Mears H-4
Mecosta H-6
Melstrand C-4
Melvin J-9
Memphis K-9
Mendon L-5
Menominee E-3
Meredith G-7
Merriman D-2
Merritt G-5
Merriweather A-10
Mesick G-5
Metz E-8
Michigamme C-3
Michigan Center L-7
Middleville K-6
Midland H-7
Mikado G-8
Milburg L-4
Milford K-8
  Vicinity ..... Pg. 110, B-2
Millbrook H-6
Millersburg E-8
Millington J-8
Mio F-8
Mineral Hills C-2
Mohawk A-2
Moline K-6
Monroe L-8
Monroe Center F-6
Montague H-4
Montrose J-8
Moorestown G-6
Moran D-7
Morenci M-7
Morley H-6
Morrice K-7
Moscow L-7
Mottville L-5
Mt. Clemens K-9
  Vicinity ..... Pg. 111, B-13
Mount Forest H-8
Mt. Morris J-8
Mount Pleasant H-6
Muir J-7
Mullett Lake E-7
Munger H-8
Munising C-5
Muskegon J-4
Muskegon Heights J-4
Muttonville K-9
Napoleon L-7
Nashville K-6
Nathan D-3
Naubinway D-6
Negaunee C-3
Newaygo J-5
New Baltimore K-9
Newberry C-6
New Buffalo L-4
New Era H-4
New Greenleaf H-9
New Haven K-9
Newport L-9
Niles L-4
Nirvana H-5
Nisula B-1
North Branch J-9
North Ironwood A-9
Northland D-3
North Muskegon J-4
Northport E-5
Northville ..... ★
  Vicinity ..... Pg. 110, E-4
Norton Shores J-4
Norwalk G-5
Norway D-2
Novi K-8
  Vicinity ..... Pg. 110, D-4
Nunica J-5
Oak Grove J-8
Oakley J-7
Oak Park ..... ★
  Vicinity ..... Pg. 111, E-8
Oakwood K-9
Ocqueoc E-8
Ogden Center M-8

Oil City H-7
Okemos K-7
Olivet K-6
Omena E-6
Onaway E-8
Onekama G-4
O'Neil F-6
Onondaga K-7
Ontonagon B-1
Orchard Lake ..... ★
  Vicinity ..... Pg. 110, B-5
Orleans J-6
Ortonville K-8
Oscoda G-9
Oskar A-2
Osseo L-7
Ossineke F-9
Otisville J-8
Otsego K-5
Otsego Lake F-7
Ottawa Lake M-8
Ovid J-7
Owosso J-7
Oxbow H-5
Oxford K-9
Ozark D-7
Painesdale B-2
Palmer C-3
Palmyra L-8
Palo J-6
Paradise C-7
Parchment K-5
Paris H-6
Parisville H-9
Parkdale G-4
Parkville L-5
Paulding C-1
Paw Paw L-5
Peck J-9
Pellston E-7
Pentwater H-4
Perkins D-4
Perrinton J-6
Perronville D-3
Perry K-7
Petersburg L-8
Petoskey E-7
Pewamo J-6
Phoenix A-2
Pickford D-7
Pierson J-6
Pigeon H-9
Pinconning H-8
Pine River H-8
Pittsford L-7
Plainfield K-8
Plainwell K-5
Pleasant Ridge ..... ★
  Vicinity ..... Pg. 111, D-9
Plymouth ..... ★
  Vicinity ..... Pg. 110, F-4
Pointe Aux Barques H-9
Polaski F-8
Pontiac K-8
  Vicinity ..... Pg. 110, A-7
Portage L-5
Port Hope H-10
Port Huron K-10
Portland J-6
Port Sanilac J-10
Posen E-8
Poseyville H-7
Potterville K-6
Powers D-3
Prescott G-8
Princeton C-3
Prudenville G-7
Pullman K-5
Quanicassee H-8
Quimby K-6
Quincy L-6
Quinnesec D-3
Raber D-8
Ralph D-3
Ramsay B-9
Randville C-3
Ransom M-7
Rattle Run K-10
Ray K-9
Reading L-6
Redford ..... ★
  Vicinity ..... Pg. 110, F-7
Reed City H-6
Reese J-8
Remus H-6
Republic C-2
Rexton D-6
Rhodes H-8
Richland K-5
Richmond K-9
Richmondville H-10
Richville J-8
Ridgeway L-8
River Rouge ..... ★
  Vicinity ..... Pg. 111, J-9
Rives Junction K-7
Roberts Landing K-10
Rochester K-9
Rock D-4
Rockford J-6
Rockland B-1
Rockwood L-9
Rodney H-6
Rogers City E-8
Rome Center L-7
Romeo K-9
Romulus ..... ★
  Vicinity ..... Pg. 110, J-5
Roosevelt Park J-5
Roscommon G-7
Roseburg J-9
Rosebush H-7
Rose City G-8
Roseville ..... ★
  Vicinity ..... Pg. 111, D-12
Rothbury H-4
Royal Oak H-4
Rudyard D-7
Rumely C-4
Ruth H-9
Saginaw J-7

Sagola D-2
St. Clair K-10
St. Clair Shores K-9
  Vicinity ..... Pg. 111, D-13
St. Ignace D-7
St. James D-6
St. Johns J-7
St. Louis J-7
Saline L-8
Samaria M-8
Sands C-3
Sandusky J-9
Sanford H-7
Sans Souci K-10
Saranac J-6
Saugatuck K-4
Sault Ste. Marie C-8
Sawyer L-4
Schaffer D-3
Schoolcraft L-5
Scottdale L-3
Scottville G-4
Sears H-6
Sebewaing H-9
Seney C-5
Sharon G-6
Shelby H-4
Shelbyville K-5
Shepherd H-7
Sherman C-5
Shingleton C-5
Shore D-4
Sidnaw C-2
Sigma F-6
Silver City A-10
Sister Lakes L-4
Six Lakes J-6
Skandia C-4
Skidway Lake G-8
Smithville J-6
Smyrna J-6
Snover J-9
Somerset L-7
Soo Junction C-6
South Boardman F-6
South Branch G-8
Southfield ..... ★
  Vicinity ..... Pg. 110, E-7
Southgate ..... ★
  Vicinity ..... Pg. 111, J-8
South Haven K-4
South Range B-1
Spring Lake J-5
Springport K-7
Spratt F-8
Spruce F-9
Stager D-2
Stalwart D-7
Standish H-8
Stanton J-6
Stanwood H-6
Stephenson E-3
Sterling G-8
Sterling Hts. ..... ★
  Vicinity ..... Pg. 111, B-10
Steuben C-4
Stevensville L-4
Stockbridge K-7
Stronach G-4
Strongs C-7
Sturgis L-5
Stutsmanville E-6
Sundell C-4
Sunfield K-6
Suttons Bay E-6
Swanson E-3
Swartz Creek J-8
Sylvester H-6
Tawas City G-8
Taylor ..... ★
  Vicinity ..... Pg. 110, J-4
Tecumseh L-8
Tekonsha L-6
Theodore D-2
Thompson D-5
Thornville J-9
Three Lakes C-2
Three Oaks L-4
Three Rivers L-5
Tiovola B-1
Tipton L-7
Tower E-7
Trenton ..... ★
  Vicinity ..... Pg. 111, L-9
Traunik C-4
Traverse City F-5
Trenary C-4
Triangle Ranch C-2
Trout Creek C-1
Trout Lake D-7
Troy K-9
  Vicinity ..... Pg. 111, B-9
Turner G-8
Tustin G-6
Twining G-8
Twin Lake J-4
Twin Lakes B-2
Ubly H-9
Union L-5
Union City L-6
Union Pier L-4
Unionville H-9
Utica K-9
  Vicinity ..... Pg. 111, A-11
Vandalia L-5
Vanderbilt F-7
Vassar J-8
Vernon K-7
Verona H-9
Vicksburg L-5
Vienna F-7
Vogel Center G-6
Vriesland K-5
Vulcan D-3
Wacousta K-7
Wadhams J-9
Wakefield B-10
Waldron M-7
Walhalla G-5
Walkerville H-5
Wallace E-3
Walled Lake ..... ★
  Vicinity ..... Pg. 110, C-4
Walton G-6
Warren ..... ★
  Vicinity ..... Pg. 111, D-11

Washington K-9
Waterford ..... ★
  Vicinity ..... Pg. 110, A-6
Waterloo K-7
Waters F-7
Watersmeet C-1
Watertown J-9
Watervliet L-4
Watson D-3
Watson K-5
Watton C-2
Waucedah D-3
Wayland K-5
Wayne ..... ★
  Vicinity ..... Pg. 110, J-3
Webberville K-7
Wells D-4
Wellston G-5
West Branch G-7
Westland ..... ★
  Vicinity ..... Pg. 110, G-5
West Olive K-4
Weston M-7
Wetmore C-4
White L-5
White Cloud H-5
Whitehall J-4
White Pine A-10
White Rock H-10
Whitney D-3
Whittemore G-8
Williamston K-7
Willis L-8
Windsor (Canada) ..... ★
  Vicinity ..... Pg. 111, J-11
Winegars H-7
Winona B-2
Wisner H-8
Wixom ..... ★
  Vicinity ..... Pg. 110, C-3
Wolf Lake G-5
Wolverine E-7
Wolverine Lake ..... ★
  Vicinity ..... Pg. 110, C-3
Woodbury K-6
Woodhaven L-9
Woodland K-6
Wyandotte ..... ★
  Vicinity ..... Pg. 111, K-8
Wyoming J-5
Yale J-9
Yalmer C-4
Ypsilanti L-8
  Vicinity ..... Pg. 110, J-2
Yuma G-5
Zeeland K-5
Zilwaukee J-8

MINNESOTA
Pages 50-51

Ada D-2
Adams L-6
Adrian L-2
Ah-gwah-ching E-4
Aitkin F-5
Akeley E-4
Albany G-4
Alberta G-2
Albert Lea L-5
Alden L-5
Alexandria G-3
Alvarado C-1
Alvwood D-5
Amboy K-5
Andover H-6
Angle Inlet A-3
Angora C-7
Annandale H-5
Anoka H-6
Appleton H-2
Arden Hills ..... ★
  Vicinity ..... Pg. 126, A-6
Argyle C-1
Arlington J-4
Ashby G-2
Askov F-7
Atwater H-4
Audubon E-2
Aurora D-7
Austin L-6
Avoca K-2
Avon G-4
Babbitt D-8
Backus E-4
Badger B-2
Bagley D-3
Ball Club D-5
Barnesville F-1
Barnum F-7
Barrett G-2
Barry G-1
Battle Lake F-3
Baudette B-4
Baxter F-4
Beardsley G-1
Beaver Bay D-9
Beaver Creek L-1
Becker H-5
Bejou D-2
Belgrade H-3
Belle Plaine J-5
Bellingham H-1
Beltrami D-1
Bemidji D-4
Bena D-5
Benedict E-4
Benson H-3
Beroun G-6
Bertha F-3
Bigelow L-2
Big Falls C-5
Big Fork C-5
Big Lake H-5
Birchdale B-5
Birchwood ..... ★
  Vicinity ..... Pg. 127, A-9
Bird Island J-4
Biwabik D-7
Blackduck D-4
Blomkest J-3
Blooming Prairie K-6

Bloomington ..... ★
  Vicinity ..... Pg. 126, F-3
Blue Earth L-5
Bluffton F-3
Bock G-5
Boy River E-5
Bovey D-5
Bowstring D-5
Boyd J-2
Braham G-6
Brainerd F-4
Branch H-6
Brandon G-3
Breckenridge F-1
Brewster L-2
Britt D-7
Brook Park G-6
Brooklyn Center ..... ★
  Vicinity ..... Pg. 126, A-4
Brooklyn Park ..... ★
  Vicinity ..... Pg. 126, A-2
Brooks C-2
Brookston E-7
Brooten H-3
Browerville F-4
Brownsdale L-6
Browns Valley G-1
Brownsville L-8
Brownton J-4
Bruno F-7
Buffalo H-5
Buffalo Lake J-4
Buhl D-6
Burtrum G-4
Butterfield K-4
Buyck C-7
Byron K-7
Caledonia L-8
Callaway E-2
Cambridge H-6
Campbell F-2
Canby J-1
Cannon Falls J-6
Canyon E-7
Carlos G-3
Carlton F-7
Cass Lake D-4
Cedar Mills J-4
Centerville K-8
Ceylon L-4
Chandler K-2
Chaska J-5
Chatfield K-7
Chickamaw Beach E-4
Chisholm D-6
Clara City H-3
Claremont K-6
Clarissa F-3
Clarkfield J-2
Clarks Grove L-6
Clearbrook D-3
Clear Lake H-5
Clearwater H-5
Clementson B-4
Climax D-1
Clinton H-1
Clitherall F-3
Clontarf H-3
Cloquet E-7
Cokato H-5
Cold Spring H-4
Coleraine D-6
Cologne J-5
Columbia Hts. ..... ★
  Vicinity ..... Pg. 126, A-4
Comfrey K-4
Comstock E-1
Cook C-7
Coon Rapids H-6
Correll H-2
Cosmos J-4
Cottage Grove J-6
Cotton E-7
Cottonwood J-2
Croftville D-10
Cromwell E-6
Crookston C-2
Crosby F-5
Crystal ..... ★
  Vicinity ..... Pg. 126, A-3
Currie K-2
Cushing F-4
Cuyuna F-5
Cyrus H-2
Dakota K-8
Dalton F-2
Danube J-3
Danvers H-2
Darfur K-4
Dassel H-4
Dawson J-2
Deer River D-5
Deerwood F-5
De Graff H-3
Delano H-5
Dent F-2
Detroit Lakes E-2
Dilworth E-1
Dodge Center K-6
Donaldson B-1
Donnelly G-2
Doran F-2
Dovray K-3
Duluth E-7
Dumont G-2
Dunnell L-4
Duquette F-7
Eagle Bend F-3
Eagle Lake K-5
East Bethel H-6
East Grand Forks C-1
East Guil Lake F-4
Easton L-5
Echo J-3
Eden Prairie ..... ★
  Vicinity ..... Pg. 126, E-1
Eden Valley H-4
Edgerton K-2
Edina ..... ★
  Vicinity ..... Pg. 126, E-3
Effie C-5

Elbow Lake G-2
Eldred D-1
Elgin K-7
Elk River H-6
Elizabeth F-2
Ellendale K-6
Ellsworth L-1
Elmer E-6
Elmore L-5
Elrosa G-3
Ely D-8
Emily E-5
Erhard F-2
Ericsburg B-6
Erksine C-2
Evansville G-3
Eveleth D-7
Eyota K-7
Fairfax J-4
Fairbault K-6
Fairmont L-4
Falcon Heights ..... ★
  Vicinity ..... Pg. 126, C-6
Farmington J-6
Farwell G-2
Felton E-1
Fergus Falls F-2
Fertile D-2
Finland D-8
Finlayson F-6
Fisher C-1
Flom E-2
Floodwood E-6
Florence K-2
Foley G-5
Forbes D-7
Forest Lake H-6
Fort Ripley F-4
Fosston D-3
Fountain K-7
Fox Home F-2
Franklin J-3
Frazee E-3
Freeport G-4
Fridley ..... ★
  Vicinity ..... Pg. 126, A-4
Frost L-5
Fulda K-3
Funkley C-4
Garfield G-3
Garrison F-5
Garvin K-2
Gary D-2
Gatzke B-3
Gaylord J-5
Gem Lake ..... ★
  Vicinity ..... Pg. 127, A-8
Geneva L-6
Georgetown E-1
Gheen C-7
Gibbon J-4
Gilbert D-7
Gilman G-5
Glencoe J-4
Glenville L-6
Glenwood G-3
Gluek H-3
Glyndon E-1
Golden Valley ..... ★
  Vicinity ..... Pg. 126, B-2
Goodhue K-7
Goodland E-6
Goodridge C-2
Good Thunder K-5
Goodview K-8
Gonvick D-3
Graceton B-4
Graceville G-2
Grand Falls J-2
Grand Marais D-9
Grand Meadow L-7
Grand Portage C-10
Grand Rapids D-6
Granby G-6
Granite Falls J-2
Grasston G-6
Greenbush B-2
Greenfield H-5
Grey Eagle G-4
Grove City H-4
Grygla C-3
Gully D-3
Guthrie E-4
Hackensack E-4
Hallock B-1
Halma B-2
Halstad D-1
Ham Lake H-6
Hancock H-2
Hanley Falls J-3
Hanska K-4
Harding F-5
Hardwick L-1
Harmony L-7
Harris G-6
Hastings J-6
Hawley E-2
Hayfield K-7
Hayward L-6
Hazel Run J-2
Hector J-4
Heidelberg J-5
Henderson J-5
Hendricks J-1
Hendrum D-1
Henning F-3
Herman G-2
Hermantown E-7
Heron Lake L-3
Hewitt F-3
Hibbing D-6
Hill City E-5
Hills L-1
Hilltop ..... ★
  Vicinity ..... Pg. 126, A-4
Hinckley G-6
Hines D-4
Hitterdal E-2
Hoffman G-2
Hokah L-8
Holland K-2
Holloway H-2
Holt C-2
Holyoke F-7

Hope K-6
Hopkins ..... ★
  Vicinity ..... Pg. 126, D-2
Houston L-8
Hovland C-10
Howard Lake H-5
Hoyt Lakes D-7
Humboldt B-1
Hutchinson J-4
Ihlen K-1
Independence H-5
Indus B-5
International Falls B-5
Inver Grove Hts. ..... ★
  Vicinity ..... Pg. 127, F-8
Iona K-2
Ironton F-5
Isabella D-8
Isanti H-6
Isle F-5
Ivanhoe J-2
Jackson L-4
Jacobson E-6
Janesville K-5
Jasper K-1
Jeffers K-3
Jenkins E-4
Johnson G-2
Jordan J-5
Karlstad B-2
Kasota K-5
Kasson K-7
Keewatin D-6
Kelliher D-4
Kellog K-7
Kelsey E-7
Kennedy B-1
Kensington G-2
Kent F-1
Kenyon K-6
Kerkhoven H-3
Kerrick F-7
Kettle River F-6
Kiester L-5
Kimball H-5
Kingston H-4
Knife River E-8
La Crescent L-8
Lafayette J-4
Lake Benton K-2
Lake Bronson B-1
Lake City J-7
Lake Crystal K-4
Lake Elmo ..... ★
  Vicinity ..... Pg. 127, B-10
Lakefield L-3
Lake George D-3
Lake Henry H-4
Lake Lillian J-4
Lake Park E-2
Lake Shore F-4
Lakeville J-6
Lake Wilson K-2
Lamberton K-3
Lancaster A-1
Lanesboro L-8
Laporte E-4
La Prairie E-6
Larsmont E-8
Lastrup G-5
Lauderdale ..... ★
  Vicinity ..... Pg. 126, B-5
Le Center K-5
Lengby D-3
Leonard D-3
LeRoy L-7
Lester Prairie J-5
LeSueur J-5
Lewiston K-8
Lilydale ..... ★
  Vicinity ..... Pg. 127, C-6
Linden Grove C-6
Lindford B-5
Lindstrom H-6
Lismore L-2
Litchfield H-4
Little Canada ..... ★
  Vicinity ..... Pg. 127, B-7
Little Falls G-4
Little Fork B-5
Little Rock G-5
Lockhart D-2
Loman B-5
Long Prairie G-4
Longville E-5
Lonsdale J-5
Lowry G-3
Lucan J-3
Lutsen D-9
Luverne L-2
Lyle L-6
Lynd K-2
Mabel L-8
Madelia K-4
Madison H-2
Madison Lake K-5
Mahnomen D-2
Mahtomedi ..... ★
  Vicinity ..... Pg. 127, A-9
Mahtowa F-6
Malmo F-6
Manchester L-5
Mankato K-4
Mantorville K-7
Maple Lake H-5
Mapleton K-5
Mapleview L-6
Maplewood ..... ★
  Vicinity ..... Pg. 127, B-8
Magnolia L-2
Marcell D-5
Margie C-5
Marietta H-1
Markham D-7
Marshall J-2
Mathtowa F-6
McGrath F-6
McGregor F-6
McIntosh D-2
Meadowlands E-7
Medford K-6
Meire Grove G-3
Melrose G-4
Menagha E-3
Mendota ..... ★
  Vicinity ..... Pg. 126, E-6

Mendota Hts. ..... ★
  Vicinity ..... Pg. 127, E-7
Mentor D-2
Middle River B-2
Milaca G-5
Milan H-2
Milroy K-2
Miltona G-3
Minneapolis ..... ★
  Vicinity ..... Pg. 126
Minneota J-2
Minnesota Lake K-5
Minnetonka ..... ★
  Vicinity ..... Pg. 126, D-1
Mizpah C-5
Montevideo H-3
Montgomery J-5
Monticello H-5
Moorhead E-1
Moose Lake F-6
Mora G-6
Morgan J-3
Morris G-2
Morton J-4
Motley F-4
Mountain Iron D-7
Mountain Lake K-3
Murdock H-3
Myrtle L-6
Nashua G-2
Nashwauk D-6
Nassau H-1
Nay-Tah-Waush D-3
Nevis E-4
New Brighton ..... ★
  Vicinity ..... Pg. 126, A-5
Newfolden B-2
New Hope ..... ★
  Vicinity ..... Pg. 126, B-2
New London H-4
New Prague J-5
New Richland K-5
New Ulm K-4
New York Mills F-3
Nicollet K-4
Nielsville D-1
Nimrod E-4
Nisswa F-4
Norcross G-2
North Branch H-6
Northfield J-6
North Mankato K-5
Northome C-5
Northrop L-4
North St. Paul ..... ★
  Vicinity ..... Pg. 127, B-9
Norwood J-5
Noyes A-1
Oakdale ..... ★
  Vicinity ..... Pg. 127, C-9
Oak Island H-1
Odessa H-1
Ogema E-2
Ogilvie G-5
Oklee C-2
Olivia J-3
Onamia G-5
Orleans A-1
Ormsby K-4
Orono ..... ★
  Vicinity ..... Pg. 127, C-9
Oronoco K-7
Orr C-7
Ortonville H-1
Osage E-3
Osakis G-3
Oslo C-1
Ottertail F-3
Outing E-5
Owatonna K-6
Palisade E-6
Parkers Prairie F-3
Park Rapids E-3
Paynesville H-4
Pelican Rapids F-2
Pencer B-3
Pennock H-3
Perley E-1
Pierz G-5
Pine City G-6
Pine Island K-7
Pine River E-4
Pine Springs ..... ★
  Vicinity ..... Pg. 127, A-9
Pinewood D-4
Pipestone K-1
Pitt B-4
Plainview K-8
Plummer C-2
Plymouth ..... ★
  Vicinity ..... Pg. 126, A-1
Ponemah C-4
Ponsford E-3
Porter J-2
Preston L-7
Princeton G-5
Prinsburg J-3
Prior Lake J-6
Proctor E-7
Quamba G-6
Randall G-4
Randolph J-6
Ray B-6
Raymond H-3
Redby C-4
Red Lake C-4
Red Lake Falls C-2
Red Wing J-7
Redwood Falls J-3
Regal H-4
Remer E-5
Renville J-3
Rice G-5
Richfield ..... ★
  Vicinity ..... Pg. 126, F-4
Richmond H-4
Richville F-2
Robbinsdale ..... ★
  Vicinity ..... Pg. 126, B-3
Rochester K-7
Rock Creek G-6
Roosevelt B-4
Roseau B-3
Rose Creek L-6
Roseville ..... ★
  Vicinity ..... Pg. 126, B-5

★ Denotes City located only on Vicinity map. City not located on State map.

Ross B-2
Rothsay F-1
Round Lake L-3
Round Prairie G-4
Royalton G-4
Rush City G-6
Rushford L-8
Russell K-2
Ruthton K-2
Rutledge F-7
Sabin E-1
Sacred Heart J-3
St. Anthony ★
Vicinity Pg. 126, B-5
St. Charles K-7
St. Clair K-5
St. Cloud H-5
St. Francis H-6
St. Hilaire C-2
St. James K-4
St. Joseph G-4
St. Louis Park ★
Vicinity Pg. 126, D-2
St. Michael H-5
St. Paul H-6
Vicinity Pg. 127
St. Paul Park ★
Vicinity Pg. 127, F-9
St. Peter K-5
St. Vincent A-1
Salol B-3
Sanborn K-3
Sandstone G-7
Sartell G-4
Sauk Centre G-4
Sauk Rapids G-4
Saum C-4
Sawyer F-6
Scanlon F-7
Schroeder D-9
Sebeka E-3
Sedan G-3
Shakopee J-5
Shelly D-1
Sherburn L-4
Shevlin D-3
Shoreview ★
Vicinity Pg. 127, A-6
Silver Bay D-9
Silver Lake J-4
Slayton K-2
Sleepy Eye K-4
Solway D-3
South International Falls B-6
South St. Paul ★
Vicinity Pg. 127, E-8
Spicer H-4
Springfield K-4
Spring Grove L-8
Spring Lake D-5
Spring Valley L-7
Squaw Lake D-5
Staples F-4
Starbuck G-3
Stephen B-1
Stewartville K-7
Storden K-3
Strandquist B-2
Strathcona B-2
Sturgeon Lake F-7
Sunberg H-3
Sunfish Lake ★
Vicinity Pg. 127, E-7
Swan River E-6
Swanville G-4
Swatara E-5
Swift B-3
Taconite D-6
Taconite Harbor D-9
Talmoon D-5
Tamarack F-6
Taylors Falls H-7
Toopi L-7
Taunton J-2
Tenney F-1
Tenstrike D-4
Terrace H-3
Thief River Falls C-2
Togo C-6
Tommald F-5
Tower C-7
Tracy K-3
Trail C-3
Trimont L-4
Trosky K-2
Truman L-4
Turtle River D-4
Twig E-7
Twin Lakes L-5
Twin Valley D-2
Two Harbors E-8
Tyler K-2
Ulen E-2
Underwood F-2
Vadnais Hts. ★
Vicinity Pg. 127, A-7
Verdi K-1
Vergas E-2
Verndale F-4
Vesta J-3
Viking C-2
Villard G-3
Vining F-3
Viola K-7
Virginia D-7
Wabasha K-7
Wabasso J-3
Waconia J-5
Wadena F-3
Wahkon F-5
Waite Park H-4
Walker E-4
Walnut Grove K-3
Waltham L-6
Wanamingo K-6
Warba E-6
Warraska B-3
Warren C-1
Warroad B-3
Waseca K-5
Waskish C-4
Watertown J-5
Waterville K-5
Watkins H-4
Watson H-2
Waubun E-2
Wawina E-6
Welcome L-4
Wells L-5
Westbrook K-3
West Concord K-6
West St. Paul ★
Vicinity Pg. 127, E-7
Wheaton G-1
Whipholt E-4
White Bear Lake ★
Vicinity Pg. 127, A-9
White Earth E-2
Wilder L-3
Willernie ★
Vicinity Pg. 127, A-9
Williams B-4
Willmar H-4
Willow River F-6
Wilton D-3
Windom K-3
Winger D-2
Winnebago L-5
Winona K-8
Wirt D-5
Winthrop J-4
Winton C-8
Wolf Lake E-3
Wolverton F-1
Woodbury J-6
Vicinity Pg. 127, D-10
Woodland G-6
Worthington L-3
Wrenshall F-7
Wyoming H-6
Zim D-7
Zimmerman H-6
Zemple D-5
Zumbro Falls K-7
Zumbrota K-7

MISSISSIPPI
Page 56

Abbeville B-5
Aberdeen C-6
Ackerman D-5
Alcorn G-2
Algoma B-5
Alligator C-3
Amory C-6
Anguilla E-3
Arcola D-3
Arkabutla A-4
Artesia D-6
Ashland A-5
Askew B-3
Avalon C-4
Avon D-2
Bailey F-6
Baird D-3
Baldwyn B-6
Banner B-5
Bassfield H-4
Batesville B-4
Baxterville H-5
Bay Saint Louis K-5
Bay Springs G-5
Beaumont H-6
Beauregard G-3
Becker C-6
Belden B-6
Belen B-3
Bellefontaine C-5
Belmont A-6
Belzoni D-3
Benoit D-2
Benton E-4
Bentonia E-3
Beulah C-2
Bigbee Valley D-6
Big Creek C-4
Biloxi K-6
Blue Mountain A-5
Bogue Chitto H-3
Bolton F-3
Booneville A-6
Boyle C-3
Brandon F-4
Braxton G-4
Brookhaven H-3
Brooklyn H-5
Brooksville D-6
Bruce C-5
Buckatunna G-6
Bude H-3
Burnsville A-6
Byhalia A-4
Caledonia C-6
Calhoun City C-5
Call Town J-6
Camden E-4
Canton E-4
Carlisle G-3
Carpenter G-3
Carriere J-5
Carrollton D-4
Carson H-4
Carthage E-4
Cary E-3
Cascilla C-4
Cedarbluff C-5
Centreville J-3
Charleston C-4
Chatawa J-4
Chatham E-2
Chunky F-6
Church Hill G-2
Clara G-6
Clarksdale B-3
Cleveland C-3
Clifftonville D-6
Clinton F-3
Coahoma B-3
Coffeeville C-4
Coila D-4
Coldwater A-4
Collins G-5
Collinsville E-6
Columbia H-4
Columbus C-6
Como B-4
Conehatta F-5
Corinth A-6
Courtland B-4
Crawford D-6
Crenshaw B-3
Crosby H-2
Cross Roads A-6
Crowder B-3
Crystal Springs G-3
Daleville E-6
Darling B-3
Darlove D-3
Decatur F-5
DeKalb E-6
Dennis A-6
Derma C-5
D'Lo G-4
Doddsville C-3
Dorsey B-6
Drew C-3
Dublin C-3
Duck Hill C-4
Dumas A-5
Duncan C-3
Dundee B-3
Durant D-4
Eastabuchie H-5
Ebenezer E-4
Ecru B-5
Eden E-3
Edinburg E-5
Edwards F-3
Electric Mills E-6
Elizabeth D-3
Ellisville G-5
Enid B-4
Enterprise F-6
Ethel D-5
Etta B-5
Eupora D-5
Falcon B-3
Falkner A-5
Fannin F-4
Fayette G-2
Fitler E-2
Flora F-3
Florence F-4
Forest F-5
Forkville F-4
Fort Adams H-2
Foxworth H-4
French Camp D-5
Friar's Pt. B-3
Fruitland Park J-5
Fulton B-6
Futheyville C-4
Gallman G-3
Gattman C-6
Gautier K-6
Georgetown G-4
Gholson E-6
Gillsburg J-3
Glen A-6
Glen Allan E-2
Glendora C-3
Gloster H-2
Golden B-6
Goodman E-4
Gore Springs C-4
Grace E-2
Grand Gulf G-2
Greenville D-2
Greenwood D-4
Grenada C-4
Gulfport K-6
Gunnison C-2
Guntown B-6
Hamilton C-6
Harperville F-5
Harriston G-2
Harrisville G-4
Hatley B-6
Hattiesburg H-5
Hazlehurst G-3
Heidelberg G-6
Helena B-3
Hermanville G-3
Hernando A-4
Hickory F-5
Hickory Flat A-5
Hilhouse C-2
Hillsboro F-5
Hiwannee G-6
Hollandale D-3
Holly Bluff E-3
Holly Springs A-5
Holcomb C-4
Homewood F-5
Horn A-3
Houston C-5
Hurley J-6
Independence A-4
Indianola D-3
Inverness D-3
Isola D-3
Itta Bena D-3
Iuka A-6
Jackson F-4
Jayess H-4
Johnstons Station H-4
Jonestown B-3
Kewanee F-6
Kilmichael D-4
Kiln K-5
Kokomo H-4
Kosciusko E-5
Lafayette Springs B-5
Lake F-5
Lakeshore K-5
Lake View ★
Vicinity Pg. 123, G-2
Lambert B-3
Lauderdale E-6
Laurel G-5
Lawrence F-5
Leaf H-6
Leakesville H-6
LeFlore C-4
Leland D-3
Lena E-4
Lessley H-2
Lexington D-4
Liberty H-3
Little Rock F-5
Long Beach K-5
Lorman G-2
Louin F-5
Louisville D-5
Louise E-3
Lucedale J-6
Lucien H-3
Ludlow F-4
Lula B-3
Lumberton J-5
Lyman K-5
Lyon B-3
Maben D-5
Macon D-6
Madden E-5
Madison F-4
Magee G-4
Magnolia H-3
Mantachie B-6
Mantee C-5
Marietta B-6
Marion F-6
Marks B-3
Martinsville G-3
Mathiston D-5
McAdams E-4
McCall Creek H-3
McCarley D-4
McComb H-3
McCondy C-5
McCool D-5
McHenry J-5
McNeill J-5
Meadville H-3
Mendenhall G-4
Meridian F-6
Merigold C-3
Metcalfe D-2
Michigan City A-5
Mineral Wells A-4
Vicinity Pg. 123, G-7
Minter City C-3
Mize G-5
Money C-3
Monticello H-4
Montpelier C-5
Montrose F-5
Mooreville B-6
Moorhead D-3
Morgan City D-3
Morgantown H-4
Morton F-4
Moselle H-5
Moss G-5
Moss Point K-6
Mound Bayou C-3
Mount Olive G-5
Myrtle A-5
Natchez H-2
Necaise J-5
Nesbit A-3
Neshoba E-5
Nettleton B-6
New Albany B-5
New Augusta H-5
New Hebron G-4
New Houlka C-5
New Site A-6
Newton F-5
Nicholson K-5
Nitta Yuma E-3
North Carrollton D-4
Noxapater E-5
Oakland C-4
Oak Vale H-4
Ocean Springs K-6
Ofahoma E-4
Okolona B-6
Olive Branch A-4
Oma G-4
Onward E-3
Osyka J-3
Oxford B-4
Pace C-2
Pachuta F-6
Paden A-6
Panther Burn E-3
Parchman C-3
Paris B-4
Pascagoula K-7
Pass Christian K-5
Paulding F-5
Paulette D-6
Pearl F-4
Pearlington K-5
Pelahatchie F-4
Perkinston J-5
Petal H-5
Pheba C-5
Philadelphia E-5
Philipp C-3
Picayune K-5
Pickens E-4
Pinkneyville J-2
Pinola G-4
Plantersville B-6
Plum Point ★
Vicinity Pg. 123, G-5
Pocahontas F-3
Pond J-2
Pontotoc B-5
Pope B-4
Poplarville J-5
Porterville E-6
Port Gibson G-2
Potts Camp A-5
Prairie C-6
Prairie Point D-6
Prentiss G-4
Preston E-6
Puckett F-4
Pulaski F-5
Purvis H-5
Quentin H-3
Quitman F-6
Raleigh G-5
Randolph B-5
Raymond F-3
Red Banks A-4
Redwood F-3
Reform D-5
Rena Lara B-2
Richland F-4
Richton H-6
Ridgecrest H-2
Ridgeland F-4
Rienzi A-6
Ripley A-5
Robinsonville A-3
Rolling Fork E-3
Rome C-3
Rosedale C-2
Roxie H-2
Ruleville C-3
Runnelstown H-6
Ruth H-4
Sallis E-4
Saltillo B-6
Sandersville G-6
Sandhill F-4
Sandy Hook H-5
Sanford H-5
Sarah B-3
Sardis B-4
Sarepta B-5
Satartia E-3
Saucier J-5
Savage A-3
Schlater D-3
Scobey C-4
Scooba E-6
Scott D-2
Sebastopol F-5
Seminary G-5
Senatobia A-4
Sessums D-6
Seypal Lake A-3
Shannon B-6
Sharon E-4
Shaw D-2
Shelby C-3
Sherard B-3
Sherman B-6
Shivers G-4
Shubuta G-6
Shuqualak D-6
Sibley H-2
Sidon D-4
Silver City E-3
Silver Creek G-4
Skene C-2
Slate Springs C-5
Slayden A-4
Sledge B-3
Smithdale H-3
Smithville B-6
Sontag G-4
Soso G-5
Southhaven A-3
Star F-4
Starkville D-6
State Line H-6
Steens C-6
Stewart D-5
Stoneville D-2
Stonewall F-6
Stovall B-3
Stringer G-5
Sturgis D-5
Summit H-3
Sumner C-3
Sumrall H-5
Sunflower D-3
Suqualena F-5
Swan Lake C-3
Swiftown D-3
Sylvarena G-5
Taylorsville G-5
Taylor B-4
Tchula D-4
Terry G-3
Thaxton B-5
Thomastown E-4
Thorn C-5
Thornton D-3
Thyatria A-4
Tie Plant C-4
Tillatoba C-4
Tinsley E-3
Tiplersville A-5
Tippo C-4
Tishomingo A-6
Toccopola B-5
Tomnolen D-5
Toomsuba F-6
Trebloc C-6
Tremont B-6
Tula B-5
Tunica A-3
Tupelo B-5
Tutwiler C-3
Tylertown H-4
Union E-5
Union Church G-3
University B-4
Utica G-3
Vaiden D-4
Valley Park E-3
Vance C-3
Vancleave J-6
Van Vleet C-5
Vardaman C-5
Vaughan E-4
Verona B-5
Vicksburg F-3
Victoria A-4
Vossburg G-6
Wade J-6
Walls A-3
Walnut A-5
Walnut Grove E-5
Walthall C-5
Washington H-2
Waterford A-4
Water Valley B-4
Waveland K-5
Waynesboro G-6
Wayside D-2
Weathersby G-4
Webb C-3
Weir D-5
Wenasoga A-6
Wesson G-3
West Gulfport J-5
West Point C-6
Wheeler A-6
Whitfield F-4
Wiggins J-6
Winona D-4
Winstonville C-3
Woodland C-5
Woodville J-2
Yazoo City E-3

MISSOURI
Pages 52-53

Adair D-5
Adrian H-2
Advance K-9
Agency F-2
Albany D-2
Alexandria D-7
Allendale D-3
Alma G-4
Altamont E-3
Alton L-7
Amoret H-2
Anderson L-2
Annapolis K-8
Anniston K-10
Appleton City H-3
Arab K-9
Arbela D-6
Arbor Terrace ★
Vicinity Pg. 140, E-5
Arbyrd M-9
Arcadia J-8
Archie H-3
Arcola J-3
Arkoe E-2
Arnold ★
Vicinity Pg. 141, L-3
Arroll K-6
Arrow Rock G-4
Asbury K-2
Ash Grove K-3
Ashland G-6
Ashley F-7
Athens D-6
Atlanta E-5
Aullville G-3
Aurora K-3
Auxvasse G-6
Ava K-5
Avalon F-4
Avondale ★
Vicinity Pg. 116, E-4
Bado K-6
Bakersfield L-6
Ballard H-3
Bardley L-7
Baring E-6
Barnard E-1
Battlefield K-4
Bay H-7
Beaufort H-7
Bellair G-5
Bella Villa ★
Vicinity Pg. 140, J-5
Bell City K-9
Belle H-6
Bellefontaine Neighbors ★
Vicinity Pg. 140, D-6
Bellerive ★
Vicinity Pg. 140, E-4
Bel Nor ★
Vicinity Pg. 140, E-4
Bel Ridge ★
Vicinity Pg. 140, E-4
Belton G-2
Bem H-7
Bendavis K-5
Bennett Springs J-5
Benton K-10
Berdell Hills ★
Vicinity Pg. 140, E-5
Berkeley ★
Vicinity Pg. 140, D-4
Bernie L-9
Bertrand K-10
Bethany D-3
Bethel E-6
Beulah J-6
Beverly Hills ★
Vicinity Pg. 140, E-5
Bevier E-5
Bible Grove D-6
Bigelow E-1
Bigspring G-7
Birch Tree K-7
Birmingham ★
Vicinity Pg. 116, E-5
Bismark J-8
Bixby J-7
Blackburn G-4
Black Jack ★
Vicinity Pg. 140, C-5
Black Walnut ★
Vicinity Pg. 140, A-3
Blodgett K-10
Bloomfield K-9
Blue Eye L-4
Blue Springs G-3
Blue Summit ★
Vicinity Pg. 117, G-5
Blythedale D-3
Bogard F-3
Bolckow E-2
Bolivar J-4
Boonesboro G-5
Boonville G-4
Bonne Terre J-8
Boshertown ★
Vicinity Pg. 140, B-2
Bosworth F-4
Bourbon H-7
Bowling Green F-7
Boynton D-4
Bradleyville L-4
Branson L-4
Braymer F-3
Brazeau J-10
Breckenridge E-3
Breckenridge Hills ★
Vicinity Pg. 140, E-3
Brentwood ★
Vicinity Pg. 141, G-4
Briar L-8
Bridgeton ★
Vicinity Pg. 140, D-3
Bridgeton Terrace ★
Vicinity Pg. 140, D-3
Brighton K-4
Brimson E-4
Brinktown H-6
Bronaugh J-2
Brookfield E-4
Broseley L-9
Brownbranch L-5
Browning E-4
Brumley H-5
Bruner K-5
Brunswick F-4
Buckhorn K-9
Bucklin E-5
Buckner G-3
Buffalo J-4
Bunker J-7
Burlington Jct. D-1
Butler H-2
Cabool K-6
Cainsville D-3
Cairo F-5
California G-5
Callao E-5
Calverton Park ★
Vicinity Pg. 140, D-4
Calwood G-6
Camdenton H-5
Cameron E-3
Campbell L-9
Canton E-7
Cape Girardeau K-9
Caplinger Mills J-3
Cardwell M-9
Carrington G-6
Carrollton F-3
Carsonville ★
Vicinity Pg. 140, E-4
Carthage K-2
Caruthersville M-10
Cascade K-9
Cassville L-3
Catherine Place J-9
Caulfield L-6
Center F-7
Centerville J-8
Centralia F-6
Chadwick L-4
Chain-O-Lakes L-3
Chaffee K-10
Chamois G-6
Champ ★
Vicinity Pg. 140, D-2
Charlack ★
Vicinity Pg. 140, E-4
Charleston K-10
Cherryville J-7
Chillicothe E-4
Chilhowee G-3
Chula E-4
Clarence E-6
Clarksdale E-2
Clarksville F-7
Clarkton L-9
Claycomo ★
Vicinity Pg. 116, D-5
Clayton ★
Vicinity Pg. 141, F-4
Clearmont D-1
Clever K-4
Clifton City G-4
Climax Springs H-4
Clinton H-3
Clyde D-2
Coal H-3
Coatsville D-5
Coldwater K-8
Cole Camp H-4
Collins J-3
Coloma F-4
Columbia G-6
Concordia G-3
Connelsville D-5
Conway J-5
Cook Station J-7
Cool Valley ★
Vicinity Pg. 140, D-4
Corning D-1
Corridon K-7
Couch L-7
Country Club Hills ★
Vicinity Pg. 140, E-5
County Life Acres ★
Vicinity Pg. 141, G-2
Craig E-1
Crane K-4
Crestwood ★
Vicinity Pg. 141, H-3
Creve Coeur ★
Vicinity Pg. 140, F-2
Crocker J-6
Cross Timbers H-4
Crystal Lake Park ★
Vicinity Pg. 141,G-2
Cuba H-7
Cul Du Sac ★
Vicinity Pg. 140, B-2
Curryville F-7
Cyrene F-7
Dadeville J-3
Dearborn F-2
Deerfield J-2
Defiance G-8
Dellwood ★
Vicinity Pg. 140, D-5
Denver D-3
Des Arc K-8
Desloge J-8
Des Peres ★
Vicinity Pg. 141, G-2
De Soto H-8
Dexter K-9
Diamond L-2
Dixon H-6
Doe Run J-8
Dongola K-9
Doniphan L-8
Doolittle J-6
Dorena L-10
Downing D-5
Drexel H-2
Drury K-5
Duenweg K-2
Duncans Bridge F-5
Dunlap E-4
Dunnegan J-4
Durham E-6
Dutchtown K-9
Eagleville D-3
East Prairie L-10
Ectonville ★
Vicinity Pg. 116, A-5
Edgar Springs J-6
Edgerton F-2
Edina E-6
Edinburg E-3
Edmundson ★
Vicinity Pg. 140, D-3
Eldon H-4
El Dorado Springs J-3
Eldridge J-5
Elkland J-4
Ellington K-7
Ellsinore K-8
Elmira F-3
Elm Point ★
Vicinity Pg. 140, B-1
Elmo D-1
Elsberry F-8
Elvins J-8
Elwood E-2
Emden E-6
Eminence K-7
Eolia F-7
Essex L-9
Esther J-8
Ethel E-5
Eudora J-4
Eunice K-6
Eureka H-8
Everton K-3
Ewing E-6
Excello F-5
Excelsior Springs F-2
Exeter L-3
Fagus L-9
Fairdealing L-8
Fairfax D-1
Fair Grove K-4
Fairmont D-6
Fair Play J-4
Fairview K-6
Farber F-6
Farley F-2
Farmersville E-4
Farmington J-9
Farrar J-9
Faucett F-2
Fayette F-5
Fayetteville G-3
Fenton ★
Vicinity Pg. 141, J-2
Ferguson ★
Vicinity Pg. 116, D-4
Ferrelview ★
Vicinity Pg. 116, B-2
Festus H-8
Fisk L-9
Filley J-3
Fillmore E-2
Flat River J-8
Flordell Hills ★
Vicinity Pg. 140, E-5
Florence G-4
Florida F-6
Florissant ★
Vicinity Pg. 140, C-4
Forbes E-1
Fordland K-4
Forest City E-1
Forest Green F-5
Foristell G-8
Forsyth L-4
Fortescue E-1
Fortuna H-5
Fountain Grove E-4
Foley G-8
Frankford F-7
Frederick Town J-9
Freeburg H-6
Fremont K-7
French Village J-9
Frontenac ★
Vicinity Pg. 141, G-2
Fulton G-6
Gainesville L-5
Galena L-3
Gallatin E-3
Galt E-4
Garden City H-3
Garwood K-8
Gatewood L-7
Gentry D-2
Gerald H-7
Gilliam F-4
Gladden J-7
Gladstone ★
Vicinity Pg. 116, D-4
Glasgow F-5
Glenaire ★
Vicinity Pg. 116, D-5
Glendale ★
Vicinity Pg. 141, G-3
Glen Echo Park ★
Vicinity Pg. 140, E-4
Glenwood D-5
Glover J-8
Golden City K-3
Goodfellow Terr. ★
Vicinity Pg. 140, E-5
Goodman K-2
Gower F-2
Graham D-1
Grain Valley G-3
Granby K-3
Grand Pass F-4
Grandin L-8
Grandview ★
Vicinity Pg. 117, L-4
Granger D-6
Grant City D-2
Grantwood Village ★
Vicinity Pg. 141, H-4

★ Denotes City located only on Vicinity map. City not located on State map.

MISSOURI

Granville — F-6
Grassy — K-9
Gray Summit — H-8
Graysville — D-5
Greencastle — D-5
Greendale — ★
  Vicinity — Pg. 140, E-4
Greenfield — K-3
Green City — D-4
Green Ridge — G-4
Greensburg — D-6
Greentop — D-5
Greenville — K-8
Greenwood — ★
  Vicinity — Pg. 117, L-7
Greer — L-7
Grovespring — K-5
Halfway — J-4
Hallsville — F-6
Halltown — K-3
Hamilton — E-3
Hanley Hills — ★
  Vicinity — Pg. 140, E-4
Hannibal — E-7
Hardin — F-3
Harris — D-4
Harrisburg — G-5
Harrisonville — G-3
Hartford — D-5
Hartville — K-5
Hartwell — H-3
Harwood — D-3
Hatfield — D-3
Hawk Point — G-7
Hayti — M-9
Hayti Heights — M-10
Hazelwood — ★
  Vicinity — Pg. 140, C-4
Helena — E-2
Hermann — G-7
Hermitage — J-4
Higbee — F-5
Higginsville — G-3
High Point — K-5
Hillsboro — H-8
Hillsdale — ★
  Vicinity — Pg. 140, E-5
Hinton — D-5
Hitt — D-5
Holcomb — L-9
Holden — G-3
Hollister — L-4
Holt — F-2
Holts Summit — G-6
Hoover — ★
  Vicinity — Pg. 116, A-2
Hopkins — D-2
Hornersville — M-9
House Springs — H-8
Houston — K-6
Houstonia — G-4
Houston Lake — ★
  Vicinity — Pg. 116, D-3
Humansville — J-3
Humphreys — E-4
Hunter — K-8
Huntleigh — ★
  Vicinity — Pg. 141, E-4
Huntsdale — G-5
Huntsville — F-5
Hurdland — E-6
Hurricane Deck — H-5
Iberia — H-6
Idlewild — K-9
Independence — G-3
  Vicinity — Pg. 117, G-6
Ironton — J-8
Jackson — J-10
Jacksonville — F-5
Jameson — E-3
Jamesport — E-3
Jamestown — G-5
Jane — L-2
Jasper — K-2
Jefferson City — H-5
Jenkins — L-3
Jennings — G-9
  Vicinity — Pg. 140, D-5
Jerico Springs — J-3
Joplin — K-2
Kahoka — D-6
Kansas City — ★
  Vicinity — Pgs. 116 & 117
Kearney — F-2
  Vicinity — Pg. 116, A-7
Kendricktown — K-3
Kennett — M-9
Kerrville — ★
  Vicinity — Pg. 116, A-2
Kewanee — L-10
Keytesville — F-5
Kidder — E-3
Kimberling City — L-4
King City — E-2
Kingston — F-3
Kinloch — ★
  Vicinity — Pg. 140, D-4
Kirksville — E-5
Kirkwood — ★
  Vicinity — Pg. 141, G-2
Kliever — G-5
Knob Noster — G-3
Knox City — E-6
Knoxville — F-3
Koshkonong — L-6
LaBelle — E-6
Laddonia — F-6
Ladue — ★
  Vicinity — Pg. 141, G-3
La Grange — E-7
Lake Lotawana — G-2
Lakenan — F-6
Lake St. Louis — G-8
Lakeshire — ★
  Vicinity — Pg. 116, D-2
Lamar — J-3
LaMonte — G-4
Lancaster — D-5
Lanton — L-6
La Plata — E-5

Lathrop — F-2
La Tour — G-3
Latty — J-8
Lawson — F-3
Leadwood — J-8
Leasburg — H-7
Lebanon — J-5
Leeper — K-8
Lee's Summit — ★
  Vicinity — Pg. 117, K-5
Leeton — H-3
Lemons — D-4
Lentner — F-6
Leonard — E-5
Lewiston — E-6
Lexington — F-3
Liberal — J-2
Liberty — F-2
  Vicinity — Pg. 116, C-5
Lilbourn — L-10
Licking — J-6
Lincoln — H-4
Linn — H-6
Linneus — E-4
Lockwood — K-3
Lone Jack — G-3
Long Lane — J-5
Longrun — L-5
Longwood — G-4
Louisiana — F-7
Lowry City — H-3
Ludlow — F-3
Luray — D-6
Macks Creek — J-5
Macon — E-5
Madison — F-6
Madrid — L-9
Malden — L-9
Manchester — G-8
  Vicinity — Pg. 141, G-1
Manes — K-5
Mansfield — K-5
Maplewood — ★
  Vicinity — Pg. 141, G-4
Marais Croche — ★
  Vicinity — Pg. 140, B-3
Marble Hills — K-9
Marcelene — E-4
March — J-4
Margona — ★
  Vicinity — Pg. 140, E-4
Marionville — K-4
Marlborough — ★
  Vicinity — Pg. 141, F-4
Marshall — F-4
Marshfield — K-5
Marthasville — G-7
Martinsburg — G-6
Mary Ridge — ★
  Vicinity — Pg. 140, E-4
Maryville — D-1
Matthews — L-10
Maxville — ★
  Vicinity — Pg. 141, L-3
Maysville — E-2
McBaine — G-5
McFall — E-2
Melbourne — E-3
Memphis — D-5
Mendon — F-4
Mercer — D-4
Mexico — F-6
Miami — F-4
Miami Station — F-4
Middle Grove — F-6
Midway — G-5
Milan — E-4
Milford — J-3
Millard — E-5
Miller — K-3
Millersburg — G-6
Mill Grove — D-4
Milo — J-2
Mineola — G-7
Miner — K-10
Mirabile — F-3
Moberly — F-5
Mokane — G-6
Moline Acres — ★
  Vicinity — Pg. 140, D-5
Monett — K-3
Monroe City — F-6
Montevallo — J-3
Montgomery City — G-7
Monticello — E-6
Montreal — J-5
Montrose — H-3
Moody — L-6
Morley — K-10
Morrisville — J-4
Mound City — E-1
Moundville — J-2
Mountain Grove — K-5
Mountain View — K-7
Mount Moriah — D-3
Mount Vernon — K-3
Mount Zion — K-5
Murphy — ★
  Vicinity — Pg. 141, K-1
Myrtle — M-6
Mystic — E-4
Nashville — K-2
Naylor — L-8
Nebo — J-5
Neelyville — L-8
Neosho — K-2
Nevada — J-2
New Bloomfield — G-6
New Boston — E-4
Newburg — J-6
New Florence — G-7
New Franklin — G-5
New Hartford — F-7
New Haven — G-7
New Hope — F-8
New London — E-7
New Madrid — L-10
New Market — F-2
New Melle — G-7
New Point — E-1
Niangua — K-5
Nixa — K-4
Noel — L-2
Norborne — F-3

Normandy — ★
  Vicinity — Pg. 140, E-5
Noser Mill — H-7
Northmoor — ★
  Vicinity — Pg. 116, D-3
Northview — K-4
Northwoods — ★
  Vicinity — Pg. 140, E-5
Norwood — ★
  Vicinity — Pg. 140, E-5
Novelty — E-5
Novinger — E-5
Oak Grove — G-3
Oak Hill — H-7
Oakland — ★
  Vicinity — Pg. 141, H-3
Oaks — ★
  Vicinity — Pg. 116, D-3
Oakview — ★
  Vicinity — Pg. 116, D-3
Oakville — ★
  Vicinity — Pg. 141, K-4
Oakwood — ★
  Vicinity — Pg. 116, D-3
Oakwood Manor — ★
  Vicinity — Pg. 116, D-4
Oakwood Park — ★
  Vicinity — Pg. 116, D-3
Ocie — L-5
Odessa — G-3
Oermann — H-8
O'Fallon — G-8
Olivette — ★
  Vicinity — Pg. 140, E-3
Olympian Village — H-8
Oran — K-10
Orchard Farm — G-8
  Vicinity — Pg. 140, A-2
Oregon — E-1
Orrick — F-3
Osage Beach — H-5
Osceola — H-3
Osgood — E-4
Oskaloosa — J-2
Otto — H-8
Overland — ★
  Vicinity — Pg. 140, E-4
Owensville — H-7
Ozark — K-4
Pacific — H-8
Pagedale — ★
  Vicinity — pg. 140, E-4
Palmyra — E-7
Paris — F-6
Parkville — ★
  Vicinity — Pg. 116, D-2
Parma — L-9
Pasadena Hills — ★
  Vicinity — Pg. 140, E-5
Pasadena Park — ★
  Vicinity — pg. 140, E-4
Patton — J-9
Pattonsburg — E-2
Peach Orchard — L-9
Peace Valley — L-6
Peculiar — G-2
Peerless Park — ★
  Vicinity — Pg. 141, H-1
Pennsboro — K-3
Perrin — E-2
Perry — F-6
Perryville — J-9
Pevely — H-8
Phelps — K-3
Philadelphia — E-6
Phillipsburg — J-5
Piedmont — K-8
Pierce City — K-3
Pilot Grove — G-4
Pilot Knob — J-8
Pine Lawn — ★
  Vicinity — Pg. 140, E-4
Pineville — L-2
Plad — J-4
Plato — J-5
Platte Woods — ★
  Vicinity — Pg. 116, D-2
Plattsburg — F-2
Pleasant Green — G-5
Pleasant Valley — ★
  Vicinity — Pg. 116, D-5
Plevna — E-6
Pocahontas — J-9
Polk — J-4
Pollock — D-4
Polo — F-3
Pomona — L-6
Poplar Bluff — L-8
Potosi — J-8
Portageville — L-10
Prairie Hill — F-5
Prairie Home — G-6
Preston — J-4
Princeton — D-4
Protem — L-5
Purdin — E-4
Purdy — L-3
Puxico — K-9
Queen City — D-5
Quincy — H-4
Qulin — L-9
Racine — K-2
Randolph — ★
  Vicinity — Pg. 116, E-5
Ravanna — D-4
Ravenwood — D-2
Raymondville — K-6
Raymore — G-2
Raytown — ★
  Vicinity — Pg. 116, H-5
Rea — E-2
Redford — K-8
Reeds Spring — L-4
Regal — F-3
Reger — E-5
Renick — F-5
Rensselaer — E-7
Republic — K-4
Reynolds — K-8
Rhineland — G-7
Rich Hill — H-2
Richland — J-5
Richmond — F-3
Richmond Heights — ★
  Vicinity — Pg. 141, G-4

Ridgedale — L-4
Ridgeway — D-3
Risco — L-9
Riverside — ★
  Vicinity — Pg. 116, E-2
Riverton — L-7
Riverview — ★
  Vicinity — Pg. 140, D-6
Rock Hill — ★
  Vicinity — Pg. 141, G-3
Rock Port — D-1
Rockville — H-3
Rocky Comfort — L-3
Rogersville — K-4
Rolla — J-6
Romance — L-5
Rombauer — K-8
Rosati — H-7
Rosebud — H-7
Rosendale — E-2
Rothville — F-4
Round Spring — K-7
Rover — L-7
Rushville — F-1
St. Ann — ★
  Vicinity — Pg. 140, D-3
St. Charles — G-8
  Vicinity — Pg. 140, C-1
Saint Clair — H-8
Saint Clement — F-7
Sainte Genevieve — H-8
St. George Village — ★
  Vicinity — Pg. 141, J-4
St. James — H-6
St. John — ★
  Vicinity — Pg. 140, E-4
St. Joseph — E-2
St. Louis —
  Vicinity — Pgs. 140 & 141
St. Marys — J-9
  Vicinity — Pg. 116, D-2
St. Patrick — E-7
Saint Robert — J-6
Salem — H-8
Saline — D-3
Salisbury — F-5
Sandy Hook — G-5
Santa Fe — F-6
Sarcoxie — K-3
Savannah — E-2
Schuerman Heights — ★
  Vicinity — Pg. 140, D-4
Scott City — K-10
Sedalia — G-4
Sedgewickville — J-9
Seligman — L-3
Senath — M-9
Seneca — L-2
Seymour — K-5
Shannondale — F-5
Shelbina — E-6
Shelbyville — E-6
Sheldon — J-3
Sheridan — D-2
Shirley — J-8
Shrewsbury — ★
  Vicinity — 141, H-4
Sikeston — K-10
Silva — K-8
Silver Lake — J-9
Simmons — K-6
Skidmore — D-1
Slater — F-4
Sleeper — J-5
Sligo — J-7
Smithton — G-4
Smithville — F-2
South Fork — L-6
South Gifford — E-5
South Lineville — D-4
Southwest City — L-2
Sparta — K-4
Spickard — E-4
Spokane — K-4
Springfield — K-4
Squires — L-5
Stanberry — E-2
State Line — M-9
Steele — M-9
Steelville — J-7
Stockton — J-3
Stoutsville — F-6
Stover — H-4
Strafford — K-4
Strasburg — G-3
Sturgeon — G-5
Sugar Creek — ★
  Vicinity — Pg. 116, H-7
Sullivan — H-7
Summersville — K-7
Sumner — F-4
Sunset Hills — ★
  Vicinity — Pg. 141, J-3
Sweet Springs — G-4
Sycamore — L-6
Sycamore Hills — ★
  Vicinity — Pg. 140, E-4
Tarkio — D-1
Tebbetts — G-6
Thayer — L-7
Theodosia — L-6
Thomas Hill — F-5
Thomasville — L-6
Thompson — F-6
Thornfield — L-5
Tiffin — J-3
Tina — F-4
Tindall — E-4
Tipton — G-5
Town & Country — ★
  Vicinity — Pg. 141, G-2
Townley — L-9
Trenton — E-4
Troy — G-8
Truesdale — G-7
Truxton — G-7
Tunas — J-4
Turney — F-2
Tuscumbia — H-5
Twin Oaks — ★
  Vicinity — Pg. 141, H-1
Tyrone — K-6
Union — H-7
Union Star — E-2
Unionville — D-4

Unity Village — ★
  Vicinity — Pg. 117, J-6
University City — ★
  Vicinity — Pg. 140, E-4
Uplands Park — ★
  Vicinity — Pg. 140, E-5
Urich — H-3
Utica — E-3
Valley Park — ★
  Vicinity — Pg. 141, H-1
Van Buren — K-8
Vandalia — F-7
Vanzant — K-6
Velda — ★
  Vicinity — Pg. 140, E-5
Velda Village Heights — ★
  Vicinity — Pg. 140, E-5
Versailles — H-5
Viburnum — J-7
Vichy — H-6
Vienna — H-6
Vinita Park — ★
  Vicinity — Pg. 140, E-4
Vulcan — K-8
Wainwright — G-6
Walnut Grove — K-4
Wardsville — H-6
Warren — E-6
Warrensburg — G-3
Warrenton — G-7
Warsaw — H-4
Warson Woods — ★
  Vicinity — Pg. 141, G-3
Washington — G-7
Wasola — L-5
Waverly — F-4
Wayland — D-6
Waynesville — J-6
Weatherby Lake — ★
  Vicinity — Pg. 116, D-2
Weaubleau — J-4
Webb City — K-2
Weber Hill — H-8
Webster Groves — ★
  Vicinity — Pg. 141, G-4
Wellston — ★
  Vicinity — Pg. 140, F-5
Wellsville — F-7
Wentworth — K-3
West Alton — ★
  Vicinity — Pg. 140, B-6
Westboro — D-1
Weston — F-1
Westphalia — H-6
West Plains — L-6
White Oak — L-9
Wilbur Park — ★
  Vicinity — Pg. 141, H-4
Wilcox — D-2
Willard — K-4
Williamsburg — G-6
Williamsville — K-8
Willow Springs — L-6
Wilson City — K-10
Winchester — ★
  Vicinity — 141, G-1
Windsor — H-4
Winfield — G-8
Winigan — E-4
Winona — K-7
Winston — F-1
Winthrop — F-1
Woodlawn — F-3
Woods Heights — F-3
Woodson Terr. — ★
  Vicinity — Pg. 140, E-4
Wooldridge — G-6
Worth — D-2
Wyaconda — D-6
Yarnell — ★
  Vicinity — Pg. 141, H-2
Yarrow — F-5
Yount — K-6
Yukon — K-6
Zalma — K-9

MONTANA
Pages 54-55

Absarokee — H-8
Acton — G-9
Alberton — E-2
Alder — H-5
Alzada — H-14
Amsterdam — ★
  Vicinity — Pg. 116, H-7
Anaconda — G-4
Angela — F-12
Antelope — B-13
Apgar — B-3
Arlee — E-3
Ashland — G-12
Augusta — D-5
Avon — F-4
Babb — B-4
Bainville — B-14
Baker — F-13
Ballantine — G-10
Basin — F-5
Bearcreek — H-8
Belfry — H-9
Belgrade — G-6
Belt — D-6
Biddle — H-13
Big Arm — D-3
Big Fork — C-3
Bighorn — G-11
Big Sandy — C-7
Big Sky — H-6
Big Timber — G-8
Billings — G-10
Birney — H-11
Black Eagle — D-6
Blackfoot — B-4
Bloomfield — E-13
Bonner — E-3
Boulder — F-5
Box Elder — C-7
Boyd — H-8
Bozeman — G-6
Bozeman Hot Springs — H-6
Brady — C-6
Bridger — H-9

Broadus — H-13
Broadview — G-9
Brockton — C-13
Brockway — D-12
Browning — B-4
Busby — H-11
Butte — H-6
Bynum — C-5
Cameron — H-6
Canyon Creek — E-5
Capitol — H-14
Carter — D-6
Cartersville — F-11
Cascade — E-6
Checkerboard — E-7
Chester — B-7
Chico Hot Springs — H-7
Chinook — B-8
Choteau — D-5
Christina — D-8
Churchill — G-6
Circle — D-12
Cleveland — C-8
Clinton — E-3
Clyde Park — G-7
Coffee Creek — D-7
Cohagen — E-11
Colstrip — G-11
Columbia Falls — B-3
Columbus — G-8
Conner — G-3
Conrad — C-5
Cooke City — H-8
Coram — C-3
Corvallis — F-3
Craig — E-5
Crane — D-14
Crow Agency — H-10
Culbertson — C-13
Custer — G-10
Cut Bank — B-5
Darby — G-3
Dayton — C-3
Dean — H-8
DeBorgia — D-2
Decker — H-11
Deer Lodge — F-4
Dell — J-4
Denton — D-7
Dillon — J-4
Divide — G-4
Dixon — D-3
Dodson — C-9
Drummond — F-4
Dupuyer — C-5
Dutton — D-5
East Glacier Park — C-4
East Helena — F-5
Edgar — H-9
Ekalaka — G-14
Elliston — F-5
Elmo — C-3
Emigrant — H-6
Ennis — H-5
Essex — C-3
Ethridge — B-5
Eureka — A-2
Fairfield — D-5
Fairview — C-14
Fallon — E-13
Fishtail — H-8
Flatwillow — E-9
Flaxville — B-12
Florence — F-3
Ford — H-12
Forestgrove — E-8
Forsyth — F-11
Fort Belknap Agency — B-9
Fort Benton — D-7
Fortine — B-2
Fort Peck — C-11
Fort Shaw — D-5
Frazer — C-12
Fenchtown — E-3
Froid — B-13
Fromberg — H-9
Four Buttes — B-12
Galen — H-5
Gallatin Gateway — H-6
Gardiner — H-7
Garneill — E-8
Garryowen — H-10
Geraldine — D-7
Geyser — D-7
Gildford — B-7
Glasgow — C-11
Glen — H-4
Glendive — E-13
Grant — H-4
Grantsdale — F-3
Grassrange — E-8
Great Falls — D-6
Hall — F-4
Hamilton — F-3
Hammond — H-13
Happy's Inn — C-2
Hardin — G-10
Harlem — B-9
Harlowtown — F-8
Harrison — H-5
Hathaway — F-11
Haugan — D-1
Havre — B-8
Helena — F-5
Helmville — E-4
Heron — C-1
Hilger — E-8
Hingham — B-7
Hinsdale — B-11
Hobson — E-7
Homestead — B-13
Hot Springs — D-2
Hungry Horse — C-3
Huntley — G-9
Huson — E-3
Hysham — F-10
Ingomar — F-10
Inverness — B-7
Jackson — H-4
Jefferson City — F-5
Joliet — H-8
Joplin — B-7
Jordan — E-11

Judith Gap — F-7
Kalispell — C-3
Kiowa — B-4
Klein — F-9
Kremlin — B-7
Lakeside — C-3
Lambert — D-13
Lame Deer — G-11
Laurel — G-9
Laurin — H-5
Lavina — F-9
Lewistown — E-8
Libby — B-1
Lima — J-4
Lincoln — E-4
Lindsay — D-13
Livingston — G-7
Lloyd — C-8
Lodge Grass — H-10
Logan — G-6
Lohman — B-8
Lolo — E-3
Lolo Hot Springs — E-2
Loma — C-7
Loring — B-10
Lothair — C-6
Malta — C-10
Manhattan — G-6
Marion — C-2
Martin City — C-3
Martinsdale — F-7
McLeod — G-7
Medicine Lake — B-13
Melrose — G-4
Melstone — F-10
Melville — G-7
Miles City — F-12
Milltown — E-3
Missoula — E-3
Moccasin — E-7
Monarch — E-6
Montague — E-7
Montana City — F-5
Moore — E-8
Musselshell — F-9
Nashua — C-11
Neihart — E-7
Niarada — C-2
Norris — H-5
Noxon — C-1
Nye — H-8
Opheim — B-11
Ovando — E-4
Pablo — C-3
Paradise — D-2
Park City — H-9
Peerless — B-12
Pendroy — C-5
Philipsburg — F-4
Plains — D-2
Plentywood — B-13
Plevna — F-14
Polebridge — B-3
Polson — D-3
Pompeys Pillar — G-10
Poplar — C-13
Port of DelBonita — B-5
Port of Morgan — B-10
Port of Opheim — B-11
Port of Raymond — A-13
Port of Whitetail — A-13
Potomac — E-4
Powderville — G-13
Power — D-5
Pray — H-7
Proctor — C-3
Radersburg — F-6
Rapelje — G-8
Ravalli — D-3
Raymond — B-13
Raynesford — D-7
Red Lodge — H-8
Redstone — B-13
Reedpoint — G-8
Reserve — B-13
Rexford — A-2
Richey — D-13
Richland — B-11
Ringling — F-7
Roberts — H-8
Rock Springs — E-12
Rollins — C-3
Ronan — D-3
Rosebud — F-12
Roundup — F-9
Roy — E-9
Rudyard — B-7
Ryegate — F-8
Saco — C-10
Saint Ignatius — D-3
Saint Mary — B-4
Saint Regis — D-2
Saltese — D-1
Santa Rita — B-5
Savage — D-13
Scoby — B-12
Seeley Lake — E-3
Shawmut — F-8
Shelby — C-6
Sheridan — H-5
Sidney — C-14
Silesia — H-9
Silver Star — H-5
Simms — D-5
Somers — C-3
Springdale — G-7
Stanford — E-7
Stevensville — F-3
Stryker — B-2
Suffolk — D-8
Sumatra — F-10
Sunburst — B-5
Sun River — D-6
Superior — D-2
Swan Lake — C-3
Sweetgrass — B-5
Tarkio — E-2
Teigen — E-9
Terry — E-13
Thompson Falls — D-2
Three Forks — G-6

Toston — F-6
Townsend — F-6
Trego — B-2
Trout Creek — C-1
Troy — B-1
Turner — B-9
Twin Bridges — H-5
Twodot — F-7
Ulm — D-6
Valier — C-5
Vaughn — D-6
Victor — F-3
Vida — C-12
Virginia City — H-5
Volborg — G-12
Walkerville — G-4
Warmsprings — F-4
Waterloo — H-5
Westby — A-14
West Glacier — B-3
West Yellowstone — J-6
Whitefish — B-3
White Hall — G-5
White Sulphur Springs — F-6
Whitetail — B-13
Wibaux — E-14
Willow Creek — G-6
Wilsall — G-7
Winifred — D-8
Winnett — E-9
Winston — F-5
Wisdom — G-4
Wise River — G-4
Wolf Creek — E-5
Wolf Point — C-12
Worden — G-10
Wyola — H-10
Yaak — B-1
Zurich — B-9

NEBRASKA
Pages 58-59

Abie — F-11
Adams — H-12
Ainsworth — D-7
Albion — E-10
Alda — G-8
Alexandria — H-11
Allen — D-3
Alliance — D-3
Alma — H-8
Almeria — D-8
Amelia — D-9
Ames — F-12
Amherst — G-8
Angora — D-3
Anoka — C-9
Anselmo — E-7
Ansley — F-8
Antioch — D-4
Arapahoe — H-7
Arcadia — F-8
Archer — F-10
Arlington — F-12
Arnold — F-7
Arthur — E-5
Ashby — D-4
Ashland — G-12
Ashton — F-9
Atkinson — D-9
Atlanta — H-8
Auburn — H-12
Aurora — G-10
Axtell — H-8
Ayr — H-9
Bancroft — E-12
Barneston — J-12
Bartlett — E-9
Bartley — H-6
Bassett — D-8
Battle Creek — E-10
Bayard — D-2
Bazile Mills — D-10
Beatrice — H-11
Beaver City — H-7
Beaver Crossing — G-11
Beemer — E-11
Belden — D-11
Belgrade — F-10
Bellevue — G-12
Belvidere — H-10
Benedict — G-10
Benkelman — H-5
Bennet — G-12
Bertrand — H-7
Berwyn — F-8
Big Springs — F-4
Bingham — D-4
Bladen — H-9
Blair — F-12
Bloomfield — D-10
Blue Hill — H-9
Blue Springs — H-11
Boelus — G-9
Boone — F-10
Boystown — F-12
Bradshaw — G-10
Brady — F-6
Brandon — G-4
Brewster — E-7
Bridgeport — E-3
Bristow — C-9
Broadwater — E-3
Brock — H-12
Broken Bow — F-8
Brownlee — D-6
Brule — F-4
Bruning — H-10
Bruno — F-11
Brunswick — D-10
Burchard — H-12
Burr — H-12
Burton — C-8
Burwell — E-8
Bushnell — F-2
Butte — C-9
Cairo — G-9
Callaway — F-7
Cambridge — H-6
Campbell — H-9
Carleton — H-10

★ Denotes City located only on Vicinity map. City not located on State map.

Carroll D-11
Cedar Bluffs F-12
Cedar Rapids F-9
Center D-10
Central City G-9
Ceresco G-12
Chadron C-3
Chambers D-9
Champion G-5
Chapman G-9
Chappell F-4
Chester J-10
Clarks F-10
Clarkson E-11
Clatonia H-11
Clay Center H-10
Clearwater D-10
Clinton C-4
Cody C-6
Coleridge D-11
Columbus F-11
Comstock F-8
Concord D-11
Cook H-12
Cornlea F-10
Cortland H-11
Cotesfield F-9
Cowles H-9
Cozad G-7
Crab Orchard H-12
Craig E-12
Crawford C-3
Creighton D-10
Creston E-11
Crete H-11
Crofton C-10
Crookston C-6
Culbertson H-6
Curtis G-6
Dakota City D-12
Dalton E-3
Danbury H-6
Dannebrog F-9
Davenport H-10
David City F-11
Dawson J-13
Daykin H-11
Decatur E-12
Denton G-11
Deshler H-10
Dewitt H-11
Dickens G-6
Diller H-11
Dix F-2
Dodge E-11
Doniphan G-9
Dorchester G-11
Douglas H-12
DuBois J-12
Dunbar G-12
Duncan F-10
Dunning E-7
Dwight G-11
Eddyville G-7
Edgar H-10
Edison H-7
Elba F-9
Elgin E-9
Eli C-5
Elk Creek H-12
Elkhorn F-12
Ellsworth D-4
Elm Creek G-8
Elmwood G-12
Elsie G-5
Elsmere D-7
Elwood G-7
Elyria E-8
Emerson D-11
Emmet D-9
Enders H-5
Endicott J-11
Ericson E-9
Eustis G-7
Exeter G-10
Fairbury H-11
Fairfield H-10
Fairmont H-10
Falls City J-13
Farnam G-7
Farwell F-9
Filley H-11
Firth H-11
Fordyce D-11
Foster D-10
Franklin H-8
Fremont F-12
Friend G-10
Fullerton F-10
Funk H-8
Gandy F-6
Garrison F-11
Geneva H-10
Genoa F-10
Gering E-2
Giltner G-9
Glenvil H-9
Goehner G-11
Gordon C-4
Gothenburg G-7
Grafton G-10
Grainton G-5
Grand Island G-9
Grant G-5
Greeley F-9
Greenwood G-12
Gresham G-11
Gretna F-12
Guide Rock J-9
Gurley F-3
Hadar E-11
Haigler H-4
Hallam H-11
Halsey E-7
Hamlet H-5
Hampton G-10
Harrisburg E-2
Harrison C-2
Hartington D-11
Harvard H-9
Hastings H-9
Hayes Center H-5
Hay Springs C-4
Hazard F-8
Hebron H-10
Hemingford D-3
Henderson G-10
Herman F-12
Hershey F-6
Hildreth H-8
Holdrege H-7
Holstein H-9
Hooper F-11
Hoskins E-11
Howells E-11
Humboldt J-12
Humphrey E-10
Huntley H-8
Hyannis D-5
Imperial G-5
Inavale J-9
Indianola H-6
Inman D-9
Ithaca F-11
Johnson H-12
Johnstown D-7
Juniata G-9
Kearney G-8
Kenesaw G-9
Kennard F-12
Keystone F-5
Kilgore C-6
Kimball F-2
Lakeside D-4
Laurel D-11
Lawrence H-9
Lebanon H-7
Leigh E-11
Lemoyne F-5
Lexington G-7
Lewellen F-4
Lewiston H-12
Liberty J-12
Lincoln G-12
Linwood F-11
Lisco E-4
Litchfield F-8
Lodgepole F-4
Long Pine D-7
Loomis H-7
Louisville G-12
Loup City F-8
Lyman D-2
Lynch C-9
Lyons E-12
Madison E-11
Madrid G-5
Magnet D-10
Malmo F-11
Marquette G-10
Martell H-11
Martinsburg D-11
Mason City F-8
Max H-5
Maxwell F-6
Maywood G-6
McCook H-6
McCool Jct. G-10
McGrew E-2
Mead F-12
Melbeta E-2
Memphis G-12
Merna F-7
Merriman C-5
Milford G-11
Miller G-8
Milligan H-10
Mills C-8
Minden H-8
Minitare E-2
Mitchell D-2
Monowi C-9
Moorefield G-6
Morrill D-2
Mullen D-6
Murray G-12
Naper C-8
Naponee J-8
Nebraska City H-12
Neligh E-10
Nelson H-10
Nenzel C-6
Newman Grove E-10
Newport D-8
Nickerson F-12
Niobrara C-10
Norden C-7
Norfolk E-10
North Bend F-11
North Loup F-9
North Platte F-6
Oak H-10
Oakdale E-10
Oakland E-12
Obert D-11
Oconto F-7
Octavia F-11
Odell J-11
Odessa G-8
Ogallala F-5
Ohiowa H-10
Omaha F-12
O'Neill D-9
Ong H-10
Ord E-8
Orleans H-8
Osceola F-10
Oshkosh E-4
Osmond D-10
Overton G-8
Oxford H-7
Page D-9
Palisade H-5
Palmer F-9
Palmyra G-12
Panama H-12
Papillion G-12
Parks H-4
Pawnee City J-12
Paxton F-5
Pender E-12
Petersburg E-10
Phillips G-9
Pickrell H-11
Pierce D-10
Pilger E-11
Plainview D-10
Platte Center F-10
Plattsmouth G-12
Pleasanton G-8
Plymouth H-11
Polk G-10
Poole G-8
Potter F-3
Prague F-11
Primrose F-9
Prosser G-9
Purdum D-7
Raeville E-10
Ragan H-8
Randolph D-11
Ravenna G-9
Raymond G-11
Red Cloud H-9
Republican City H-8
Reynolds J-11
Richland F-11
Rising City F-11
Riverdale G-8
Riverton H-9
Roca G-11
Rockville F-9
Rogers F-11
Rosalie E-12
Rose D-8
Roseland H-9
Rushville C-4
Ruskin H-9
St. Edward F-10
St. Helena C-11
St. Libory G-9
St. Mary H-12
St. Paul F-9
Santee C-10
Sargent E-8
Saronville H-9
Schuyler F-11
Scotia F-9
Scottsbluff D-2
Scribner F-11
Seneca D-6
Seward G-11
Shelby F-10
Shelton G-8
Shickley H-10
Sholes D-11
Shubert H-12
Sidney F-3
Silver Creek F-10
Smithfield G-7
Snyder E-11
South Bend G-12
South Souix City D-12
Spalding E-9
Sparks C-7
Spencer C-9
Springfield G-12
Springview C-8
Stamford H-7
Stanton E-11
Staplehurst G-11
Stapleton F-6
Steele City J-11
Steinauer H-12
Stockville G-6
Strang H-10
Stratton H-5
Stromsburg F-10
Stuart D-8
Summerfield J-12
Sumner G-8
Surprise G-10
Superior J-10
Sutherland F-6
Sutton H-10
Syracuse H-12
Table Rock H-12
Tarnov F-10
Taylor E-8
Tecumseh H-12
Tekamah E-12
Terrytown D-2
Thayer G-10
Thedford E-6
Thurston E-12
Tilden E-10
Tobias H-11
Trenton H-6
Trumbull G-9
Tryon E-6
Uehling E-12
Ulysses G-11
Unadilla G-12
Upland H-8
Valentine C-7
Valley F-12
Valparaiso G-11
Venango G-4
Verdel C-10
Verdigre D-10
Verdon J-13
Waco G-10
Wahoo F-12
Wakefield D-11
Wallace G-5
Walthill E-12
Wauneta H-5
Wausa D-10
Waverly G-12
Wayne D-11
Weeping Water G-12
Weissert F-8
Wellfleet G-6
Western H-11
Westerville F-8
West Point E-11
Whitman D-5
Whitney C-3
Wilber H-11
Wilcox H-8
Willow Island G-7
Wilsonville H-7
Winnebago D-12
Winnetoon D-9
Winside E-11
Winslow F-12
Wisner E-11
Wolbach F-9
Wood Lake C-7
Wood River G-9
Wymore J-12
Wynot C-11
York G-10

NEVADA
Pages 60-61

Adaven H-8
Alamo K-8
Arden M-8
Armogosa Valley L-6
Ash Springs K-8
Austin F-5
Babbitt G-3
Baker G-10
Battle Mountain D-5
Beatty L-6
Beowawe D-6
Blue Diamond M-8
Boulder City M-9
Bunkerville L-10
Cactus Springs L-7
Caliente J-9
Cal Nev Ari N-9
Carlin D-7
Carp K-9
Carson City F-2
Carver's G-5
Caselton J-9
Charleston Park M-7
Cherry Creek E-8
Coaldale H-4
Cold Spring F-4
Contact B-9
Cottonwood Cove N-9
Crescent Valley D-6
Crystal Bay F-1
Currant G-8
Currie E-9
Dayton F-2
Deeth C-8
Denio A-3
Denio Junction A-3
Dixie Valley E-4
Duckwater G-7
Dyer J-4
Eagle Picher Mine D-3
East Ely F-9
East Las Vegas M-9
Elgin K-9
Elko D-7
Ely F-9
Empire D-2
Eureka F-7
Fallon F-3
Fernley F-2
Frenchman F-3
Gabbs G-4
Gardnerville G-1
Genoa F-1
Gerlach D-2
Glenbrook F-1
Glendale L-9
Golconda C-5
Goldfield J-5
Gold Hill F-2
Gold Point K-5
Goodsprings M-8
Halleck C-8
Hawthorne H-3
Hazen F-2
Henderson M-9
Hiko J-8
Imlay D-4
Incline Village F-1
Indian Springs L-8
Ione G-5
Jack Creek B-7
Jackpot B-9
Jarbridge B-8
Jean N-8
Jiggs D-8
Kimberly F-8
Lage's E-9
Lamoille D-8
Las Vegas M-8
Laughlin N-9
Lee D-8
Lee Canyon L-7
Lida K-5
Logandale L-9
Lovelock E-3
Lund G-9
Luning H-4
Major's Place G-9
Manhattan H-5
Mason G-2
McDermitt A-5
McGill F-9
Mercury L-7
Mesquite L-10
Midas C-6
Middle Gate F-4
Mill City D-4
Mina H-4
Minden G-1
Moapa L-9
Montello C-10
Mountain City B-7
Mt. Charleston M-7
Nelson N-9
Nixon E-2
North Las Vegas M-9
Nyala H-7
Oasis C-9
Oreana D-4
Orovada B-4
Overton L-9
Owyhee B-7
Pahrump M-7
Panaca J-9
Paradise Valley B-5
Pioche J-9
Preston G-8
Rachel J-7
Reno F-1
Rhyolite L-6
Round Mt. G-5
Ruby Valley D-8
Ruth F-9
Sandy N-8
Schurz G-3
Scotty's Jct. K-5
Searchlight N-8
Shantytown E-8
Silver City F-2
Silverpeak J-4
Silver Springs F-2
Sloan M-8
Smith G-2
S. Lake Tahoe G-1
Sparks F-1
Stateline G-1
Stewart F-1
Stillwater F-3
Sulphur C-3
Sutcliffe E-1
Thousand Springs C-9
Tonopah H-5
Tuscarora C-7
Unionville D-4
Ursine J-10
Valmy D-6
Verdi F-1
Virginia City F-2
Wabuska F-2
Wadsworth E-2
Warm Springs H-6
Washoe F-1
Weed Heights G-2
Wellington G-2
Wells C-9
Winnemucca C-4
Yerington G-2
Zephyr Cove F-1

NEW HAMPSHIRE
Page 57

Albany E-6
Allenstown H-6
Alstead H-3
Alton G-6
Alton Bay G-6
Amherst J-5
Andover G-5
Antrim J-4
Ashland F-5
Ashuelot K-3
Atkinson J-6
Auburn J-6
Barnstead H-6
Bartlett E-6
Bath E-4
Bayside J-7
Bedford J-5
Belmont G-5
Bennington J-4
Berlin D-6
Bethlehem D-5
Boscawen H-5
Bow Center H-5
Bradford H-4
Brentwood J-6
Bretton Woods E-5
Bridgewater G-5
Bristol G-5
Brookfield G-6
Brookline K-5
Canaan G-4
Canaan Center F-4
Candia J-6
Carroll D-5
Cascade D-6
Center Conway E-6
Center Ossipee F-6
Charlestown H-3
Cheever F-4
Chester J-6
Chesterfield J-3
Chichester H-5
Chocorua F-6
Cilleyville G-4
Claremont G-4
Coburn G-6
Cold River H-3
Colebrook B-5
Columbia B-5
Concord H-5
Contoocook H-5
Conway E-6
Coos Junction D-5
Cornish City G-3
Cornish Flat G-3
Crystal C-6
Danbury G-5
Davisville H-5
Derry J-6
Derry Village J-6
Dixville B-6
Dorchester F-4
Dover H-7
Drewsville H-3
Dummer C-6
Dublin J-4
Durham H-7
East Alton G-6
E. Barrington H-6
E. Grafton G-4
E. Hebron F-5
E. Kingston J-6
E. Lempster H-3
East Madison F-6
Easton E-5
Eaton Center F-6
Effingham Falls F-6
Elkins G-4
Epping J-6
Epsom H-6
Epsom Four Corners H-6
Errol C-6
Exeter J-7
Fabyan D-5
Farmington G-7
Fish Market G-4
Francestown H-4
Francestown J-4
Franconia D-5
Franklin G-5
Freedom F-6
Fremont J-6
Gerrish H-5
Gilford G-6
Gilmans Corner F-4
Gilmanton G-6
Gilsum J-4
Glendale G-5
Glen E-6
Glen House D-6
Goffs Falls J-5
Goffstown J-5
Gorham D-6
Goshen H-4
Gossville H-6
Grafton G-4
Grantham G-4
Greenfield J-4
Greenville K-4
Groton F-4
Groveton C-5
Guild G-4
Guildhall C-5
Hampton J-7
Hampton Beach J-7
Hancock J-4
Hanover F-4
Hanover Center F-4
Harts Location E-6
Haverhill E-4
Hebron F-4
Henniker H-5
Hillsborough H-4
Hinsdale J-3
Hooksett H-6
Hopkinton H-5
Hoyts Corner H-4
Hudson K-6
Jackson D-6
Jaffrey J-4
James City H-6
Jefferson D-5
Jefferson Highlands D-5
Keene J-3
Kensington J-7
Kidderville B-5
Kingston J-6
Laconia G-5
Lancaster D-5
Laskey Corner G-7
Leavitts Hill H-6
Lebanon G-4
Lincoln E-5
Lisbon E-4
Little Boars Head J-7
Littleton D-5
Loudon H-5
Lower Shaker Village G-4
Lower Village J-3
Lyme F-4
Lyme Center F-4
Madbury H-7
Madison F-6
Manchester J-5
Marlborough J-4
Marlow H-4
Martin J-6
Mason K-5
Meadows D-5
Meredith F-5
Meriden G-3
Merrimack J-5
Milan C-6
Milford J-5
Milton G-6
Mirror Lake F-6
Monroe D-4
Mont Vernon J-5
Moultonborough F-6
Moultonville F-6
Mt. Sunapee H-4
Munsonville J-4
Nashua K-5
New Durham G-6
New Hampton G-5
New Ipswich K-4
New London G-4
New Rye H-6
Newbury H-4
Newfields J-7
Newington H-7
Newmarket J-7
Newport H-4
Newton J-7
N. Chatham D-6
N. Conway E-6
N. Grantham G-4
N. Groton F-5
N. Haverhill E-4
N. Rochester H-6
N. Sandwich F-6
N. Stratford C-5
N. Sutton H-4
N. Wakefield G-6
N. Woodstock E-5
Northfield G-5
Northumberland C-5
Northwood H-6
Orford F-4
Orfordville F-4
Ossipee F-6
Otterville G-4
Pages Corner H-5
Parker J-5
Pearls Corner H-5
Pelham K-6
Pembroke H-6
Penacook H-5
Peterborough J-4
Pinardville J-5
Pittsburg B-6
Pittsfield H-6
Plainfield G-3
Plaistow J-6
Plymouth F-5
Portsmouth H-7
Potter Place G-4
Province Lake F-7
Randolph D-6
Raymond J-6
Redstone E-6
Richmond K-3
Rindge K-4
Riverton G-5
Rochester H-6
Rollinsford H-7
Rumney F-5
Rye J-7
Rye Beach J-7
Salem J-6
Salisbury H-5
Sanbornton G-5
Sanbornville G-7
Sandwich F-6
Seabrook Beach J-7
Sharon J-4
Shelburne D-6
Somersworth H-7
S. Acworth H-4
S. Cornish G-3
S. Danbury G-5
S. Stoddard J-4
S. Weare J-5
Spofford J-3
Springfield G-4
Squantum J-4
Stark C-6
Stewartstown B-6
Stoddard H-4
Stratford C-5
Stratham J-7
Sunapee G-4
Surry J-3
Sutton H-4
Swiftwater E-4
Tamworth F-6
Temple J-4
Thornton F-5
Tilton G-5
Trapshire H-3
Troy J-4
Tuftonboro F-6
Twin Mountain D-5
Union G-6
Wakefield G-7
Wallis Sands J-7
Walpole J-3
Warner H-5
Warren F-4
Washington H-4
Weare J-5
Wendell H-4
Wentworth F-4
Wentworths Location B-6
West Campton F-5
West Lebanon F-4
West Milan C-6
Westmoreland J-3
W. Ossipee F-6
West Rindge K-4
W. Rumney F-5
W. Stewartstown B-5
W. Swanzey J-4
Westville J-6
Willey House E-5
Wilton J-5
Winchester J-3
Windsor H-4
Winnisquam G-5
Wolfeboro G-6
Wonalancet F-6
Woodman F-7
Woodsville E-4

NEW JERSEY
Page 31

Absecon G-5
Adelphia D-6
Albion ★
Vicinity Pg. 137, G-9
Allamuchy B-4
Allendale ★
Vicinity Pg. 128, D-4
Almonesson ★
Vicinity Pg. 137, G-8
Alloway G-3
Alpha C-3
Alpine B-6
Vicinity Pg. 128, F-7
Anderson B-4
Andover B-4
Annandale C-4
Ardena D-6
Asbury Park D-6
Ashland ★
Vicinity Pg. 137, F-9
Atco F-4
Atlantic City G-5
Atlantic Highlands D-6
Atsion F-4
Auburn F-2
Audubon ★
Vicinity Pg. 137, E-8
Augusta A-4
Aura F-3
Avalon H-4
Avenel ★
Vicinity Pg. 129, P-1
Bakersville D-5
Baptistown C-3
Barnegat F-6
Barnegat Light F-6
Barrington ★
Vicinity Pg. 137, F-8
Batsto F-4
Bayonne C-6
Vicinity Pg. 129, M-4
Bay Point H-3
Bayville F-6
Beach Haven G-6
Bedminster C-4
Belleville ★
Vicinity Pg. 129, J-3
Bellmawr ★
Vicinity Pg. 137, F-8
Belmar E-6
Bergenfield ★
Vicinity Pg. 128, F-6
Berlin F-4
Vicinity Pg. 137, G-10
Bernardsville C-4
Beverly ★
Vicinity Pg. 137, B-10
Blackwood ★
Vicinity Pg. 137, G-8
Bloomfield ★
Vicinity Pg. 129, J-3
Bloomingdale ★
Vicinity Pg. 128, D-1
Bloomsbury C-3
Boonton B-5
Bordentown E-4
Bradley Beach E-6
Branchville A-4
Bridgeboro ★
Vicinity Pg. 137, C-9
Bridgeport F-2
Vicinity Pg. 137, G-4
Bridgeton G-3
Brielle E-6
Bringantine G-5
Brooklawn ★
Vicinity Pg. 137, E-7
Browns A-5
Brunswick D-5
Buena G-4
Burlington E-4
Vicinity Pg. 137, B-11
Burrs Mills E-4
Butler B-5
Buttzville B-4
Byram D-3
Caldwell ★
Vicinity Pg. 129, H-1
Camden E-4
Vicinity Pg. 137, D-7
Cape May J-4
Cape May Point J-4
Carlls Corner G-3
Carlstadt B-6
Carteret C-6
Vicinity Pg. 129, P-2
Cassville E-5
Cedar Bridge E-6
Cedar Grove ★
Vicinity Pg. 128, G-2
Cedarville G-3
Centerton ★
Vicinity Pg. 137, C-10
Centre Grove G-3
Chambers Corners E-4
Chatham C-5
Cherry Hill E-4
Vicinity Pg. 137, E-8
Chester B-5
Churchtown F-2
Cinnaminson E-3
Vicinity Pg. 137, C-9
Clark ★
Vicinity Pg. 129, N-1
Clarksboro ★
Vicinity Pg. 137, G-6
Clarksburg D-5
Clarksville D-5
Clayton F-3
Clementon ★
Vicinity Pg. 137, G-9
Clermont H-4
Cliffside ★
Vicinity Pg. 129, J-6
Clifton B-6
Vicinity Pg. 128, H-3
Closter ★
Vicinity Pg. 128, E-7
Coffins Corner ★
Vicinity Pg. 137, F-9
Cohansey G-3
Cold Springs J-3
Colesville A-4
Collingswood ★
Vicinity Pg. 137, E-8
Colonia ★
Vicinity Pg. 129, P-1
Conovertown G-5
Cranbury D-5
Cranford ★
Vicinity Pg. 129, M-1
Cresskill B-6
Vicinity Pg. 128, F-7
Croton C-4
Culvers Inlet A-4
Cumberland G-4
Dayton D-5
Deacons ★
Vicinity Pg. 137, C-11
Deerfield G-3
Delair ★
Vicinity Pg. 137, D-8
Delanco ★
Vicinity Pg. 137, B-9
Delaware B-3
Delaware City F-2
Delmont H-4
Dennisville H-4
Denville B-5
Deptford ★
Vicinity Pg. 137, F-7
Dias Creek H-3
Dividing Creek H-3
Dorothy G-4
Dover B-5
Downstown G-4
Dukes Bridge F-5
Dumont ★
Vicinity Pg. 128, F-6
East Orange ★
Vicinity Pg. 129, J-3
East Rutherford ★
Vicinity Pg. 129, H-4
Eatontown D-6
Egg Harbor G-5
Eldora H-4
Elizabeth C-6
Vicinity Pg. 129, M-3
Ellisburg ★
Vicinity Pg. 137, E-9
Elm F-4
Elmwood Park ★
Vicinity Pg. 128, G-4
Elwood G-4
Emerson ★
Vicinity Pg. 128, E-6
Englewood B-6
Vicinity Pg. 128, G-7
Erskine A-6
Vicinity Pg. 128, B-1
Estell Manor G-4
Estellville G-4
Everittstown C-3
Evesboro ★
Vicinity Pg. 137, E-10
Ewanville E-4
Fairfield B-6
Vicinity Pg. 128, G-1

★ Denotes City located only on Vicinity map. City not located on State map.

NEW MEXICO
Pages 62-63

NEW YORK
Pages 64-67

★ Denotes City located only on Vicinity map. City not located on State map.

NEW YORK

Barneveld....H-14
Batavia....J-6
Bath....M-8
Bayberry....J-11
Bay Park....★
  Vicinity....Pg.131,P-12
Bay Ridge....★
  Vicinity....Pg.129,M-6
Bay Shore....E-6
  Vicinity....Pg.131,M-19
Bayside....★
  Vicinity....Pg.129,K-10
Bayville....D-5
  Vicinity....Pg.131,H-14
Beacon....A-3
Bedford....B-4
  Vicinity....Pg.130,B-13
Bedford Hills....★
  Vicinity....Pg.130,A-11
Belle Rose....★
  Vicinity....Pg.131,M-11
Bellmore....★
  Vicinity....Pg.131,N-15
Belmont....N-6
Bemus Pt.....N-2
Benson Hurst....★
  Vicinity....Pg.129,N-7
Bergen....J-6
Bethpage....D-5
  Vicinity....Pg.131,M-15
Big Flats....N-9
Binghamton....N-12
Black River....E-13
Blasdell....K-4
Blauve....★
  Vicinity....Pg.128,C-7
Bloomingburgh....A-2
Bloomingdale....C-18
Bloomington....N-18
Bolivar....N-6
Boonville....G-14
Bovsen Bay....J-11
Brasher Falls....B-16
Brentwood....D-6
Brewerton....★
  Vicinity....Pg.131,L-19
Brewster....B-4
Briarcliff Manor....C-4
Bridgehampton....C-9
Bridgewater....K-14
Brightwaters....★
  Vicinity....Pg.131,M-19
Broadalbin....J-18
Brockport....J-6
Brocton....M-2
Bronx....★
  Vicinity....Pg.128,G-8
Bronxville....★
  Vicinity....Pg.128,F-9
Brooklyn....★
  Vicinity....Pg.129,L-9
Brownville....E-12
Brushton....B-16
Buchanan....B-3
Buffalo....K-4
Burdett....M-10
Burke....B-17
Cairo....M-18
Caledonia....K-7
Calverton....D-7
Cambridge....J-20
Camden....H-13
Camillus....J-11
Canajoharie....K-16
Canandaigua....K-8
Canarisie....★
  Vicinity....Pg.129,M-8
Canaseraga....M-7
Canastota....J-13
Candor....M-11
Canisteo....M-7
Canton....C-14
Cape Vincent....E-11
Carle Place....★
  Vicinity....Pg.131,M-13
Carthage....E-13
Cassadaga....M-2
Castile....L-6
Castleton Corners....★
  Vicinity....Pg.129,N-4
Castleton-on-Hudson....L-18
Castorland....F-13
Cato....J-10
Catskill....M-19
Cattaraugus....M-4
Cayuga....K-10
Cayuga Hts.....L-11
Cazenovia....K-13
Cedarhurst....★
  Vicinity....Pg.131,P-11
Celoron....N-2
Central Square....H-12
Centereach....D-6
Center Ilsip....★
  Vicinity....Pg.131,K-20
Center Moriches....D-7
Central Nyack....★
  Vicinity....Pg.128,B-7
Central Valley....B-3
Champlain....A-19
Chappaqua....B-4
  Vicinity....Pg.130,A-9
Chateaugay....A-17
Chatham....M-19
Chaumont....E-12
Chelsea....★
  Vicinity....Pg.129,N-3
Cherry Creek....M-3
Cherry Valley....K-16
Chester....B-2
Chittenango....J-13
Churchville....J-7
Claverack....M-19
Clarence Center....J-4
Clayton....J-14
Clayville....J-14
Cleveland....H-12
Clifton Knolls....K-18
Clifton Springs....K-9
Clinton....J-14
Clintondale....O-18
Clyde....J-10
Cobleskill....K-17
Cohocton....L-8

Cohoes....K-19
Cold Brook....H-15
Cold Spring....J-11
Coldspring Harbor....★
  Vicinity....Pg.131,J-15
College Point....★
  Vicinity....Pg.129,J-9
Colonie....K-18
Commack....★
  Vicinity....Pg.131,K-18
Coney Island....★
  Vicinity....Pg.129,P-10
Constableville....G-14
Constantia....H-12
Cooperstown....K-15
Copenhagen....F-13
Copiague....★
  Vicinity....Pg.131,N-17
Coram....D-6
Corfu....K-5
Corinth....H-18
Corning....N-9
Cornwall-on-Hudson....A-3
Cortland....L-11
County Knolls....J-18
Coxsackie....M-19
Croghan....F-14
Croton-on-Hudson....B-3
Cuba....N-5
Cutchogue....C-8
Dannemora....B-18
Dansville....L-7
Deer Park....D-5
  Vicinity....Pg.131,L-18
Deferiet....E-13
Delanson....K-17
Delevan....K-5
Delhi....M-15
Delmar....L-18
Depew....K-4
Deposit....N-14
DeRuyter....K-13
DeWitt....J-11
Dexter....E-12
Dobbs Ferry....C-3
  Vicinity....Pg.128,D-8
Dolgeville....J-16
Dongan Hills....★
  Vicinity....Pg.129,N-4
Dover Plains....O-19
Dresden....L-9
Dryden....L-11
Dundee....L-9
Dunkirk....M-2
Earlville....K-13
E. Aurora....K-4
E. Bloomfield....K-8
E. Glenville....J-18
E. Hampton....C-9
E. Hills....★
  Vicinity....Pg.131,K-13
E. Irvington....★
  Vicinity....Pg.128,C-9
E. Ilsip....★
  Vicinity....Pg.131,M-20
E. Meadow....D-4
  Vicinity....Pg.131,M-14
E. New York....★
  Vicinity....Pg.129,L-8
E. Northport....D-5
  Vicinity....Pg.131,J-17
E. Quogue....D-8
E. Randolph....N-3
E. Rochester....J-8
E. Rockaway....★
  Vicinity....Pg.131,P-12
East View....★
  Vicinity....Pg.128,C-9
Eastchester....★
  Vicinity....Pg.128,E-9
Eatons Neck....C-7
Eden....L-4
Edgemere....★
  Vicinity....Pg.129,N-10
Edwards....D-14
Elba....J-6
Elbridge....J-11
Ellenville....O-17
Ellicottville....M-4
Ellisburg....F-12
Elma....K-4
Elmira Hts.....N-10
Elmont....★
  Vicinity....Pg.131,N-12
Elwood....★
  Vicinity....Pg.131,J-17
Endicott....N-12
Endwell....N-12
Esperance....K-17
Evans Mills....E-12
Fabius....K-12
Fair Harbor....★
  Vicinity....Pg.131,P-20
Fair Haven....H-10
Fairport....J-8
Falconer....N-3
Farmingdale....★
  Vicinity....Pg.131,M-16
Farnham....L-3
Far Rockaway....★
  Vicinity....Pg.129,N-10
Fayetteville....J-12
Fernwood....H-19
Fillmore....M-6
Firthcliffe....A-3
Fishkill....A-3
Flatbush....★
  Vicinity....Pg.129,M-7
Fleischmanns....N-17
Florida....B-2
Flower Hill....★
  Vicinity....Pg.131,K-12
Flushing....★
  Vicinity....Pg.129,K-9
Fonda....J-17
Forest Hills....★
  Vicinity....Pg.129,K-9
Forestville....M-3
Ft. Ann....G-19
Ft. Edward....H-19
Ft. Johnson....J-17
Ft. Plain....K-16
Ft. Salonga....★
  Vicinity....Pg.130,H-18

Frankfort....J-15
Franklin....M-15
Franklin Square....★
  Vicinity....Pg.131,N-12
Franklinville....M-5
Freeport....E-4
  Vicinity....Pg.131,P-14
Frewsburg....N-3
Friendship....N-6
Fulton....H-11
Fultonville....K-17
Gainesville....L-6
Galway....J-18
Gang Mills....N-9
Garden City Park....★
  Vicinity....Pg.131,M-12
Geneseo....K-7
Geneva....K-9
Gilbertsville....L-14
Glasco....N-18
Glen Cove....D-4
  Vicinity....Pg.131,M-12
Glenham....A-3
Glen Park....E-12
Glens Falls....H-19
Glenville....★
  Vicinity....Pg.128,C-9
Glenwood Landing....★
  Vicinity....Pg.131,K-13
Gloversville....J-17
Goldens Bridge....B-4
Goshen....B-2
Gouverneur....D-14
Gowanda....M-3
Grand View....★
  Vicinity....Pg.128,C-8
Granville....G-20
Great Kills....★
  Vicinity....Pg.129,P-4
Great Neck....D-4
  Vicinity....Pg.131,K-11
Greece....J-7
Greene....M-13
Green Island....K-19
Green Lawn....★
  Vicinity....Pg.131,J-7
Greenport....C-9
Greenvale....★
  Vicinity....Pg.131,K-13
Greenville....★
  Vicinity....Pg.128,D-9
Greenwich....J-20
Greenwood Lake....B-2
Groton....L-11
Hadley....H-19
Hagaman....J-18
Halesite....★
  Vicinity....Pg.131,H-16
Hamburg....L-4
Hamilton....K-14
Hammond....C-13
Hammondsport....M-8
Hampton Bays....D-8
Hancock....O-15
Hannibal....H-10
Harriman....B-2
Harrison....C-4
  Vicinity....Pg.130,G-11
Harrisville....E-14
Hartsdale....★
  Vicinity....Pg.128,D-10
Hastings-On-Hudson....★
  Vicinity....Pg.128,D-8
Hauppage....★
  Vicinity....Pg.131,K-20
Haverstraw....B-3
Haviland....★
  Vicinity....Pg.131,P-12
Hawthorne....★
  Vicinity....Pg.128,B-9
Hempstead....E-4
Herkimer....J-15
Hermon....C-14
Herrings....E-13
Heuvelton....C-14
Hewlett....★
  Vicinity....Pg.131,P-12
Hicksville....D-5
  Vicinity....Pg.131,L-14
Highland....O-18
Highland Falls....B-3
Highland Mills....B-3
Hillburn....C-2
  Vicinity....Pg.128,B-3
Hillcrest....★
  Vicinity....Pg.128,B-5
Hillside Lake....P-19
Hilton....H-7
Hobart....M-16
Holbrook....D-6
Holcomb....K-8
Holland....L-5
Holland Patent....H-14
Holley....J-6
Homer....L-11
Honeoye Falls....K-7
Hoosick Falls....J-20
Hornell....M-7
Horseheads....N-10
Houghton....M-6
Hudson....M-19
Hudson Falls....H-19
Hugenot....★
  Vicinity....Pg.129,Q-2
Hugenot Park....★
  Vicinity....Pg.129,Q-3
Hunter....M-17
Huntington....D-5
  Vicinity....Pg.131,J-16
Huntington Sta.....D-5
  Vicinity....Pg.131,K-17
Hurley....O-18
Hyde Park....O-18
Ilion....J-15
Interlaken....L-10
Inwood....★
  Vicinity....Pg.131,Q-11
Irvington....★
  Vicinity....Pg.128,C-8
Iselin....★
  Vicinity....Pg.129,P-1
Island Park....★
  Vicinity....Pg.131,P-13
Islip....D-6
  Vicinity....Pg.131,M-20

Islip Terrace....★
  Vicinity....Pg.131,L-20
Ithaca....M-11
Jackson Hts.....★
  Vicinity....Pg.129,K-8
Jamaica....★
  Vicinity....Pg.129,L-10
Jamesport....C-8
Jamestown....N-3
Jefferson Hts.....M-18
Jefferson Valley....B-4
Jeffersonville....O-15
Jericho....D-5
  Vicinity....Pg.131,L-14
Johnson City....N-12
Johnstown....J-17
Jordan....J-11
Keeseville....C-19
Kenmore....K-3
Kerhonkson....O-17
Keuka Park....L-9
Kinderhook....M-19
Kings Park....★
  Vicinity....Pg.131,J-19
Kings Point....★
  Vicinity....Pg.129,K-10
Kingston....N-18
Kiryas Joel....B-2
Lackawanna....K-4
Lacona....G-12
Ladentown....★
  Vicinity....Pg.128,A-5
Lake Carmel....A-4
Lake Erie Beach....L-3
Lake George....G-19
Lake Luzerne....H-18
Lake Placid....D-18
Lake Ronkonkoma....D-6
Lakewood....N-2
Lancaster....K-4
Lansing....L-11
Larchmont....★
  Vicinity....Pg.130,H-14
Latham....K-18
Lattingtown....★
  Vicinity....Pg.131,J-12
Laurens....L-15
Leicester....K-6
LeRoy....K-6
Levittown....E-5
  Vicinity....Pg.131,M-15
Lewiston....J-3
Liberty....O-16
Lido Beach....★
  Vicinity....Pg.131,Q-13
Lima....K-7
Lime Lake....M-5
Limestone....N-4
Lincoln Park....N-18
Lindenhurst....E-5
  Vicinity....Pg.131,N-17
Lisle....M-12
Little Falls....J-16
Little Valley....M-4
Liverpool....J-11
Livingston Manor....O-16
Livonia....K-7
Lloyd Harbor....D-5
  Vicinity....Pg.131,H-17
Lockport....J-4
Locust Valley....★
  Vicinity....Pg.131,H-13
Lodi....L-10
Long Beach....E-4
  Vicinity....Pg.131,Q-12
Long Island....★
  Vicinity....Pg.129,K-7
Lorenz Pk.....M-19
Lowville....F-14
Lynbrook....★
  Vicinity....Pg.131,P-11
Lyndonville....H-5
Lyons....J-9
Lyons Falls....G-14
Macedon....J-8
Machias....M-5
Madison....K-14
Mahopac....B-4
Malone....B-17
Mamaroneck....C-4
  Vicinity....Pg.130,G-11
Manchester....J-8
Manlius....J-12
Manhattan....★
  Vicinity....Pg.129,J-7
Mannsville....F-12
Manorhaven....★
  Vicinity....Pg.131,J-11
Marathon....M-12
Marcellus....J-11
Margaretville....N-16
Marlboro....P-18
Massapequa....E-5
  Vicinity....Pg.131,N-15
Massena....B-15
Mastic....D-7
Mastic Beach....D-7
Mattituck....C-8
Mattydale....J-11
Maybrook....A-2
Mayfield....M-2
Mayville....M-2
McGraw....L-12
Mechanicville....K-19
Medford....D-6
Medina....J-5
Melville....★
  Vicinity....Pg.131,L-16
Menands....K-19
Meridian....J-11
Merrick....★
  Vicinity....Pg.131,P-14
Merriewold Lake....B-2
Mexico....H-11
Middleburgh....L-17
Middle Hope....P-18
Middleport....J-5
Middletown....A-2
Middleville....J-15
Milburn....★
  Vicinity....Pg.129,K-1
Milford....K-15
Millbrook....O-19

Millerton....N-19
Millport....M-10
Milton....O-18
Minetto....H-11
Mineville....E-19
Minoa....J-12
Mohawk....J-15
Monroe....B-2
Monsey....C-3
  Vicinity....Pg.128,B-5
Montauk....C-10
Montgomery....A-2
Monticello....P-16
Montour Falls....M-10
Mooers....A-19
Moravia....L-11
Morris....L-15
Morristown....C-13
Morrisville....K-13
Mt. Ivy....★
  Vicinity....Pg.128,A-6
Mount Kisco....★
  Vicinity....Pg.130,B-11
Mount Morris....L-7
Mount Vernon....D-4
  Vicinity....Pg.128,F-9
Munnsville....J-13
Munsey Park....★
  Vicinity....Pg.131,L-13
Nanuet....C-3
Napanoch....O-17
Naples....L-8
Nassau....L-19
Nassau Shores....★
  Vicinity....Pg.131,O-16
Nelliston....J-16
New Berlin....L-14
New Brighton....★
  Vicinity....Pg.129,M-5
New Cassel....★
  Vicinity....Pg.131,L-14
New Drop....★
  Vicinity....Pg.129,P-4
New Drop Beach....★
  Vicinity....Pg.129,P-4
New Hartford....J-14
New Hyde Park....★
  Vicinity....Pg.131,M-12
New Paltz....O-18
New Rochelle....D-4
  Vicinity....Pg.128,F-9
New Suffolk....C-8
New York City....E-3
  Vicinity....Pgs.128-131
New York Mills....J-14
Newark....J-9
Newark Valley....M-12
Newburgh....A-3
Newfane....H-4
Newport....H-15
Niagara Falls....J-3
Nichols....N-11
Norfolk....B-15
N. Collins....L-3
N. Haven....C-9
N. Hornell....M-7
Northport....★
  Vicinity....Pg.131,H-17
N. Sea....C-9
N. Syracuse....J-12
N. Tareytown....★
  Vicinity....Pg.128,B-9
N. Tonawanda....J-4
Northville....H-17
Norwich....L-13
Norwood....B-15
Noyack....C-9
Nunda....L-6
Nyack....C-3
  Vicinity....Pg.128,B-7
Oak Beach....★
  Vicinity....Pg.131,P-19
Oakfield....J-5
Ocean Beach....E-6
  Vicinity....Pg.131,P-20
Oceanside....★
  Vicinity....Pg.131,P-13
Odessa....M-10
Ogdensburg....B-13
Olcott....H-4
Old Forge....G-14
Olean....N-5
Oneida....J-13
Oneida Castle....J-13
Oneonta....M-15
Orangeburg....★
  Vicinity....Pg.128,C-6
Orange Lake....A-3
Orchard Park....K-4
Oriskany....J-14
Oriskany Falls....J-14
Ossining....C-4
Otego....M-14
Otisville....A-1
Ovid....L-10
Owego....N-11
Oxford....M-13
Oyster Bay....★
  Vicinity....Pg.131,J-14
Ozone Park....★
  Vicinity....Pg.129,L-9
Painted Post....N-9
Palatine Bridge....K-16
Palisades....★
  Vicinity....Pg.131,J-13
Palmyra....J-9
Panama....N-2
Parish....H-12
Patchogue....D-6
Pawling....B-4
Pearl River....C-3
  Vicinity....Pg.128,C-6
Peconic....C-8
Peekskill....B-3
Pelham....★
  Vicinity....Pg.128,G-9
Pelham Manor....★
  Vicinity....Pg.128,G-10
Penn Yan....L-9
Perry....K-6
Perrysburg....M-3
Peru....C-19
Phelps....K-9

Philadelphia....E-13
Philmont....M-19
Phoenix....H-11
Piermont....★
  Vicinity....Pg.128,C-8
Pike....L-6
Pine Bush....P-17
Pine Hill....N-17
Pine Plains....N-19
Pittsford....J-8
Plainview....D-5
  Vicinity....Pg.131,L-16
Plattsburgh....B-19
Pleasantville....★
  Vicinity....Pg.128,B-9
Pocantico Hills....★
  Vicinity....Pg.128,B-9
Poestenkill....K-20
Poland....H-15
Pomona....B-3
Port Byron....J-10
Port Dickinson....N-13
Port Ewen....N-18
Port Henry....E-19
Port Jefferson....C-6
Port Jervis....B-1
Port Leyden....G-14
Port Richmond....★
  Vicinity....Pg.129,M-3
Port Washington....★
  Vicinity....Pg.131,K-12
Portville....N-5
Potsdam....B-15
Poughkeepsie....O-19
Pound Ridge....★
  Vicinity....Pg.130,B-13
Prospect....H-14
Pulaski....G-12
Queens....★
  Vicinity....Pg.129,K-10
Queens Village....★
  Vicinity....Pg.129,K-10
Quogue....D-8
Ramapo....★
  Vicinity....Pg.128,B-3
Randolph....N-3
Ransomville....J-4
Ravena....L-18
Red Creek....H-10
Red Hook....N-19
Red Oaks Mill....O-19
Remsen....H-15
Rensselaer....L-19
Rensselaer Falls....C-14
Rhinebeck....N-19
Richburg....N-6
Richfield Springs....K-15
Richmond Hill....★
  Vicinity....Pg.129,L-9
Richmondville....L-17
Richville....C-14
Ridge....D-7
Riverdale....★
  Vicinity....Pg.130,A-20
Riverhead....D-8
Riverside....N-9
Rochester....J-7
Rockville Centre....★
  Vicinity....Pg.131,P-12
Rocky Pt.....C-7
Roessleville....K-18
Rome....J-13
Roosevelt....★
  Vicinity....Pg.131,N-14
Roselyn Hts.....★
  Vicinity....Pg.131,L-12
Rosendale....O-18
Rossville....★
  Vicinity....Pg.129,P-3
Round Lake....J-19
Rouses Point....A-19
Ruby....N-18
Rushville....K-9
Rye....C-4
  Vicinity....Pg.130,G-12
Rye Brook....C-4
Sackets Harbor....F-12
Sag Harbor....C-9
St. James....★
  Vicinity....Pg.131,J-20
St. Lawrence Park....D-12
St. Johnsville....J-16
Salamanca....N-4
Salem....H-20
Sand Ridge....J-11
Sandy Creek....G-12
Sands Point....★
  Vicinity....Pg.131,J-11
San Remo....★
  Vicinity....Pg.131,J-19
Saranac Lake....D-18
Saratoga Springs....J-18
Saugerties....N-18
Savona....M-9
Sayville....D-6
Scarsdale....C-4
  Vicinity....Pg.128,E-9
Schaghticoke....K-19
Schenectady....K-18
Schenevus....L-15
Schoharie....L-17
Schuylerville....J-19
Scotch Town....A-2
Scotia....K-18
Scottsville....J-7
Sea Cliff....★
  Vicinity....Pg.131,J-13
Seaford....★
  Vicinity....Pg.131,N-15
Selden....D-6
Seneca Falls....K-10
Sharon Springs....K-16
Shelter Island....C-9
Sherburne....L-14
Sherman....N-1
Sherrill....J-13
Shirley....D-7
Shortsville....K-8
Sidney....M-14
Silver Creek....L-3
Silver Springs....L-6
Sinclairville....M-2
Skaneateles....K-11
Sloatsburg....C-2
Smithtown....D-6
  Vicinity....Pg.131,J-20

Smyrna....L-13
Sodus....H-9
Sodus Point....H-9
Solvay....J-11
Sound Beach....C-6
S. Corning....N-9
S. Fallsburg....O-16
S. Glens Falls....H-19
Southampton....D-9
Southold....C-8
S. Nyack....★
  Vicinity....Pg.128,C-8
Southport....N-10
Sparkill....★
  Vicinity....Pg.128,D-7
Speculator....G-17
Spencer....N-11
Spencerport....J-7
Springfield Gdns.....★
  Vicinity....Pg.129,L-10
Springs....C-9
Spring Valley....★
  Vicinity....Pg.128,B-5
Stamford....M-16
Stannards....N-6
Star Lake....D-15
Staten Island....★
  Vicinity....Pg.129,O-5
Stillwater....J-19
Stony Brook....D-6
  Vicinity....Pg.130,H-20
Stony Pt.....B-3
Stottville....M-19
Suffern....C-2
  Vicinity....Pg.128,B-3
Sylvan Beach....J-13
Syosset....★
  Vicinity....Pg.131,K-15
Syracuse....J-12
Tall Man....★
  Vicinity....Pg.128,B-4
Tannersville....M-17
Tappan....★
  Vicinity....Pg.128,D-7
Tarrytown....C-4
  Vicinity....Pg.128,C-8
Theresa....D-13
Thomaston....★
  Vicinity....Pg.131,L-11
Thornwood....★
  Vicinity....Pg.128,B-10
Ticonderoga....F-19
Tivoli....N-18
Tonawanda....J-4
Town Line....K-5
Troy....K-19
Trumansburg....L-10
Trumbull....★
  Vicinity....Pg.130,A-20
Tuckahoe....D-8
Tully....K-12
Tupper Lake....D-17
Turin....F-14
Tuxedo Park....B-2
Unadilla....M-14
Union Springs....K-10
Union Vale....★
  Vicinity....Pg.131,N-19
Unionville....B-15
Upper Nyack....★
  Vicinity....Pg.128,B-7
Utica....J-15
Valhalla....★
  Vicinity....Pg.128,C-10
Valatie....L-19
Valley Cottage....★
  Vicinity....Pg.128,B-7
Valley Falls....J-20
Valley Stream....★
  Vicinity....Pg.131,P-11
Van Etten....N-11
Vernon....J-14
Verona....J-12
Victor....K-8
Village of the Branch....★
  Vicinity....Pg.131,J-20
Viola....★
  Vicinity....Pg.128,B-5
Vista....★
  Vicinity....Pg.130,B-14
Voorheesville....L-18
Waddington....B-14
Wading River....C-7
Wainscott....A-2
Walden....A-2
Walker....N-15
Walton....M-14
Walton Park....B-2
Wampsville....J-13
Wantagh....★
  Vicinity....Pg.131,N-15
Wappinger Falls....A-3
Warrensburg....G-19
Warsaw....K-6
Warwick....B-2
Washingtonville....A-2
Waterford....K-19
Waterloo....K-9
Watertown....E-13
Waterville....K-14
Watervliet....K-19
Watkins Glen....M-10
Waverly....N-11
Wayland....L-7
Webster....J-8
Weedsport....J-10
Wellsburg....N-10
Wellsville....N-6
W. Babylon....★
  Vicinity....Pg.131,M-18
Westbury....★
  Vicinity....Pg.131,M-13
W. Carthage....E-13
West End....L-15
W. Nyack....★
  Vicinity....Pg.128,B-7
West Point....B-3

W. Sand Lake....L-19
W. Winfield....K-15
Westfield....M-2
Westhampton....D-8
Westons Mills....N-5
Westport....D-19
White Plains....C-4
  Vicinity....Pg.128,D-10
Whitehall....G-20
Whitesboro....J-14
Whitney Point....M-12
Willard....L-10
Williamsville....K-4
Williston Park....★
  Vicinity....Pg.131,L-13
Willow Brook....★
  Vicinity....Pg.129,N-4
Wilson....H-4
Windsor....N-13
Winthrop....B-16
Witherbee....E-19
Wolcott....H-10
Woodbury....★
  Vicinity....Pg.131,K-15
Woodhull....N-8
Woodridge....O-17
Woodstock....N-18
Wyadanch....★
  Vicinity....Pg.131,L-17
Wyoming....K-6
Yonkers....D-3
  Vicinity....Pg.128,F-8
Yorkshire....L-5
Yorktown Hts.....B-4
Yorkville....J-13
Youngstown....H-3
Zena....N-18

NORTH CAROLINA
Pages 68-69

Abbotsburg....E-10
Aberdeen....D-8
Acre....C-12
Ahoskie....A-12
Albemarle....D-7
Allen....E-9
Allenton....D-2
Andrews....D-12
Arapahoe....C-12
Archdale....C-7
Arcola....B-10
Arlington....B-6
Asheboro....C-8
Asheville....C-4
Ashland....B-12
Askin....D-12
Aurora....C-12
Avery Creek....D-4
Avon....C-14
Ayden....C-11
Badin....D-7
Bagley....C-10
Bald Creek....C-4
Baldwin....B-5
Bandana....C-4
Banner Elk....B-5
Barco....A-13
Bat Cave....D-4
Battleboro....B-11
Bear Creek....C-9
Beaufort....D-13
Belgrade....D-11
Belhaven....C-12
Belmont....D-6
Benson....C-9
Bertha....A-13
Bessemer City....D-6
Bethel....B-11
Bethel Hill....A-9
Beulaville....D-11
Biscoe....C-8
Black Mtn.....C-4
Bladenboro....E-9
Blowing Rock....B-5
Bolton....E-10
Boone....B-5
Bottom....A-6
Bowdens....D-10
Brasstown....D-2
Brevard....D-4
Bridgeton....D-12
Bridgewater....C-5
Brookford....C-5
Brooks Cross Roads....B-6
Brooksdale....B-9
Browns Summit....B-8
Brunswick....F-10
Bryson City....D-2
Bulade....B-4
Bullock....B-10
Bunn....B-10
Burch....E-6
Burgaw....E-10
Burlington....B-8
Burnsville....C-4
Buxton....C-14
Calypso....D-10
Camden....A-13
Cameron....D-9
Candor....D-8
Canto....C-3
Canton....D-3
Carolina Beach....F-11
Carr....B-8
Carrboro....B-9
Cary....C-9
Casville....B-8
Cedar Creek....D-10
Celo....C-4
Centerville....B-10
Chadbourn....F-10
Chapel Hill....B-9
Charlotte....D-6
Cherokee....D-3
Cherryville....D-5
China Grove....C-7
Chinquapin....D-11
Chocowinity....C-12
Churchland....C-7
Cid....C-7

168 Index

NORTH CAROLINA OHIO

Clemmons B-7
Clinton D-10
Coinjock A-13
College B-8
Colerain B-12
Columbia B-13
Columbus D-5
Comfort D-11
Concord D-7
Conover C-6
Conway A-12
Corapeake A-12
Cornelius D-6
Council E-10
Cove Creek C-3
Creedmoor B-9
Creswell B-13
Culberson D-1
Currie E-10
Dalton B-7
Darden B-12
Davidson C-6
Deep Gap B-5
Del Mar Beach E-12
Delway E-10
Dillsboro D-3
Dobson B-7
Dortches B-10
Dudley D-10
Dunn D-9
Durham B-9
E. Flat Rock D-4
E. Lake B-13
E. Laport D-3
Eastwood D-8
Eden A-8
Edenhouse B-12
Edenton B-12
Edneyville D-4
Edward C-12
Ela D-3
Eldorado C-7
Elizabeth City A-13
Elizabethtown E-10
Elkin B-6
Elk Park B-4
Emery D-8
Enfield B-11
Englehard C-13
Erwin D-9
Essex B-10
Ether C-8
Evergreen E-9
Fair Bluff E-9
Fairmont E-9
Faison D-10
Farmville C-11
Fayetteville D-9
Flat Rock D-4
Fletcher D-4
Folkston E-11
Fontana Village D-2
Forest City D-5
Fountain C-11
Franklin D-2
Freeman E-10
Fremont C-10
Frisco C-14
Fuquay-Varina C-9
Garysburg A-11
Gastonia D-6
Gates A-12
Gilkey D-5
Glade Valley B-6
Glenview B-10
Glenwood C-5
Gneiss D-3
Godwin D-9
Gold Rock B-11
Goldsboro C-10
Goldston C-8
Gordonton B-9
Graingers C-11
Grandy A-13
Granite Falls C-5
Granite Quarry C-7
Grantham D-10
Greensboro B-8
Greenville C-11
Grifton C-11
Grissettown F-10
Guilford B-8
Halifax A-11
Hamlet D-8
Hampstead E-11
Harbinger B-14
Harmony B-6
Harrells E-10
Harrisburg D-7
Hatteras C-14
Hazelwood D-3
Heartsease B-11
Henderson B-10
Hendersonville D-4
Henrietta D-5
Hertford B-12
Hester B-9
Hickory C-6
High Hampton D-3
High Point B-7
Hightowers B-8
Hill Crest D-8
Hillsborough B-9
Hiwassee D-2
Hoffman D-8
Holly Ridge E-11
Hollyville (Cash Corner) D-12
Hope Mills D-9
House C-11
Hubert D-11
Hudson C-5
Ingleside B-10
Jackson A-11
Jacksonville D-11
Jamesville B-12
Jarvisburg A-13
Jason C-11
Johns E-9
Jonas Ridge C-5
Jonesville B-6
Julian B-8
Justice B-10
Kannopolis C-7
Keener D-10
Kelley E-10
Kenansville D-11
Kenly C-10
Kernersville B-7
Kinston D-11
Kilkenny B-13
King B-7
Kings Mountain D-5
Kittrell B-10
Kuhns D-12
Kure Beach F-11
La Grange D-11
Lake Landing C-13
Lakeview D-9
Landis C-7
Lauada D-3
Laurinburg E-9
Lawrence B-11
Leasburg A-9
Leechville C-13
Lemon Springs C-8
Lenoir C-5
Level Cross B-8
Lewis A-9
Lexington C-7
Liberia B-10
Liberty B-8
Lilesville D-8
Lillington C-9
Lincolnton D-5
Linwood C-7
Lizzie C-11
Logan D-5
Long Beach F-10
Longview C-5
Louisburg B-10
Lucama C-11
Luck C-3
Lumber Bridge D-9
Lumberton E-9
Madison B-7
Magnolia D-10
Maco E-10
Maiden C-6
Mamie B-13
Manns Harbor B-13
Manteo B-14
Maple Hill E-11
Marble D-2
Marion C-5
Marshall C-3
Mars Hill C-4
Marshville D-7
Marston D-8
Matthews D-7
Mayodan B-8
Maysville D-12
Maxton E-9
Mesic C-12
Micro C-10
Middleburg A-10
Midway B-12
Midway Park E-11
Mineola C-12
Mint Hill D-7
Mocksville C-7
Moncure C-9
Monroe D-7
Monticello B-8
Mooresville C-7
Morehead City D-12
Morgans Corner A-13
Morganton C-5
Mountain Home D-4
Mount Airy A-7
Mount Gilead D-7
Mount Holly D-6
Mount Mourne C-6
Mount Olive D-11
Mulberry B-6
Murfreesboro A-11
Nags Head B-14
Nakina F-10
Nantahala D-2
Nashville B-10
New Bern D-12
New Holland C-13
Newport D-12
New Salem D-7
Newton C-6
Newton Grove D-10
Norlina A-10
Norman D-8
Northside B-9
North Wilkesboro B-6
Oakboro D-7
Oak Grove B-9
Ocracoke D-14
Old Dock F-10
Old Fort C-4
Olivia D-9
Osceola B-8
Oswalt C-6
Oteen C-4
Otto D-3
Oxford B-9
Pantego C-12
Patterson B-5
Peachtree D-2
Pembroke E-9
Penderlea E-10
Picks B-9
Pilot Mountain B-7
Pinebluff D-8
Pineola B-5
Pine Ridge A-6
Pinetops B-11
Pineville D-6
Pinnacle B-7
Pireway F-10
Pittsboro C-9
Pleasant Hill A-11
Plymouth B-12
Point Harbor B-13
Polkton D-7
Pollockville D-11
Portsmouth D-13
Postell D-1
Powells Point B-14
Raeford D-9
Rainbow Springs D-2
Raleigh B-10
Randleman C-8
Ranger D-1
Ransomville C-12
Rapids A-11
Red Hill B-4
Red Springs E-9
Reidsville B-8
Rheasville A-11
Rich Square B-11
Roanoke A-11
Robbinsville D-2
Robersonville B-11
Rockingham D-8
Rocky Mount B-11
Rodanthe B-14
Roduco A-12
Roper B-12
Roseboro D-10
Rose Hill D-10
Roseneath B-11
Rougemont B-9
Rowan E-10
Rowland E-9
Roxboro A-9
Ruth D-5
Rutherfordton D-4
Saint Pauls E-9
Salemburg D-10
Salisbury C-7
Salvo B-14
Sanford C-9
Schley B-9
Scotland Neck B-11
Scott's Hill E-11
Scranton C-13
Seabreeze F-11
Seagrove C-8
Selma C-10
Shallotte F-10
Sharpsburg B-11
Shelby D-5
Siler City C-8
Sioux B-4
Skyland C-4
Smith F-10
Smithfield C-10
Snow Hill C-11
Southern Pines D-9
Southport F-10
Sparta A-6
Spencer C-7
Spindale D-5
Spivey's Corner D-10
Spout Springs D-9
Spring Hill B-11
Spring Lake D-9
Spruce Pine C-5
Stanhope B-10
Stanleyville B-7
State Road B-6
Statesville C-6
Steeds C-8
Stocksville C-4
Stokes C-12
Stony Point C-6
Stovall A-10
Stratford A-6
Sunbury A-12
Supply F-10
Surl B-9
Surf City E-11
Swann C-9
Swannanoa C-4
Swansboro E-12
Swiss C-4
Sylva D-3
Tabor City E-9
Tapoco D-2
Tarboro B-11
Tar Heel E-9
Taylorsville C-6
Teachey D-10
Thomasville C-7
Tillery B-11
Tin City E-10
Tobaccoville B-7
Toluca C-5
Topsail Beach E-11
Topton D-2
Tramway C-9
Trenton D-11
Trent Woods D-12
Trust C-3
Tryon D-4
Turnersburg C-6
Tuxedo D-4
Twin Oaks A-6
Unaka D-2
Ulah C-8
Uwharrie C-8
Valdese C-5
Vanceboro C-11
Vander D-9
Violet D-1
Wadesboro E-7
Wadeville D-7
Wagram D-9
Wake Forest B-10
Wallace E-11
Wallburg B-7
Walnut C-4
Walnut Creek C-11
Wanchese B-14
Wards Corner E-10
Warsaw D-10
Waves B-14
Waynesville D-3
Weldon A-11
Wendell C-10
Wesser D-2
Westminster D-5
West Onslow Beach E-12
Whalebone B-14
White Lake E-10
White Oak E-10
Whiteville F-10
Wilbar B-6
Wilkesboro B-6
Willard E-10
Williamsburg B-8
Williamston B-12
Wilmar C-12
Wilmington F-10
Wilmington Beach F-11
Wilson C-11
Windsor B-12
Wingate D-7
Winnabow F-10
Winston-Salem B-7
Winterville C-11
Winton A-12
Wise A-10
Wise Fork D-11
Woodland B-11
Woodley B-13
Wrightsville Beach F-11
Yadkinville B-7
Yancyville A-8
Yeatesville C-12
Youngsville B-10
Zebulon B-10

NORTH DAKOTA
Pages 70-71

Abercrombie G-10
Adams C-9
Alamo B-2
Alexander C-2
Alice F-9
Almont F-5
Alsen B-8
Ambrose A-3
Amenia F-10
Amidon F-2
Anamoose D-6
Aneta D-9
Antler A-5
Ardoch C-9
Arena E-6
Argusville E-10
Arnegard C-2
Arthur E-10
Ashley G-7
Ayr E-9
Balfour D-6
Balta C-6
Bantry B-6
Barney G-10
Barton B-6
Bathgate B-9
Battleview B-3
Beach E-1
Belcourt B-7
Belfield E-2
Benedict D-5
Bergen C-6
Berlin G-8
Berthold C-5
Berwick C-6
Beulah E-4
Binford D-8
Bisbee B-7
Bismarck F-5
Bottineau B-6
Bowbells B-4
Bowdon D-7
Bowman G-2
Braddock F-6
Breien F-5
Bremen D-7
Brinsmade C-7
Brocket C-8
Bucyrus G-3
Buffalo F-9
Burlington C-5
Butte D-6
Buxton E-10
Calio B-8
Calvin B-8
Cando C-7
Carbury B-6
Carpio B-4
Carrington E-8
Carson F-4
Casselton F-9
Cathay D-7
Cavalier B-9
Cayuga G-9
Center E-4
Charlson C-3
Chaseley E-7
Churchs Ferry C-7
Cleveland F-7
Clifford E-10
Colehabor D-5
Colfax G-10
Columbus A-3
Conway C-9
Cooperstown E-9
Courtenay E-8
Crary D-8
Crosby A-2
Crystal B-9
Davenport F-9
Dawson F-6
Dazey E-9
Deering C-5
Des Lacs C-4
Devils Lake C-8
Dickey G-8
Dickinson E-3
Dodge E-4
Donnybrook B-4
Douglas D-5
Drake D-6
Drayton C-10
Driscoll F-6
Dunn Center D-3
Dunseith B-7
Dwight G-10
Edgeley G-8
Edinburg C-9
Edmore C-8
Egeland B-8
Elgin F-4
Ellendale G-8
Elliott G-9
Emerado D-9
Enderlin F-9
Epping C-2
Esmond C-7
Fairdale C-8
Fairfield E-3
Farimont H-10
Fargo F-10
Fessenden D-7
Fingal F-9
Finley E-9
Flasher F-5
Flaxton B-4
Flora D-7
Forbes G-7
Fordville C-9
Forman G-9
Fort Ransom G-9
Fortuna A-2
Forest River C-9
Fort Yates G-5
Freda G-5
Fredonia G-7
Fullerton G-8
Gackle F-7
Galesburg E-9
Gardena B-6
Garrison D-5
Gascoyne G-2
Gilby C-9
Gladstone E-3
Glenburn B-5
Glenfield E-8
Glen Ullin F-4
Golden Valley E-4
Golva E-1
Goodrich D-6
Grafton C-9
Grand Forks D-10
Grandin E-10
Grano B-4
Granville C-6
Grassy Butte D-3
Great Bend G-10
Grenora B-2
Gwinner G-9
Hague G-6
Haley G-2
Halliday D-4
Hamberg D-7
Hampden B-8
Hankinson H-10
Hanks B-2
Hannaford E-9
Hannah B-8
Harlow C-7
Harvey D-7
Harwood F-10
Hatton D-9
Havana H-9
Haynes G-3
Hazelton F-6
Hazen E-4
Hebron E-4
Heimdal D-7
Hensel B-9
Hettinger G-3
Hillsboro E-10
Hoopie C-9
Hope E-9
Hunter E-9
Hurdsfield E-6
Inkster C-9
Jamestown F-8
Joliette B-10
Jud F-8
Karlsruhe C-6
Kathryn F-9
Kenmare B-4
Kensal E-8
Kief D-6
Killdeer D-3
Kindred F-10
Knox C-7
Kramer B-6
Kulm G-8
Lakota D-8
La Moure G-8
Landa B-5
Langdon B-8
Lansford B-5
Larimore D-9
Larson A-3
Lawton C-8
Leeds C-7
Leal E-8
Lehr G-7
Leith G-4
Leonard F-9
Lidgerwood G-9
Lignite A-4
Lincoln F-5
Linton G-6
Lisbon G-9
Litchville F-8
Loma B-8
Loraine B-5
Ludden H-9
Maddock D-7
Makoti C-4
Mandan F-5
Manfred D-7
Manning E-3
Mantador G-10
Manvel D-9
Marion F-8
Marmarth F-2
Marshall E-4
Martin D-6
Max D-5
Maxbass B-5
Mayville E-9
Maza C-7
McClusky D-6
McHenry D-8
McVille D-9
Medina F-7
Medora E-2
Melville E-8
Mercer D-5
Merricourt G-7
Michigan D-9
Milnor G-9
Milton B-9
Minnewaukan D-7
Minot C-5
Minto C-9
Mohall B-5
Monango G-8
Montpelier F-8
Mooreton G-10
Mountain B-9
Mott F-3
Munich B-8
Mylo B-7
Napoleon F-7
Neche B-9
Nekoma B-8
Newburg B-6
New England F-3
New Leipzig F-4
New Rockford D-8
New Salem F-5
New Town C-4
Nome F-9
Noonan A-3
Northwood D-9
Oakes G-9
Oberon D-7
Omemee B-6
Oriska F-9
Osnabrock B-8
Overly B-6
Page E-9
Palermo B-4
Park River C-9
Parshall C-4
Pekin D-8
Pembina B-9
Perth B-7
Petersburg D-9
Pettibone E-7
Pick City D-4
Pillsbury E-9
Pingree E-8
Pisek C-9
Plaza C-4
Portal A-3
Portland E-10
Powers Lake B-3
Prairie Rose F-10
Raleigh G-5
Rawson C-2
Ray B-3
Reeder G-3
Regan E-6
Regent F-3
Rhame G-2
Richardton F-4
Riverdale D-5
Riverside F-10
Robinson E-7
Rock Lake B-7
Rogers E-9
Rolette B-7
Rolla B-7
Ross C-4
Rugby C-7
Ruso D-5
Rutland H-9
Ryder C-4
St. Thomas B-10
Sanborn F-8
Sarles B-8
Sawyer C-5
Scranton G-2
Selfridge G-5
Sentinel Butte E-2
Sharon D-9
Sheilds G-5
Sheldon F-9
Sherwood A-4
Sheyenne D-8
Sibley E-9
S. Heart E-3
Spring Brook C-2
Spiritwood F-8
Spiritwood Lake F-8
Solen G-5
Souris B-6
Stanley C-3
Stanton E-5
Starkweather C-8
Steele F-6
Strasburg G-6
Streeter F-7
Surrey C-5
Sydney F-8
Sykeston E-7
Tappen F-7
Thompson D-10
Tioga B-3
Tolley B-4
Tolna D-8
Tower City F-9
Towner C-6
Turtle Lake D-5
Tuttle E-6
Underwood D-5
Upham B-6
Valley City F-9
Velva C-5
Venturia G-7
Verendrye C-6
Verona G-9
Voltaire C-6
Wahpeton G-10
Walcott G-10
Wales B-8
Walhalla B-9
Walseth B-5
Warwick D-8
Washburn E-5
Watford City C-3
Wellsburg D-7
West Fargo F-10
Westfeild G-6
Westhope B-5
Wheelock C-3
White Earth B-3
Wildrose B-3
Williston C-2
Willow City B-6
Wilton E-5
Wimbledon E-8
Wing E-6
Wishek G-7
Wolford C-7
Woodworth E-7
Wyndmere G-9
York C-7
Ypsilanti F-8
Zap E-4
Zeeland G-6

OHIO
Pages 72-73

Abbeyville ★
Vicinity Pg. 107, J-1
Aberdeen N-3
Ada G-3
Addyston M-1
Adena J-9
Akron G-8
Albany M-7
Alexandria J-5
Alger H-3
Alliance G-9
Amanda L-5
Amberley ★
Vicinity Pg. 106, C-5
Amelia M-2
Amesville L-7
Amherst F-6
Amsterdam H-9
Andover E-10
Anna J-2
Ansonia J-1
Antwerp F-1
Apple Creek H-7
Arcadia F-4
Arcanum K-1
Archbold E-2
Arlington G-3
Ashland G-6
Ashley H-5
Ashtabula D-9
Ashville K-5
Athalia O-6
Athens M-7
Attica G-5
Aurora F-8
Avon Lake E-7
Bainbridge M-4
Baltic H-8
Baltimore K-6
Barberton G-8
Barnesville K-9
Barnsourg ★
Vicinity Pg. 106, B-2
Batavia M-2
Bath ★
Vicinity Pg. 107, K-4
Bath Ctr. ★
Vicinity Pg. 107, K-4
Bay View E-5
Beach City H-8
Beallsville K-9
Beavercreek K-3
Beaverdam G-3
Bedford ★
Vicinity Pg. 107, E-6
Bedford Heights ★
Vicinity Pg. 107, E-6
Beechwood ★
Vicinity Pg. 107, C-6
Bellaire K-10
Bellbrook L-2
Belle Center H-3
Belle Valley K-8
Bellefontaine H-3
Bellevue F-5
Bellville H-6
Belmont J-9
Beliot G-9
Belpre M-8
Belvis ★
Vicinity Pg. 106, A-2
Bennetts Corner ★
Vicinity Pg. 107, H-2
Berea F-7
Vicinity Pg. 107, F-1
Bergholz H-9
Berlin Heights F-6
Bethel N-2
Bethseda J-9
Bettsville F-4
Beverly L-8
Bexley J-5
Blanchester M-3
Bloomdale F-4
Bloomingburg L-4
Bloomville G-5
Blue Ash ★
Vicinity Pg. 107, B-6
Blufton G-3
Bolivar H-8
Botkins H-2
Botzum ★
Vicinity Pg. 107, K-5
Bowersville L-3
Bowling Green F-3
Bradford J-2
Bradner F-4
Brecksville ★
Vicinity Pg. 107, F-4
Bremen K-6
Brewster H-8
Bridgetown ★
Vicinity Pg. 106, D-1
Brilliant J-10
Broadview Heights ★
Vicinity Pg. 107, G-4
Brook Park ★
Vicinity Pg. 107, E-2
Brooklyn ★
Vicinity Pg. 107, D-3
Brunswick F-7
Vicinity Pg. 107, J-2
Bryan E-2
Buchtel L-7
Bucyrus G-5
Burton E-9
Butler H-6
Byesville K-8
Cadiz J-9
Cairo G-2
Caldwell K-8
Caledonia H-5
Cambridge J-8
Camden L-1
Campbell F-10
Canal Fulton G-8
Canal Winchester K-5
Canfield G-9
Canton G-8
Cardington H-5
Carey G-4
Carroll K-5
Carrollton H-9
Castalia F-5
Cedarville K-3
Celina H-1
Centerburg J-5
Centerville L-2
Chardon E-9
Chauncey L-7
Chesapeake O-6
Chesterhill L-7
Cheviot M-1
Vicinity Pg. 106, D-2
Chillicothe L-5
Christianburg J-3
Cincinnati M-2
Vicinity Pg. 106, H-2
Circleville L-5
Clarington K-9
Clarksville L-3
Cleveland E-7
Vicinity Pg. 107, C-5
Cleveland Heights E-8
Vicinity Pg. 107, C-6
Cleves L-1
Clyde F-5
Coal Grove O-6
Coalton M-6
Coldwater H-1
Columbiana G-10
Columbus J-4
Columbus Grove G-2
Conneaut D-10
Continental F-2
Convoy G-1
Coolville M-7
Corning L-7
Cortland F-10
Coshocton J-7
Covington J-2
Craig Beach G-9
Crestline G-5
Cridersville H-2
Crooksville K-7
Crown City O-6
Cuyahoga Falls F-8
Vicinity Pg. 107, K-7
Cygnet F-3
Dalton G-8
Danville H-6
Darbydale K-4
Dayton K-2
De Graff J-3
Delaware J-4
Delphos G-2
Delta E-3
Dennison H-9
Deshler F-3
Dover H-8
Dresden J-7
Dublin J-4
Dunkirk G-3
Dunlap ★
Vicinity Pg. 106, A-1
E. Canton G-9
E. Cleveland E-8
Vicinity Pg. 107, B-5
E. Liverpool H-10
E. Palestine G-10
E. Sparta H-8
Eastlake E-8
Eaton K-1
Edgerton E-1
Edison H-5
Edon E-1
Eldorado K-1
Elmore E-4
Elmwood Place ★
Vicinity Pg. 106, C-4
Elyria F-7
Emery ★
Vicinity Pg. 107, D-6
Englewood K-2
Enon K-3
Euclid E-8
Vicinity Pg. 107, A-6
Evendale ★
Vicinity Pg. 106, B-5
Everett ★
Vicinity Pg. 107, J-5
Fairfax ★
Vicinity Pg. 106, D-6
Fairfield L-1
Fairlawn F-8
Fairport Harbor D-9
Fairview Park ★
Vicinity Pg. 107, D-1
Fayette E-2
Felicity N-2
Findlay G-3
Flushing J-9
Forest G-4
Forest Park M-1
Vicinity Pg. 106, A-3
Fostoria F-4
Frankfort L-4
Franklin L-2
Frazeysburg J-7
Fredericktown H-6
Fredericksburg H-7
Freeport J-8
Fremont F-5
Ft. Jennings G-2
Ft. Loramie J-2
Ft. Recovery H-1
Ft. Shawnee H-2
Gahanna J-5
Galion H-5
Gallipolis N-7
Gambier J-9
Garfield Hts. F-8
Vicinity Pg. 107, D-5
Garrettsville F-9
Geneva D-9
Geneva-on-the-Lake D-9
Genoa E-4
Georgetown N-3
Germantown L-1
Gettysburg J-1
Ghent ★
Vicinity Pg. 107, K-4
Gibsonburg F-4
Girard F-10
Glandorf G-2

★ Denotes City located only on Vicinity map. City not located on State map.

OHIO

Glendale .... ★
  Vicinity ...... Pg. 106, A-4
Glouster L-7
Gnaddenhutten J-8
Golf Manor .... ★
  Vicinity ...... Pg. 106, C-5
Grafton F-7
Grand Rapids F-3
Granger .... ★
  Vicinity ...... Pg. 107, K-3
Granville J-6
Gratis L-1
Green Springs F-5
Greenfield L-4
Greenhills .... ★
  Vicinity ...... Pg. 106, A-3
Greenville J-1
Greenwich G-6
Grove City K-4
Groveport K-5
Hamden M-6
Hamersville N-3
Hamilton L-1
Hamler F-3
Hanover J-6
Hanoverton G-9
Harrison M-1
Harrod H-3
Hartville G-9
Hayesville G-7
Heath K-6
Hebron K-6
Hicksville N-3
Higginsport N-3
Hilliard J-4
Hillsboro M-3
Hinckley .... ★
  Vicinity ...... Pg. 107, H-2
Hiram F-9
Holgate F-2
Holland E-3
Hopedale J-9
Hubbard F-10
Huber Hts K-2
Hudson .... ★
  Vicinity ...... Pg. 107, F-8
Huron F-6
Independence ..... ★
  Vicinity ...... Pg. 107, F-4
Ironton O-6
Jackson M-5
Jackson Center ... H-3
Jacksonville L-7
Jamestown L-3
Jefferson E-9
Jeffersonville ... L-4
Jeromesville G-7
Jewett J-9
Johnstown J-6
Junction City K-6
Kalida G-2
Kellys Island E-6
Kent F-8
Kenton H-3
Kettering K-2
Killbuck H-7
Kingston L-5
Kirkersville K-6
Kirtland Hills ... E-9
Lagrange F-7
Lakemore G-8
Lakeview H-3
Lakewood .... ★
  Vicinity ...... Pg. 107, C-1
Lancaster L-5
Laurelville L-5
Lebanon L-2
Leesburg L-4
Leetonia G-10
Leipsic G-3
Lewisburg K-1
Lexington H-6
Liberty Center ... E-2
Lima G-2
Lincoln Heights .. ★
  Vicinity ...... Pg. 106, B-4
Lindsey F-5
Lisbon G-10
Lockland .... ★
  Vicinity ...... Pg. 106, B-4
Lodi G-7
Logan L-6
London K-4
Lorain E-6
Lordstown F-9
Loudonville H-6
Louisville G-9
Loveland M-2
Lowell L-8
Lowellville G-10
Luckey F-4
Lynchburg M-3
Lyndhurst .... ★
  Vicinity ...... Pg. 107, B-7
Lyons E-2
Macedonia F-8
  Vicinity ...... Pg. 107, G-6
Madei .... ★
  Vicinity ...... Pg. 106, C-6
Madeira M-2
Madison E-9
Magnolia H-8
Malta K-7
Malvern H-9
Manchester N-3
Mansfield H-6
Maple Hts .... ★
  Vicinity ...... Pg. 107, E-5
Marblehead E-5
Marietta L-8
Marimet .... ★
  Vicinity ...... Pg. 106, D-6
Marion H-4
Martins Ferry J-10
Martinsville L-3
Marysville J-4
Mason L-2
Massillon G-8
Maumee E-3
Mayfield Hts ..... ★
  Vicinity ...... Pg. 107, B-7
McArthur M-6
McClure F-3
McConnelsville ... L-7

McGuffey H-3
Mechanicsburg J-4
Medina F-7
  Vicinity ...... Pg. 107, K-1
Mentor E-8
Mentor-on-the-Lake E-8
Metamora E-3
Miamisburg L-2
Middle Point G-2
Middleburg Hts ... ★
  Vicinity ...... Pg. 107, F-2
Middleport M-7
Middletown L-1
Midvale H-8
Midway K-4
Milan F-6
Milford M-2
Milford Center ... J-4
Millersburg H-7
Millersport K-6
Millville L-1
Mineral City H-8
Minerva G-9
Minerva Park J-5
Mingo Junction ... J-10
Minster H-2
Monroe L-2
Monroeville F-5
Montpelier E-1
Moraine K-2
Moreland Hills ... ★
  Vicinity ...... Pg. 107, D-7
Morrow L-2
Mt. Blanchard G-4
Mt. Gilead H-5
Mt. Healthy ...... ★
  Vicinity ...... Pg. 106, B-3
Mt. Orab M-3
Mt. Sterling K-4
Mt. Vernon H-6
Mt. Victory H-4
Murray City L-7
Napoleon F-2
Navarre H-8
Nelsonville L-6
Nevada G-5
Newark J-6
New Boston N-5
New Bremen H-2
New Carlisle K-3
Newcomerstown ... J-8
New Concord K-7
New Holland L-4
New Knoxville H-2
New Lebanon K-1
New Lexington L-7
New London G-6
New Madison K-1
New Maramoras L-9
New Miami .... ★
New Middletown ... G-10
New Paris K-1
New Philadelphia . H-8
New Richmond N-2
New Straitsville . L-6
Newton Falls F-9
Newton M-2
New Vienna L-3
New Washington ... G-5
New Waterford G-10
Niles F-10
N. Baltimore F-3
N. Bend M-1
N. Canton G-8
N. College Hill .. ★
  Vicinity ...... Pg. 106, C-3
N. Fairfield F-6
N. Kingsville D-10
N. Lewisburg J-4
N. Olmsted .... ★
  Vicinity ...... Pg. 107, E-1
N. Ridgeville F-7
N. Royalton ...... ★
  Vicinity ...... Pg. 107, F-3
Northwood E-4
Norton G-8
Norwalk F-6
Norwood .... ★
  Vicinity ...... Pg. 106, D-4
Oak Harbor E-5
Oak Hill N-6
Oakwood K-2
Oakwood F-2
Obetz K-5
Ohio City G-1
Ontario H-6
Orange .... ★
  Vicinity ...... Pg. 107, D-7
Orangeville F-10
Oregon E-4
Orrville G-7
Orwell E-9
Ottawa G-3
Ottawa Hills E-3
Ottoville G-2
Owensville M-2
Oxford L-1
Painesville E-9
Pandora G-3
Parma .... ★
  Vicinity ...... Pg. 107, E-3
Parma Hts .... ★
  Vicinity ...... Pg. 107, E-2
Pataskala K-6
Paulding F-1
Payne G-1
Pemberville F-4
Peninsula .... ★
  Vicinity ...... Pg. 107, J-5
Pepper Pike ...... ★
  Vicinity ...... Pg. 107, C-7
Perry E-9
Perrysburg E-4
Perrysville H-6
Philo K-7
Pickerington K-5
Piketon M-5
Pioneer E-1
Piqua J-2
Plain City J-4
Pleasant Hill J-2
Pleasantville K-6
Plymouth G-6

Poland G-10
Pomeroy M-7
Port Clinton E-5
Portsmouth N-5
Port Washington . J-8
Powhatan Point ... K-9
Proctorville O-6
Prospect H-4
Put-in-Bay E-5
Quaker City K-8
Quincy J-3
Racine M-7
Ravenna F-9
Reading M-2
  Vicinity ...... Pg. 106, B-5
Remsen Corners ... ★
  Vicinity ...... Pg. 107, K-3
Republic F-5
Reynoldsburg K-5
Richfield .... ★
  Vicinity ...... Pg. 107, K-5
Richmond H-9
Richmond Hts ..... ★
  Vicinity ...... Pg. 107, A-6
Rio Grande N-6
Ripley N-3
Rising Sun F-4
Rittman G-7
Roaming Shores ... E-10
Rock Creek E-9
Rockford H-1
Rocky River ...... ★
  Vicinity ...... Pg. 107, C-1
Rome N-4
Roseville K-7
Rossford E-4
Rushsylvania H-3
Russells Point ... H-3
Rutland M-7
Sabina L-3
Sagamore Hills ... ★
  Vicinity ...... Pg. 107, G-5
St. Bernard ...... ★
  Vicinity ...... Pg. 106, D-4
St. Clairsville .. J-9
St. Henry H-1
St. Marys H-2
St. Paris J-3
Salem G-9
Salineville H-9
Sandusky E-6
Sarahsville K-8
Sardinia N-3
Scio H-9
Seaman N-4
Sebring G-9
Senecaville K-8
Seven Hills ...... ★
  Vicinity ...... Pg. 107, E-4
Seven Mile L-1
Seville G-7
Shadyside K-9
Shaker Hts .... ★
  Vicinity ...... Pg. 107, C-6
Sharonville ...... ★
  Vicinity ...... Pg. 106, B-4
Shawnee L-7
Sheffield Lake ... E-6
Shelby G-6
Sherwood F-1
Shiloh G-6
Shreve H-7
Sidney J-2
Silverton .... ★
  Vicinity ...... Pg. 106, C-5
Sinking Springs .. N-4
Smithfield J-10
Solon .... ★
  Vicinity ...... Pg. 107, E-7
Somerset K-6
S. Bloomfield K-5
S. Charleston K-3
S. Euclid .... ★
  Vicinity ...... Pg. 107, B-6
S. Lebannon L-2
S. Point O-6
S. Russell F-9
S. Webster N-5
S. Zanesville K-7
Spencer F-7
Spencerville G-2
Springboro L-2
Springdale .... ★
  Vicinity ...... Pg. 106, A-4
Springfield K-3
Spring Valley L-3
Steubenville H-10
Stockport L-7
Stoutsville L-5
Stow .... ★
  Vicinity ...... Pg. 107, K-7
Strasburg H-8
Streetsboro F-9
Strongsville ..... ★
  Vicinity ...... Pg. 107, G-1
Struthers G-10
Stryker E-2
Sugarcreek H-8
Sugar Grove L-6
Sunbury J-5
Swanton E-3
Sycamore G-4
Sylvania E-3
Syracruse M-7
Tallmadge G-8
Terrace Park M-2
Thornville K-6
Tiffin G-4
Tiltonsville J-10
Tipp City K-2
Toledo E-3
Toronto H-10
Trenton L-1
Trimble L-7
Trotwood K-2
Troy J-2
Tuscarawas H-8
Twinsburg .... ★
  Vicinity ...... Pg. 107, G-7
Uhrichsville H-8
Union K-2
Union City J-1
University Hts ... ★
  Vicinity ...... Pg. 107, C-6

Upper Arlington .. K-4
Upper Sandusky ... G-4
Urbana J-3
Urbancrest K-4
Utica J-6
Valley View ...... ★
  Vicinity ...... Pg. 107, E-5
Vandalia K-2
Van Wert G-1
Vermilion F-6
Versailles J-2
Vinton M-6
Wadsworth F-6
Wakeman E-4
Walbridge E-4
Wapokoneta H-2
Warren F-10
  Vicinity ...... Pg. 107, D-6
Warrensville Hts . ★
  Vicinity ...... Pg. 107, D-6
Warsaw J-7
Washington C.H. .. L-4
Washingtonville .. G-10
Waterville E-3
Wauseon E-3
Waverly M-5
Wayne F-4
Waynesburg G-9
Waynesfield H-3
Waynesville L-3
Wellington F-6
Wellston M-6
Wellsville H-9
W. Alexandria K-1
W. Carrollton K-2
W. Farmington F-9
W. Jefferson K-4
W. Lafayette J-8
W. Liberty J-3
W. Mansfield H-4
W. Milton K-2
W. Richfield ..... ★
  Vicinity ...... Pg. 107, H-4
W. Salem G-7
W. Union N-4
W. Unity E-2
Westerville J-5
Westfield Center . G-7
Weston F-3
Whitehall K-5
Whitehouse E-3
Wickliffe .... ★
  Vicinity ...... Pg. 107, A-7
Wilkesville M-6
Willard G-5
Williamsburg M-2
Williamsport L-4
Willoughby E-8
Willoughby Hills . E-8
Willowick E-8
Willshire H-1
Wilmington L-3
Winchester N-3
Windham F-9
Wintersville H-10
Woodlawn .... ★
  Vicinity ...... Pg. 106, B-4
Woodsfield K-9
Woodville F-4
Wooster G-7
Worthington J-5
Wyoming .... ★
  Vicinity ...... Pg. 106, B-4
Xenia L-3
Yellow Springs ... K-3
Yorkville J-10
Youngstown F-9
Zanesville K-7

OKLAHOMA

OKLAHOMA
Pages 74-75

Achille J-11
Acme G-8
Ada G-10
Adair C-13
Addington H-9
Afton C-13
Agra E-10
Agawam G-9
Ahloso G-11
Albany H-11
Albert F-8
Albion G-13
Alcorn C-9
Alderson G-12
Aledo E-7
Alex F-9
Alfalfa F-8
Aline D-8
Allen G-11
Alpers G-9
Altus G-6
Alva C-7
Amber F-9
Ames D-8
Amorita B-8
Anadarko F-8
Angora E-6
Antioch G-9
Apache F-8
Apperson C-10
Apple E-13
Aqua Park E-7
Arapaho E-7
Arcadia E-9
  Vicinity ...... Pg. 133, A-7
Ardmore H-10
Arnett D-6
Asher F-10
Ashland H-11
Atoka H-11
Atwood F-11
Avant C-11
Avard C-7
Avery E-10
Aydelotte F-7
Babbs D-7
Bache G-12
Bado F-12
Bald Hill E-12

Balko C-5
Barr D-8
Barnesdall C-11
Bartlesville C-11
Battiest G-13
Baum H-11
Bearden F-11
Beckett (Sunray) . G-9
Beggs E-11
Belva C-7
Bellvue E-11
Bengal G-13
Bennington H-11
Bernice C-13
Bessie E-7
Bethany .... ★
  Vicinity ...... Pg. 133, D-2
Bethel H-13
Bethel Acres F-10
Big Cabin C-13
Big Cedar G-13
Bigheart C-11
Billings C-9
Binger F-8
Bison D-9
Bixby D-12
Blackburn D-11
Blackwell C-10
Blair G-6
Blanchard F-9
Blocker F-12
Blue H-11
Boise City C-2
Boley E-11
Bomar J-10
Boswell H-11
Bowlegs F-11
Box E-13
Boyd C-4
Boynton E-12
Brace C-7
Braggs E-12
Braman C-9
Bray G-9
Briartown F-13
Bridgeport E-8
Brinkman E-6
Bristow E-11
Broken Arrow D-11
Broken Bow H-13
Bromide H-10
Brooksville F-10
Brown H-11
Brushy E-13
Bryant F-11
Bryan's Corner ... C-4
Buffalo C-6
Bugtussle E-13
Burbank C-10
Burlington C-8
Burmah E-7
Burns Flat F-7
Butler E-7
Byars G-10
Byng G-10
Byron C-8
Cabiniss F-11
Cache G-8
Caddo H-11
Calvin F-11
Camargo D-7
Cameron F-13
Canadian F-12
Canadian Club City F-9
Caney H-11
Canton D-8
Capron C-8
Carleton C-8
Carmen C-8
Carnegie F-8
Carney E-10
Carrier D-8
Carter F-6
Cartwright J-10
Cashion E-9
Castle E-11
Caston F-13
Catesby C-6
Catoosa D-12
Cedar Springs D-8
Cement F-8
Centrahoma G-11
Centralia C-12
Ceres D-10
Cestos D-7
Chandler E-10
Chattanooga G-7
Checotah E-12
Chelsea C-12
Cherokee C-8
Cheyenne E-6
Chester D-7
Chickasha F-8
Childers C-12
Chilocco B-10
Chisney F-10
Chloeta C-13
Chockie G-11
Choska E-12
Chouteau D-12
Cimmarron City ... E-9
Claremore C-12
Clayton G-12
Clemscott H-9
Cleo Springs D-8
Cleveland D-11
Clinton E-7
Cloud Chief F-7
Coalgate G-11
Cogar F-8
Colbert J-11
Cole F-9
Coleman H-11
Collinsville C-11
Colony F-7
Comanche H-9
Commerce B-13
Connerville G-11
Cookietown G-8
Cookson E-13
Cooperton G-7
Cora C-7
Cordell F-7
Corinne F-13

Corn E-7
Cornish H-9
Corum H-8
Council Hill E-12
Courtney H-9
Covington D-9
Cowden F-8
Coweta D-12
Cowlington F-13
Coyle E-10
Crawford E-6
Crekola E-12
Crescent E-9
Creta F-6
Criner F-9
Cromwell F-11
Crowder F-12
Curtis C-7
Cushing D-10
Custer City E-7
Cyril G-8
Dacoma C-8
Daisy G-12
Dale F-10
Darwin H-12
Davenport E-10
Davidson H-7
Davis G-10
Deer Creek C-9
Delaware C-12
Del City .... ★
  Vicinity ...... Pg. 133, F-5
Dempsey E-6
Denman F-13
Depew E-11
Devol H-7
Dewey C-12
Dibble F-9
Dickson H-10
Dill City F-7
Dillworth B-10
Dougherty G-10
Douglas D-9
Dover E-9
Doxey E-6
Drake H-10
Driftwood C-8
Drummond D-8
Drumright E-11
Duke G-6
Dunbar G-12
Duncan G-9
Durant H-11
Durham E-6
Durwood H-10
Dustin F-11
Eagle City D-8
Eakly F-8
Earlsboro F-10
Eason F-10
Eddy C-9
Edith C-7
Edmond .... ★
  Vicinity ...... Pg. 133, A-5
Eldon D-13
Eldorado G-6
Elgin G-8
Elk City E-6
Elmer G-6
Elmore City G-9
Elmwood C-5
El Reno E-8
Emet H-10
Empire City G-8
Enid D-9
Enterprise F-12
Erick E-6
Eubanks G-12
Eufaula F-12
Eva C-3
Fairfax C-10
Fairland C-13
Fairmont D-9
Fair Oaks D-11
Fallis E-10
Fallon H-12
Fanshawe F-13
Fargo D-6
Fairview D-8
Farris H-12
Faxon G-8
Fay E-7
Featherston F-12
Felt C-2
Fewell G-13
Fillmore H-11
Finley H-12
Fittstown G-10
Fitzhugh G-10
Fleetwood H-9
Fletcher G-8
Floris B-5
Folsom H-11
Foraker C-11
Forest Park ...... ★
  Vicinity ...... Pg. 133, D-5
Forgan B-5
Ft. Arbuckle G-9
Ft. Cobb F-8
Fort Gibson E-13
Ft. Supply C-6
Ft. Towson H-13
Forty One F-7
Foss E-7
Four Corners E-10
Fowler H-9
Fox H-9
Foyil C-12
Francis G-10
Frederick G-7
Freedom C-7
Gage D-6
Gans F-13
Gansel D-9
Garber D-9
Garvin H-13
Gate B-5
Gay H-12
Gene Autry H-10
Gerty F-11
Gibbon B-8
Gideon H-12
Glencoe D-10

Glenpool D-11
Glover H-13
Goldsby F-9
Goltry C-8
Goodland H-12
Goodwell C-3
Goodwin E-13
Gore E-13
Gotebo F-7
Gould G-6
Gowan G-12
Gracemont F-8
Grady H-9
Grainola B-11
Grandfield H-7
Granite F-6
Grant H-12
Gray D-6
Grayson E-12
Green D-13
Greenfield E-8
Griggs C-3
Grove C-13
Guthrie E-9
Guymon C-3
Haileyville G-12
Hall Park F-9
Hammon E-7
Hardy B-10
Harden City G-10
Hardesty C-4
Harmon D-6
Harrah E-10
Hartshorne G-12
Haskell E-12
Hastings H-8
Hayward D-9
Haywood F-12
Hayworth J-13
Headrick G-7
Healdton H-9
Heavener G-13
Hectorville E-12
Helena C-8
Hennepin G-9
Hennessey D-9
Henryetta E-11
Hickory G-10
Higgins G-12
Hillsdale C-8
Hinton E-8
Hitchcock D-8
Hitchita E-12
Hobart F-7
Hodgen G-13
Hoffman E-12
Holdenville F-11
Hollis G-6
Hollister H-7
Hollow C-12
Homestead D-8
Hominy D-11
Honobia G-13
Hooker B-4
Hoot Owl D-13
Hopeton C-7
Hough B-3
Howe F-13
Hucmac D-7
Hugo H-12
Hulah B-11
Hulbert D-13
Hulen G-8
Humphreys G-7
Hunter C-9
Huskey H-12
Hydro E-8
Idabel H-13
Indiahoma G-7
Indianola F-12
Ingersoll C-8
Inola D-12
Isabella D-8
Jay C-13
Jefferson C-9
Jennings D-11
Jenks D-11
Jesse G-11
Jet C-8
Joy G-10
Kansas D-13
Karma J-11
Karns E-8
Kellyville E-11
Kendrick E-10
Kenton B-2
Keota F-13
Keyes C-3
Kiamichi G-12
Kildare C-10
Kingfisher E-9
Kingston H-10
Kinta F-13
Kiowa G-11
Knowles C-5
Konawa F-10
Kosoma H-12
Kremlin C-9
Kusa E-12
Lacey D-8
Lahoma D-8
Lake Valley F-9
Lambert C-8
Lamont C-9
Lane H-11
Langston E-10
Latta G-10
Laverne C-6
Lawton G-8
Leach D-13
Lebanon H-10
Leedey E-6
Lehigh G-11
Lela D-10
Lenapah C-12
Leon H-9
Leonard E-12
Lequire F-13
Letitia E-13
Lewisville F-8
Lexington F-9
Lima F-10

Lincoln G-6
Lindsay G-9
Little Axe F-10
Little City H-10
Loco H-9
Locust Grove D-12
Lone Grove H-9
Lone Wolf F-7
Longdale D-8
Lookeba F-8
Lookout B-7
Lovedale C-6
Loveland H-7
Lovell D-9
Loyal D-8
Lucien D-9
Lugert F-7
Lula G-11
Luther E-10
Lyman C-6
Lyons E-13
Macomb F-10
Madill H-10
Manchester B-9
Mangum F-6
Manitou G-7
Mannsville H-10
Maramac D-10
Marietta H-10
Marland C-10
Marlow G-9
Marshall D-9
Martha G-6
Mason E-11
Maud F-10
Mayfield G-6
Maysville G-9
Mazie D-12
McAlester F-12
McBride H-13
McCurtain F-13
McLoud E-10
McQueen G-6
McWillie C-9
Mead H-10
Medford C-9
Meeker E-10
Meers G-8
Meno D-8
Meridian E-10
Messer H-12
Miami B-13
Middleberg F-9
Milburn H-11
Milfay E-11
Mill Creek G-10
Millerton H-13
Milo H-10
Milton F-13
Minco F-9
Mocane B-5
Moore .... ★
  Vicinity ...... Pg. 133, H-5
Mooreland C-7
Morris E-12
Morrison D-10
Mounds D-11
Mountain View F-7
Moyers H-12
Mulhall D-9
Muskogee E-12
Mustang E-9
Mutual D-7
Narcissa C-13
Nardin C-9
Nash C-8
Nashoba G-13
Navina E-9
Nebo H-12
New Alluwe C-12
Newcastle F-9
  Vicinity ...... Pg. 133, J-2
New Eucha D-13
Newkirk C-10
New Prue D-11
New Ringold H-13
New Tulsa D-11
New Woodville H-10
Nichols Hills .... ★
  Vicinity ...... Pg. 133, C-4
Nida H-11
Ninnekah F-9
Noble F-9
Norge F-9
Norman F-9
  Vicinity ...... Pg. 133, K-5
North Miami B-13
Nowata C-12
Numa C-9
Oakland H-10
Oaks D-13
Oakwood D-7
Ochelata C-12
Octavia G-13
Oilton D-11
Okeene D-8
Okemah E-11
Okesa C-11
Okfuskee E-11
Oklahoma City E-9
  Vicinity ...... Pg. 133
Okmulgee E-12
Oleta H-12
Olney G-11
Olustee G-6
Omega E-8
Onapa F-12
Oneta D-12
Oologah C-12
Optima C-4
Ord H-12
Orienta D-8
Orion D-7
Osage D-11
Overbrook H-10
Owasso D-12
Owen B-11
Ozark F-12
Paden E-11
Panama F-13
Panola F-13

★ Denotes City located only on Vicinity map. City not located on State map.

OKLAHOMA

Paoli ... G-10
Patton ... C-13
Pauls Valley ... G-10
Pawhuska ... C-11
Pawnee ... D-10
Payne ... G-9
Pearson ... F-10
Pearsonia ... C-11
Peggs ... D-13
Peoria ... B-13
Perkins ... D-10
Perry ... D-10
Pershing ... C-11
Petersburg ... H-9
Pharoah ... F-11
Phillips ... G-11
Picher ... B-13
Pierce ... F-10
Pink ... E-11
Pittsburg ... G-12
Plainview ... C-7
Platter ... H-11
Poccasset ... F-8
Ponca City ... C-10
Pond Creek ... C-9
Pontotoc ... G-11
Pooleville ... G-9
Porter ... E-12
Porum ... F-13
Poteau ... F-13
Prague ... E-10
Preston ... E-11
Proctor ... D-13
Pryor ... D-12
Pumpkin Center ... G-8
Purdy ... G-9
Purcell ... F-9
Putnam ... E-7
Pyramid Corners ... C-12
Quapaw ... B-13
Quay ... D-10
Quinlan ... C-7
Quinton ... F-12
Ralston ... C-10
Ramona ... C-11
Randlett ... H-8
Rankin ... E-6
Ratliff City ... G-9
Rattan ... H-12
Rayla ... H-10
Reagan ... H-10
Red Bird ... E-12
Redden ... H-12
Red Oak ... F-13
Red Rock ... D-10
Reed ... F-6
Remus ... F-10
Renfrow ... B-9
Retrop ... F-7
Reydon ... E-6
Rhea ... E-7
Richards Spur ... G-8
Ringling ... H-9
Ringold ... H-12
Ringwood ... D-8
Roberta ... H-11
Rocky ... F-7
Roff ... G-10
Roll ... E-6
Roosevelt ... G-7
Rose ... D-13
Rosedale ... G-10
Rosston ... C-6
Rubottom ... H-9
Rush Springs ... G-9
Ryan ... H-9
Saddle Mountain ... G-7
St. Louis ... F-10
Salem ... F-12
Salina ... D-13
Sallisaw ... E-13
Salt Fork ... C-9
Sand Springs ... D-11
Sapulpa ... D-11
Sasakwa ... F-10
Sayre ... F-6
Scraper ... D-13
Scullin ... G-10
Seiling ... D-7
Selman ... C-6
Seminole ... F-11
Sentinel ... F-7
Shamrock ... E-11
Sharon ... D-6
Shattuck ... D-6
Shawnee ... E-11
Shidler ... C-11
Skedee ... D-10
Skiatook ... D-11
Slapout ... C-6
Slaughterville ... F-10
Slick ... E-11
Smithville ... G-13
Snow ... G-12
Snyder ... G-7
Soper ... H-12
Southard ... D-8
South Coffeyville ... C-12
Sparks ... E-10
Spear ... H-12
Spencer ... ★
  Vicinity ... Pg. 133, D-6
Sperry ... D-11
Spiro ... F-13
Springer ... H-10
Stanley ... G-12
Steadman ... G-11
Stecker ... F-8
Sterling ... G-8
Stidham ... F-12
Stigler ... F-13
Stillwater ... D-10
Stilwell ... E-13
Stonewall ... F-11
Stratford ... G-10
Stringtown ... G-11
Strong City ... E-6
Stroud ... E-11
Stuart ... G-11
Sturgis ... B-3
Sugden ... H-9
Sulphur ... G-10
Summerfield ... G-13
Summit ... E-12

Sumner ... D-10
Sumpter ... C-10
Surrey Hills ... B-3
Sweetwater ... E-6
Sylvian ... F-11
Tabler ... F-9
Taft ... E-12
Tahlequah ... E-13
Talala ... C-12
Talihina ... G-13
Tallant ... C-11
Taloga ... D-7
Tangier ... C-6
Tatums ... G-9
Taylor ... H-8
Tecumseh ... F-10
Tegarden ... C-7
Temple ... H-8
Terlton ... D-11
Terral ... H-9
Texola ... F-6
The Village ... ★
  Vicinity ... Pg. 133, C-3
Thomas ... E-7
Three Sands ... C-10
Tiawah ... D-12
Tipton ... G-7
Tonkawa ... C-9
Trail ... D-7
Tribbey ... E-10
Trousdale ... E-10
Troy ... E-10
Tryon ... E-10
Tullahassee ... E-12
Tulsa ... D-11
Tupelo ... G-11
Turpin ... C-4
Tushka ... H-11
Tuskahoma ... G-13
Tuttle ... F-9
Tyrone ... B-4
Unger ... H-12
Union City ... F-8
Utica ... H-11
Valley Brook ... ★
  Vicinity ... Pg. 133, F-5
Valliant ... H-13
Vanoss ... G-10
Velma ... G-9
Vera ... C-12
Verdigris ... D-12
Vernon ... F-11
Vian ... E-13
Vici ... D-7
Vinita ... C-12
Vinson ... F-6
Wade ... H-11
Wadena ... H-12
Wagoner ... E-12
Wakita ... C-9
Wanette ... F-10
Wapanucka ... H-11
Warner ... E-12
Warr Acres ... ★
  Vicinity ... Pg. 133, C-2
Warren ... C-7
Warwick ... E-10
Waters ... H-8
Watonga ... E-8
Watova ... C-12
Watson ... G-13
Waukomis ... D-8
Waurika ... H-9
Wayne ... F-10
Waynoka ... C-7
Weatherford ... E-7
Webb City ... C-10
Webbers Falls ... E-12
Welch ... C-13
Weleetka ... F-11
Wellston ... E-11
Welty ... E-11
Westport ... D-11
Wetumka ... F-11
Wewoka ... F-11
Wheeless ... E-4
White Bead ... G-9
Whitefield ... F-12
White Oak ... C-8
Whitesboro ... G-13
Willis ... J-10
Willow ... H-6
Wilson ... H-9
Wolco ... C-11
Woodford ... H-10
Woodlawn Park ... ★
  Vicinity ... Pg. 133, D-2
Woodward ... D-7
Wyandotte ... C-13
Wynona ... C-11
Wynnewood ... G-10
Yale ... D-10
Yeager ... F-11
Yewed ... C-8
Yost ... D-10
Yuba ... J-11
Yukon ... E-9
Zeb ... E-13
Zena ... C-13

OREGON
Pages 76-77

Adair Village ... D-3
Adams ... C-9
Adel ... J-7
Adrian ... G-11
Agness ... H-1
Albany ... D-3
Algoma ... J-5
Alicel ... C-10
Allegany ... F-2
Alpine ... E-3
Alsea ... E-3
Amity ... D-3
Andrews ... J-9
Antelope ... D-6
Applegate ... J-2
Arlington ... C-7
Arock ... H-10
Ashland ... J-3

Ashwood ... D-6
Astoria ... A-3
Athena ... C-9
Aurora ... C-4
Austin ... E-9
Azalea ... H-3
Bandon ... G-1
Banks ... B-3
Barlow ... C-4
Bay City ... B-2
Beatty ... J-5
Beaver ... C-2
Beaverton ... ★
  Vicinity ... Pg. 139, E-2
Belknap Springs ... E-4
Bend ... E-5
Biggs ... C-6
Birkenfeld ... B-3
Blalock ... C-7
Blodgett ... D-3
Blue River ... E-4
Boardman ... B-8
Bonanza ... J-5
Bonneville ... C-5
Boring ... C-5
Breitenbush Springs ... D-5
Bridge ... G-2
Bridgeport ... E-10
Brightwood ... C-5
Broadbent ... G-2
Brogan ... F-10
Brookings ... J-1
Brooks ... D-4
Brothers ... F-6
Brownsville ... E-4
Burns ... G-8
Burns Jct. ... J-9
Butte Falls ... H-4
Buxton ... B-3
Camas Valley ... G-2
Camp Sherman ... E-5
Canby ... C-4
Cannon Beach ... B-2
Canyon City ... E-9
Canyonville ... H-2
Carlton ... C-3
Cascade Locks ... C-5
Cascadia ... E-4
Cave Junction ... J-2
Cecil ... C-7
Central Point ... J-3
Charleston ... G-1
Chemult ... G-5
Cheshire ... E-3
Chiloquin ... H-5
Christmas Valley ... G-6
Clackamas ... ★
  Vicinity ... Pg. 139, G-6
Clatskanie ... A-3
Cloverdale ... C-2
Coburg ... E-3
Colton ... C-4
Columbia City ... B-3
Condon ... D-7
Coos Bay ... F-1
Coquille ... G-2
Corvallis ... D-2
Cottage Grove ... F-3
Council Crest ... ★
  Vicinity ... Pg. 139, E-4
Cove ... D-10
Crabtree ... D-4
Crane ... G-9
Crawfordsville ... E-4
Crescent ... G-5
Crescent Lake ... G-4
Creswell ... F-3
Culp Creek ... F-3
Culver ... E-5
Curtin ... F-3
Cushman ... E-2
Dairy ... J-5
Dallas ... D-3
Days Creek ... H-2
Dayton ... C-3
Dayville ... E-8
Dee ... C-5
Deer Island ... B-3
Denmark ... G-1
Depoe Bay ... D-2
Detroit ... D-4
Dexter ... F-3
Diamond ... H-9
Diamond Lake ... G-4
Dillard ... G-2
Disston ... F-4
Donald ... C-4
Dorena ... F-4
Drain ... F-3
Drew ... H-3
Drewsey ... F-9
Dufur ... C-6
Dunes City ... F-2
Durkee ... E-10
Eagle Creek ... C-4
Eagle Point ... J-3
Eastside ... G-2
Echo ... C-8
Eddyville ... D-3
Elgin ... C-10
Elk City ... D-2
Elkton ... F-2
Elmira ... E-3
Enterprise ... C-11
Estacada ... C-5
Eugene ... E-3
Falls City ... D-3
Falls Creek ... F-3

Gates ... D-4
Gaylord ... H-2
Gearhart ... A-2
Gervais ... C-4
Gibbon ... C-9
Gilchrist ... G-5
Gladstone ... ★
  Vicinity ... Pg. 139, G-6
Glenada ... E-2
Glendale ... H-2
Gleneden Beach ... D-2
Glenwood ... B-3
Glide ... B-4
Goble ... B-4
Gold Beach ... H-1
Gold Hill ... H-3
Goshen ... F-3
Government Camp ... C-5
Grand Ronde ... C-3
Grangeville ... D-9
Granite ... D-9
Grants Pass ... H-2
Grass Valley ... C-6
Gravelford ... D-2
Greenhorn ... E-9
Gresham ... C-5
Haines ... D-10
Halfway ... D-11
Halsey ... E-3
Hamilton ... D-8
Hammond ... A-2
Hampton ... F-6
Harbor ... J-1
Hardman ... D-7
Harlan ... D-2
Harper ... F-10
Harrisburg ... E-3
Hebo ... C-2
Heceta Beach ... E-2
Helix ... C-8
Heppner ... C-8
Hereford ... E-10
Hermiston ... B-8
Hillsboro ... C-3
Hillsdale ... ★
  Vicinity ... Pg. 139, E-4
Hines ... G-8
Holley ... E-4
Homestead ... D-11
Hood River ... B-5
Horton Blachly ... E-3
Hoskins ... D-3
Hubbard ... C-4
Idanha ... D-5
Ideyld Park ... G-3
Imbler ... C-10
Imnaha ... C-11
Independence ... D-3
Ione ... B-8
Ironside ... B-8
Irrigon ... B-8
Island City ... C-10
Jacksonville ... J-3
Jamieson ... F-10
Jasper ... E-3
Jefferson ... D-3
Jewell ... B-3
John Day ... E-9
Jordan Valley ... H-10
Joseph ... D-11
Junction City ... E-3
Juntura ... F-9
Kamela ... C-9
Keizer ... C-3
Keno ... J-4
Kent ... C-6
Kerby ... J-2
Kernville ... D-2

Kimberly ... D-7
King City ... ★
  Vicinity ... Pg. 139, ★
Kings Valley ... D-3
Kinzua ... D-7
Klamatch Agency ... H-4
Klamath Falls ... J-4
Lacomb ... D-4
Lafayette ... C-3
La Grande ... D-9
Lakecreek ... J-3
Lake Grove ... ★
  Vicinity ... Pg. 139, G-5
Lakeside ... F-1
Lakeview ... J-7
Langlois ... G-1
LaPine ... G-5
Lawen ... G-9
Leaburg ... E-4
Lebanon ... D-4
Lexington ... C-7
Lincoln Beach ... D-2
Lincoln City ... D-2
Lonerock ... D-7
Long Creek ... E-8
Lorane ... F-3
Lostine ... C-11
Lowell ... F-4
Lyons ... D-4
Mabel ... E-3
Madras ... E-6
Malin ... K-5
Manning ... B-4
Manzanita ... B-2
Mapleton ... E-2
Marcola ... E-4
Marion ... D-4
Maupin ... C-6
Mayville ... D-7
McCredie Springs ... F-4
McKenzie Bridge ... E-5
McMinnville ... C-3
McNary ... B-8
Meacham ... C-9
Medford ... J-3
Medical Springs ... D-10
Merlin ... H-2
Merrill ... K-5
Metolius ... E-6
Metzger ... ★
  Vicinity ... Pg. 139, F-2
Midland ... J-5
Mikkalo ... C-7
Mill City ... D-4
Millersburg ... D-3
Millican ... F-6
Milo ... H-3
Milton Freewater ... B-10
Milwaukie ... ★
  Vicinity ... Pg. 139, F-5
Minam ... C-10
Mission ... C-9
Mist ... B-3
Mitchell ... E-7
Mohawk ... E-4
Mohler ... B-2
Molalla ... C-4
Monmouth ... D-3
Monroe ... E-3
Monument ... D-8
Moro ... C-6
Mosier ... C-6
Mt. Angel ... C-4
Mount Hood ... C-5
Mount Vernon ... E-8
Mulino ... C-4
Multnomah ... ★
  Vicinity ... Pg. 139, E-3
Murphy ... J-2
Myrtle Creek ... G-3
Myrtle Point ... G-1
Nehalem ... B-3
Neskowin ... C-2
Netarts ... C-2
Newberg ... C-3
New Bridge ... E-11
Newport ... D-2
New Princeton ... G-9
North Bend ... B-4
North Plains ... B-4
North Powder ... D-10
Norway ... G-1
Noti ... E-3
Nyssa ... F-11
Oak Grove ... ★
  Vicinity ... Pg. 139, G-5
Oakland ... G-2
Oakridge ... F-4
O'Brien ... J-2
Oceanside ... C-2
Odell ... C-5
Olene ... J-5
Ontario ... F-11
Ophir ... H-1
Oregon City ... C-4
Otis ... C-2
Otter Rock ... D-2
Owyhee ... F-11
Pacific City ... C-2
Paisley ... H-6
Parkdale ... C-5
Paulina ... F-7
Pendleton ... C-8
Philomath ... D-3
Phoenix ... J-3
Pilot Rock ... C-9
Pinehurst ... J-4
Pistol River ... J-1
Pleasant Hill ... F-4
Pleasant Valley ... E-10
Plush ... J-7
Portland ... ★
  Vicinity ... Pg. 139
Port Orford ... H-1
Post ... E-7
Powell Butte ... E-6
Powers ... H-2
Prairie City ... E-9
Prescott ... B-4
Prineville ... E-6
Prospect ... H-4
Rainier ... B-4
Raleigh Hills ... ★
  Vicinity ... Pg. 139, E-3

Redmond ... E-5
Reedsport ... F-2
Remote ... G-2
Rhododendron ... C-5
Richland ... E-11
Riddle ... G-2
Riley ... G-8
Ritter ... D-8
Riverton ... G-1
Rockaway ... B-2
Rogue River ... H-3
Rome ... H-10
Roseburg ... G-3
Rufus ... C-7
Saginaw ... F-3
St. Helens ... B-4
St. Paul ... C-3
Salem ... D-3
Scappoose ... B-3
Scio ... D-4
Scottsburg ... F-2
Seal Rock ... D-2
Seaside ... A-2
Selma ... J-2
Seneca ... E-8
Shady Cove ... H-3
Shaniko ... D-6
Sheaville ... H-11
Shedd ... E-3
Sheridan ... C-3
Sherwood ... D-7
Siletz ... D-2
Siltcoos ... F-2
Silver Lake ... G-6
Silverton ... D-4
Sisters ... E-5
Sitkum ... G-2
Sixes ... G-1
Skipanon ... A-2
Sodaville ... E-4
S. Beach ... D-2
Sprague River ... J-5
Spray ... D-7
Springfield ... E-3
Stanfield ... B-8
Stayton ... D-4
Sublimity ... D-4
Summer Lake ... H-6
Summerville ... C-10
Sumner ... G-2
Sumpter ... E-9
Sunriver ... F-5
Sutherlin ... G-3
Sweethome ... E-4
Swisshome ... E-2
Tangent ... D-3
Telocaset ... D-10
Tenmile ... G-2
Terrebonne ... E-5
The Dalles ... C-6
Tidewater ... E-2
Tiernan ... E-2
Tigard ... ★
  Vicinity ... Pg. 139, F-2
Tillamook ... C-3
Tiller ... H-3
Toledo ... D-2
Trail ... H-3
Troutdale ... C-5
Troy ... B-10
Tualatin ... C-4
Tygh Valley ... C-6
Umapine ... B-9
Umatilla ... B-8
Umpqua ... G-2
Union ... D-10
Union Creek ... H-3
Unity ... E-10
Vale ... F-10
Valley Falls ... J-6
Valsetz ... D-2
Veneta ... E-3
Vernonia ... B-3
Vida ... E-4
Waldport ... E-2
Walker ... F-3
Wallowa ... C-10
Walterville ... E-4
Walton ... E-3
Wamic ... C-6
Warm Springs ... D-5
Warren ... B-4
Wasco ... C-7
Waterloo ... E-4
Wedderburn ... H-1
Wemme ... C-5
Westfall ... F-10
West Fir ... F-4
West Linn ... C-4
West Slope ... ★
  Vicinity ... Pg. 139, ★
West Stayton ... D-3
Wheeler ... B-2
White City ... J-3
Whitney ... E-9
Wilbur ... G-3
Wilderville ... J-2
Williamina ... C-3
Williams ... J-3
Willow Creek ... F-10
Wilsonville ... C-4
Winchester ... G-3
Winchester Bay ... F-1
Winston ... H-2
Wolfcreek ... H-2
Wonder ... J-2
Woodburn ... C-3
Woodson ... A-3
Yachats ... E-2
Yamhill ... C-3
Yoncalla ... F-3
Zigzag ... C-5

PENNSYLVANIA
Pages 78-79

Abington ... ★
  Vicinity ... Pg. 137, A-7
Adamstown ... H-11
Addison ... K-4
Akron ... H-11
Alba ... C-9
Albion ... C-1
Alburtis ... G-12
Aliquippa ... G-2
Allentown ... G-13
Altoona ... G-6
Ambler ... H-13
Ambridge ... G-2
Andalusia ... ★
  Vicinity ... Pg. 137, B-9
Archbald ... D-12
Ardmore ... ★
  Vicinity ... Pg. 136, C-5
Arendtsville ... J-8
Armagh ... G-4
Ashley ... E-11
Ashville ... G-5
Aston Mills ... ★
  Vicinity ... Pg. 136, C-10
Athens ... C-10
Audubon ... ★
  Vicinity ... Pg. 136, A-3
Austin ... D-7
Avalon ... ★
  Vicinity ... Pg. 138, B-1
Avis ... E-8
Avonmore ... G-3
Baden ... G-2
Baldwin ... ★
  Vicinity ... Pg. 138, F-2
Bally ... H-12
Bangor ... F-13
Barnesboro ... G-5
Barren Hill ... ★
  Vicinity ... Pg. 136, A-4
Bath ... G-13
Battle's Cor. ... ★
  Vicinity ... Pg. 136, D-3
Beaver ... G-1
Beaver Falls ... F-1
Beavertown ... F-8
Bedford ... J-6
Beech Creek ... E-8
Bellefonte ... F-7
Belle Vernon ... J-2
Bellevue ... ★
  Vicinity ... Pg. 138, B-2
Bellwood ... G-6
Ben Avon ... ★
  Vicinity ... Pg. 138, A-1
Bendersville ... J-8
Benson ... H-5
Benton ... E-10
Berlin ... J-4
Berrysburg ... G-9
Berwick ... E-10
Bessemer ... F-1
Bethayres ... ★
  Vicinity ... Pg. 137, A-8
Bethel Park ... ★
  Vicinity ... Pg. 138, G-2
Bethlehem ... G-13
Big Beaver ... F-1
Big Run ... H-5
Birdsboro ... H-12
Blain ... H-7
Blairsville ... G-4
Black Horse ... ★
  Vicinity ... Pg. 136, E-3
Blakely ... D-12
Blooming Valley ... C-2
Bloomsburg ... F-10
Blossburg ... C-8
Bolivar ... H-4
Booth Cors. ... ★
  Vicinity ... Pg. 136, F-2
Boothwyn ... ★
  Vicinity ... Pg. 136, F-3
Boswell ... H-4
Bowmanstown ... F-12
Boyertown ... H-12
Brackenridge ... G-3
Bradford ... C-5
Brentwood ... ★
  Vicinity ... Pg. 138, F-2
Briar Creek ... E-11
Bridgeport ... ★
  Vicinity ... Pg. 136, A-4
Bridgeville ... ★
  Vicinity ... Pg. 138, F-1
Bristol ... J-14
Broad Top City ... H-6
Brockway ... E-5
Brookhaven ... ★
  Vicinity ... Pg. 136, F-3
Brookville ... E-4
Broomall ... ★
  Vicinity ... Pg. 136, ★
Brownstown ... H-11
Brownsville ... J-2
Bruin ... E-3
Bryn Athyn ... ★
  Vicinity ... Pg. 137, A-8
Bryn Mawr ... ★
  Vicinity ... Pg. 136, C-4
Burgettstown ... H-1
Burlington ... C-10
Burnham ... G-8
Butler ... F-3
California ... J-2
Callensburg ... F-3
Gallery ... F-2
Cambridge Springs ... D-2
Canonsburg ... H-2
Canton ... C-9
Carbondale ... D-12
Carlisle ... H-9
Carmichaels ... J-2
Carnegie ... ★
  Vicinity ... Pg. 138, D-2
Carrolltown ... G-5
Carroll Valley ... K-8
Cassandra ... H-5
Castle Shannon ... ★
  Vicinity ... Pg. 138, E-2
Catasauqua ... G-12

Catawissa ... F-10
Centerville ... C-3
Centerville ... J-2
Central City ... J-5
Centre Hall ... F-7
Chadds Ford ... ★
  Vicinity ... Pg. 136, D-5
Chalfont ... H-13
Chambersburg ... H-7
Charleroi ... H-2
Charlestown ... ★
  Vicinity ... Pg. 136, A-1
Chelsea ... ★
  Vicinity ... Pg. 136, F-2
Cheltenham ... ★
  Vicinity ... Pg. 137, B-7
Cherry Tree ... G-5
Chester ... J-13
Chester Heights ... J-12
Chester Hill ... F-6
Cheyney ... ★
  Vicinity ... Pg. 136, D-2
Chicora ... F-3
Clairton ... H-2
Clarendon ... C-4
Clarion ... E-4
Clarks Summit ... D-12
Clearfield ... F-6
Cleona ... H-10
Clifton Heights ... ★
  Vicinity ... Pg. 136, E-3
Clintonville ... E-3
Clymer ... G-4
Coalport ... G-5
Coatesville ... J-12
Cochranton ... D-2
Cokeburg ... H-2
Collegeville ... H-13
Colling Dale ... ★
  Vicinity ... Pg. 136, E-5
Columbia ... J-10
Concordville ... ★
  Vicinity ... Pg. 136, E-3
Conneautville ... C-1
Connellsville ... J-3
Connoquenessing ... F-2
Conshohocken ... ★
  Vicinity ... Pg. 136, B-4
Conway ... G-2
Conyngham ... F-11
Coopersburg ... G-13
Cooperstown ... D-3
Coraopolis ... G-2
Cornwall ... H-10
Cornwells Hts. ... ★
  Vicinity ... Pg. 137, B-9
Corry ... C-3
Corsica ... E-4
Coudersport ... C-7
Crafton ... ★
  Vicinity ... Pg. 138, D-2
Cranesville ... C-1
Creekside ... G-4
Cresson ... G-5
Cressona ... G-11
Croydon ... ★
  Vicinity ... Pg. 137, B-10
Crozierville ... ★
  Vicinity ... Pg. 136, E-3
Curwensville ... F-6
Dalesford ... ★
  Vicinity ... Pg. 136, D-3
Dallastown ... J-10
Dalton ... D-12
Danville ... F-10
Darby ... J-13
  Vicinity ... Pg. 136, E-5
Darling ... ★
  Vicinity ... Pg. 136, E-2
Darlington Cors. ... ★
  Vicinity ... Pg. 136, D-1
Dauphin ... H-9
Dayton ... F-4
Delaware Water Gap ... F-13
Delta ... K-11
Denver ... H-11
Derry ... H-4
Devon ... ★
  Vicinity ... Pg. 136, D-3
Devoult ... ★
  Vicinity ... Pg. 136, B-1
Dickson City ... D-12
Dillsburg ... J-9
Dilworthton ... ★
  Vicinity ... Pg. 136, D-3
Donegal ... J-4
Donora ... H-2
Dormont ... ★
  Vicinity ... Pg. 138, E-2
Dover ... J-9
Downingtown ... J-12
Doylestown ... H-13
Drexel Hill ... ★
  Vicinity ... Pg. 136, E-5
Driftwood ... E-6
DuBois ... E-5
Duboistown ... E-9
Dudley ... H-6
Dunbar ... J-3
Duncannon ... H-9
Duncansville ... H-6
Dunmore ... D-12
Dushore ... D-10
Eagles Mere ... D-10
Eagleville ... ★
  Vicinity ... Pg. 136, A-3
E. Berlin ... J-9
E. Brady ... F-3
E. Butler ... F-3
E. Greenville ... H-12
E. Lansdowne ... ★
  Vicinity ... Pg. 136, D-5
Easton ... G-13
E. Stroudsburg ... F-13
E. Washington ... H-2
Eau Claire ... E-3
Ebensburg ... G-5
Economy ... G-2
Eddystone ... ★
  Vicinity ... Pg. 136, F-4
Edge Hill ... ★
  Vicinity ... Pg. 137, A-6
Edgely ... ★
  Vicinity ... Pg. 137, A-11

★ Denotes City located only on Vicinity map. City not located on State map.

PENNSYLVANIA

SOUTH DAKOTA

PENNSYLVANIA

Edgemont ... ★
  Vicinity ... Pg. 136, D-3
Edinboro ... C-2
Edwardsville ... E-11
Elam ... ★
  Vicinity ... Pg. 136, F-1
Eldred ... C-6
Elizabethtown ... H-10
Elizabethville ... G-10
Elkins Park ... ★
  Vicinity ... Pg. 137, B-7
Elkland ... C-8
Ellport ... F-2
Ellwood City ... F-2
Elwyn ... ★
  Vicinity ... Pg. 136, E-3
Emlenton ... D-3
Emmaus ... G-12
Emporium ... D-6
Ephrata ... H-11
Erie ... B-2
Etna ... ★
  Vicinity ... Pg. 138, B-5
Evans City ... F-2
Everett ... J-6
Everson ... J-3
Factoryville ... D-12
Fairchance ... J-3
Fairfield ... K-8
Fairview ... B-2
Falls Creek ... E-5
Farrell ... E-1
Faust Cor. ... ★
  Vicinity ... Pg. 136, A-2
Fawn Grove ... K-10
Feasterville ... ★
  Vicinity ... Pg. 137, A-9
Fitzwatertown ... ★
  Vicinity ... Pg. 137, A-7
Five Points ... ★
  Vicinity ... Pg. 137, A-6
Fleetwood ... G-12
Flemington ... E-8
Florida Park ... ★
  Vicinity ... Pg. 136, D-3
Ford City ... F-3
Ford Cliff ... F-3
Forest City ... D-12
Forksville ... D-10
Fort Washington ... ★
  Vicinity ... Pg. 136, A-6
Forty Fort ... E-12
Fountain Hill ... G-13
Frackville ... F-11
Franklin ... D-3
Franklin Park ... G-2
Freeburg ... G-9
Freeland ... F-12
Freemansburg ... G-13
Freeport ... G-3
Galeton ... C-8
Gallitzin ... G-5
Garden City ... ★
  Vicinity ... Pg. 136, F-4
Garrett ... J-4
Geistown ... H-5
General Greene ... ★
  Vicinity ... Pg. 136, C-1
Gettysburg ... K-9
Girard ... B-2
Gladwyne ... ★
  Vicinity ... Pg. 136, C-5
Glen Campbell ... F-5
Glen Hope ... F-6
Glen Mills ... ★
  Vicinity ... Pg. 136, E-2
Glenolden ... ★
  Vicinity ... Pg. 136, E-5
Glen Riddle ... ★
  Vicinity ... Pg. 136, E-3
Glen Rock ... K-10
Glenside ... ★
  Vicinity ... Pg. 137, A-8
Gordon ... G-11
Goshenville ... ★
  Vicinity ... Pg. 136, C-1
Gradyville ... ★
  Vicinity ... Pg. 136, D-2
Grampian ... F-5
Gratz ... G-10
Great Bend ... C-12
Greencastle ... K-7
Greensboro ... K-2
Greensburg ... H-3
Greentree ... ★
  Vicinity ... Pg. 138, E-2
Greenville ... D-1
Grenloch ... ★
  Vicinity ... Pg. 136, C-1
Grove City ... E-2
Gulph Mills ... ★
  Vicinity ... Pg. 136, B-4
Halifax ... G-9
Hallstead ... C-12
Hamburg ... G-11
Hanover ... K-9
Harmony ... F-2
Harrisburg ... H-9
Harrisville ... E-2
Harveys Lake ... D-11
Hastings ... G-5
Hatboro ... H-13
Haverford ... ★
  Vicinity ... Pg. 136, C-5
Havertown ... ★
  Vicinity ... Pg. 136, D-4
Hawley ... D-13
Hawthorn ... F-4
Hazleton ... F-11
Heidelberg ... ★
  Vicinity ... Pg. 138, E-1
Herndon ... G-9
Hershey ... H-10
Hollidaysburg ... H-6
Homer City ... G-4
Homestead ... ★
  Vicinity ... Pg. 138, E-6
Honesdale ... D-13
Honey Brook ... J-12
Hooversville ... H-5
Hop Bottom ... C-12
Hopewell ... H-7
Houston ... H-2
Houtzdale ... F-6

Howeville ... ★
  Vicinity ... Pg. 136, B-2
Hughesville ... E-10
Hummelstown ... H-10
Huntingdon ... G-7
Hydetown ... C-3
Hyndman ... K-5
Indiana ... G-4
Indian Lake ... J-5
Industry ... G-1
Ingram ... ★
  Vicinity ... Pg. 138, D-1
Irvona ... F-5
Ithan ... ★
  Vicinity ... Pg. 136, C-4
Jackson Center ... E-2
Jacobus ... J-10
Jamestown ... H-3
Jeanette ... ★
Jeddo ... H-2
Jefferson ... H-2
Jeffersonville ... ★
  Vicinity ... Pg. 136, A-4
Jennerstown ... H-4
Jermyn ... D-12
Jersey Shore ... E-9
Jim Thorpe ... F-12
Johnsonburg ... D-5
Johnson's Cors. ... ★
  Vicinity ... Pg. 136, F-1
Johnstown ... H-5
Jonestown ... H-10
Kane ... C-5
Karns City ... F-3
Kennett Square ... J-12
Kimberton ... ★
  Vicinity ... Pg. 136, A-1
King of Prussia ... ★
  Vicinity ... Pg. 136, B-3
Kingston ... E-12
Kistler ... H-7
Kittanning ... F-3
Knox ... E-3
Knoxville ... C-8
Kulpmont ... F-10
Kutztown ... G-12
Lake City ... B-2
Lake Pine ... ★
  Vicinity ... Pg. 137, F-11
Lancaster ... J-11
Landisburg ... H-8
Lanesboro ... C-12
Langhorne ... H-14
Lansdale ... H-13
Lansdowne ... ★
  Vicinity ... Pg. 136, D-5
Lansford ... F-12
Laporte ... D-10
Latrobe ... H-4
Laureldale ... H-12
Lebanon ... H-10
Leesport ... G-11
Lehighton ... F-12
Lemoyne ... H-9
Lenni ... ★
  Vicinity ... Pg. 136, E-3
Leopard ... ★
  Vicinity ... Pg. 136, C-3
Le Raysville ... C-11
Levittown ... F-9
Lewisburg ... G-8
Lewistown ... G-8
Liberty ... D-9
Ligonier ... H-4
Lilly ... H-5
Lima ... ★
  Vicinity ... Pg. 136, E-3
Lincoln ... H-11
Linesville ... C-1
Linwood ... ★
  Vicinity ... Pg. 136, F-3
Lititz ... J-11
Little Meadows ... C-11
Littlestown ... K-9
Liverpool ... G-9
Llanerch ... ★
  Vicinity ... Pg. 136, D-5
Locksley ... ★
  Vicinity ... Pg. 136, D-2
Lock Haven ... E-8
Logantown ... E-8
Loretto ... G-5
Lower Burrell ... F-5
Lumber City ... F-5
Lykens ... G-10
McKean ... B-2
McKeesport ... H-2
Mckee's Rocks ... ★
  Vicinity ... Pg. 138, C-2
McSherrytown ... K-9
McVeytown ... G-7
Macungie ... G-12
Mahanoy City ... F-11
Manchester ... J-10
Manheim ... J-11
Manns Choice ... J-5
Manoa ... ★
  Vicinity ... Pg. 136, D-4
Mansfield ... C-9
Marcus Hook ... ★
  Vicinity ... Pg. 136, F-3
Marianna ... J-2
Marietta ... J-10
Marion Center ... F-4
Markham ... ★
  Vicinity ... Pg. 136, E-2
Marklesburg ... H-6
Marpel ... ★
  Vicinity ... Pg. 136, D-4
Mars ... G-2
Martinsburg ... H-6
Marysville ... H-9
Masontown ... J-2
Matamoras ... D-14
Meadville ... C-2
Mechanicsburg ... H-9
Media ... ★
  Vicinity ... Pg. 136, E-3
Melrose Park ... ★
  Vicinity ... Pg. 137, B-7
Mercer ... E-2
Mercersburg ... K-7
Meshoppen ... D-11
Meyersdale ... K-4

Middleburg ... G-9
Middletown ... H-10
  Vicinity ... Pg. 137, A-10
Midland ... G-1
Mifflinburg ... F-9
Mifflinton ... G-8
Milbourne ... ★
  Vicinity ... Pg. 136, D-5
Milesburg ... F-7
Milford ... E-14
Mill Creek ... H-7
Millersburg ... G-9
Millerstown ... G-9
Millersville ... J-10
Mill Hall ... E-8
Millheim ... F-8
Mill Village ... C-2
Milltown ... ★
  Vicinity ... Pg. 136, D-1
Milton ... F-9
Minersville ... G-11
Mohnton ... H-11
Monaca ... G-2
Monessen ... H-2
Monroe ... C-10
Monroeville ... G-3
  Vicinity ... Pg. 138
Mont Alto ... J-8
Mont Clare ... ★
  Vicinity ... Pg. 136, A-2
Montoursville ... E-9
Montrose ... C-12
Moosic ... D-12
Morrisville ... H-14
Morton ... ★
  Vicinity ... Pg. 136, E-4
Moscow ... E-12
Mt. Carmel ... F-10
Mt. Holly Springs ... J-9
Mt. Jewett ... C-5
Mt. Joy ... J-10
Mt. Lebanon ... ★
  Vicinity ... Pg. 138, E-2
Mt. Oliver ... ★
  Vicinity ... Pg. 138, E-4
Mt. Pleasant ... H-3
Mt. Pocono ... E-13
Mt. Union ... H-7
Mountville ... J-10
Mt. Wolf ... J-10
Murrysville ... G-3
Myerstown ... H-11
Nanticoke ... E-11
Nanty-Glo ... G-5
Narberth ... J-13
  Vicinity ... Pg. 136, D-5
Nazareth ... G-13
Nescopeck ... F-11
New Albany ... D-10
New Alexandria ... H-3
New Beaver ... F-1
New Berlin ... F-9
New Bethlehem ... F-4
New Brighton ... F-1
New Buffalo ... G-9
Newburg ... F-5
Newburg ... J-8
New Castle ... F-1
New Centerville ... J-4
  Vicinity ... Pg. 136, B-3
New Cumberland ... H-9
New Eagle ... H-2
New Florence ... H-4
New Freedom ... K-10
New Holland ... J-11
New Kensington ... G-3
New Lebanon ... D-2
New Milford ... C-12
New Oxford ... J-9
New Paris ... J-5
New Philadelphia ... G-11
Newport ... ★
Newportville ... ★
  Vicinity ... Pg. 137, A-10
Newry ... H-6
New Salem ... J-2
New Stanton ... H-3
Newton Square ... ★
  Vicinity ... Pg. 136, D-3
Newville ... H-8
New Wilmington ... E-1
Nicholson ... D-12
Norristown ... J-13
  Vicinity ... Pg. 136, A-4
Northampton ... G-12
N. Apollo ... G-3
N. Belle Vernon ... H-3
N. East ... B-3
Northumberland ... F-10
N. Wales ... H-13
N. York ... J-10
Norwood ... ★
  Vicinity ... Pg. 136, E-5
Nuangola ... E-11
Oakford ... ★
  Vicinity ... Pg. 137, A-9
Oakland ... C-12
Oakmont ... G-3
Oaks ... ★
  Vicinity ... Pg. 136, D-4
Ogden ... ★
  Vicinity ... Pg. 136, F-2
Ogontz ... ★
  Vicinity ... Pg. 137, B-7
Ohiopyle ... J-3
Oil City ... D-3
Old Forge ... D-12
Olyphant ... D-12
Orangeville ... E-10
Orbisonia ... H-7
Orrstown ... J-8
Orwigsburg ... G-11
Osceola Mills ... F-6
Oswego ... C-7
Oxford ... K-11
Painters Crossroads ... ★
  Vicinity ... Pg. 136, E-1
Palmerton ... F-12
Palmyra ... H-10
Paoli ... ★
  Vicinity ... Pg. 136, C-2
Parker ... E-3
Parkesburg ... J-11
Parkside ... ★
  Vicinity ... Pg. 136, E-1

Patton ... G-5
Paxtang ... G-9
Pen Argyl ... F-13
Pennsburg ... H-12
Penn Valley ... ★
  Vicinity ... Pg. 136, C-5
Penn Wynne ... ★
  Vicinity ... Pg. 136, D-5
Perkasie ... H-13
Perkiomen ... ★
  Vicinity ... Pg. 136, A-2
Petersburg ... G-7
Petrolia ... F-3
Philadelphia ... J-13
  Vicinity ... Pg. 136 & 137
Philipsburg ... F-6
Philmont ... ★
  Vicinity ... Pg. 137, A-8
Phoenixville ... H-12
Pickering ... ★
  Vicinity ... Pg. 136, A-2
Picture Rocks ... E-10
Pillow ... G-9
Pine Grove ... G-10
Pittsburgh ... G-2
  Vicinity ... Pg. 138
Pittston ... E-12
Planebrook ... ★
  Vicinity ... Pg. 136, B-1
Platea ... B-2
Pleasantville ... H-5
Pleasantville ... D-3
Plum ... G-3
Plumville ... F-4
Plymouth Meeting ... ★
  Vicinity ... Pg. 136, A-5
Point Marion ... K-2
Polk ... D-2
Portage ... H-5
Port Allegany ... C-6
Port Carbon ... G-11
Portersville ... F-2
Port Kennedy ... ★
  Vicinity ... Pg. 136, A-3
Portland ... F-13
Port Matilda ... F-7
Port Royal ... G-8
Pottstown ... H-12
Pottsville ... G-11
Prospect Park ... ★
  Vicinity ... Pg. 136, E-4
Punxsutawney ... F-4
Quakertown ... H-13
Quarryville ... J-11
Radnor ... ★
  Vicinity ... Pg. 136, C-4
Rainsburg ... J-5
Ramey ... F-6
Reading ... H-11
Red Hill ... H-13
Red Lion ... J-10
Renovo ... E-7
Reynoldsville ... E-5
Rices Landing ... J-2
Richland ... H-11
Richlandtown ... G-13
Ridgway ... D-5
Ridley Park ... ★
  Vicinity ... Pg. 136, E-4
Rimersburg ... E-3
Ringtown ... F-11
Riverside ... F-10
Roaring Spring ... H-6
Robesonia ... H-11
Rockdale ... ★
  Vicinity ... Pg. 136, A-3
Rockledge ... ★
  Vicinity ... Pg. 137, B-7
Rockwood ... J-4
Rocky Hill ... ★
  Vicinity ... Pg. 136, D-1
Rome ... C-11
Roseto ... F-13
Rose Tree ... ★
  Vicinity ... Pg. 136, D-3
Rose Valley ... ★
  Vicinity ... Pg. 136, E-3
Roseville ... C-9
Roslyn ... ★
  Vicinity ... Pg. 137, A-7
Rosslyn Farms ... ★
  Vicinity ... Pg. 138, D-1
Rouseville ... D-3
Royersford ... H-12
Rural Valley ... F-4
Rutledge ... ★
  Vicinity ... Pg. 136, B-2
Saegertown ... C-2
St. Clair ... G-11
St. Davids ... ★
  Vicinity ... Pg. 136, C-4
St. Marys ... D-5
St. Petersburg ... E-3
Salisbury ... K-4
Salladasburg ... E-9
Saltillo ... H-7
Saltsburg ... G-3
Sandy Lake ... D-2
Sankertown ... G-5
Saxonburg ... F-3
Saxton ... H-6
Sayre ... C-10
Scalp Level ... H-5
Schuylkill Haven ... G-11
Scottdale ... H-3
Scranton ... D-12
Selinsgrove ... F-9
Sellersville ... H-13
Seward ... H-4
Shade Gap ... H-7
Shamokin ... F-10
Shamokin Dam ... F-9
Sharon ... E-1
Sharon Hill ... ★
  Vicinity ... Pg. 136, E-5
Sharpsburg ... ★
  Vicinity ... Pg. 138, B-5
Shenandoah ... F-11
Shickshinny ... E-11
Shillington ... H-11
Shinglehouse ... C-6
Shippensburg ... J-8
Shippenville ... E-3
Shirleysburg ... H-7

Shoemakersville ... G-11
Shrewsbury ... K-10
Slatington ... F-12
Sligo ... E-3
Slippery Rock ... E-2
Smethport ... C-6
Smithfield ... K-3
Snow Shoe ... F-7
Somerset ... J-4
Souderton ... H-13
S. Bethlehem ... F-4
S. Coatesville ... J-12
S. Connellsville ... J-3
S. Fork ... H-5
S. Renovo ... E-7
S. Waverly ... B-10
S. Williamsport ... E-9
Spangler ... G-5
Spartansburg ... C-3
Springboro ... C-1
Spring City ... H-12
Spring Grove ... J-9
State College ... F-7
Stillwater ... E-10
Stoneboro ... D-2
Stoystown ... H-5
Strasburg ... J-11
Strattanville ... E-4
Strausstown ... G-11
Stroudsburg ... F-13
Sugarcreek ... D-3
Sugar Grove ... B-4
Sugartown ... ★
  Vicinity ... Pg. 136, C-2
Summerhill ... H-5
Summerville ... E-4
Summit Hill ... F-12
Sunbury ... F-10
Susquehanna ... C-12
Swarthmore ... ★
  Vicinity ... Pg. 136, E-4
Swedeland ... ★
  Vicinity ... Pg. 136, B-4
Swedesburg ... ★
  Vicinity ... Pg. 136, A-4
Sykesville ... F-5
Sylvania ... B-9
Tamaqua ... F-12
Tanguy ... ★
  Vicinity ... Pg. 136, D-2
Tatamy ... G-13
Taylor ... D-12
Telford ... H-13
Temple ... H-11
Terre Hill ... H-11
Thompsontown ... G-9
Thornburg ... ★
  Vicinity ... Pg. 138, D-1
Thornton ... ★
  Vicinity ... Pg. 136, E-1
Three Springs ... H-7
Tidioute ... C-4
Tioga ... C-9
Tionesta ... D-3
Titusville ... D-3
Topton ... G-12
Towanda ... C-10
Tower City ... G-10
Townville ... C-3
Trainer ... ★
  Vicinity ... Pg. 136, F-3
Trappe ... H-12
Tremont ... G-10
Trevose ... ★
  Vicinity ... Pg. 137, A-9
Trooper ... ★
  Vicinity ... Pg. 136, A-3
Troutville ... F-5
Troy ... C-10
Tullytown ... H-14
Tunkhannock ... D-11
Twin Oaks ... ★
  Vicinity ... Pg. 136, F-3
Tyrone ... G-6
Union City ... C-3
Union Dale ... C-11
Uniontown ... J-3
Upper Darby ... ★
  Vicinity ... Pg. 136, D-5
Ursina ... K-4
Valley Forge ... ★
  Vicinity ... Pg. 136, B-2
Valleystore ... ★
  Vicinity ... Pg. 136, B-2
Vandergrift ... G-3
Venango ... C-2
Village Green ... ★
  Vicinity ... Pg. 136, F-3
Villanova ... ★
  Vicinity ... Pg. 136, C-4
Walnutport ... G-12
Wampum ... F-1
Ward ... ★
  Vicinity ... Pg. 136, C-4
Warren ... C-4
Washington ... H-2
Waterford ... C-2
Watsontown ... E-9
Wattsburg ... B-3
Waymart ... D-13
Wayne ... ★
  Vicinity ... Pg. 136, C-3
Waynesboro ... K-8
Waynesburg ... J-2
Weatherly ... F-12
Wellersburg ... K-5
Wellsboro ... C-9
Wellsville ... J-9
Wesleyville ... B-2
W. Brownsville ... J-2
W. Chester ... J-12
W. Conshohocken ... ★
  Vicinity ... Pg. 136, B-4
W. Easton ... G-13
Westfield ... C-8
W. Grove ... K-12
W. Hazleton ... F-11
W. Homestead ... ★
  Vicinity ... Pg. 138, E-5
W. Kittanning ... F-3
W. Liberty ... F-2
W. Middlesex ... E-1

W. Mifflin ... G-3
  Vicinity ... Pg. 138, G-5
Westmont ... H-5
W. Newton ... H-3
Westover ... G-5
W. Pittston ... E-12
W. Reading ... H-11
West View ... ★
  Vicinity ... Pg. 138, A-2
W. York ... J-10
Wheatland ... E-1
Whitehall ... G-12
  Vicinity ... Pg. 138, G-4
White Haven ... E-12
White Horse ... ★
  Vicinity ... Pg. 136, C-2
White Marsh ... ★
  Vicinity ... Pg. 136, A-6
White Oak ... H-3
Wilkes-Barre ... E-12
Wilkinsburg ... G-3
Williamsburg ... G-6
Williams Cors. ... ★
  Vicinity ... Pg. 136, A-2
Williamsport ... E-9
Williamstown ... G-10
Willow Grove ... ★
  Vicinity ... Pg. 137, A-7
Willstown ... ★
  Vicinity ... Pg. 136, B-1
Wilson ... G-13
Wilsons Cors. ... ★
  Vicinity ... Pg. 136, A-1
Windber ... H-5
Wind Gap ... F-13
Winterstown ... K-10
Womelsdorf ... H-11
Woodbury ... F-3
Worthington ... F-3
Wrightsville ... J-10
Wyalusing ... D-11
Wynnewood ... ★
  Vicinity ... Pg. 136, C-5
Wyola ... ★
  Vicinity ... Pg. 136, C-3
Wyomissing ... H-11
Yardley ... H-14
Yeadon ... J-13
  Vicinity ... Pg. 136, E-5
York ... J-10
York Springs ... J-9
Youngsville ... C-4
Youngwood ... H-3
Zelienople ... F-2

RHODE ISLAND
Page 30

Bristol ... E-7
Central Falls ... C-6
Charlestown ... F-6
Coventry ... D-6
Cranston ... D-6
East Greenwich ... D-6
East Providence ... D-6
Exeter ... D-6
Greenville ... D-6
Hope Valley ... E-6
Jamestown ... E-7
Johnston ... D-6
Kenyon ... E-6
Middletown ... E-7
Nasonville ... D-6
Narragansett Pier ... E-7
Newport ... E-7
Pawtucket ... C-6
Perryville ... E-6
Point Judith ... F-7
Portsmouth ... D-7
Primrose ... D-6
Providence ... D-6
Tiverton ... D-7
Warren ... D-7
Warwick ... D-6
Wakefield ... E-6
Washington ... D-7
Westerly ... F-6
Wickford ... E-7
Woonsocket ... C-6
Wyoming ... E-6

SOUTH CAROLINA
Pages 68-69

Abbeville ... F-4
Adams Run ... J-7
Aiken ... G-5
Allendale ... J-6
Almeda ... J-6
Anderson ... F-4
Andrews ... H-8
Antreville ... F-4
Appleton ... H-6
Awendaw ... J-8
Aynor ... F-9
Ballentine ... G-6
Bamberg ... H-6
Barnwell ... H-6
Batesburg ... G-6
Bayboro ... F-9
Belton ... F-4
Bennettsville ... D-8
Bethany ... E-6
Bethera ... H-8
Bethune ... E-7
Bishopville ... F-7
Blacksburg ... D-5
Blackville ... H-6
Blenheim ... E-8
Bonneau ... H-8
Bowman ... H-7
Boykin ... F-7
Bradley ... F-4
Branchville ... H-6
Brownsville ... E-8
Buffalo ... E-5
Bullock Creek ... E-5
Cades ... G-8
Cainhoy ... H-8
Calhoun Falls ... F-4

Camden ... F-7
Canadys ... H-7
Carlisle ... E-6
Cassatt ... F-7
Cayce ... G-6
Centenary ... F-9
Central ... E-3
Chappells ... F-5
Charleston ... J-8
Chelsea ... K-7
Cheraw ... E-8
Chesnee ... D-5
Chester ... E-6
Chesterfield ... E-8
Clarkshill ... G-5
Clemson ... E-3
Cleveland ... D-4
Clinton ... F-5
Clio ... D-8
Clover ... D-6
Columbia ... G-6
Conway ... F-9
Cool Spring ... F-9
Cresent Beach ... G-10
Cross ... H-7
Cross Hill ... F-5
Cross Keys ... E-5
Dalzell ... F-7
Darlington ... F-8
Debidue Beach ... H-9
De Kalb ... D-7
Denmark ... H-6
Dillon ... E-9
Dovesville ... F-8
Due West ... F-4
Edgefield ... G-5
Edgemoor ... E-6
Edisto Beach ... J-8
Edisto Island ... J-7
Effingham ... H-6
Ehrhardt ... H-6
Elgin ... F-6
Enoree ... E-5
Estill ... J-6
Eureka ... G-5
Evcrgreen ... F-8
Fair Bluff ... F-9
Fairfax ... J-6
Florence ... F-8
Floydale ... F-9
Folly Beach ... J-8
Forest Beach ... K-7
Foreston ... G-7
Fork ... F-9
Fork Shoals ... E-4
Fort Mill ... D-6
Fountain Inn ... E-5
Frogmore ... K-7
Furman ... J-6
Gaffney ... D-5
Galivants Ferry ... F-9
Garden City ... G-9
Gardens Corner ... J-7
Garnett ... J-6
Georgetown ... G-9
Givhans ... H-8
Goose Creek ... H-8
Gourdin ... G-8
Gownesyville ... D-4
Grays ... J-6
Great Falls ... E-6
Greeleyville ... G-8
Green Pond ... J-7
Green Sea ... F-9
Greenville ... E-4
Greenwood ... F-4
Greer ... E-4
Hampton ... J-6
Hanahan ... H-8
Hannah ... G-8
Hardeeville ... K-6
Hartsville ... F-7
Hemingway ... G-8
Hickory Grove ... D-5
Hickory Grove ... F-9
Hilda ... H-6
Hilton Head Island ... K-7
Holly Hill ... H-7
Honea Path ... F-4
Horrell Hill ... F-6
Indiantown ... G-8
Inman ... D-4
Islandton ... H-7
Isle of Palms ... J-8
Iva ... F-4
Jackson ... H-5
Jacksonboro ... J-7
James Island ... J-8
Jamestown ... H-8
Jefferson ... E-7
Jenkinsville ... F-6
Joanna ... F-5
Johnston ... G-5
Jonesville ... E-5
Kershaw ... E-7
Kingsburg ... G-8
Kingstree ... G-8
Kitchings Mill ... H-6
Kline ... H-6
Lake City ... G-8
Lamar ... F-7
Lancaster ... E-7
Landrum ... D-4
Lane ... G-8
Latta ... E-8
Laurens ... E-5
Leeds ... E-5
Lena ... J-6
Lesslie ... E-6
Lewis ... F-6
Lexington ... G-6
Liberty ... E-4
Litchfield Beach ... G-9
Little River ... F-10
Lockhart ... E-5
Lodge ... H-6
Loris ... F-9

Mayesville ... F-7
McBee ... E-7
McColl ... E-8
McConnells ... E-6
McCormick ... G-4
Moncks Corner ... H-8
Mt. Carmel ... F-4
Mount Pleasant ... J-8
Mountville ... F-5
Mullins ... F-9
Myrtle Beach ... G-10
Newberry ... F-5
New Ellenton ... H-5
New Prospect ... D-5
Nichols ... F-9
Nine Times ... E-3
Ninety Six ... F-5
Nixonville ... F-9
North ... G-6
North Augusta ... G-5
North Charleston ... J-8
Oatland ... G-9
Oats ... F-8
Olanta ... F-8
Olar ... H-6
Orangeburg ... G-6
Outland ... G-9
Owdoms ... G-5
Pacolet ... E-5
Pageland ... E-7
Pamplico ... G-8
Parksville ... G-5
Patrick ... E-7
Pauline ... E-5
Pawleys Island ... G-9
Pelion ... G-6
Pendleton ... E-4
Pickens ... E-4
Pineland ... J-6
Pineville ... G-8
Pinewood ... F-7
Plum Branch ... G-4
Pontiac ... F-6
Prosperity ... F-5
Pumpkintown ... E-4
Rantowles ... J-8
Rhems ... G-8
Richburg ... E-6
Ridgeland ... J-6
Ridgeway ... F-6
Rock Hill ... D-6
Rockton ... F-6
Rockville ... J-8
Rocky Bottom ... E-3
Ruby ... E-7
St. George ... H-7
St. Matthews ... G-7
St. Stephen ... G-8
Salem ... E-3
Salters ... G-8
Saluda ... F-5
Sampit ... H-8
Sandy Run ... G-6
Santee ... G-7
Sardinia ... G-7
Sardis ... F-8
Scranton ... G-8
Seneca ... E-3
Sharon ... E-5
Simpson ... F-6
Simpsonville ... E-4
Smallwood ... F-6
Smoaks ... H-7
Society Hill ... E-8
Spartanburg ... D-5
Springfield ... G-6
Starr ... F-4
Stateburg ... F-7
Steedman ... G-6
Stoney Hill ... F-5
Stuckey ... G-8
Sullivans Island ... J-8
Summerton ... G-7
Summerville ... H-7
Sumter ... F-7
Surfside Beach ... G-9
Switzerland ... K-6
Sycamore ... H-6
Tamassee ... E-3
Tatum ... E-8
Taylor ... F-6
Tillman ... J-6
Timmonsville ... F-8
Tradesville ... E-7
Travelers Rest ... E-4
Trenton ... G-5
Trio ... G-8
Troy ... G-5
Union ... E-5
Varnville ... J-6
Wagener ... G-6
Walhalla ... E-3
Walterboro ... J-7
Ward ... G-5
Wateree ... F-7
Wells ... H-7
Westminster ... E-3
Westville ... E-7
Whitehall ... J-8
Whitmire ... E-5
Willington ... G-4
Williston ... H-5
Winnsboro ... F-6
Woodruff ... E-5
Yauhannah ... G-9
Yemassee ... J-7
Yonges Island ... J-8
York ... D-6

SOUTH DAKOTA
Pages 80-81

Aberdeen ... D-8
Agar ... E-7
Akaska ... D-7
Albee ... E-11
Alcester ... J-11
Alexandria ... H-9
Alpena ... G-9
Altamont ... D-11
Andover ... D-10
Arlington ... F-11
Armour ... H-9

★ Denotes City located only on Vicinity map. City not located on State map.

SOUTH DAKOTA   TEXAS

SOUTH DAKOTA

Artas C-7
Artesian G-9
Ashton E-9
Astoria F-11
Aurora G-11
Badger F-10
Bancroft F-10
Batesland H-4
Belle Fourche E-2
Belvidere G-5
Beresford J-11
Bison D-4
Blunt F-7
Bonesteel J-8
Bowdle D-7
Bradley E-10
Brandon H-11
Brandt F-11
Brentford E-9
Bristol D-10
Britton C-10
Broadland F-9
Brookings F-11
Bruce F-10
Bryant F-10
Buffalo C-3
Buffalo Gap G-2
Burke J-8
Bushnell F-11
Butler D-10
Camp Crook C-2
Canova G-10
Canton H-11
Carthage G-10
Castle Rock D-3
Cavour F-9
Centerville J-10
Central City F-2
Chamberlain G-8
Chelsea E-8
Claire City C-10
Clark E-10
Clear Lake F-11
Colman G-11
Colton G-11
Columbia D-9
Conde E-9
Corona D-11
Corsica H-9
Cottonwood G-4
Cresbard E-8
Crooks H-11
Custer G-2
Dallas H-7
Dante J-9
Deadwood F-2
Dell Rapids G-11
Delmont H-9
De Smet F-9
Dimock H-9
Doland H-9
Dolton H-10
Draper G-6
Dupree E-5
Eagle Butte E-5
Eden D-10
Edgemont H-2
Elk Point K-11
Elm Springs F-3
Emery H-10
Enning E-4
Erwin F-10
Estelline F-10
Ethan H-9
Eureka D-8
Fairburn G-3
Fairfax J-8
Fairpoint E-3
Fairview J-11
Faith E-4
Farmer G-10
Farmingdale F-3
Faulkton E-8
Flandreau G-11
Florence E-10
Fort Pierre F-6
Fort Thompson G-7
Frankfort E-9
Frederick C-9
Freeman H-10
Fruitdale E-2
Fulton G-9
Gann Valley G-8
Garden City E-10
Garretson H-11
Gary E-11
Gayville J-10
Geddes H-8
Gettysburg E-7
Glad Valley D-5
Glenham D-7
Goodwin E-11
Greenwood H-9
Gregory H-7
Grenville D-10
Groton D-9
Hammer C-10
Harrisburg H-11
Harrold F-7
Hayti F-10
Hecla C-9
Hermosa G-3
Herreid C-7
Herrick H-7
Highmore F-8
Hill City G-2
Hillsview D-8
Hitchcock F-9
Hosmer D-8
Hot Springs H-2
Houghton C-9
Hoven D-7
Howard G-10
Howes E-4
Huron F-9
Interior G-4
Ipswich D-8
Irene J-10
Iroquois F-9
Isabel D-5
Java D-7
Jefferson K-11
Kadoka G-5
Kennebec G-7

Keystone G-2
Kranzburg E-11
Kimball G-8
Kyle H-4
Lake Andes J-8
Lake City D-10
Lake Preston F-10
Lane G-9
Langford D-10
La Plant D-6
Lead F-2
Lebanon E-7
La Bolt E-11
Lemmon C-4
Lennox H-11
Leola D-8
Letcher G-9
Lily E-10
Little Eagle C-6
Lodgepole C-4
Long Lake C-8
Long Valley H-5
Lowry D-7
Loyalton D-8
Ludlow C-3
Madison G-11
Marion H-10
Martin H-4
Marty H-9
Marvin E-11
Maurine D-4
McIntosh C-5
McLaughlin C-6
Meadow D-4
Mellette E-9
Menno H-10
Midland G-5
Milbank E-11
Miller F-8
Mission H-6
Mitchell H-9
Mobridge D-6
Monroe H-10
Montrose H-10
Morristown C-5
Mound City C-7
Mount Vernon H-9
Mud Butte D-3
Murdo G-6
Naples D-5
New Effington C-11
Newell E-2
Nisland E-2
North Sioux City K-11
Northville D-8
Nunda G-11
Oacoma G-7
Oelrichs H-2
Oglala H-3
Okreek H-6
Oldham G-10
Olivet H-10
Onaka D-8
Onida E-7
Orient E-8
Ortley D-10
Parker H-10
Parkston H-9
Parmelee H-5
Philip G-5
Pierpont D-10
Pierre F-6
Pine Ridge H-3
Plainview E-4
Plankinton G-9
Platte H-8
Pollock C-7
Prairie City C-3
Presho G-6
Pringle G-2
Provo H-2
Pukwana G-8
Quinn G-4
Ramona G-10
Rapid City F-3
Raymond E-9
Redfield E-8
Redig D-2
Ree Heights F-8
Reliance G-7
Reva C-3
Revillo E-11
Ravinia J-9
Ridgeview E-6
Rockerville F-2
Rockham E-8
Roscoe D-8
Rosebud H-5
Roslyn D-10
Roswell G-9
St. Charles J-8
Saint Francis H-5
Saint Lawrence F-8
Salem H-10
Scenic G-3
Scotland J-10
Selby D-7
Seneca E-7
Shadehill C-4
Sherman H-11
Sioux Falls H-11
Sisseton D-11
Sorum D-3
South Shore E-10
Spearfish F-2
Springfield J-9
Stephan F-8
Stickney H-9
Stockholm E-11
Stoneville E-4
Stratford D-9
Sturgis F-2
Summit D-11
Tabor J-9
Tea H-11
Timber Lake D-6
Tolstoy D-7
Toronto E-11
Trail City D-6
Trent G-11
Tripp H-9
Tulare F-9
Turton E-9
Tyndall J-9

Union Center E-3
Veblen C-10
Verdon D-9
Vermillion J-10
Vienna F-10
Viborg H-10
Vilas G-10
Vivian G-6
Wagner J-9
Wakonda J-10
Wall G-4
Wallace E-10
Ward G-11
Wasta F-4
Watertown E-10
Waubay D-10
Webster D-10
Wessington F-8
Wessington Springs G-8
Westport D-9
Wetonka D-8
Wewela J-7
White F-11
White Butte C-4
White Lake G-8
White River H-6
Whitewood E-2
Willow Lake F-10
Wilmot D-11
Winner H-7
Wood H-6
Woonsocket G-9
Worthing H-11
Yale F-9
Yankton J-10
Zeona D-3

TENNESSEE
Page 42

Adams F-6
Adamsville J-4
Aetna H-5
Alamo H-3
Alcoa H-11
Alexandria G-8
Algood G-8
Allardt G-13
Allens H-3
Allisona H-7
Allred G-9
Alpine E-8
Altamont H-8
Alto H-7
Arlington J-2
　Vicinity Pg. 123, A-10
Arrington H-7
Arthur F-11
Ashland City G-6
Ashport H-2
Athens H-10
Atoka H-2
Atwood G-4
Bailey ★
　Vicinity Pg. 123, A-9
Banner Hill G-13
Bartlett J-2
　Vicinity Pg. 123, C-7
Bean Station G-12
Beardstown H-5
Beersheba Springs H-8
Belfast J-7
　Vicinity Pg. 123, F-8
Bell Buckle H-7
Belle Meade G-6
Bells H-3
Belvidere J-7
Benjestown ★
　Vicinity Pg. 123, B-3
Benton J-10
Bethel H-3
Bethel Sprgs. H-4
Big Sandy G-4
Big Spring H-8
Birchwood J-9
Blaine G-11
Bluff City F-13
Bogota G-3
Bolivar J-3
Bolton ★
　Vicinity Pg. 123, A-9
Bon Air H-8
Braden H-2
Bradford G-4
Brattontown F-8
Brentwood G-6
Briceville G-10
Brighton H-2
Bristol F-13
Brownsville H-3
Bruceton G-4
Brunswick ★
　Vicinity Pg. 123, B-8
Bulls Gap G-12
Bumpus Mills F-5
Burlison H-2
Burns G-6
Busby J-5
Byrdstown F-9
Calhoun J-10
Camden G-5
Camelot F-12
Capleville ★
　Vicinity Pg. 123, C-6
Carlisle G-5
Carthage G-8
Caryville G-10
Cash Point J-6
Cedar Hill F-6
Celina F-9
Center Point J-6
Centertown H-7
Centerville H-6
Central G-3
Chapel Hill H-7
Charleston J-10
Charlotte G-6
Chattanooga J-8
Church Hill F-12
Clarksburg H-4
Clarksville F-5
Cleveland J-9

Clifton J-5
Clinton G-10
Coalmont H-8
Collegedale J-9
Collierville J-2
　Vicinity Pg. 123, F-10
Collinwood J-5
Columbia H-6
Cookeville G-8
Copperhill J-10
Cordova ★
　Vicinity Pg. 123, D-8
Cornersville J-6
Cosby H-11
Covington H-2
Crab Orchard H-9
Cross Plains F-6
Crossville H-9
Culleoka H-6
Cumberland City G-5
Cumberland Furnace G-6
Cumberland Gap F-11
Dancyville H-3
Dandridge G-12
Darden H-4
Dayton H-9
Decatur H-9
Decaturville H-4
Decherd J-8
Deerfield H-8
Dibrell H-8
Dickson G-5
Dixon Springs G-8
Donelson G-7
Dover F-5
Dowalltown H-8
Doyle H-8
Dresden G-4
Ducktown J-10
Dunlap J-9
Dyer G-3
Dyersburg G-3
Eads ★
　Vicinity Pg. 123, C-10
Eagan F-10
East Ridge J-9
Eastview J-4
Egypt ★
　Vicinity Pg. 123, B-6
Elizabethton G-13
Elkton J-6
Ellandale ★
　Vicinity Pg. 123, B-7
Elmore Park ★
　Vicinity Pg. 123, C-7
Elora J-7
Elza G-10
Englewood J-10
Erin G-5
Erwin G-13
Estill Springs J-7
Ethridge J-6
Etowah H-10
Evensville H-9
Fairview G-6
Farmington H-6
Fayetteville J-6
Finley G-3
Fisherville ★
　Vicinity Pg. 123, D-10
Flatcreek J-7
Flat Woods H-5
Flynns Lick F-9
Forbus F-9
Forest Hill ★
　Vicinity Pg. 123, F-8
Franklin H-6
Friendship G-3
Fruitvale H-3
Gadsden H-3
Gainesboro G-8
Gallatin G-7
Gates H-3
Gatlinburg H-11
Germantown J-2
　Vicinity Pg. 123, E-8
Gibson H-4
Gildfield ★
　Vicinity Pg. 123, A-8
Gilt Edge H-2
Gleason G-4
Goodlettsville G-7
Goodspring J-6
Gordonsville G-8
Grand Junction J-3
Grandview H-9
Gray G-13
Graysville H-9
Greenback H-10
Greenbrier G-6
Greeneville G-12
Greenfield G-4
Grimsley H-9
Habersham H-8
Haletown J-8
Halls G-3
Halls Crossroads G-11
Hampshire H-5
Hampton G-13
Hanging Limb G-9
Harriman H-9
Harrogate F-11
Hartsville G-7
Henderson H-4
Hendersonville G-7
Henning H-2
Henry G-4
Hickory Valley J-3
Highland J-5
Hohenwald H-5
Hornsby J-3
Humboldt H-3
Hunter G-13
Huntingdon G-4
Huntland J-7
Huntsville G-10
India G-3
Inglewood ★
Iron City J-5
Isham G-10
Ivydell G-10
Jacksboro G-10
Jackson H-3
Jamestown G-9

Jasper J-8
Jefferson City G-11
Jellico F-10
Joelton G-6
Johnson City G-13
Jonesboro G-12
Kelso J-7
Kenton G-3
Kingsport F-13
Kingston H-10
Kirkland H-10
Knoxville G-11
Kodak H-11
Kyles Ford F-12
Laager J-8
Lafayette F-8
LaFollette G-10
Lake City G-10
Lake Mont H-10
Lakesite J-9
Lancing G-9
Lascassas H-7
Latham G-4
La Vergne G-7
Lawrenceburg J-5
Leach G-4
Lebanon G-7
Leipers Fork H-6
Lenoir City H-10
Lenow ★
　Vicinity Pg. 123, C-9
Lewisburg H-6
Lexington H-4
Liberty G-8
Linden H-5
Livingston G-8
Lobelville H-5
Locke J-2
Loretto J-5
Loudon H-10
Lucy ★
　Vicinity Pg. 123, A-6
Luttrell G-11
Lutts J-5
Lyles H-6
Lynchburg J-7
Lynnville J-6
Madisonville H-10
Malesus H-3
Manchester H-7
Martin G-4
Maryville H-10
Mascot G-11
Mason H-2
Maury City H-3
Mayland G-8
Maynardville G-10
McBurg J-6
McEwen G-5
McKenzie G-4
McMinnville H-8
Medina H-3
Medon H-3
Memphis J-1
　Vicinity Pg. 123
Mercer H-3
Michie J-4
Middleton J-3
Midtown H-10
Milan H-4
Milledgeville J-4
Milligan College G-13
Millington J-2
Minor Hill J-6
Miston G-2
Monteagle J-8
Monterey G-9
Morgantown H-9
Morris Chapel H-4
Morrison H-7
Morristown G-11
Moscow J-3
Mosheim G-12
Moss F-8
Mound City ★
　Vicinity Pg. 123, C-2
Mountain City F-14
Mount Carmel F-12
Mount Crest H-9
Mt. Juliet G-7
Mt. Olive H-11
Mt. Pleasant H-6
Mount Vernon J-10
Munford H-2
Murfreesboro H-7
Nashville G-6
Newbern G-3
New Hope K-8
New Johnsonville G-5
New Market G-11
Newport G-12
New River G-10
New Tazwell G-11
New Town J-6
Niota H-10
Nixon J-4
Norris G-10
Nunnelly H-5
Oakdale G-9
Oakland J-2
Oak Ridge G-10
Oakville ★
　Vicinity Pg. 123, F-5
Oakwood F-5
Obion G-3
Ocoee J-10
Olivehill J-4
Oliver Springs G-10
Oneida F-10
Palmer J-8
Palmyra F-5
Paris G-4
Parrottsville G-12
Parsons H-4
Peakland H-6
Pegram G-6
Petersburg J-7
Philadelphia H-10
Pigeon Forge H-11
Pikeville H-9
Pinewood H-5

Pinson H-4
Pisgah ★
　Vicinity Pg. 123, D-8
Pleasant Hill G-8
Pleasant View G-6
Portland F-7
Providence G-7
Pulaski J-6
Puncheon Camp G-11
Puryear F-4
Quebeck H-8
Raleigh ★
　Vicinity Pg. 123, B-6
Ramer J-4
Ramsey ★
　Vicinity Pg. 123, A-4
Raus J-7
Red Bank J-8
Red Boiling Springs F-8
Riceville J-10
Rickman G-8
Ridgely G-3
Ridgetop G-6
Ridley H-6
Ripley H-2
Rives G-3
Roan Mountain G-13
Robbins G-9
Rockwood H-9
Rocky Fork G-12
Rogersville G-12
Rossville J-2
Russellville G-12
Rutherford G-3
Rutledge G-11
Sadlersville F-6
Saint Bethlehem F-5
Saint Joseph J-5
Sale Creek J-9
Saltillo H-4
Sandy Hook H-6
Sardis H-4
Savannah J-4
Scotts Hill H-4
Selmer J-4
Sevierville H-11
Sewanee J-8
Shady Valley F-13
Sharon G-3
Shackleett ★
Shelby Farms ★
　Vicinity Pg. 123, D-7
Shelbyville H-7
Shooks Gap H-11
Shop Springs G-8
Signal Mtn. J-8
Siloam F-7
Silver Point G-8
Skiem H-8
Smartt H-8
Smithville G-8
Smyrna G-7
Sneedville G-11
Soddy-Daisy J-9
Somerville J-2
South Carthage G-8
S. Dyersburg G-3
South Fulton F-3
South Pittsburg J-8
Sparta H-8
Spencer H-8
Spring City H-9
Spring Creek H-3
Spring Hill H-6
Spring Lake ★
　Vicinity Pg. 123, B-7
Springfield F-6
Stanton H-3
Sullivan Gdn. F-12
Summertown J-5
Sunbright G-10
Surgoinsville F-12
Sweetwater H-10
Taft J-6
Tarpley H-10
Tazewell G-11
Tellico Plains J-10
Tennessee Ridge G-5
Timberlake H-4
Tiptonville F-3
Toone J-3
Tracy City J-8
Trenton G-3
Trezevant G-4
Triune H-7
Troy G-3
Tullahoma J-7
Tusculum G-13
Unicoi G-13
Union City F-3
Union Grove J-9
Victoria J-8
Viola H-8
Violet J-10
Vonore H-10
Waco J-6
Walden J-8
Walkertown H-8
Walling H-8
Wartburg G-10
Wartrace H-7
Watertown G-8
Waverly G-5
Waynesboro J-5
Webber City J-5
Westmoreland F-7
West Shiloh J-4
White Bluff G-6
White House F-7
White Pine G-12
Whiteville J-3
Whitesburg G-12
Whitwell J-8
Wilder G-8
Williston J-3
Willette F-8
Winchester J-7
Winfield F-10
Woodbury H-8
Woodland Mills F-3
Woodstock ★
　Vicinity Pg. 123, B-5
Wyanoke ★
　Vicinity Pg. 123, E-1

Wynnburg G-3

TEXAS
Pages 90-94

Abernathy E-7
Abilene G-10
Acala J-2
Ackerly F-9
Acme D-10
Adamsville G-12
Addicks ★
　Vicinity Pg. 109, A-8
Addison ★
　Vicinity Pg. 108, C-3
Adrian C-7
Agua Dulce P-12
Aiken E-8
Alanreed C-9
Alba F-15
Albany G-11
Aldine L-15
Aledo F-12
Alexander H-12
Algoa M-16
Alice P-12
Allamoore J-3
Allen F-14
Alleyton L-14
Allison B-10
Almeda ★
　Vicinity Pg. 114, F-5
Alpine K-5
Altair M-14
Alto H-16
Alvarado G-13
Alvin M-16
Alvord E-12
Amarillo C-7
Ambrose E-14
Ames L-16
Amherst D-7
Anderson K-15
Andrews G-8
Angleton M-16
Anna E-14
Annona F-16
Anson G-10
Antelope F-12
Anthony H-2
Anton E-7
Apple Springs J-16
Aquilla H-13
Aransas Pass P-13
Archer City E-11
Argyle F-12
Arlington G-13
　Vicinity Pg. 108, A-3
Arp G-16
Artesia Wells P-10
Arthur City E-15
Asherton N-10
Ashtola C-8
Aspermont F-9
Athens G-15
Atlanta F-17
Aubrey E-13
Austin L-13
Austonio J-15
Austwell N-14
Avery F-16
Avinger F-17
Avoca G-10
Azle F-12
Baclift M-16
Bagwell E-16
Bailey E-15
Baird G-10
Bakersfield K-7
Balch Springs ★
　Vicinity Pg. 109, E-11
Ballinger H-10
Balmorhea J-5
Bangs H-11
Barclay J-13
Bardwell H-14
Barksdale L-10
Barnhart J-8
Barrett ★
　Vicinity Pg. 115, B-9
Barry H-14
Barstow J-4
Bartlett K-13
Bassett F-16
Bastrop L-13
Batesville N-10
Batson L-16
Bay City M-15
Baytown ★
　Vicinity Pg. 115, D-10
Beasley M-15
Beaumont L-17
Beckville G-17
Bedias K-15
Bee Cave L-12
Beeville N-13
Belfalls J-13
Bellaire ★
　Vicinity Pg. 114, E-4
Bellevue E-12
Bellmead H-13
Bells E-14
Bellville L-14
Belmont M-13
Belton J-13
Ben J-3
Ben Arnold J-14
Benavides P-11
Ben Bolt P-12
Benbrook ★
　Vicinity Pg. 108, G-1
Benchley K-14
Benjamin F-10
Bertram K-12
Bettie F-16
Big Lake J-7
Big Sandy G-16
Big Spring H-8

Big Wells N-10
Birthright F-15
Bishop Q-12
Bivins F-17
Black D-7
Blackwell H-9
Blanco L-11
Blanket H-11
Bleakwood K-18
Bledsoe E-6
Blessing N-15
Blewett M-9
Blooming Grove H-14
Bloomington N-13
Blossom E-15
Bluegrove E-12
Blue Mound ★
　Vicinity Pg. 108, C-3
Blue Ridge F-14
Bluff Dale G-12
Bluffton K-11
Boca Chica S-13
Boerne L-11
Bogata F-15
Boling M-14
Bomarton F-10
Bon Wier K-18
Booker A-9
Boonsville F-12
Borger B-8
Boston F-17
Bovina D-6
Bowie E-12
Boxelder F-16
Boyd F-13
Boys Ranch Tacosa B-7
Brackettville M-9
Bradshaw H-10
Brady J-10
Brandon H-14
Brashear F-15
Brazoria M-15
Breckenridge G-11
Bremond J-14
Brenham L-14
Brice D-9
Bridge City L-18
Bridgeport F-12
Briggs K-12
Briscoe B-9
Broaddus J-17
Brookeland J-17
Brookesmith H-11
Brookshire L-15
Brookside Village ★
　Vicinity Pg. 114, F-6
Bronco E-6
Bronson J-17
Bronte H-9
Brownfield E-7
Brownsboro G-15
Brownsville S-13
Brownwood H-11
Brundage N-10
Bruni Q-11
Bryan K-14
Bryson F-12
Buchanan Dam K-12
Buckholts K-13
Buda L-12
Buffalo J-14
Buffalo Gap G-10
Bula E-6
Bulcher E-13
Bullard H-16
Buna K-17
Bunavista B-8
Bunker Hill ★
　Vicinity Pg. 114, D-3
Burkburnett E-11
Burke J-16
Burkeville K-18
Burleson G-13
　Vicinity Pg. 108, G-3
Burlington J-14
Burnet K-12
Burton L-14
Bushland C-7
Bustamanete R-11
Butler K-13
Byers E-12
Bynum H-14
Cactus A-7
Caddo G-11
Caddo Mills F-14
Caldwell K-14
Calf Creek J-10
Call L-17
Calliham N-12
Calvert K-14
Camden K-16
Cameron K-14
Camilla K-16
Camp Verde L-11
Camp Wood L-10
Campbell F-15
Campbellton N-12
Canadian B-9
Canton G-15
Canutillo H-1
Canyon C-7
Caps G-10
Carbon G-11
Carlos K-15
Carlsbad H-9
Carlton H-12
Carmine L-14
Carmona J-16
Carrizo Springs N-10
Carrollton ★
　Vicinity Pg. 109, A-8
Carta Valley L-9
Carthage H-17
Cash F-14
Castle Hills ★
　Vicinity Pg. 111, M-11
Castroville M-11
Cat Spring L-14
Catarina N-10
Cayuga H-15
Cedar Creek L-13
Cedar Hill ★
　Vicinity Pg. 109, F-7

★ Denotes City located only on Vicinity map. City not located on State map.

TEXAS

Cedar Park ... K-12
Celeste ... F-14
Celina ... F-14
Center ... H-17
Centerville ... J-15
Chandler ... G-15
Channelview ... L-16
 Vicinity ... Pg. 115, D-6
Channing ... B-7
Chapman ... H-16
Chapman Ranch ... P-13
Chappell Hill ... L-15
Charco ... N-13
Charlotte ... N-11
Cherokee ... J-11
Cherry Spring ... K-11
Chester ... J-17
Chico ... F-12
Childress ... D-9
Chillicothe ... D-10
Chilton ... J-13
China ... L-17
Chireno ... J-17
Chisos Basin ... M-5
Chriesman ... K-14
Christine ... N-11
Christoval ... J-9
Circle Back ... E-6
Cisco ... G-11
Cistern ... L-13
Clairette ... H-12
Clairemont ... F-9
Clarendon ... C-9
Clarksville ... E-16
Claude ... C-8
Clayton ... H-17
Clear Lake Shore ... ★
 Vicinity ... Pg. 115, C-6
Cleburne ... G-13
Cleveland ... K-16
Clifton ... H-13
Clint ... H-1
Clodine ... ★
 Vicinity ... Pg. 114, D-2
Cloverleaf ... ★
 Vicinity ... Pg. 115, D-6
Clute ... M-16
Clyde ... G-10
Coahoma ... G-8
Cockrell Hill ... ★
 Vicinity ... Pg. 109, D-8
Coffey City ... H-15
Coldspring ... K-16
Coleman ... H-10
College Station ... K-14
Colleyville ... ★
 Vicinity ... Pg. 108, B-5
Collinsville ... F-13
Colmesneil ... K-17
Colorado City ... G-9
Columbus ... L-14
Comanche ... H-11
Combes ... R-12
Comfort ... L-11
Commerce ... F-15
Como ... F-15
Comstock ... L-8
Concan ... M-10
Concepcion ... Q-12
Concord ... H-16
Conroe ... K-16
Converse ... L-11
Conway ... C-8
Cooks Point ... K-14
Cookville ... F-16
Coolidge ... H-14
Cooper ... F-15
Coppell ... ★
 Vicinity ... Pg. 108, B-3
Copperas Cove ... J-12
Cornudas ... H-3
Corpus Christi ... P-13
Corrigan ... J-16
Corsicana ... H-14
Coryell City ... J-13
Cost ... M-13
Cotton Center ... E-7
Cottonwood ... H-11
Cottonwood ... K-15
Cotulla ... N-10
Country Campus ... K-16
Coy City ... N-12
Coyanosa ... J-5
Crabbs Praire ... K-15
Crandall ... G-14
Crane ... J-6
Cranvills Gap ... H-13
Crawford ... J-13
Cresson ... G-13
Crews ... H-10
Crockett ... J-15
Crosby ... L-15
Crosby ... ★
 Vicinity ... Pg. 115, B-9
Crosbyton ... E-8
Cross ... K-15
Cross Cut ... H-11
Cross Plains ... H-11
Cross Roads ... H-15
Crow ... G-16
Crowell ... E-10
Crowley ... H-13
 Vicinity ... Pg. 108, F-3
Crystal City ... N-10
Cuero ... M-13
Cumby ... F-14
Cushing ... H-16
Cut & Shoot ... K-15
Cypress ... L-15
 Vicinity ... Pg. 114, A-1
Cypress Mill ... K-12
Dabney ... M-9
Dacosta ... N-14
Daingerfield ... F-16
Daisetta ... K-16
Dalhart ... B-7
Dalingworth Gardens ... ★
 Vicinity ... Pg. 108, G-5
Dallas ... G-14
 Vicinity ... Pg. 109
Damon ... M-15
Danbury ... M-15
Danevang ... M-15
Darrouzett ... A-9

Dawn ... C-7
Dawson ... H-14
Dayton ... K-16
Deanville ... K-14
De Berry ... G-17
Decatur ... F-13
Deer Park ... L-16
 Vicinity ... Pg. 115, E-8
De Kalb ... E-16
De Leon ... H-11
Dell City ... H-3
Del Rio ... M-8
Del Valle ... L-13
Denison ... E-14
Denton ... F-13
Denver City ... F-6
Deport ... E-15
Derby ... N-11
Desdemona ... H-12
DeSoto ... ★
 Vicinity ... Pg. 109, F-8
Devers ... L-17
Devine ... M-11
Dew ... H-15
Dewalt ... ★
 Vicinity ... Pg. 114, G-3
De Wees ... M-12
Deweyville ... K-18
D'Hanis ... M-10
Diana ... G-16
Diboll ... J-16
Dickens ... E-9
Dickinson ... L-15
Dilley ... N-11
Dime Box ... K-14
Dimmitt ... D-7
Dinero ... P-12
Divot ... M-11
Dobbin ... K-15
Dodd City ... E-15
Donie ... J-14
Donna ... S-12
Doole ... J-10
Doss ... K-11
Doucette ... K-17
Dougherty ... E-8
Douglass ... H-16
Douglassville ... F-17
Dozier ... C-9
Draw ... F-8
Dripping Springs ... L-12
Driscoll ... P-12
Dryden ... L-7
Dublin ... H-12
Dumas ... B-7
Dumont ... E-9
Duncanville ... ★
 Vicinity ... Pg. 109, F-7
Dundee ... E-11
Dunlay ... M-11
Dunn ... G-9
Dyersdale ... ★
 Vicinity ... Pg. 115, B-7
Eagle Lake ... M-14
Eagle Pass ... N-9
Early ... H-11
Earth ... D-7
East Columbia ... M-15
Easterly ... J-14
Easton ... G-16
Eastland ... G-11
Edcouch ... R-12
Eddy ... J-13
Eden ... J-10
Edgecliff ... ★
 Vicinity ... Pg. 108, E-3
Edinburg ... R-12
Edmonson ... D-7
Edna ... M-14
Edom ... G-15
Edroy ... P-13
Egypt ... M-14
El Campo ... M-14
Elbert ... F-11
Eldorado ... K-9
Electra ... E-11
Elgin ... K-13
Elkhart ... H-15
El Lago ... ★
 Vicinity ... Pg. 115, E-8
Ellinger ... L-14
Elmendort ... M-12
Elmo ... G-15
El Paso ... H-2
Elsa ... R-12
El Sauz ... R-10
Elysian Fields ... G-17
Elwood ... L-14
Emory ... F-14
Encinal ... N-10
Encino ... Q-12
Energy ... H-12
Enloe ... F-15
Ennis ... G-14
Enochs ... E-6
Escobas ... Q-11
Estelline ... D-9
Etoile ... J-16
Etter ... B-7
Euless ... ★
 Vicinity ... Pg. 108, C-6
Eureka ... H-14
Eustace ... G-15
Evadale ... K-17
Evant ... J-12
Evergreen ... K-16
Everman ... ★
 Vicinity ... Pg. 108, F-3
Fabens ... H-1
Fairbanks ... ★
 Vicinity ... Pg. 114, C-3
Fairfield ... H-15
Fairlie ... F-15
Fairview ... G-8
Fairy ... H-12
Falcon ... R-11
Falfurias ... Q-12
Falls City ... N-12
Fannin ... N-13
Farmers Branch ... ★
 Vicinity ... Pg. 109, B-8
Farmersville ... F-14

Farrar ... J-14
Farwell ... D-6
Fayetteville ... L-14
Fedor ... K-13
Fentress ... L-13
Ferris ... G-14
 Vicinity ... Pg. 109, D-8
Fieldton ... E-7
Fink ... E-14
Finney ... E-9
Flatonia ... M-13
Flomot ... D-8
Florence ... J-12
Floresville ... M-12
Flower Mound ... ★
 Vicinity ... Pg. 108, A-6
Floydada ... E-8
Fluvanna ... F-8
Follett ... A-10
Forest ... J-16
Forestburg ... F-13
Forest Hill ... ★
 Vicinity ... Pg. 108, E-4
Forney ... G-14
Fort Clark Springs ... M-9
Fort Davis ... K-5
Fort Gates ... J-12
Fort Griffin ... G-10
Fort Hancock ... J-2
Fort McKavett ... K-10
Fort Stockton ... J-6
Fort Worth ... G-13
 Vicinity ... Pg. 108
Fostoria ... K-16
Fowlerton ... N-11
Frame Switch ... K-13
Francitas ... N-14
Frankel City ... G-6
Franklin ... J-14
Franklin ... H-15
Fransworth ... A-9
Fred ... K-17
Fredericksburg ... L-11
Fredonia ... K-11
Freeport ... N-16
Freer ... P-11
Fresno ... ★
 Vicinity ... Pg. 114, G-4
Friday ... J-16
Friendswood ... ★
 Vicinity ... Pg. 115, G-8
Friona ... D-6
Frisco ... F-13
Fritch ... B-8
Frio Town ... M-10
Frost ... H-14
Fruitland ... F-12
Fruitvale ... G-15
Fulbright ... F-16
Fulshear ... P-13
Fulton ... P-13
Funston ... G-10
Gail ... G-8
Gainesville ... F-13
Galena Park ... ★
 Vicinity ... Pg. 115, D-7
Galveston ... M-16
Ganado ... M-14
Garden City ... H-8
Gardendale ... H-7
Garden Valley ... G-15
Garland ... G-14
 Vicinity ... Pg. 109, B-10
Garrison ... H-17
Garwood ... M-14
Gary ... H-17
Gatesville ... J-12
Gause ... K-14
Geneva ... J-17
George West ... N-12
Georgetown ... K-13
Geronimo ... L-12
Giddings ... L-13
Gilchrist ... M-17
Gillett ... M-12
Gilliland ... E-10
Gilmer ... G-16
Girard ... F-9
Girvin ... J-7
Gladewater ... G-16
Glazier ... B-9
Glenn ... E-9
Glen Cove ... H-10
Glen Flora ... M-14
Glen Heights ... ★
 Vicinity ... Pg. 109, G-8
Glenrio ... C-6
Glen Rose ... H-13
Glenwood ... G-16
Glidden ... L-14
Gober ... F-15
Godley ... G-13
Goldsboro ... H-10
Goldthwaite ... J-11
Goliad ... N-13
Gonzales ... M-13
Goodland ... E-6
Goodlett ... D-10
Goodnight ... C-8
Goodrich ... K-16
Gordon ... G-12
Gordonville ... E-13
Goree ... F-10
Gorman ... H-11
Gouldbusk ... J-10
Graford ... F-12
Graham ... F-11
Granbury ... G-12
Grandfalls ... J-6
Grand Prairie ... G-13
 Vicinity ... Pg. 109, D-6
Grand Saline ... G-15
Granger ... K-13
Grapeland ... J-15
Grapevine ... ★
 Vicinity ... Pg. 108, B-6
Grassland ... F-8
Grayback ... E-11
Greenville ... F-15
Grit ... K-10
Groesbeck ... J-14
Groom ... C-8
Groves ... L-17
Groveton ... J-16
Gruver ... A-8

Gun Barrel City ... G-14
Gustine ... H-12
Guthrie ... E-9
Guy ... M-15
Hagansport ... F-15
Hale Center ... E-7
Halletsville ... M-14
Hallsville ... G-16
Haltom City ... ★
 Vicinity ... Pg. 108, C-3
Hamilton ... H-12
Hamlin ... G-9
Hamshire ... L-17
Happy ... C-7
Hardin ... L-16
Hargill ... R-12
Harker Heights ... J-13
Harleton ... G-16
Harlingen ... S-13
Harper ... L-10
Harrold ... E-11
Hart ... D-6
Hartley ... B-7
Harwood ... L-13
Haskell ... F-10
Haslet ... ★
 Vicinity ... Pg. 108, A-3
Hatchel ... H-10
Hawkins ... G-16
Hawley ... G-10
Hearne ... K-14
Hebbronville ... Q-11
Hebron ... ★
 Vicinity ... Pg. 109, A-8
Hedley ... C-9
Hedwig Village ... ★
 Vicinity ... Pg. 114, D-3
Helena ... N-13
Hemleigh ... G-9
Hemphill ... J-18
Hempstead ... L-14
Henderson ... G-16
Henrietta ... E-12
Hereford ... C-6
Hext ... K-10
Hye ... L-11
Hico ... H-12
Hidalgo ... S-12
Higgins ... B-10
Highbank ... J-14
Highland Park ... ★
 Vicinity ... Pg. 109, C-9
Highlands ... L-16
 Vicinity ... Pg. 115, C-9
Hillister ... K-17
Hillsboro ... H-13
Hillshire Village ... ★
 Vicinity ... Pg. 114, C-4
Hilltop Lakes ... J-14
Hindes ... N-11
Hitchcock ... M-16
Hobson ... N-12
Hochheim ... M-13
Hockley ... L-15
Holland ... K-13
Holliday ... E-11
Hondo ... M-11
Honey Grove ... F-15
Horizon City ... H-1
Houston ... L-16
Howe ... F-14
Hubbard ... H-14
Hull ... K-16
 Vicinity ... Pg. 115, A-6
Huckabay ... G-12
Hudson ... J-16
Hughes Springs ... F-16
Humble ... L-16
 Vicinity ... Pg. 115, A-6
Hunt ... L-10
Hunters Creek ... ★
 Vicinity ... Pg. 114, D-4
Huntington ... J-17
Huntsville ... K-15
Hurlwood ... E-7
Hurst ... ★
 Vicinity ... Pg. 108, C-5
Hutto ... K-13
Hutchins ... G-14
 Vicinity ... Pg. 109, E-10
Idalou ... E-7
Imperial ... J-6
Inadal ... G-9
Independence ... L-14
Indian Gap ... J-12
Industry ... L-14
Inez ... N-13
Ingleside ... P-13
Ingram ... L-11
Iola ... K-15
Iowa Park ... E-11
Ira ... G-8
Iraan ... J-7
Iredell ... H-12
Ireland ... J-12
Irving ... G-13
 Vicinity ... Pg. 109, C-7
Italy ... H-13
Itasca ... H-13
Jacinto City ... ★
 Vicinity ... Pg. 115, D-7
Jacksboro ... F-12
Jacksonville ... H-15
Jamestown ... J-18
Jarrell ... K-13
Jasper ... J-17
Jayton ... F-9
Jean ... F-11
Jefferson ... F-16
Jermyn ... F-12
Jersey Village ... ★
 Vicinity ... Pg. 114, B-3
Jewett ... J-14
Joaquin ... H-17
Johnson City ... L-12
Johntown ... F-16
Jolly ... E-11
Jollyville ... K-12
Jonesboro ... J-12
Jonesville ... G-17
Jones Creek ... M-15
Jonestown ... K-12
Joshua ... G-13
Jourdanton ... N-11

Joy ... E-12
Junction ... K-10
Justiceburg ... F-8
Justin ... F-13
Kalgary ... F-8
Kamay ... E-11
Kanawha ... E-16
Karnack ... G-17
Karnes City ... N-12
Katemcy ... K-10
Katy ... L-15
Kaufman ... G-14
Keene ... G-13
Keller ... ★
 Vicinity ... Pg. 108, B-4
Kellerville ... C-9
Kemp ... G-14
Kempner ... J-12
Kendleton ... M-15
Kenedy ... N-12
Kennard ... J-16
Kennedale ... ★
 Vicinity ... Pg. 108, E-4
Kenney ... L-14
Kent ... J-4
Kerens ... H-14
Kermit ... H-5
Kerrville ... L-11
Kilgore ... G-16
Killeen ... J-13
Kingsland ... K-11
Kings Mill ... C-8
Kingston ... F-14
Kinwood ... Q-12
 Vicinity ... Pg. 114, B-6
Kirbyville ... K-17
Kirkland ... D-9
Kirvin ... H-14
Kleburg ... ★
 Vicinity ... Pg. 109, E-11
Knickerbocker ... J-9
Knippa ... M-10
Knott ... G-7
Knox City ... F-10
Kopperl ... H-13
Kosse ... J-14
Kountze ... K-17
Kress ... D-7
Krum ... F-13
Kyle ... L-12
Ladonia ... F-15
Lago Vista ... K-12
La Grange ... L-13
La Grulla ... R-11
Laguna Park ... H-13
La Joya ... R-11
Lake Hills ... M-11
Lake Jackson ... M-15
Lakeport ... G-16
Lakeside ... ★
 Vicinity ... Pg. 108, C-1
Lakeside Village ... H-13
Lakeview ... D-9
Lake Worth ... ★
 Vicinity ... Pg. 108, C-2
La Marque ... M-16
Lamesa ... G-7
Lamkin ... H-12
Lampasas ... J-12
Lancaster ... G-14
 Vicinity ... Pg. 109, F-9
Lane City ... M-15
Laneville ... L-7
La Paloma ... S-13
La Porte ... L-16
 Vicinity ... Pg. 115, E-9
La Pryor ... N-9
Laredo ... R-12
La Reforma ... D-6
Lariat ... C-8
Lark ... H-15
Lasara ... R-12
Latexo ... J-15
La Vernia ... M-12
La Villa ... R-12
Lavon ... F-14
La Ward ... N-14
Lawn ... H-10
League City ... L-15
Leakey ... L-10
Leander ... K-12
Leesburg ... F-16
Leesville ... M-13
LeFors ... C-9
Leggett ... K-16
Lehman ... E-6
Leigh ... G-17
Lela ... C-9
Lelia Lake ... C-9
Leming ... M-12
Lenorah ... G-7
Leon Springs ... L-11
Leon Valley ... L-11
Leona ... J-15
Leonard ... F-14
Levelland ... E-7
Lewisville ... F-13
 Vicinity ... Pg. 109, A-6
Lexington ... K-13
Liberty ... L-16
Liberty Hill ... K-12
Libscomb ... A-9
Lincoln ... L-13
Lincoln City ... ★
 Vicinity ... Pg. 114, C-5
Linden ... E-16
Lindsay ... F-13
Lingleville ... H-12
Liverpool ... M-16
Livingston ... K-16
Llano ... K-11
Lobo ... J-3
Lockett ... E-10
Lockhart ... L-13
Lockney ... E-8
Lodi ... F-17
Loma Alta ... L-8
Lomax ... H-8

Lometa ... J-11
London ... K-10
Lone Star ... F-16
Long Branch ... H-16
Long Mott ... N-14
Longview ... G-16
Longworth ... G-9
Loop ... F-7
Lopeno ... R-11
Loraine ... G-9
Lorena ... J-13
Lorenzo ... E-8
Los Angeles ... N-11
Los Fresnos ... S-13
Lott ... J-13
Louise ... M-14
Lovelady ... J-15
Loving ... F-11
Loyal Valley ... K-11
Lubbock ... F-8
Luckenbach ... L-11
Lueders ... G-10
Lufkin ... J-16
Luling ... L-13
Lumberton ... K-17
Lutie ... C-9
Lydia ... F-16
Lyford ... R-13
Lynchburg ... ★
 Vicinity ... Pg. 115, D-9
Lyons ... K-14
Lytle ... M-11
Mabank ... G-15
Madisonville ... J-15
Magnolia ... M-15
Malakoff ... H-15
Malta ... F-16
Mankins ... E-11
Mansfield ... G-13
 Vicinity ... Pg. 108, F-4
Maple ... E-6
Marathon ... K-5
Marble Falls ... K-12
Marfa ... K-4
Marietta ... F-17
Marion ... M-12
Markham ... M-15
Marlin ... J-14
Marquez ... J-15
Mart ... J-14
Martindale ... L-13
Martinez ... M-12
Martinsville ... M-17
Marynael ... G-9
Mason ... K-11
Masterson ... C-8
Matador ... E-8
Matagorda ... N-15
Mathis ... P-12
Maud ... F-17
Mauriceville ... L-17
Maxwell ... L-13
May ... H-11
Maydelle ... H-16
Maypearl ... G-13
McAdoo ... E-8
McAllen ... S-12
McCamey ... J-7
McCauley ... G-9
McDade ... L-13
McFaddin ... N-13
McGregor ... J-13
McKinney ... F-14
McLean ... C-9
McLeod ... F-17
McMahan ... L-13
McNary ... J-2
McQueeney ... L-12
Meadow ... F-7
Medicine Mound ... D-10
Medina ... L-11
Megargel ... F-11
Melissa ... F-14
Melrose ... H-17
Melvin ... J-10
Memphis ... D-9
Menard ... K-10
Mendoza ... L-13
Mentone ... J-4
Mercedes ... S-12
Meridian ... H-13
Merkel ... G-9
Mertzon ... J-8
Mesquite ... G-14
 Vicinity ... Pg. 109, C-11
Mexia ... H-14
Miami ... B-9
Mico ... M-11
Midland ... H-7
Midlothian ... G-14
Midway ... J-18
Milam ... J-18
Milano ... K-14
Miles ... H-9
Milford ... G-13
Millersview ... J-10
Millett ... N-11
Millican ... K-14
Mineola ... G-15
Mineral ... N-11
Mineral Wells ... G-12
Mingus ... G-12
Mirando City ... Q-11
Mission ... S-12
Missouri City ... ★
 Vicinity ... Pg. 114, F-3
Mobeetie ... B-9
Moline ... J-12
Monahans ... H-6
Monkstown ... E-15
Monroe ... E-17
Montague ... E-12
Montalba ... H-15
Mont Belvieu ... L-16
Monte Alto ... S-12
Montgomery ... K-15
Monthalia ... M-13
Moody ... J-13
Moore ... M-11
Mooring ... K-14
Moran ... G-11
Morgan ... H-13
Morgan Mill ... G-12

Morgans Point ... ★
 Vicinity ... Pg. 115, E-10
Morton ... E-6
Morton Valley ... G-11
Morse ... B-8
Moscow ... J-16
Moss Hill ... K-16
Moulton ... M-13
Mountain Home ... L-11
Mount Calm ... H-13
Mt. Enterprise ... H-16
Mt. Pleasant ... F-16
Mount Sylvan ... G-16
Mount Vernon ... F-16
Muenster ... E-13
Muldoon ... L-14
Muleshoe ... D-6
Mullin ... J-11
Munday ... F-10
Murchison ... G-15
Murphy ... ★
 Vicinity ... Pg. 109, A-11
Myrtle Springs ... G-15
Nacogdoches ... H-16
Nada ... M-14
Naples ... F-16
Nash ... E-17
Natalia ... M-11
Navasota ... K-15
Nazareth ... D-7
Neches ... H-15
Nederland ... K-17
Needmore ... E-6
Needville ... M-15
Nelsonville ... L-14
Newark ... ★
 Vicinity ... Pg. 108, A-1
New Baden ... J-14
New Boston ... E-17
New Braunfels ... L-12
New Caney ... L-16
Newcastle ... F-11
New Deal ... E-7
New Gulf ... M-15
New "O" Home ... F-7
New London ... G-16
Newman ... H-2
Newsome ... F-16
New Summerfield ... H-16
Newton ... K-18
New Ulm ... L-14
New Waverly ... K-15
Nixon ... M-13
Nocona ... E-12
Nolan ... G-9
Nolanville ... J-13
Nome ... K-17
Noodle ... G-10
Nordheim ... N-13
Normandy ... N-9
Normangee ... J-15
Normanna ... N-13
North Houston ... ★
 Vicinity ... Pg. 114, B-4
Northcrest ... H-13
Northfield ... D-9
N. Richland Hills ... ★
 Vicinity ... Pg. 108, C-4
North Zulch ... K-14
Norton ... H-9
Notrees ... H-6
Novice ... H-10
Nuevo Laredo ... P-10
Nursery ... N-13
Oakalla ... J-12
Oak Hill ... L-12
Oakland ... M-14
Oakwood ... H-15
O'Brien ... F-10
Odell ... D-10
Odem ... P-13
Odessa ... H-7
O'Donnell ... F-7
Oglesby ... J-13
Oilton ... Q-11
Oklaunion ... E-11
Old Boston ... E-17
Old Center ... H-17
Olden ... G-11
Oldenburg ... L-14
Old Glory ... F-9
Old Ocean ... M-15
Olivia ... N-14
Olmito ... S-13
Olney ... F-11
Olton ... D-7
Omaha ... F-16
Oplin ... H-10
Orange ... K-18
Orange Grove ... P-12
Ore City ... F-16
Orla ... H-5
Osceola ... H-13
Ottine ... M-13
Ovalo ... H-10
Overton ... G-16
Ovilla ... ★
 Vicinity ... Pg. 109, G-8
Ozona ... K-8
Paducah ... E-9
Paige ... L-13
Paint Rock ... J-10
Palacios ... N-14
Palestine ... H-15
Palmer ... G-14
Palo Pinto ... G-12
Pampa ... B-8
Pandora ... M-13
Panhandle ... C-8
Panna Maria ... M-12
Panola ... G-17
Paradise ... F-13
Paris ... E-15
Parnell ... D-9
Pasadena ... L-16
 Vicinity ... Pg. 115, E-8
Patroon ... H-17
Pattison ... L-15
Pattonville ... E-16
Pawnee ... N-12
Paxton ... H-17
Pearland ... ★
 Vicinity ... Pg. 115, G-6

Pearsall ... N-11
Pecos ... J-5
Peggy ... N-12
Penelope ... H-14
Penwell ... H-6
Perrin ... F-12
Perry ... J-13
Perryton ... A-9
Petersburg ... E-8
Petrolia ... E-12
Pettus ... N-12
Petty ... E-15
Pflugerville ... K-13
Phalba ... G-15
Pharr ... S-12
Phillips ... B-7
Pilot Point ... E-14
Pinehurst ... K-17
Pineland ... J-17
Pine Mills ... G-16
Pioneer ... H-11
Pipecreek ... L-11
Pittsburg ... F-16
Placedo ... N-14
Plains ... F-6
Plainview ... D-7
Plano ... F-14
 Vicinity ... Pg. 109, A-10
Plantersville ... K-15
Plata ... L-4
Pleasanton ... N-12
Point Comfort ... N-14
Pollok ... J-16
Ponder ... F-13
Ponta ... H-16
Pontotoc ... K-11
Poolville ... F-12
Port Aransas ... P-13
Port Arthur ... L-17
Port Bolivar ... L-16
Port Isabel ... S-13
Portland ... P-13
Port Lavaca ... N-14
Port Mansfield ... R-13
Port Neches ... L-17
Port "O" Connor ... N-14
Posey ... F-8
Post ... F-8
Poteet ... M-12
Poth ... M-12
Pottsboro ... E-14
Pottsville ... H-12
Powderly ... E-15
Powell ... H-14
Poynor ... H-15
Prairie Dell ... K-13
Prairie Hill ... H-14
Prairie Lea ... L-13
Prairie View ... L-15
Premont ... Q-12
Presidio ... L-4
Priddy ... H-11
Princeton ... F-14
Pritchett ... G-16
Proctor ... H-12
Prosper ... F-14
Purmela ... J-12
Putnam ... G-11
Pyote ... H-5
Quail ... C-9
Quanah ... D-10
Queen City ... F-17
Quemado ... N-9
Quitaque ... D-8
Quinlan ... G-14
Quitman ... G-15
Ralls ... E-8
Ramirez ... Q-12
Randado ... Q-11
Randlett Charlie ... E-12
Randolph ... F-14
Ranger ... G-11
Rankin ... J-7
Ratcliff ... J-16
Ravenna ... E-14
Rayland ... E-10
Raymondville ... R-12
Raywood ... L-17
Reagan ... J-14
Realitos ... Q-11
Red Oak ... M-4
Red Oak ... ★
 Vicinity ... Pg. 109, G-9
Red Springs ... E-10
Reese Vill. ... E-7
Refugio ... P-13
Reklaw ... H-16
Reno ... E-15
Retta ... ★
 Vicinity ... Pg. 108, F-5
Rhome ... F-13
Ricardo ... Q-12
Rice ... H-14
Richardson ... ★
 Vicinity ... Pg. 109, A-9
Richland ... G-14
Richland Hills ... ★
 Vicinity ... Pg. 108, C-3
Richland Springs ... J-11
Richmond ... L-15
Richwood ... M-16
Ridge ... J-14
Riesel ... J-13
Rio Grande City ... R-11
Riomendina ... M-11
Rios ... Q-12
Rio Vista ... H-13
Rising Star ... H-11
River Oaks ... ★
 Vicinity ... Pg. 108, D-2
Riverside ... K-15
Riviera ... Q-12
Robbins ... J-15
Roans Prairie ... K-15
Robert Lee ... H-9
Robinson ... J-13
Robstown ... P-13
Roby ... G-9
Rochelle ... J-11
Rochester ... F-10
Rockdale ... K-14

Rock Island M-14
Rockland J-17
Rockport P-13
Rocksprings L-9
Rockwood J-10
Roganville K-17
Rogers K-13
Roma R-10
Romayor K-16
Roosevelt K-9
Ropesville F-7
Rosanky L-13
Roscoe G-9
Rosebud J-14
Rosenberg M-15
Rosevine J-17
Rosser G-14
Rotan G-9
Round Mountain K-12
Round Rock K-13
Round Top L-14
Rowena H-9
Roxton E-15
Royalty J-6
Rule F-10
Runge M-12
Rusk H-16
Rye K-16
Sabinal M-10
Sacul H-16
Sadler E-14
Sagerton F-10
Saginaw G-13
Vicinity Pg. 108, C-2
Saint Hedwig M-12
Saint Jo E-13
Salado J-13
Salt Falt H-3
Samnorwood C-9
San Angelo J-9
San Antonio M-12
San Augustine H-17
San Benito S-13
Sanderson L-6
Sandia P-12
San Diego P-12
San Elizario H-1
Sanford B-8
San Felipe L-14
San Gabriel K-13
Sanger F-13
San Isidro R-11
San Juan R-12
San Marcus L-12
San Patricio P-12
San Perlita R-13
San Saba J-11
Sansom Park ★
Vicinity Pg. 108, C-2
Santa Anna H-11
Santa Elena R-12
Santa Fe L-15
Santa Maria S-12
Santa Rosa S-12
San Ygnacio Q-10
Saragosa J-5
Saratoga K-17
Sarita Q-12
Satsuma ★
Vicinity Pg. 114, B-2
Savoy E-14
Schertz M-12
Schulenburg L-14
Scotland E-11
Scottsville G-17
Scranton G-11
Seabrook L-16
Vicinity Pg. 115, G-10
Seadrift N-14
Seagraves F-6
Sealy L-14
Sebastian R-13
Sebastopol K-16
Segovia K-10
Seguin M-12
Seminole G-6
Seven Points G-14
Seymour E-11
Shafter L-4
Shallowater E-7
Shamrock C-9
Shannon F-12
Sheffield K-7
Shelbyville H-17
Sheldon L-16
Vicinity Pg. 115, C-8
Sheridan M-14
Sherman F-14
Sherwood J-9
Shiner M-13
Shiro K-15
Shoreacres ★
Vicinity Pg. 115, F-10
Sidney H-11
Sierra Blanca J-3
Silsbee K-17
Silver H-9
Silverton D-8
Simms F-16
Singleton K-15
Sinton P-13
Sisterdale L-11
Sivells Bend E-13
Skellytown B-8
Skidmore P-12
Slaton F-8
Slidell F-13
Slocum H-16
Smada ★
Vicinity Pg. 114, F-3
Smiley M-13
Smyer E-7
Snook K-14
Snyder G-8
Spade E-7
Spanish Fort E-12
Speaks M-14
Spearman B-8
Spicewood K-12
Splendora L-16
Spofford M-9
Spring L-15
Spring Branch L-11
Springlake D-7
Springtown F-12
Spring Valley ★
Vicinity Pg. 114, C-3
Spur F-9
Spurger K-17
Socorro H-1
Somerset M-11
Somerville K-14
Sonora K-9
Sour Lake L-17
South Bend F-11
South Houston ★
Vicinity Pg. 115, E-7
Southlake ★
Vicinity Pg. 108, B-5
Southland F-7
South Padre Island S-13
South Plains D-8
Stafford M-15
Vicinity Pg. 114, F-3
Stamford F-10
Stanton H-7
Star J-12
Stephenville H-12
Sterling City H-8
Stinnett B-8
Stockdale M-12
Stoneburg E-12
Stoneham K-15
Stonwall L-11
Stowell L-16
Stratford A-7
Strawn G-11
Streetman H-14
Study Butte M-5
Sudan E-7
Swenson F-9
Sugar Land L-15
Vicinity Pg. 114, F-2
Sullivan City S-11
Sulphur Sprs. F-15
Summerfield D-6
Sumner E-15
Sundown F-6
Sunray B-8
Sunset F-13
Sutherland Springs M-12
Sweeny M-15
Sweetwater G-9
Swenson F-9
Taft P-13
Tahoka F-7
Talco F-16
Talpa H-10
Tankersley J-9
Tarpley M-10
Tarzan G-7
Tatum G-17
Taylor K-13
Taylor Lake Village ★
Vicinity Pg. 115, F-9
Teague H-14
Telegraph K-9
Telephone E-15
Telferner N-14
Tell D-9
Temple J-13
Tenaha H-17
Tenneseee Colony H-15
Tennyson H-9
Terlingua M-5
Terrell G-14
Texhoma A-8
Texarkana F-17
Texas City M-16
Texico D-6
Texline A-6
Thackerville E-13
Thalia E-10
The Grove J-13
The Woodlands L-15
Thomaston M-13
Thorndale K-13
Thornton J-14
Thorp Spring G-12
Thrall K-13
Three Rivers N-12
Throckmorton F-10
Tilden N-11
Timpson H-17
Tioga F-13
Tira F-15
Tivoli N-14
Toco E-15
Tolar G-12
Tomball K-15
Tool H-14
Topsey J-12
Tornillo H-1
Toyah J-5
Trammells ★
Vicinity Pg. 114, G-4
Trent G-10
Trenton F-14
Trickham H-11
Trinidad H-15
Trinity J-15
Troup H-16
Troy J-13
Tucker H-15
Tuleta N-13
Tulia D-7
Turkey D-9
Turnertown H-16
Tuscola H-10
Twin Sisters L-12
Twitty C-9
Tye G-10
Tyler G-16
Tynan P-12
Umbarger C-7
Unity E-15
Universal City M-12
University Park ★
Vicinity Pg. 109, C-9
Utopia M-10
Uvalde M-10
Valentine K-4
Valera H-10
Valley Mills H-13
Valley Spring K-11
Valley View F-13
Van G-15
Van Alstyne F-14
Vancourt J-9
Vanderbilt N-14
Vanderpool L-10
Van Horn J-3
Van Vleck M-15
Vashti E-12
Vealmoor G-8
Vega C-7
Venus G-13
Vera E-10
Vernon E-11
Vincent G-8
Victoria N-13
Vidor L-17
Village Mills K-17
Voca J-11
Voss H-10
Votaw K-17
Waco H-13
Wadsworth N-15
Waelder L-13
Waka A-9
Wake Village E-17
Wall J-9
Waller L-15
Wallis L-14
Walnut Sprs. H-13
Warda L-13
Warren K-17
Washington K-15
Waskom G-17
Water Valley H-9
Watson K-12
Watsonville ★
Vicinity Pg. 108, F-5
Waxahachie G-14
Wayside D-8
Weatherford G-12
Weaver F-15
Webster M-16
Vicinity Pg. 115, G-8
Weches J-16
Weesatche N-13
Weimar L-14
Weinort F-10
Welch F-7
Wellington D-9
Wellman F-7
Wells J-16
Weslaco S-12
West H-13
Westbrook G-8
West Columbia M-15
Westfield L-15
Westhoff M-13
Westlake ★
Vicinity Pg. 108, A-4
West Orange K-17
West Point L-13
Westover F-11
Westworth ★
Vicinity Pg. 108, D-2
Wharton M-15
Wheeler C-9
Wheeler G-15
Wheelock K-14
White Deer C-8
Whiteface E-6
Whiteflat E-9
Whitehouse G-15
Whitesboro E-13
Whitewright F-14
Whitharral E-7
Whitney H-13
Whittsett N-12
Wichita Falls E-12
Wickett H-5
Wiergate J-18
Wildorado C-7
Wildwood K-17
Wilkinson F-16
Willow City K-11
Wills Point G-15
Wilmer ★
Vicinity Pg. 109, F-10
Wilson F-8
Wimberly L-12
Winchell J-11
Windhorst E-12
Windom F-15
Winfield F-16
Wingate H-9
Wink H-5
Winnie L-17
Winnsboro F-15
Winona G-16
Winters H-9
Wolfe City F-15
Wolfforth F-7
Woodlake J-16
Woodland Hills ★
Vicinity Pg. 109, F-8
Woodlawn G-17
Woodsboro P-13
Woodson F-11
Woodville K-17
Woodward N-10
Wortham H-14
Yoakum M-13
Yorktown M-13
Zephyr H-11
Zavalla J-17
Zorn L-12

UTAH
Pages 82-83

Abraham H-4
Adamsville K-4
Alpine F-5
Alta F-6
Altamont F-8
Alton L-4
American Fork F-5
Aneth M-9
Angle K-5
Annabella J-5
Antimony K-5
Arcadia F-8
Aurora J-5
Austin J-5
Axtell H-5
Vicinity Pg. 142, E-2
Bear River City D-5
Beaver K-4
Benjamin F-5
Bennion ★
Vicinity Pg. 142, F-5
Beryl K-2
Bicknell K-5
Birdseye G-6
Black Rock J-4
Blanding L-9
Bluebell F-8
Blue Cr. C-4
Bluff L-9
Bluffdale F-5
Vicinity Pg. 142, J-5
Bonanza G-9
Boneta F-7
Boulder K-6
Bountiful E-5
Brian Head L-4
Bridgeland F-8
Brigham City D-5
Bryce Canyon L-4
Burmester E-4
Burrville J-5
Cannonville L-4
Carbonville G-7
Castle Dale H-7
Castle Rock E-6
Cave C-5
Cedar City L-3
Cedar Cr. C-3
Cedar Fort F-5
Cedar Springs D-4
Cedarview F-8
Center Cr. F-6
Centerfield H-5
Centerville E-5
Central L-2
Central J-5
Charleston F-6
Chester H-5
Circleville K-4
Cisco J-9
Clarkston C-5
Clawson H-6
Clear Creek C-3
Clearcreek G-6
Clearfield E-5
Cleveland H-7
Clinton D-5
Clover F-4
Coalville E-6
Colton G-7
Copperton ★
Vicinity Pg. 142, G-2
Corinne D-5
Cornish C-5
Cottonwood ★
Vicinity Pg. 142, F-7
Cove Fort J-4
Crescent Jct. J-8
Croydon E-6
Delle E-4
Delta H-4
Deseret H-4
Desert Mound L-3
Dividend G-5
Draper F-5
Vicinity Pg. 142, H-6
Duchesne F-8
Dugway F-4
East Carbon City H-8
Eastland L-9
Echo E-6
Elberta G-5
Elmo H-7
Elsinore J-5
Emery J-6
Enoch L-3
Enterprise L-2
Ephraim H-6
Escalante L-5
Etna D-2
Eureka G-5
Fairfield F-5
Fairview G-6
Farmington E-5
Faust F-4
Fayette H-5
Ferron H-6
Fillmore H-5
Fish Sprs. G-3
Flowell H-4
Fort Duchesne F-8
Fountain Green G-5
Francis F-6
Freedom G-5
Fremont J-6
Fremont Jct. J-6
Fruit Hts. E-5
Fruitland F-7
Fry Canyon L-8
Gandy G-2
Garden City C-6
Garrison H-2
Geneva F-5
Genola G-5
Glendale L-4
Glenwood J-5
Goshen G-5
Granger ★
Vicinity Pg. 142, D-4
Granite ★
Vicinity Pg. 142, G-7
Grantsville E-4
Green Lake E-9
Green River H-8
Greenville K-4
Greenwich J-5
Greenwood H-4
Grouse Cr. C-2
Gunlock L-2
Gunnison H-5
Gusher F-8
Hamilton Fort L-3
Hanksville K-7
Hanna F-7
Harrisville D-5
Hatch L-4
Hatton J-4
Heber City F-6
Helper G-7
Henefer E-6
Henrieville L-5
Herriman ★
Vicinity Pg. 143, J-4
Hiawatha H-6
Hinckley H-4
Holden H-5
Vicinity Pg. 143, E-7
Honeyville D-4
Hot Springs D-5
Howell C-4
Hoytsville E-6
Huntington H-7
Huntsville D-6
Hurricane M-3
Hyde Park C-5
Hyrum D-5
Indianola G-6
Ioka F-8
Iron Sprs. L-3
Ivins M-2
Jensen F-9
Jericho G-5
Joseph J-4
Junction K-4
Kamas F-6
Kanab M-4
Kanarraville L-3
Kanosh J-4
Kaysville E-5
Kearns ★
Vicinity Pg. 143, E-3
Kelton D-4
Kenilworth G-7
Kimball Jct. E-6
Kingston K-5
Knolls E-3
Koosharem J-5
Lake Pt. Jct. E-4
Lake Shore F-5
Laketown C-6
Lake View F-5
Lapoint F-8
La Sal K-9
La Sal Jct. K-9
La Verkin L-3
Leamington H-5
Leeds L-3
Leeton F-8
Lehl F-5
Leota F-9
Levan G-5
Lewiston C-5
Loa J-6
Logan D-5
Long Valley Jct. L-4
Low E-3
Lucin D-2
Lund K-3
Lyman J-6
Lynndyl G-4
Maeser F-9
Magna E-5
Vicinity Pg. 143, E-2
Manderfield J-4
Manila E-8
Manti H-6
Mantua D-5
Mapleton F-6
Marion E-6
Marysvale J-4
Mayfield H-6
Meadow J-4
Meadowville C-6
Mendon D-5
Mexican Hat M-8
Midvale E-5
Vicinity Pg. 143, F-6
Midway F-6
Milburn G-6
Milford J-3
Mills H-5
Minersville K-3
Moab J-9
Modena K-2
Molen H-6
Mona G-5
Monarch F-8
Monroe J-5
Montezuma Creek M-9
Monticello K-9
Moore H-6
Morgan E-5
Moroni G-6
Mountain Home F-7
Mt. Carmel L-4
Mt. Carmel Jct. L-4
Mt. Emmons F-7
Mt. Pleasant G-6
Murray E-5
Vicinity Pg. 143, E-6
Myton F-8
Naples F-9
Neola F-8
Nephi G-5
Newcastle L-2
New Harmony L-3
Newton C-5
N. Logan C-6
N. Salt Lake E-5
Oak City H-5
Oakley E-6
Oasis H-4
Ogden D-5
Orangeville H-6
Orderville L-4
Orem F-6
Ouray F-9
Panguitch K-4
Paradise D-5
Paragonah K-4
Park City E-6
Park Valley C-3
Parowan K-4
Payson G-5
Peoa E-6
Perry Willard D-5
Pigeon Hollow Jct. H-6
Pine Valley L-2
Pintura L-3
Plain City D-5
Pleasant Grove F-6
Pleasant View D-5
Plymouth C-5
Portage C-5
Porterville E-6
Price G-7
Providence D-5
Provo F-6
Randolph D-6
Redmond H-5
Richfield J-5
Richmond C-5
Riverdale D-5
Riverton F-5
Vicinity Pg. 143, H-5
Rockville M-3
Roosevelt F-8
Rosette C-3
Rowley Jct. E-4
Roy D-5
Ruby's Inn L-4
St. George M-3
St. John Sta. F-4
Salem G-5
Salina H-5
Salt Lake City E-5
Vicinity Pg. 143
Sandy F-5
Vicinity Pg. 143, G-6
Santa Clara M-2
Santaquin G-5
Scipio H-5
Scofield G-6
Sevier J-4
Shivwits M-2
Sigurd J-5
Silver City G-5
Smithfield C-5
Smyths J-3
Snowville C-4
Soldier Summit G-7
South Jordan F-5
Vicinity Pg. 143, G-5
S. Ogden D-5
S. Salt Lake E-5
Vicinity Pg. 143, D-6
Spanish Fork F-6
Spring City H-6
Springdale M-3
Spring Lake G-5
Springville F-6
Sterling H-6
Stockton F-4
Summit K-3
Summit Pt. K-9
Sunnyside G-8
Sunset E-5
Sutherland H-4
Syracuse E-5
Tabiona F-7
Talmage F-7
Taylors ★
Vicinity Pg. 143, E-5
Teasdale K-6
Terrace D-6
Thistle G-6
Thompson H-9
Toquerville L-3
Torrey K-6
Tooele F-4
Trementon D-5
Tropic L-5
Trout Cr. G-2
Tucker G-6
Vicinity Pg. 143, E-5
Upalco F-8
Uvada L-2
Venice J-5
Vernon F-4
Vernal F-9
Veyo L-2
Virgin M-3
Wahsatch D-6
Wales H-5
Wanship E-6
Washington M-3
Washington Terrace D-5
Wellington G-7
Wellsville D-5
Wendover E-2
West Jordan F-5
Vicinity Pg. 143, F-4
West Point D-4
W. Valley City E-5
Whiterocks F-8
Widtsoe Jct. K-5
Woodruff D-6
Woods Cross E-5
Woodside H-8
Zane K-2

VERMONT
Page 57

Addison E-1
Albany C-3
Alburg B-1
Arlington J-1
Ascutney G-3
Averill B-5
Bakersfield B-2
Barnard F-3
Barnet D-4
Barton C-4
Barre E-3
Bellows Falls H-3
Bennington J-1
Bethel F-3
Bloomfield C-5
Bomoseen G-1
Bondville H-2
Broadford Cen. E-3
Bradford E-4
Brandon F-2
Brattleboro J-3
Bridgewater Cors. G-3
Bridport E-1
Bristol E-1
Brunswick Sprs. C-5
Burlington D-1
Cambridge J-1
Cambridgeport H-3
Canaan B-5
Castleton G-1
Charlotte D-1
Chelsea E-3
Chester H-3
Chester Depot H-3
Chimney Cor. C-2
Concord D-4
Cornwall E-1
Danby G-2
Dorset H-1
E. Barnet D-4
E. Berkshire B-2
E. Braintree E-3
E. Brookfield E-3
E. Burke C-4
E. Calais D-3
E. Dorset H-2
E. Fairfield C-2
E. Granville E-2
E. Hardwick D-3
E. Haven C-4
Eden C-3
Eden Mills C-3
Enosburg Falls B-2
Essex Jct. D-2
Fairfax C-2
Fair Haven G-1
Ferrisburg E-1
Franklin B-2
Gassetts H-3
Gaysville F-3
Glover C-4
Grafton H-3
Grand Isle C-1
Granville E-2
Groton E-4
Guilford J-3
Hancock F-2
Hardwick D-3
Heartwellville J-2
Hinesburg D-2
Holland B-4
Hubbardton F-1
Hyde Park C-3
Irasburg B-3
Irasville E-2
Island Pond C-4
Jacksonville J-2
Jamaica H-2
Jay B-3
Jeffersonville C-2
Johnson C-3
Lake Elmore C-3
Leicester F-2
Londonderry H-2
Lowell B-3
Ludlow G-2
Lunenberg D-5
Lyndon D-4
Lyndonville C-4
Maidstone C-5
Manchester H-2
Manchester Cen. H-2
Marlboro J-2
Marshfield D-3
Middlebury E-1
Middletown Sprs. G-1
Mill Village E-3
Montgomery Cen. B-3
Montpelier D-3
Morgan Cen. B-4
Morrisville C-3
Newbury E-4
Newfane J-2
New Haven E-1
New Haven Jct. E-1
Newport D-4
N. Concord D-4
N. Dorset H-2
Northfield E-2
N. Hero C-1
N. Hyde Park C-3
N. Rupert H-1
Norton B-4
Norwich F-3
Orleans B-4
Orwell F-1
Passumpsic D-4
Pawlet H-1
Pittsfield F-2
Pittsford F-2
Pittsford Mills F-2
Plainfield D-3
Plymouth Union G-2
Plymouth C-2
Pompanoosuc F-3
Post Mills F-3
Poultney G-1
Pownal J-1
Pownal Cen. J-1
Putney J-3
Quechee G-3
Randolph F-2
Richford B-2
Richmond D-2
Ripton E-2
Rochester F-2
Robinson F-2
Royalton F-2
Roxbury E-2
Rutland G-2
St. Albans B-2
St. George D-2
St. Johnsbury D-4
Searsburg J-2
Shafsbury J-1
Shafsbury Cen. J-2
Shelburne D-1
Sherburne Cen. G-2
Shoreham F-1
Simonsville H-2
S. Burlington D-1
S. Dorset G-2
S. Hinesburg D-2
S. Shafsbury J-2
S. Starksboro E-2
S. Royalton F-3
S. Ryegate E-4
S. Walden D-3
S. Wallingford G-2
Springfield H-3
Starksboro E-2
Stockbridge F-2
Sudbury F-1
Sunderland J-2
Swanton B-1
Ticonderoga F-1
Townshend J-2
Troy B-3
Tunbridge F-3
Tyson G-2
Underhill Flats D-2
Union Village F-3
Vergennes E-1
Vershire E-3
Waits River E-3
Waitsfield E-2
Wallingford G-2
Waltham E-1
Wardsboro J-2
Warren E-2
Waterbury D-2
Waterbury Ctr. D-3
Washington E-3
Websterville E-3
Wells G-1
W. Addison E-1
W. Braintree E-2
W. Bridgewater G-2
W. Burke C-4
W. Danville D-3
W. Dover J-2
W. Fairlee F-3
Westfield B-3
W. Groton E-3
W. Hartford F-3
W. Rutland G-2
W. Topsham E-3
W. Wardsboro J-2
White River Jct. G-3
Whiting F-1
Wilmington J-2
Windsor G-3
Wolcott C-3
Woodbury D-3
Woodford J-1
Woodstock G-3
Worcester D-3

VIRGINIA
Pages 46-47

Abingdon J-3
Accomac G-12
Achilles H-11
Adner G-11
Afton F-8
Alberta H-9
Alexandria D-10
Vicinity Pg. 148, H-4
Altavista H-7
Amelia G-9
Amherst G-7
Amissville E-9
Amonate H-4
Annandale ★
Vicinity Pg. 148, H-2
Appalachia H-2
Appomattox G-8
Ark G-11
Arlington E-10
Vicinity Pg. 148, F-3
Arvonia G-8
Ash Grove ★
Vicinity Pg. 148, E-1
Ashland G-10
Atkins H-4
Augusta Springs F-7
Austinville H-5
Axton J-6
Bacons Castle H-11
Bailey's Crossroads ★
Vicinity Pg. 148, G-3
Ballsville G-9
Banco E-9
Bandy H-4
Barhamsville G-11
Bassett J-6
Bassett Forks J-6
Bastian H-4
Bavon G-11
Bayview H-12
Bedford G-7
Belle Haven G-12
Vicinity Pg. 148, J-4
Belleview ★
Vicinity Pg. 148, E-1
Bells Crossroads F-9
Bensley G-10
Bent Creek G-8
Bentonville E-9
Bergton E-8
Berryville D-9
Beverly Forest ★
Vicinity Pg. 148, J-2
Big Island G-7
Big Stone Gap H-2
Birchleaf H-3
Bishop H-4
Blacksburg H-5
Blackstone H-9
Blackwater J-2
Blairs J-7
Bland H-5
Bluefield H-4
Bluemont D-9
Blue Ridge G-6
Bon Air G-9
Boones Mill H-6
Boswells Tavern F-9
Bowling Green F-10
Boyce D-9
Boydton J-8
Boykins J-10
Branchville J-10
Brandy Station E-9
Bremo Bluff G-9
Bridgewater F-7
Brightwood E-9
Bristol J-3
Broadford H-4
Broadway E-8
Brodnax J-9
Brookneal H-7
Brownsburg F-7
Brunswick J-9
Buchanan G-7
Buckingham G-8
Bucknell Manor ★
Vicinity Pg. 148, J-4
Buena Vista G-7
Burgess F-11
Burke ★
Vicinity Pg. 148, J-1
Burkeville H-9
Burn Chimney H-6
Burrowsville H-10
Callaghan G-6
Callands H-7
Callao F-11
Calverton E-10
Cana J-5
Cape Charles G-12
Capron J-10
Carmel Church F-10
Carson H-10
Cartersville G-9
Castlewoods H-3
Catalpa E-9
Catlett E-9
Cave Spring H-6
Cedar Bluff H-4
Central Garage G-10
Chapel Acres ★
Vicinity Pg. 148, K-1
Charles City G-10
Charlotte Court House H-8
Charlottesville F-8
Chase City J-8
Chatham J-7
Cheriton G-12
Chesapeake J-12
Chester G-10
Chilhowie H-4
Chincoteague F-13
Christiansburg H-6
Chula G-9
Clarksville J-8
Claudville J-6
Clear Brook D-9
Clifford G-7
Clifton Forge F-6
Clinchco H-3
Clinchport J-2
Clintwood H-3
Clover H-8
Cluster Springs J-8
Cochran H-9
Cody H-8
Coeburn H-2
Colleen G-8
Collierstown G-7
Collinsville J-6
Colonial Beach F-10
Colonial Heights H-10
Concord G-7
Copper Hill H-6
Courtland J-10
Covesville F-8
Covington G-6
Craigsville F-7
Crewe H-9
Cross Junction C-9
Crows G-6
Crozet F-8
Cuckoo F-9
Cullen H-8
Culpeper E-9
Cumberland G-9
Dale E-9
Dale City E-10
Damascus J-3
Danieltown H-9
Dante H-3
Danville J-7
Davenport H-3
Dawn F-10
Dayton E-7
Deerfield F-7
Deltaville G-11
De Witt H-9
Dickensonville H-3
Dillwyn G-8
Dinwiddie H-9
Disputanta H-10
Dixie F-9
Doswell F-10
Dot J-2
Drakes Branch H-8
Draper H-5
Drewryville J-10
Dryden J-2
Dublin H-5
Dugspur H-5
Dumfries E-10
Dungannon H-2
Eagle Rock G-6
Earls H-9
Earnham F-11
Eastville G-12
Edgerton J-9
Edinburg E-8
Elberon H-11
Elkton E-8
Elliston H-6
Elon G-7
Emporia J-10
Esmont G-8
Evington H-7
Ewing J-1
Exmore G-12
Fairfax ★
Vicinity Pg. 148, G-1

★ Denotes City located only on Vicinity map. City not located on State map.

VIRGINIA

Fairfield F-7
Falling Spring F-6
Falls Church ★
  Vicinity Pg. 148, G-2
Falmouth E-10
Fancy Gap J-5
Farmville G-9
Ferrum H-6
Fieldale J-6
Fincastle G-6
Finchley F-8
Fishersville F-8
Flint Hill E-9
Floyd H-6
Forest G-7
Forestville E-8
Fork Union G-8
Fort Blackmore H-3
Franconia ★
  Vicinity Pg. 148, J-3
Franklin J-11
Fredericksburg E-10
Fries J-5
Front D-9
Fulks Run E-8
Galax J-5
Gate City J-2
Georges Tavern G-9
Gladehill H-6
Glade Spring H-4
Glasgow G-7
Gloucester G-11
Gloucester Point H-11
Golansville F-10
Goochland G-9
Gordonsville E-9
Gore D-8
Goshen F-7
Great Falls ★
  Vicinity Pg. 148, D-1
Green Bay H-9
Greenway ★
  Vicinity Pg. 148, E-1
Gretna H-7
Grottoes F-8
Groveton ★
  Vicinity Pg. 148, J-4
Grundy H-3
Guilford G-12
Gum Spring G-9
Gwynn G-11
Hague F-11
Halifax H-8
Hampton H-11
Hanover G-10
Harrisonburg E-8
Hartwood E-10
Hayfield D-9
Haynesville F-11
Healings Springs F-6
Hermitage F-8
Herndon D-10
Hickory J-11
Hightown E-7
Hillsboro D-9
Hillsville H-5
Hollins G-6
Homeville H-10
Honaker H-3
Hopewell G-10
Howardsville G-8
Independence J-4
Iron Gate G-7
Irvington G-11
Ivanhoe J-5
Ivor H-11
Jamestown H-11
Jarratt J-10
Jetersville G-9
Jewell Ridge H-4
Johnsontown G-12
Jonesville J-2
Jordan Mines G-6
Keene F-8
Keene Mill Heights ★
  Vicinity Pg. 148, J-1
Keller G-12
Kenbridge H-9
Kenmore ★
  Vicinity Pg. 148, E-1
Keysville H-9
Kilmarnock G-11
King George F-10
Kings Park ★
  Vicinity Pg. 148, H-1
Kiptopeke H-12
Konnarock J-4
Lacey Spring E-8
La Crosse J-9
Ladysmith F-10
Lancaster G-11
Langley ★
  Vicinity Pg. 148, E-2
Laurel Fork J-5
Lawrenceville J-9
Lebanon H-3
Leesburg D-10
Leesville F-7
Lerty F-10
Lexington G-7
Lightfoot G-11
Lignum E-9
Littleton H-10
Lively G-11
Locust E-9
Long Island H-7
Louisa F-9
Lovettsville D-9
Low Moor G-6
Lucketts D-9
Lunenburg H-9
Luray E-8
Lyells F-11
Lynchburg G-7
Madison E-9
Madison Heights G-7
Madison Mills E-9
Manakin G-9
Manassas E-9
Manassas Park E-10
Manquin G-10
Mappsville F-12
Marion H-4
Markham D-9
Marshall D-9
Martinsville J-6
Massies Mill F-7
Mathews G-11
Matoaca H-10
Maurertown D-8
Mauzy E-8
McDowell E-7
McGaheysville E-8
McKenney H-9
McLean ★
  Vicinity Pg. 149, E-3
McLean Hamlet ★
  Vicinity Pg. 149, E-1

Meadows of Dan H-6
Mechanicsville G-10
Meherrin H-8
Melfa G-12
Middleburg D-10
Middletown D-8
Midland E-9
Midlothian G-9
Milford F-10
Millboro F-7
Millboro Springs F-7
Millers Tavern F-11
Mineral F-9
Moneta H-7
Monroe G-7
Montebello F-7
Monterey E-7
Montpelier F-10
Montross F-10
Morrisville E-9
Moseley G-9
Mt. Airy H-7
Mt. Jackson E-8
Mt. Olive D-8
Mt. Rush G-8
Mt. Sidney F-7
Mustoe F-7
Narrows G-5
Naruna H-8
Nassawadox G-12
Natural Bridge G-7
Nellysford F-8
Nelsonia G-12
New Baltimore E-9
Newbern H-5
New Castle G-6
New Kent G-10
New Market E-8
New Post F-10
Newport G-6
Newport News H-11
Nickelsville J-3
Nora H-3
Norfolk H-12
North Garden F-8
Norton H-2
Nottoway H-9
Nuttsville G-11
Oak Hall F-12
Oak Wood H-3
  Vicinity Pg. 149, K-3
Oakville G-8
Oilville G-9
Onancock G-12
Opal E-9
Orange E-9
Paint Bank G-6
Painter G-12
Palmer Springs J-9
Palmyra F-9
Paris D-9
Parksley G-12
Patrick Springs J-5
Pearisburg G-5
Pembroke H-5
Pendaw ★
  Vicinity Pg. 149, J-4
Penhook H-7
Pennington Gap H-2
Penn Laird E-8
Petersburg H-10
Pimmit Hills ★
  Vicinity Pg. 149, F-2
Piney River G-7
Pocahontas H-4
Pohick Estates ★
  Vicinity Pg. 149, K-1
Poplar Camp H-5
Poplar Hill H-5
Poquoson H-11
Port Republic F-8
Port Royal F-10
Portsmouth H-11
Post Oak F-9
Pound H-2
Prospect Hill ★
  Vicinity Pg. 149, F-7
Providence Forge G-11
Prince George H-10
Princess Anne J-12
Pulaski H-5
Pungoteague G-12
Purcellville D-9
Purdy G-7
Quantico E-10
Quinby G-12
Radford H-5
Raven H-3
Rawley Springs E-7
Red Ash H-3
Red Oak J-8
Rehoboth H-8
Remington E-9
Renan H-7
Reston D-10
Rice H-9
Richlands H-3
Richmond G-10
Ridgeway J-6
Rileyville E-8
Riner H-5
Rixeyville E-9
Roanoke H-6
Rochelle F-9
Rockbridge Baths F-7
Rockville G-9
Rocky Gap H-5
Rocky Mount H-6
Rose Hill J-1
Roseland G-8
Round Hill D-9
Roxbury G-10
Ruckersville F-9
Rural Retreat H-4
St. Paul H-3
St. Stephens Church G-11
Salem G-6
Saltville H-4
Saxis G-11
Scotland H-11
Scottsville G-8
Sebrell J-10
Shadewell F-8
Shady Oak ★
  Vicinity Pg. 149, D-1
Shipman G-8
Shenandoah E-8
Sheppards G-8
Sinking Creek G-6
Skippers J-10
Snell F-10
S. Boston J-8
South Hill J-9
Spencer J-6
Sperryville E-9
Spotsylvania F-9
Springfield ★
  Vicinity Pg. 149, J-2

Spring Grove H-11
Spring Hill ★
  Vicinity Pg. 149, E-1
Sprouses Corner G-8
Stacy G-3
Stafford E-10
Stanardsville F-8
Stanley E-8
Stanleytown J-6
Staunton F-7
Steeles F-7
Stephens City D-9
Stevensburg E-9
Stevensville G-11
Stewartsville H-6
Stony Creek H-10
Strasburg D-8
Stuart J-6
Sturats Draft F-7
Studley G-10
Sugar Grove J-4
Sunbright J-2
Surry H-11
Sussex H-10
Sweet Chalybeate G-6
Sydnorsville H-6
Sylvatus H-5
Syria E-8
Talleysville G-10
Tappahannock F-10
Tasley G-12
Tavern F-7
Tazewell H-4
Templeman F-11
The Falls H-9
Thornburg F-10
Timberville E-8
Trenholm G-9
Triangle E-10
Troutdale J-4
Troutville G-6
Turbeville J-8
Unionville F-9
Upperville D-9
Urbanna G-11
Vanderpool E-7
Vansant H-3
Vera G-8
Verona F-7
Victoria H-9
Vienna ★
  Vicinity Pg. 149, F-1
Village F-11
Villboro E-9
Vinton H-6
Virginia F-8
Virginia Beach H-12
Va. Mineral Springs G-6
Volens H-8
Wachapreague G-12
Wakefield H-10
Warm Springs F-6
Warrenton E-9
Warsaw F-11
Washington E-9
Waverly H-10
Waynesboro F-8
Waynewood ★
  Vicinity Pg. 149, K-4
Weber City J-2
Wellington ★
  Vicinity Pg. 149, K-4
West Augusta F-7
W. Point G-11
Whaleyville J-11
White Marsh H-11
Wilderness E-9
Williamsburg H-11
Williamstown F-7
Willis H-6
Willis Wharf G-12
Wilsons H-9
Winchester D-9
Windsor H-11
Windsor Estates ★
  Vicinity Pg. 149, J-3
Wingina G-8
Winterpock G-10
Wise H-3
Woodbridge E-10
Woodlawn J-5
Woods Mill G-8
Woodstock D-8
Woodville E-9
Wylliesburg H-8
Wytheville H-5
Yorktown H-11
Zion X-Roads E-9
Zuni H-11

WASHINGTON
Pages 84-85

Aberdeen D-3
Acme A-5
Addy B-10
Adna E-4
Aeneas B-9
Agnew B-4
Airway Heights D-10
Aladdin B-11
Albion E-11
Alder E-5
Alderwood Maner ★
  Vicinity Pg. 147, A-4
Alger B-5
Allen B-5
Allyn C-4
Almira D-9
Aimota F-11
Aloha D-3
Altoona F-3
Alpha E-4
Amanda Park D-2
American River E-6
Anacortes B-4
Anatone G-11
Appledale B-10
Appleyard D-7
Arden C-7
Ardenvoir C-7
Ariel D-4
Arlington B-5
Arizna B-10
Ashford E-5
Asotin G-11
Auburn C-5
Ayer F-9
Baring C-6
Basin City E-8
Battle Ground E-4
Batum E-9
Bay Center E-3
Bay City E-2
Belfair C-4

Bellevue ★
  Vicinity Pg. 147, E-6
Bellingham A-5
Benge E-10
Benton City F-8
Beverley D-7
Bickleton G-7
Big Lake B-5
Bingen G-6
Birch Bay A-4
Black Diamond D-5
Blaine A-4
Bluecreek B-10
Bluestem D-10
Blyn B-4
Bossburg B-10
Bothell ★
  Vicinity Pg. 147, B-5
Boundary A-11
Boyds B-10
Brady D-3
Bremerton C-4
Brewster C-8
Bridgeport C-8
Brier ★
  Vicinity Pg. 147, A-4
Brinnon C-4
Brooklyn E-3
Brush Prairie E-4
Buckeye C-11
Buckley D-5
Bucoda E-4
Buena F-7
Burbank G-9
Burien ★
  Vicinity Pg. 147, H-3
Burnett D-5
Byrne Mawr ★
  Vicinity Pg. 147, H-5
B Z Corner G-5
Camano B-4
Camas G-4
Camden C-11
Canby D-10
Carbonado E-5
Carborg C-4
Carlisle D-3
Carlton B-7
Carnation C-5
Carrolls G-4
Carson G-5
Cashmere D-7
Castle Rock F-3
Cathlamet F-3
Cedar Falls D-5
Cedarville E-3
Cedonia C-10
Center C-4
Centralia E-4
Chattaroy C-11
Chehalis E-4
Chelan C-7
Cheney D-10
Chesaw A-9
Chewelah C-10
Chinook F-2
Cinebar E-5
Clallam Bay B-3
Clarkston F-11
Clayton C-11
Cle Elum D-6
Cliffdell E-6
Cloverland G-11
Clyde F-2
Coalfield ★
  Vicinity Pg. 147, H-6
Colbert D-11
Coles Corner C-6
Colfax E-10
College Place G-9
Colton F-11
Columbia Beach C-5
Colville B-10
Conconully B-8
Concrete B-6
Connell E-9
Conway B-5
Cook G-5
Copalis Beach D-2
Copalis Crossing D-3
Corfu E-8
Cosmopolis D-3
Couada C-10
Cougar F-4
Coulee City D-9
Coulee Dam C-9
Coupeville B-4
Cowiche E-6
Crescent D-10
Crescent Bar D-8
Creston D-9
Crystal Mountain E-6
Cunningham E-9
Curlew A-9
Custer A-4
Dalkena C-11
Danville A-10
Darrington B-6
Davenport D-10
Dayton G-10
Deer Harbor A-4
Deer Park C-10
Del Rio C-8
Deming A-5
Desert Aire F-8
Des Moines ★
  Vicinity Pg. 147, K-3
Dewatto D-4
Diamond D-10
Diamond Lake C-11
Disautel B-8
Discovery Bay B-4
Dixie G-10
Dodge F-10
Doe Bay A-4
Doty E-3
Downs D-10
Dryden D-7
Duluth G-4
Dungeness B-4
Du Pont C-4
Dusty F-10
Eastgate ★
  Vicinity Pg. 147, F-6
Easton D-6
Eastsound A-4
East Wenatchee D-7
Eatonville E-5
Echo B-10
Edison B-5
Edmonds ★
  Vicinity Pg. 147, A-4
Edwall D-10
Elbe E-5
Eldon D-4

Electric City C-8
Elk C-11
Ellensburg E-7
Ellisford B-8
Elma D-3
Elmer City C-9
Elwha B-3
Endicott E-10
Entiat D-7
Enumclaw D-5
Ephrata D-8
Ethel E-4
Eureka F-9
Evans B-10
Everett C-5
Everson A-5
Ewan E-10
Fairbanks E-11
Fairfield D-11
Fairholm B-3
Fall City D-5
Farmington E-11
Farris C-7
Federal Way D-5
Felida G-4
Ferndale A-9
Ferry F-9
Finley G-8
Fircrest ★
Ford C-10
Forks C-2
Four Lakes D-11
Frances E-3
Freeland C-5
Freeman D-11
Friday Harbor B-4
Gardiner B-4
Garfield E-11
Garland Mineral Sprs. C-6
George D-8
Gig Harbor C-4
Gilmer E-4
Glacier A-5
Glade F-9
Gleed E-7
Glenoma E-5
Glenwood G-6
Glifford B-10
Gold Bar C-6
Goldendale G-6
Goose Prairie E-6
Gould City F-10
Govan C-9
Grand Coulee C-9
Grandview F-7
Granger F-7
Granite Falls C-5
Grayland E-2
Grays River F-3
Greenacres D-11
Green Mountain G-4
Greenwater D-6
Grisdale D-3
Gromore F-6
Grotto C-6
Guemes B-4
Hamilton B-5
Hartline D-8
Harrah F-7
Harrington D-9
Hatton E-9
Havillah A-9
Hay F-10
Heart O' The Hills B-3
Holden B-7
Holly C-4
Home D-4
Hoodsport D-3
Hooper F-10
Hoquiam D-2
Hover G-8
Hunters C-10
Husum G-5
Hyak D-6
Illia F-10
Ilwaco F-2
Index C-6
Incheium B-10
Ione B-11
Issaquah D-5
Juanita ★
  Vicinity Pg. 147, J-2
Joyce B-3
Kahlotus E-9
Kalaloch C-2
Kalama F-4
Kamilche D-3
Kapowsin D-5
Keller C-9
Kelso E-4
Kenmoore ★
  Vicinity Pg. 147, B-4
Kennewick G-8
Kent ★
  Vicinity Pg. 147, J-4
Kettle Falls B-10
Keyport C-4
Keystone B-4
Kingston C-4
Kiona ★
  Vicinity Pg. 147, A-1
Kirkland ★
  Vicinity Pg. 147, D-5
Kittitas D-7
Klaus E-4
Klickitat G-6
Klipsan Beach E-2
Kooskooskie G-10
Krupp D-9
La Center G-4
Lacey D-4
La Conner B-4
La Crosse F-10
La Grande E-5
Lakebay C-4
Lake Stevens C-5
Lamona D-9
Lamont E-10
Langley C-5
La Push C-2
Latah D-11
Laurel A-5
Laurier A-10
Lawrence D-6
Leavenworth C-7
Lebam E-3
Lester D-6
Lexington F-4
Liberty D-7
Lilliwaup D-4
Lind E-9
Littlerock E-4
Long Beach F-2
Longbranch C-4
Longmire E-5
Longview F-4

Loomis B-8
Loon Lake C-11
Lopez B-4
Lowden G-9
Low Gap E-8
Lucerne B-7
Lummi A-4
Lyle G-6
Lyman B-5
Lynden A-5
Lynnwood ★
  Vicinity Pg. 147, A-3
Mabton F-7
Malaga D-7
Malden E-10
Malo A-9
Malone E-4
Malott B-8
Mansfield C-8
Maple Beach A-4
Maple Falls A-5
Maplewood ★
  Vicinity Pg. 147, H-5
Marblemount B-6
Marcellus D-10
Marcus B-10
Markham E-3
Marshall D-11
Marys Corner E-4
Marysville C-5
Matlock D-3
Mattawa F-7
Mazama B-7
McCleary D-4
McKenna E-5
McMurray B-5
Medical Lake D-10
Menlo E-3
Mercer Island ★
  Vicinity Pg. 147, F-4
Mesa F-8
Metaline A-11
Metaline Falls A-11
Methow C-7
Mica D-11
Midway A-9
Milan C-11
Miles C-10
Millwood D-11
Mineral E-5
Moclips D-2
Mohler D-10
Molson A-8
Monahans D-7
Monitor D-7
Monroe C-5
Monse C-8
Monte Cristo C-6
Montesano E-3
Morton F-5
Moses Lake E-8
Mossyrock E-4
Moulton G-4
Mount Baker A-6
Mt. Hope D-11
Mountlake C-5
Mountlake Terrace ★
  Vicinity Pg. 147, A-3
Mt. Vernon B-5
Moxee City F-7
Mukilteo C-5
Naches E-7
Nahcotta E-2
Napauine E-4
Naselle F-3
Neah Bay B-2
Neilton D-3
Nemah E-3
Nespelem C-9
Newcastle ★
  Vicinity Pg. 147, G-5
Newhalem B-6
Newport C-11
Nighthawk A-8
Ninemile Falls D-10
Ninemile C-9
Nooksack A-5
Normandy Park ★
  Vicinity Pg. 147, J-2
North Cove E-2
Northport A-10
Oakesdale E-11
Oak Harbor B-4
Oakville E-3
Ocean City D-2
Ocean Park E-2
Ocean Shore D-2
Odessa D-9
Ohop E-5
Okanogan B-8
Old Toroda B-9
Olga A-4
Olympia D-4
Omak B-8
Onalaska E-4
Orcas A-4
Orchards G-4
Orient A-10
Orin B-10
Orondo D-7
Oroville A-8
Orting D-5
Oso B-5
Ostrander F-4
Othello E-8
Outlook F-7
Oyhut D-2
Oysterville E-2
Ozette B-2
Packwood D-5
Pacific ★
  Vicinity Pg. 147, K-4
Pacific Beach D-2
Page E-9
Palisades D-8
Palouse E-11
Paradise E-5
Park A-5
Park Rapids B-11
Parker F-7
Pasco F-9
Pateros C-8
Paterson G-8
Pe Ell E-3
Penawawa F-10
Peshastin D-7
Piedmont B-3
Pine City E-10
Pioneer G-4
Plain C-7
Plaza D-11
Pleasant View F-9
Plymouth G-8
Point Roberts A-4
Pomeroy F-10
Port Angeles B-4
Port Gamble C-4
Port Orchard C-4
Port Townsend B-4
Potlatch D-3
Poulsbo C-4
Prescott G-9
Prosser F-8
Pullman F-11
Puyallup D-5
Queets C-2
Quilcene C-4
Quinault D-3
Quincy D-8
Rainier E-4
Ralston E-10
Randle E-5
Raymond E-3
Reardan D-10
Redmond C-5
Renton ★
  Vicinity Pg. 147, H-4
Republic B-9
Rice B-10
Richland G-8
Richland Highlands ★
  Vicinity Pg. 147, B-3
Richmond Beach ★
  Vicinity Pg. 147, B-3
Ridgefield G-4
Rimrock E-6
Riparia F-10
Ritzville E-9
Riverside B-8
Robe C-5
Roche Harbor B-4
Rochester E-3

Rockford D-11
Rock Island D-7
Rockport B-6
Rogersburg G-11
Rollingbay ★
  Vicinity Pg. 147, D-1
Ronald D-6
Roosevelt G-7
Rosalia E-11
Rosburg F-3
Roslyn D-6
Roxboro E-9
Roy D-5
Royal City E-8
Rulo C-8
Ryderwood E-3
St. Andrews D-8
St. John E-11
Salmon La Sac D-6
Sappho B-2
Sara G-4
Satus F-7
Sawyer F-7
Schrag E-9
Seabeck C-4
Seattle C-5
  Vicinity Pg. 147
Seaview F-2
Sedro Woolley B-5
Sekiu B-2
Selah E-7
Seltice E-11
Sequim B-4
Shelton D-3
Silvana B-5
Silver Creek E-4
Silverdale C-4
Silverton C-6
Skamokawa F-3
Skykomish C-6
Smyrna E-8
Snohomish C-5
Snoqualmie D-5
Soap Lake D-8
Soda Springs E-6
South Bend E-3
South Cle Elum D-6
South Point C-4
Southworth D-4
Spangle D-11
Spirit B-11
Spokane D-11
Sprague D-10
Springdale C-10
Spring Valley E-11
Staircase C-3
Stanwood B-5
Starbuck G-10
Startup C-6
Stehekin B-7
Steilacoom D-5
Stella F-4
Steptoe E-11
Stevenson G-5
Sudbury D-9
Sultan C-6
Sumas A-5
Summit D-5
Sumner D-5
Sunnyside F-7
Sutico E-3
Tacoma D-5
Taholah D-2
Tampico F-6
Teanaway D-6
Tekoa E-11
Tenino E-4
Terrace E-11
Thornton E-11
Tieton E-6
Tiger B-11
Tillicum ★
Tokeland E-2
Tokio D-9
Toledo E-4
Tonasket B-9
Tono E-4
Toppenish F-7
Toroda A-9
Touchet G-9
Toutle F-4
Trout Lake G-6
Tumtum C-10
Tumwater E-4
Twisp C-7
Tyler D-10
Union D-4
Union Gap F-7
Uniontown F-11
Usk C-11
Utsalady B-4
Vader F-4
Valley C-10
Vancouver G-4
  Vicinity Pg. 139, A-5
Vantage E-7
Van Zandt A-5
Vashon Heights ★
  Vicinity Pg. 148, G-1
Vashon Island ★
  Vicinity Pg. 148, K-1
Vaughn C-4
Vernita E-8
Wahkiacus G-6
Wakefield C-8
Waitsburg G-10
Walla Walla G-10
Wallula G-9
Wapato F-7
Warden E-8
Washougal G-4
Washtucna E-9
Waterville D-8
Wauconda B-9
Waukon D-10
Waverly D-11
Wawawai F-11
Wellpinit C-10
Wenatchee D-7
Wenatchee Heights D-7
Westport E-2
West Richland F-8
Wheeler D-8
Whites D-3
White Salmon G-6
White Swan F-6
Wickersham B-5
Wilbur C-9
Wiley City F-7
Wilkeson D-5
Willada E-10
Willapa E-3
Willard G-5
Wilson Creek D-8
Winlock E-4
Winona E-10

★ Denotes City located only on Vicinity map. City not located on State map.

Winslow ... C-4
Winthrop ... B-8
Wishram ... G-6
Wishram Heights ... G-6
Withrow ... D-8
Woodinville ... ★
  Vicinity ... Pg. 148, B-6
Woodland ... G-4
Yacolt ... G-4
Yakima ... F-7
Yelm ... E-4
Zillah ... F-7

WEST VIRGINIA
Pages 46-47

Adrian ... D-6
Alderson ... G-5
Alma ... C-5
Ansted ... F-5
Arbovale ... F-6
Arnett ... F-4
Arnoldsburg ... E-4
Athens ... G-5
Baker ... D-8
Ballard ... G-5
Barboursville ... E-3
Barrackville ... C-6
Barrett ... F-4
Beatrice ... F-4
Beckley ... G-5
Belington ... D-6
Belle ... F-4
Berkeley Springs ... C-9
Bethany ... B-6
Bethlehem ... B-5
Beverly ... D-6
Bigbend ... D-4
Big Creek ... F-4
Birch River ... E-5
Bismark ... D-7
Blacksville ... C-6
Blair ... F-4
Bluefield ... H-4
Boyer ... F-6
Bradshaw ... G-3
Brandywine ... E-7
Bridgeport ... D-6
Brooks ... G-5
Buckhannon ... D-6
Buffalo ... E-3
Burnsville ... E-5
Burton ... C-6
Cabins ... D-7
Calvin ... E-5
Camden ... D-5
Cameron ... C-5
Canfield ... D-6
Chapmanville ... F-3
Charleston ... E-4
Charles Town ... D-9
Charmco ... F-5
Chattaroy ... G-3
Cheat Bridge ... E-6
Chester ... A-6
Chloe ... E-5
Cisco ... D-4
Clarksburg ... D-5
Clay ... E-5
Clendenin ... E-4
Cleveland ... E-5
Clifftop ... G-4
Crab Orchard ... G-4
Craigsville ... F-5
Creston ... D-4
Dailey ... E-6
Davis ... D-7
Dellslow ... C-6
Diana ... E-5
Dixie ... F-4
Dunbar ... E-4
Dunlow ... F-3
Droop ... F-6
Eccles ... G-5
Echo ... F-3
Edray ... F-6
Eleanor ... E-4
Elizabeth ... D-4
Elkins ... D-6
Elkview ... E-4
Elkwater ... E-6
Ellenboro ... D-5
Elton ... F-5
Etam ... D-7
Evans ... E-4
Fairmont ... C-6
Fayetteville ... F-4
Fellowsville ... D-6
Follansbee ... B-6
Forest Hill ... G-5
Fort Ashby ... C-8
Fort Gay ... F-3
Frankford ... F-5
Franklin ... E-7
Frost ... F-6
Gallipolis Ferry ... E-3
Galloway ... D-6
Gary ... G-4
Gauley Bridge ... F-4
Gilbert ... G-3
Gilmer ... D-5
Glenville ... D-5
Grafton ... D-6
Grantsville ... D-4
Hamlin ... F-3
Harper ... E-7
Harrisville ... D-4
Hastings ... C-5
Heaters ... E-5
Hico ... F-5
Hinton ... G-5
Hodgesville ... D-6
Hogsett ... E-3
Holden ... F-3
Huntington ... E-3
Hurricane ... E-3
Huttonsville ... E-6
Iaeger ... G-4
Intermont ... D-8
Inwood ... D-9
Jane Lew ... D-5
Jeffery ... F-4
Jesse ... G-4
Julian ... F-4
Junction ... D-8
Kenna ... E-4

Kenova ... E-3
Kermit ... G-3
Keyser ... C-8
Keystone ... G-4
Kiahsville ... F-3
Kingwood ... C-6
Larew ... C-7
Lavalette ... E-3
Lehew ... D-8
Leopold ... D-5
Lesage ... E-3
Lewisburg ... G-5
Liberty ... D-4
Lindside ... G-5
Littleton ... C-5
Lizemores ... E-5
Logan ... F-4
Looneyville ... E-4
Lost City ... D-8
Lost Creek ... D-5
Lost River ... D-8
Lumberport ... C-6
Maben ... F-4
Mace ... E-6
Madison ... F-4
Man ... G-4
Mannington ... C-6
Marlinton ... F-6
Marmet ... E-4
Martinsburg ... C-9
Mason ... D-3
Matewan ... G-3
Maysville ... D-7
McMechen ... B-5
Middlebourne ... C-5
Mill Point ... F-6
Mineralwells ... D-4
Moorefield ... D-8
Morgantown ... C-7
Moundsville ... C-5
Mountain ... C-5
Mt. Alto ... D-4
Mt. Gay ... F-3
Mt. Hope ... F-4
Mt. Storm ... D-7
Mt. Zion ... D-4
Mullens ... F-4
Needmore ... D-8
Nestorville ... D-6
New Cumberland ... A-5
New Martinsville ... C-5
Nimitz ... F-5
Nitro ... E-4
North Fork ... G-4
Norton ... D-6
Oak Flat ... E-7
Oak Hill ... F-5
Oceana ... G-4
Old Fields ... D-8
Orgas ... E-4
Paden City ... C-5
Palestine ... D-4
Parkersburg ... D-4
Parsons ... D-7
Petersburg ... D-8
Peterstown ... G-5
Philippi ... D-6
Pickaway ... G-5
Pike ... D-5
Pineville ... G-4
Pipestem ... G-5
Pleasant Dale ... D-8
Point Pleasant ... D-3
Porterfield ... D-4
Prince ... F-5
Princeton ... G-4
Purgitsville ... D-8
Racine ... E-4
Rainelle ... F-5
Rand ... E-4
Ranger ... F-3
Ravenswood ... D-4
Reader ... C-5
Red Creek ... D-7
Reedy ... D-4
Renick ... F-5
Richwood ... E-5
Rio ... C-8
Ripley ... D-4
Riverton ... E-7
Robson ... E-4
Rock Camp ... G-5
Rock Cave ... E-6
Roderfield ... G-4
Romney ... C-8
Roncevert ... G-5
St. Albans ... E-4
St. Marys ... C-5
Salem ... D-5
Seneca Rocks ... D-7
Seth ... F-4
Shinnston ... C-6
Sistersville ... C-5
Slanesville ... C-8
Slate ... D-4
Smithers ... F-4
Smithfield ... C-5
Smithville ... D-5
Sophia ... G-4
South Charleston ... E-4
Spencer ... D-4
Springfield ... C-8
Stickney ... G-5
Summersville ... E-5
Sweetland ... F-3
Switzer ... G-3
Thomas ... D-7
Thornwood ... E-7
Troy ... D-5
Tunnelton ... D-6
Union ... G-5
Valley Bend ... E-6
Valley Head ... E-6
Van ... F-4
Vienna ... D-4
Walkersville ... D-6
Waneta ... D-6
War ... G-4
Washburn ... D-5
Wayne ... F-3
Webster Springs ... E-6
Weirton ... B-5
Welch ... G-4
Wellsburg ... B-5
W. Hamlin ... F-3
Weston ... D-5
Westover ... C-6
W. Union ... D-5
Wheeling ... B-5

White Sulphur Springs ... F-6
Whitesville ... F-4
Williamson ... G-3
Williamstown ... C-4
Wymer ... D-7

WISCONSIN
Pages 86-87

Abbotsford ... F-5
Adams ... H-5
Adell ... J-8
Algoma ... G-8
Alma ... G-2
Alma Center ... G-4
Altoona ... G-3
Amery ... F-2
Aniwa ... F-6
Antigo ... F-6
Appleton ... G-7
Arcadia ... H-3
Argyle ... K-5
Arlington ... J-6
Ashland ... D-4
Ashwaubenon ... G-7
Athens ... F-5
Auburndale ... G-5
Augusta ... G-4
Baldwin ... F-2
Balsam Lake ... E-2
Bangor ... H-4
Baraboo ... J-5
Barneveld ... K-5
Barron ... F-3
Bay City ... G-2
Bayfield ... D-4
Bayside ... ★
  Vicinity ... Pg. 125, A-6
Beadstown ... J-4
Beaver Dam ... J-7
Belgium ... J-8
Belleville ... K-6
Belmont ... K-5
Beloit ... L-6
Benton ... L-5
Berlin ... H-6
Big Bend ... ★
  Vicinity ... Pg. 125, H-1
Birchwood ... E-3
Birnamwood ... F-6
Black Creek ... G-7
Black Earth ... K-5
Black River Falls ... G-4
Blair ... H-3
Blanchardville ... K-5
Bloomer ... F-3
Bloomington ... K-4
Bonduel ... G-7
Boscobel ... K-4
Boyceville ... F-2
Boyd ... G-4
Brandon ... J-7
Brillion ... H-8
Brodhead ... L-6
Brokaw ... F-6
Brookfield ... ★
  Vicinity ... Pg. 125, D-3
Brown Deer ... ★
  Vicinity ... Pg. 125, A-5
Browntown ... L-6
Bruce ... F-3
Buena Park ... ★
  Vicinity ... Pg. 125, K-1
Buffalo ... H-2
Burlington ... K-8
Butler ... ★
  Vicinity ... Pg. 125, C-3
Butternut ... E-4
Cable ... D-3
Caddy Vista ... ★
  Vicinity ... Pg. 125, J-6
Cadott ... F-3
Caledonia ... ★
  Vicinity ... Pg. 125, K-8
Calhoun ... ★
  Vicinity ... Pg. 125, E-3
Cambra ... J-6
Cambridge ... K-7
Cameron ... F-3
Campbellsport ... J-8
Camp Douglas ... H-5
Carrollville ... ★
  Vicinity ... Pg. 125, K-6
Cascade ... J-8
Casco ... G-8
Cashton ... J-4
Cassville ... K-4
Catawba ... E-4
Cazenovia ... J-5
Cecil ... G-7
Centuria ... E-2
Chetek ... F-3
Chilton ... H-8
Chippewa Falls ... F-3
Clayton ... F-2
Clear Lake ... F-2
Cleveland ... H-8
Clinton ... L-7
Clintonville ... G-7
Cochrane ... H-3
Colby ... G-5
Coleman ... F-8
Colfax ... F-3
Colgate ... ★
  Vicinity ... Pg. 125, A-1
Coloma ... H-6
Columbus ... J-7
Coon Valley ... J-4
Cornell ... F-4
Couderay ... E-3
Crandon ... E-7
Crivitz ... F-8
Cross Plains ... K-6
Cuba City ... L-4
Cudahy ... ★
  Vicinity ... Pg. 125, F-6
Cumberland ... E-2
Dane ... J-6
Darien ... L-7
Darlington ... K-5
Deerfield ... J-6
DeForest ... J-6
Delavan ... L-7
Denmark ... G-8
De Pere ... G-8

De Soto ... J-4
Dickeyville ... L-4
Dodgeville ... K-5
Dorchester ... F-5
Dousman ... K-7
Dresser ... F-1
Duplainville ... ★
  Vicinity ... Pg. 125, D-2
Durand ... G-2
Durham ... ★
  Vicinity ... Pg. 125, H-3
Eagle ... K-7
Eagle River ... E-6
Eastman ... K-4
East Troy ... K-7
Eau Claire ... G-3
Eden ... J-7
Edgar ... F-5
Edgerton ... K-6
Egg Harbor ... F-9
Elano ... F-6
Eleva ... G-3
Elkhart Lake ... H-8
Elk Mound ... G-3
Ellsworth ... G-2
Elm Grove ... ★
  Vicinity ... Pg. 125, D-3
Elmwood ... G-2
Elroy ... J-5
Embarrass ... G-7
Ephraim ... F-9
Ettrick ... H-3
Evansville ... K-6
Exeland ... E-3
Fairchild ... G-4
Fall Creek ... G-4
Fall River ... J-6
Fennimore ... K-4
Ferryville ... K-4
Fifield ... E-5
Florence ... E-7
Fontana ... L-7
Footville ... K-6
Forestville ... G-8
Fort Atkinson ... K-7
Fountain City ... H-3
Fox Lake ... J-6
Fox Point ... ★
  Vicinity ... Pg. 125, B-6
Francis Creek ... H-8
Franklin ... K-8
Franksville ... ★
  Vicinity ... Pg. 125, K-8
Frederick ... E-2
Fremont ... H-7
Friendship ... H-6
Fussville ... ★
  Vicinity ... Pg. 125, B-3
Galesville ... H-3
Gays Mills ... J-4
Genoa ... J-4
Genoa City ... L-7
Germantown ... J-8
  Vicinity ... Pg. 125, A-3
Gillett ... F-7
Gilman ... F-4
Glendale ... ★
  Vicinity ... Pg. 125, B-6
Glen Flora ... E-4
Glenwood City ... F-2
Goerkes Corner ... ★
  Vicinity ... Pg. 125, C-2
Grafton ... J-8
Granton ... G-5
Grantsburg ... E-1
Granville ... ★
  Vicinity ... Pg. 125, A-4
Gratiot ... L-5
Green Bay ... G-8
Greendale ... ★
  Vicinity ... Pg. 125, G-6
Greenfield ... ★
  Vicinity ... Pg. 125, F-5
Green Lake ... H-6
Greenwood ... G-4
Gresham ... G-7
Hales Corners ... ★
  Vicinity ... Pg. 125, G-5
Hammond ... F-2
Hancock ... H-6
Hartford ... J-8
Hartland ... K-8
Haugen ... E-2
Hawkins ... E-4
Hayward ... E-3
Hazel Green ... L-5
Heafford Jct. ... E-5
Highland ... K-5
Hilbert ... H-8
Hillsboro ... J-4
Holmen ... H-3
Horicon ... J-7
Hortonville ... G-7
Howard ... G-8
Howards Grove ... H-8
Hudson ... F-1
Hurley ... D-5
Husher ... ★
  Vicinity ... Pg. 125, J-6
Independence ... G-3
Ingram ... E-4
Iola ... G-6
Iron River ... C-3
Jackson ... J-8
Janesville ... K-6
Jefferson ... K-7
Johnson Creek ... K-7
Junction City ... G-5
Juneau ... J-7
Kaukauna ... H-8
Kendall ... H-5
Kennan ... E-4
Kenosha ... K-8
Keshena ... F-7
Kewaskum ... J-8
Kewaunee ... G-9
Kiel ... H-8
Kimberly ... H-8
Kingston ... J-6
Lac Du Flambeau ... D-5
La Crosse ... H-3
Ladysmith ... F-3
La Farge ... J-4

Lake Delton ... J-5
Lake Geneva ... L-7
Lake Nebagamon ... D-3
Lancaster ... K-4
Lannon ... ★
  Vicinity ... Pg. 125, B-2
Lena ... G-8
Linden ... K-5
Little Chute ... G-7
Livingston ... K-4
Lodi ... J-6
Loganville ... J-5
Lomira ... J-7
Lone Rock ... J-5
Loyal ... G-4
Lublin ... F-4
Luck ... E-2
Luxemburg ... G-8
Madison ... K-6
  Vicinity ... Pg. 125
Manawa ... G-7
Manitowish ... D-5
Manitowoc ... H-9
Marathon City ... F-5
Marcy ... ★
  Vicinity ... Pg. 125, C-2
Marinette ... F-8
Marion ... G-7
Markesan ... J-6
Marshall ... K-6
Marshfield ... G-5
Mason ... D-4
Mauston ... H-5
Mayville ... J-7
Mazomanie ... J-6
Medford ... F-4
Meeker ... ★
  Vicinity ... Pg. 125, A-2
Mellen ... D-4
Melrose ... H-4
Melvina ... H-4
Menasha ... H-7
Menomonee ... F-8
Menomonee Falls ... J-7
  Vicinity ... Pg. 125, A-2
Menomonie ... G-2
Mequon ... ★
  Vicinity ... Pg. 125, A-5
Merrill ... F-6
Merrillan ... G-4
Middleton ... K-6
Milltown ... E-2
Milton ... K-7
Milwaukee ... K-8
  Vicintiy ... Pg. 125
Mineral Point ... K-5
Minong ... D-3
Mishicot ... H-8
Mondovi ... G-3
Monico ... E-6
Monona ... ★
  Vicinity ... Pg. 125, E-2
Monroe ... L-5
Montello ... H-6
Montfort ... K-5
Monticello ... K-5
Montreal ... D-4
Mosinee ... G-5
Mount Horeb ... K-6
Mount Sterling ... J-4
Mukwonago ... K-7
Muscoda ... K-4
Muskego ... ★
  Vicinity ... Pg. 125, G-2
Necedah ... H-5
Neenah ... H-7
Neillsville ... G-4
Nekoosa ... H-5
Nelson ... G-2
Nelsonville ... G-6
Neshkoro ... H-6
New Berlin ... ★
  Vicinity ... Pg. 125, E-2
New Glarus ... K-6
New Holstein ... H-8
New London ... G-7
New Richmond ... F-2
Niagara ... E-8
North Cape ... ★
  Vicinity ... Pg. 125, K-4
North Fond du Lac ... H-7
North Hudson ... F-1
Norwalk ... H-4
Oak Creek ... ★
  Vicinity ... Pg. 125, G-6
Oakwood ... ★
  Vicinity ... Pg. 125
Oconomowoc ... K-7
Oconto ... F-8
Oconto Falls ... G-8
Orfordville ... K-6
Omro ... H-7
Onalaska ... H-3
Oostburg ... J-8
Ontario ... J-4
Osceola ... F-2
Oshkosh ... H-7
Osseo ... G-3
Owen ... F-4
Oxford ... H-6
Paddock Lake ... K-7
Palmyra ... K-7
Pardeeville ... J-6
Park Falls ... E-4
Park Ridge ... ★
  Vicinity ... Pg. 125, C-1
Patch Grove ... K-4
Paynesville ... ★
  Vicinity ... Pg. 125, H-5
Pembine ... E-8
Pepin ... G-2
Peshtigo ... F-8
Pewaukee ... ★
  Vicinity ... Pg. 125, C-1
Phelps ... E-6
Phillips ... E-5
Pigeon Falls ... G-3
Pittsville ... H-5
Plain ... J-5
Plainfield ... H-6
Platteville ... K-4
Plover ... H-6
Plum City ... G-2
Plymouth ... H-8
Poplar ... C-3
Portage ... J-6
Port Edwards ... H-5
Port Washington ... J-8
Potosi ... K-4
Pound ... F-8
Prairie du Chien ... K-4

Prairie du Sac ... J-5
Prentice ... E-5
Prescott ... G-1
Princeton ... H-6
Prospect ... ★
  Vicinity ... Pg. 125, B-2
Pulaski ... G-8
Racine ... K-8
Radisson ... E-3
Randolph ... J-7
Rawson ... ★
  Vicinity ... Pg. 125
Raymond ... ★
  Vicinity ... Pg. 125, K-4
Redgranite ... H-6
Reedsburg ... J-5
Reedsville ... H-8
Rhinelander ... E-6
Rib Lake ... F-5
Rice Lake ... E-3
Richland Center ... J-5
Ridgeland ... F-2
Ridgeway ... K-5
Rio ... J-6
Ripon ... H-7
River Falls ... G-1
River Hills ... ★
  Vicinity ... Pg. 125, A-6
Riverside ... D-2
Roberts ... F-2
Rosendale ... H-7
Rosholt ... G-6
Rothschild ... G-5
Saint Francis ... ★
  Vicinity ... Pg. 125, A-2
St. Francis ... K-8
Saint Martins ... ★
  Vicinity ... Pg. 125, G-4
Sauk City ... J-5
Saukville ... J-8
Schofield ... F-6
Seymour ... G-7
Sharon ... L-7
Shawano ... G-7
Sheboygan ... H-9
Sheboygan Falls ... H-8
Sheldon ... F-3
Shell Lake ... E-2
Shiocton ... G-7
Shorewood ... ★
  Vicinity ... Pg. 125, L-5
Shullsburg ... L-5
Siren ... E-2
Sister Bay ... F-9
Soldiers Grove ... J-4
Solon Springs ... D-3
Somerset ... F-1
South Milwaukee ... K-8
  Vicinity ... Pg. 125, G-6
Sparta ... H-4
Spencer ... G-4
Spooner ... E-3
Spring Green ... J-5
Spring Valley ... G-2
Stanley ... G-4
Stevens Point ... G-5
Stockbridge ... H-7
Stockholm ... G-2
Stoddard ... J-4
Stoughton ... K-6
Stratford ... G-5
Strum ... G-3
Sturgeon Bay ... F-8
Sturtevant ... ★
  Vicinity ... Pg. 125, J-6
Sun Prairie ... K-7
Superior ... C-2
Suring ... F-7
Sussex ... ★
  Vicinity ... Pg. 125, B-1
Tabor ... ★
  Vicinity ... Pg. 125, J-6
Taylors Falls ... E-1
Tess Corners ... ★
  Vicinity ... Pg. 125, K-5
Theresa ... J-7
Thompsonville ... ★
  Vicinity ... Pg. 125, J-7
Thorp ... G-4
Tichigan ... ★
  Vicinity ... Pg. 125, J-1
Tigerton ... G-7
Tomah ... H-4
Tomahawk ... E-6
Tony ... F-4
Trempealeau ... H-3
Turtle Lake ... E-2
Two Rivers ... H-9
Union Center ... J-5
Union Church ... ★
  Vicinity ... Pg. 125, A-4
Union Grove ... K-8
Unity ... G-5
Valders ... H-8
Vernon ... ★
  Vicinity ... Pg. 125, J-1
Verona ... K-6
Vesper ... G-5
Village of Superior ... C-2
Viroqua ... J-4
Walworth ... L-7
Washburn ... C-3
Waterford ... K-7
Waterloo ... K-6
Watertown ... J-7
Waukesha ... K-8
  Vicinity ... Pg. 125, E-1
Waunakee ... K-6
Waupaca ... G-7
Waupun ... J-7
Wausau ... F-6
Wausaukee ... F-8
Wautoma ... H-6
Wauwatosa ... K-8
  Vicinity ... Pg. 125, D-4
Webster ... E-2
West Allis ... ★
  Vicinity ... Pg. 125, E-4
West Bend ... J-8
Westby ... J-4
Westfield ... H-6
West Granville ... ★
  Vicinity ... Pg. 125, B-5
West Milwaukee ... ★
  Vicinity ... Pg. 125, E-5
Weyauwega ... G-6
Whitefish Bay ... ★
  Vicinity ... Pg. 125, C-6

Whitehall ... G-3
White Lake ... F-7
Whitelaw ... H-8
Whitewater ... K-7
Whiting ... G-6
Wild Rose ... G-6
Willow Creek ... ★
  Vicinity ... Pg. 125, A-2
Wind Pt. ... K-8
Winneconne ... H-7
Winter ... E-4
Wisconsin Dells ... J-6
Wisconsin Rapids ... G-5
Withee ... F-4
Wittenberg ... G-6
Wonewoc ... J-5
Woodruff ... E-6
Wrightstown ... G-8
Wyeville ... H-5
Wyocena ... J-6
Yuba ... J-4

WYOMING
Pages 88-89

Acme ... B-8
Afton ... E-2
Aladdin ... B-12
Albany ... H-12
Albin ... H-12
Alcova ... F-8
Alpine Jct. ... E-3
Alva ... B-11
Arapahoe ... F-6
Arminto ... E-8
Arlington ... H-9
Arvada ... C-9
Atlantic City ... F-5
Auburn ... F-3
Baggs ... H-9
Bairoil ... G-7
Banner ... B-8
Bar Nunn ... F-9
Barnum ... D-8
Basin ... C-7
Beckton ... B-8
Bedford ... E-3
Beulah ... B-12
Big Horn ... B-8
Big Piney ... G-4
Big Sandy ... F-5
Big Trails ... D-8
Bill ... E-10
Bitter Creek ... H-6
Bondurant ... E-4
Border ... H-10
Bosler ... F-5
Boulder ... F-5
Boxelder ... F-10
Buffalo ... C-8
Burgess Jct. ... B-7
Burlington ... C-6
Burns ... H-12
Burntfork ... J-4
Burris ... E-5
Byron ... B-6
Calpet ... F-3
Canyon Jct. ... B-2
Carpenter ... J-12
Carter ... J-4
Casper ... F-9
Centennial ... H-11
Cheyenne ... J-11
Chugwater ... H-11
Clareton ... D-11
Clark ... B-5
Clearmont ... B-9
Cody ... B-5
Cokeville ... G-3
Colony ... B-12
Cora ... F-4
Cowley ... B-6
Creston Jct. ... H-8
Crowheart ... E-5
Daniel ... F-4
Dayton ... B-8
Deaver ... B-6
Devil's Tower Jct. ... B-11
Diamondville ... H-3
Dixon ... J-7
Douglas ... F-10
Dubois ... E-5
E. Thermopolis ... D-7
Eden ... H-5
Edgerton ... E-9
Egbert ... J-12
Elk Basin ... B-6
Elk Mountain ... H-9
Elkol ... H-4
Elmo ... H-9
Emblem ... C-6
Encampment ... J-9
Ethete ... F-6
Etna ... E-3
Evanston ... J-3
Evansville ... F-9
Farson ... G-5
Fairview ... E-3
Federal ... J-11
Fort Bridger ... J-4
Fort Laramie ... H-11
Fort Washakie ... F-5
Four Corners ... C-12
Fountain ... B-6
Freedom ... E-3
Frontier ... H-3
Garland ... B-6
Gillette ... C-10
Glendo ... F-11
Glenrock ... F-10
Granger ... H-4
Grass Creek ... D-6
Green River ... H-5
Greybull ... C-7
Grover ... F-3
Guernsey ... G-11

Hyattville ... C-8
Iron Mountain ... H-11
Jackson ... E-3
Jay Em ... G-12
Jeffrey City ... G-7
Jenny Lake ... D-3
Kaycee ... D-9
Keeline ... F-11
Kelly ... E-4
Kemmerer ... H-3
Kinnear ... E-6
Kirby ... D-7
La Barge ... G-4
La Grange ... H-12
Lake Jct. ... B-2
Lamont ... G-8
Lance Creek ... E-11
Lander ... F-6
Laramie ... J-10
Leiter ... B-9
Lightning Flat ... B-11
Linch ... D-9
Lingle ... G-12
Little America ... J-4
Lonetree ... J-4
Lost Cabin ... E-8
Lost Springs ... F-11
Lovell ... B-6
Lucerne ... D-7
Lusk ... F-12
Lyman ... J-4
Lysite ... E-7
Madison Jct. ... B-3
Mammoth Springs Jct. ... B-3
Manderson ... C-7
Manville ... F-11
Marbleton ... F-4
Mayoworth ... D-8
McFadden ... H-10
McKinnon ... J-4
Medicine Bow ... H-9
Meeteetse ... C-6
Merna ... F-4
Midwest ... E-9
Millburne ... J-4
Mills ... F-9
Moneta ... E-7
Moorcroft ... C-11
Moose ... D-3
Moran Jct. ... D-4
Morrisey ... D-11
Morton ... E-6
Mountain Home ... J-10
Mountain View ... J-4
Mule Cr. Jct. ... E-12
Natrona ... E-9
Newcastle ... D-12
New Haven ... C-12
Node ... F-12
Norris Jct. ... B-3
Old Faithful ... C-3
Opal ... H-4
Orin ... F-10
Orpha ... F-10
Osage ... D-12
Oshoto ... B-11
Otto ... C-7
Pahaska ... C-4
Parkman ... B-8
Pavillion ... E-6
Piedmont ... J-4
Pine Bluffs ... J-12
Pinedale ... F-5
Pine Tree Jct. ... D-10
Pitchfork ... D-5
Point of Rocks ... H-6
Powder River ... F-8
Powell ... B-6
Quealy ... J-5
Ralston ... B-6
Ranchester ... B-8
Rawlins ... H-8
Recluse ... B-10
Redbird ... E-12
Red Desert ... H-6
Reliance ... H-6
Reno Jct. ... D-10
Riverside ... J-9
Riverton ... F-7
Robertson ... J-4
Rochelle ... D-11
Rock River ... H-10
Rock Springs ... H-5
Rocky Point ... B-11
Rozet ... C-10
Sage ... H-3
St. Stephens ... F-6
Saratoga ... J-9
Savageton ... D-10
Savery ... J-9
Shawnee ... F-11
Shell ... C-7
Sheridan ... B-8
Shirley Basin ... G-9
Shoshoni ... E-7
Sinclair ... H-8
Slater ... H-11
Smoot ... F-3
South Pass City ... F-5
Spotted Horse ... B-10
Story ... C-8
Sundance ... C-12
Sunrise ... G-11
Superior ... H-6
Sussex ... D-9
Table Rock ... H-7
Ten Sleep ... D-8
Teton Village ... D-3
Thayne ... E-3
Thermopolis ... D-6
Tie Siding ... J-10
Torrington ... G-12
Tower Jct. ... B-4
Turnerville ... F-3
Uccross ... C-11
Upton ... C-11
Urie ... J-4
Valley ... C-5
Van Tassell ... F-12
Veteran ... G-12
Walcott ... H-9
Waltman ... E-8
Wamsutter ... H-7
W. Thumb Jct. ... C-3
Weston ... B-10
Wheatland ... G-11
Wilson ... D-3
Woods Landing ... J-10
Worland ... D-7
Wright ... D-10
Yoder ... H-12

★ Denotes City located only on Vicinity map. City not located on State map.